Noise & Hearing Conservation Manual

Noise & Hearing Conservation Manual

edited by

Elliott H. Berger, M.S., Mem. INCE
Manager, Acoustical Engineering
E-A-R Division, Cabot Corporation
Indianapolis, Indiana

W. Dixon Ward, Ph.D.
Professor
Departments of Communication Disorders and Otolaryngology
University of Minnesota
Minneapolis, Minnesota

Jeffrey C. Morrill, M.S.
President
Impact Hearing Conservation, Inc.
Kansas City, Missouri

Larry H. Royster, Ph.D.
Professor
Department of Mechanical Engineering
North Carolina State University
Raleigh, North Carolina

American Industrial Hygiene Association
Akron, Ohio

Printed in USA
ISBN #0932627-21-8

Copies available:
American Industrial Hygiene Association
475 Wolf Ledges Parkway
Akron, OH 44311-1087

Authors

Elliott H. Berger, M.S., Mem. INCE
Manager, Acoustical Engineering
E-A-R Division, Cabot Corp.
Indianapolis, IN 46268-0898

Robert D. Bruce, E.E., Mem. INCE
Principal Consultant
Hoover Keith & Bruce Inc.
Houston, TX 77082-2632

Allen L. Cudworth, Sc.D.
Vice President
Liberty Mutual Insurance Co.
Boston, MA 02118-0001

John J. Earshen, M.S.
Vice President and Chief Scientist
Metrosonics, Inc.
Rochester, NY 14692-3075

Kathy A. Foltner, M.A., CCC-A
President
Audio-Vestibular Testing Center, Inc.
Lansing, MI 48912-2104

Jeffrey C. Morrill, M.S.
President
Impact Hearing Conservation, Inc.
Kansas City, MO 64111-2736

Paul B. Ostergaard, S.M., P.E.
President
Ostergaard Associates
Caldwell, NJ 07006-5384

Julia D. Royster, Ph.D., CCC-A/SLP
President
Environmental Noise Consultants, Inc.
Cary, NC 27511-0144

Larry H. Royster, Ph.D.
Professor
Department of Mechanical Engineering
North Carolina State University
Raleigh, NC 27695-7910

Alice H. Suter, Ph.D.
Consultant in Industrial Audiology
1501 Red Oak Drive
Silver Spring, MD 20910-1550

Edwin H. Toothman, B.S.
Director, Occupational Health
Bethlehem Steel Corp.
Bethlehem, PA 18016-7699

W. Dixon Ward, Ph.D.
Professor
Departments of Communication Disorders and Otolaryngology
University of Minnesota
Minneapolis, MN 55414-3287

Foreword

This edition of the American Industrial Hygiene Association's *Industrial Noise Manual* represents the latest information on the subject. It is intended to serve not only those with a technical interest in noise, but also those who need a better understanding of the subject.

Many persons have worked diligently to develop this manual. The American Industrial Hygiene Association is pleased to acknowledge their contributions in making this revised edition possible.

We hope that as you read and apply the information contained herein, everyone will benefit through better control of exposure to noise.

Howard L. Kusnetz
President
American Industrial Hygiene
Association

Preface

The first edition of this Manual appeared in 1958. The fact that it was soon out of print was indicative of the broad interest in noise, which was not limited to only the industrial hygiene community. A second edition produced in 1966, was reprinted in 1970. It was revised and issued as a third edition in 1975. Early in 1982 the AIHA Noise Committee, under the chairmanship of Carl Bohl, began work on the fourth edition. In 1984 Jeffrey Morrill assumed the chair of that committee and appointed Elliott Berger as the project coordinator. The manuscript was completed in the spring of 1986.

The fourth edition, rather than being a revision of the previous Manual, represents totally new material, with the exception of Chapters 2, 12, and 13, which have been extensively updated and revised. New chapters have been added in critical areas such as audiometric data base analysis, education and motivation, and visual evaluation of the outer ear. Also, the title has been changed from the *Industrial Noise Manual* to the *Noise and Hearing Conservation Manual*, to better reflect the scope and content of the new book.

Each chapter was primarily written by the author(s) with whose name(s) it is associated. Prior to editing, the chapters received journal-like peer reviews from the editors as well as the following contributors: Carl Bohl, Dennis Driscoll, John Earshen, Donald Gasaway, Donald Joseph, Warren Kundert, Paul Ostergaard, Julia Royster, Alice Suter, William Thornton, and Vern Tubergen. Subsequently, the editors wheedled, cajoled, persuaded, and nit-picked to strive for accuracy, clarity, completeness, and consistency (both within and between chapters). Where disagreements existed, the authors' preferences were honored.

In conformance with professional society guidelines, an effort was made to metricize this text. However, current engineering and construction practice still makes wide use of the English system. Therefore, to improve the applicability of the Manual, Chapter 2 (Physics of Sound) uses both English and metric units, and Chapter 12 (Engineering Controls), which deals extensively with engineering, architectural, and construction elements, uses only English units.

In this edition of the Manual we have attempted to be comprehensive in scope, yet practical in content. The purpose was to prepare a thoroughly referenced and indexed text that could bring novice industrial hygienists up to speed and direct them towards additional sources of information while still providing a standard handbook for the more experienced practitioners. At the same time, due to the multidisciplinary nature of noise and hearing conservation (hygiene, safety, engineering, acoustics, medical, audiological, behavioral, and legal), we feel the Manual should prove useful to those beyond the industrial hygiene community. And it is our hope that it will also serve as a principal or supplemental textbook for courses and seminars on hearing conservation and the effects of noise on man.

We extend our sincerest appreciation to the staff of the American Industrial Hygiene Association for their excellent and extensive efforts in the preparation of this manuscript for publication. Additionally, we wish to acknowledge the administrative assistance of Loretah Rowland and the cover design by Anne Mueller.

With those introductory remarks we present to you the fourth edition.

August, 1986

Elliott Berger
Dix Ward
Jeff Morrill
Larry Royster

Contents

Authors v
Foreword vii
Preface ix
List of Abbreviations xiii

1. **Hearing Conservation** 1
Alice H. Suter
2. **Physics of Sound** 19
Paul B. Ostergaard
3. **Sound Measurement: Instrumentation and Noise Descriptors** 37
John J. Earshen
4. **Noise Surveys and Data Analysis** 97
Larry H. Royster, Elliott H. Berger, and Julia Doswell Royster
5. **Anatomy and Physiology of the Ear: Normal and Damaged Hearing** 177
W. Dixon Ward
6. **Auditory Effects of Noise** 197
W. Dixon Ward
7. **Visual Evaluation of the External Ear and Eardrum** 217
Kathy A. Foltner
8. **Hearing Measurement** 233
Jeffrey C. Morrill
9. **Audiometric Data Base Analysis** 293
Julia Doswell Royster and Larry H. Royster
10. **Hearing Protection Devices** 319
Elliott H. Berger
11. **Education and Motivation** 383
Larry H. Royster and Julia Doswell Royster
12. **Engineering Controls** 417
Robert D. Bruce and Edwin H. Toothman
13. **Workers' Compensation** 523
Allen L. Cudworth

Appendices

I. **Department of Labor Occupational Noise Exposure Standard** 537
II. **Annotated Listing of Noise and Hearing Conservation Films and Videotapes** 563

Subject Index 585

Abbreviations

AAO	American Academy of Otolaryngology--Head and Neck Surgery
AAOO	American Academy of Opthalmology and Otolaryngology
ACGIH	American Conference of Governmental Industrial Hygienists
ADBA	Audiometric Data Base Analysis
AIHA	American Industrial Hygiene Association
AMA	American Medical Association
ANSI	American National Standards Institute
ARTL	Age-Related Threshold Level
ASA	Acoustical Society of America, and also (until 1966) American Standards Association
ASHA	American Speech-Language-Hearing Association
ASTM	American Society for Testing and Materials
BC	Bone and Tissue Conduction
CAOHC	Council for Accreditation in Occupational Hearing Conservation
CFR	Code of Federal Regulation
df	Degrees of Freedom
DI	Directivity Index
DIS	Draft International Standard
DOD	Department of Defense
ENT	Ear, Nose and Throat
EPA	Environmental Protection Agency
FDA	Food and Drug Administration
FECA	Federal Employees' Compensation Act
FFT	Fast Fourier Transform
GLR	Graphic Level Recorder
HCA	Hearing Conservation Amendment
HCP	Hearing Conservation Program
HI	Hearing Impairment
HPD	Hearing Protection Device
HL	Hearing Level
HTL	Hearing Threshold Level
HVAC	Heating, Ventilating and Air Conditioning
IEC	International Electrotechnical Commission
IL	Insertion Loss
INEP	Industrial-Noise-Exposed Population
INIPTS	Industrial-Noise-Induced Permanent Threshold Shift
ISO	International Organization for Standardization
LCL	Lower Confidence Limit
MAF	Minimum Audible Field
MAP	Minimum Audible Pressure

ME	Medical Evidence
MSHA	Mine Safety and Health Administration
NBS	National Bureau of Standards
NINEP	Non-Industrial-Noise-Exposed Population
NIOSH	National Institute for Occupational Safety and Health
NIPTS	Noise-Induced Permanent Threshold Shift
NR	Noise Reduction
NRR	Noise Reduction Rating
OSHA	Occupational Safety and Health Administration
PE	Pressure Equalization
PEL	Permissible Exposure Limit
PNC	Preferred Noise Criteria
PTS	Permanent Threshold Shift
REAT	Real-Ear Attenuation at Threshold
RMS	Root Mean Square
SD	Speech Discrimination
SEL	Sound Exposure Level
SIL	Speech Interference Level
SLM	Sound Level Meter
SN	Serial Number
SNO	Survey Number
SPL	Sound Pressure Level
STL	Sound Transmission Loss
STS	Standard Threshold Shift
TL	Transmission Loss
TR	Transmissibility
TTS	Temporary Threshold Shift
TWA	Time-Weighted Average
UCL	Upper Confidence Limit
WC	Workers' Compensation
WCL	Workers' Compensation Law

Noise and Hearing Conservation Manual, edited by
E.H. Berger, W.D. Ward, J.C. Morrill and L.H. Royster

1 Hearing Conservation

Alice H. Suter

Contents

	Page
Historical Background	1
Need for Hearing Conservation	4
Summary of the Effects of Noise	6
Hearing Loss	6
Communication Interference	7
Effects on Performance	8
Annoyance	8
Noise and Stress Diseases	9
The Solution: Hearing Conservation Programs	9
Noise Measurement	10
Engineering and Administrative Controls	11
Audiometric Testing	12
Hearing Protection	13
Employee Training and Education	14
Record Keeping	15
Program Evaluation	15
Summary	16
References	16

Historical Background

People have been aware of the connection between noise and hearing loss for centuries. W. Burns cites cases of noise-induced hearing loss described by C. H. Parry in 1825 (Burns, 1973). One involved a British admiral "being almost entirely deaf for fourteen days following the firing of eighty broadsides from his ship, H. M. S. *Formidable,* in the year 1782." Before the industrial revolution these cases were limited to pursuits like battle (where the consequences were often much worse) and blacksmithing. The advent of the industrial revolution changed noise-induced hearing loss from an infrequent occupational condition to an epidemic. As Burns points out, one of the chief early contributers was the coal-fired steam engine with the necessity for riveted boilers. Thus, a name which still survives today was given to the condition: boilermaker's ear.

Despite general recognition of the fact that noise causes hearing loss, little was done about it until recent decades. As with so many occupational hazards, workers felt that it was a necessary price to pay in an industrial job, and managers felt that the noise was either too difficult or too costly to control. Toward the end of World War II the American hearing-impaired population was swelled by returning soldiers with noise-induced hearing losses. Consequently, the armed services established "aural rehabilitation" centers, and the field of audiology was created (Newby, 1964). Later, the armed services developed programs to protect servicemen against the hazards of noise; the earliest of these was embodied in Air Force Regulation 160-3 in 1956 (Air Force, 1956).

Worker compensation for hearing loss must also have influenced the development of hearing conservation programs. For a long time most worker compensation statutes did not cover noise-induced hearing loss either because it was not considered an accident or because it did not result in lost earnings. Then in 1948 the New York Court of Appeals awarded compensation to a hearing-impaired worker who demonstrated no loss of earnings, and in 1950 a similar decision was handed down in Wisconsin (Newby, 1964).

Around this time researchers began to investigate the relationship between hearing loss and various noise parameters such as level, frequency, and duration. In 1950, K. D. Kryter published one of the earliest and most famous monographs describing safe and hazardous noise exposures (Kryter, 1950), followed shortly by Rosenblith and Stevens (1953) and a committee of the American Standards Association (1954). During the 1960s a number of important hearing loss studies took place. Between 1960 and 1965 W. L. Baughn studied the effects of noise on more than 6000 automobile workers (Baughn, 1973). Although Baughn's data and method of analysis were not actually published until 1973, they had considerable influence on damage-risk criteria and standard setting at the time. Another important study was conducted by W. Burns and D. W. Robinson who investigated the effects of noise on 759 industrial workers between 1963 and 1968 (Burns and Robinson, 1970). A third study of 792 industrial workers was conducted by the National Institute for Occupational Safety and Health (NIOSH, 1973) between 1968 and 1972. In the fourth large study W. Passchier-Vermeer combined and analyzed the data from ten different noise and hearing loss studies of American and European origin, with a total subject population of about 4600 (Passchier-Vermeer, 1968). The results of these studies have had considerable influence on consensus as well as governmental noise standards in U.S. and abroad.

With a growing body of information on the effects of noise on hearing (and the consequences of long-term, unprotected exposure) there also was a growing impetus to control workers' exposures through regulation. Few employers had established company-wide hearing conservation programs, with the notable exception of the DuPont Company, which has had a program since 1956 (Pell, 1972), and many of the companies served by Employers Mutual Liability Insurance Company of Wausau, due to the crusading efforts of Roger Maas in the late 1950s and 1960s.

The first U.S. Government noise regulation was issued by the Department of Labor in 1969 under the authority of the Walsh-Healey Public Contracts Act (U.S. Dept. Labor, 1969). The standard applied only to employers who contracted with the U.S. Government. Most of the standard's provisions are still in effect at the time this manual is being prepared. It specifies a maximum permissible A-weighted exposure level (PEL) of 90 decibels (dB) over an 8-hour day, with a 5-dB increase allowed for every halving of exposure duration. Thus the PEL is not a single level, but a time-weighted average level experienced throughout the day. A ceiling for impulse noise is set at a peak sound pressure level of 140 dB, but this ceiling appears to be advisory rather than mandatory because the standard reads "should" instead of "shall" in this case. Noise levels are to be measured on the A-weighting network of a sound level meter set to "slow" meter response. Whenever the permissible exposure levels are exceeded, employers must issue hearing protection (and employees must wear them), and must reduce the exposure to within permissible levels by feasible engineering controls. The standard also had a provision for "continuing, effective hearing conservation programs" whenever the permissible levels were exceeded, which has since been modified.

The Walsh-Healey noise standard became applicable to all of general industry in 1971 after the enactment of the Occupational Safety and Health Act. Soon after, the Occupational Safety and Health Administration (OSHA) embarked on the process of revision. NIOSH published a criteria document in 1972 as a result of its noise study, recommending reduction of the PEL to 85 dBA after OSHA had studied the technical feasibility of such a reduction (NIOSH, 1972). OSHA established a formal advisory committee in 1973, and issued a proposed revision in 1974 (OSHA, 1974). The proposal retained the 90-dBA PEL but gave detailed requirements for "hearing conservation programs" (meaning primarily noise monitoring, audiometry, and hearing protection) to be initiated at a time-weighted average level of 85 dBA. During two sets of public hearings and years of deliberation, two major issues emerged: 1) whether the PEL should be set at 85 or 90 dBA, and 2) whether OSHA should continue to require engineering and administrative controls as the primary means of compliance or allow hearing protectors to be given equal footing.

By 1981 OSHA still had not made a decision on these two issues, but seeing the need for more precise and detailed requirements for hearing conservation programs, the agency amended the noise standard with detailed hearing conservation requirements on January 16, 1981 (OSHA, 1981a). These requirements replaced the one-sentence statement about hearing conservation programs in the Walsh-Healey noise standard.*

Due to a change of administration just four days later, the effective date of the amendment was delayed while the regulation underwent an extensive review. Portions were made effective (and therefore enforceable) in August 1981 (OSHA, 1981 b) and OSHA held public hearings on those provisions that were still stayed in March of 1982. The entire amendment was issued in final form on March 8, 1983 (OSHA, 1983). The noise standard as amended (29 CFR 1910.95 (a) through (p) with appendices) is reprinted in Appendix I of this manual as it appears in the Code of Federal Regulations in 1983. The reader should be aware, however, that OSHA standards are always subject to amendment or revision, and that the noise standard discussed and reprinted in this manual may be changed before the manual is revised. For a thorough explanation of the amendment's requirements, including the agency's rationale for each section and a description of the costs and benefits of compliance, the reader should consult the preambles of each action in the Federal Register (OSHA, 1981 a and b, and 1983).

These documents are usually available free of charge from the OSHA Publications Office, U.S. Department of Labor, Washington, DC 20210.

Need for Hearing Conservation

According to a recent Environmental Protection Agency (EPA) report prepared by the acoustical consulting firm, Bolt, Beranek and Newman, there are more than nine million Americans exposed to daily average occupational noise levels above 85 dBA (EPA, 1981). Most of these workers are in the production industries, but other occupations also can be noisy. The breakdown is in Table 1.1.

---

*It is important to note that the 90-dBA PEL and the requirement for engineering or administrative controls have only been supplemented, not superseded, by the hearing conservation amendment.

TABLE 1.1

Workers Exposed to Daily Average Noise Levels Exceeding 85 dBA by Employment Area

Agriculture	323,000
Mining	400,000
Construction	513,000
Manufacturing and Utilities	5,124,000
Transportation	1,934,000
Military	976,000
Total	9,270,000

From *Noise in America,* EPA Report No. 550/9-81-101.

In the preamble to the hearing conservation amendment (January 1981) OSHA estimated that the total cost of the hearing conservation amendment would be $270 million. This cost covers 5.5 million workers in general industry and maritime occupations. On the basis of a NIOSH study (NIOSH, 1975), OSHA estimated that approximately $16 million per year was currently being spent on hearing conservation programs (OSHA, 1981a) (excluding expenditures for engineering controls). This estimate indicates that 83% of the 5.5 million workers in general industry and maritime were not being covered, and were in need of hearing conservation services. Similar conclusions can be drawn from another NIOSH study, the National Occupational Health Survey (NIOSH, 1978). The situation is likely to be even less favorable for construction and mine workers, who are not covered by the hearing conservation amendment, and for transportation and agricultural workers, most of whom are not covered by any federal noise standard.

Although noise-induced hearing loss is one of the most common occupational "diseases" or conditions, it is often underrated because there are no visible effects and, except in very rare cases, there is no pain. There is only a gradual, progressive loss of communication with family and friends, and loss of sensitivity to the environment. Good hearing is always taken for granted until it is lost.

The course of noise-induced hearing loss is insidious. The first sign is usually that other people do not seem to speak as clearly as they used to. The noise-exposed person will have to ask others to repeat and grows annoyed with their lack of consideration. As the loss becomes worse, the hearing-impaired person will begin to withdraw from social situations.

Church, movies, and parties lose their attraction and the individual will choose to stay at home. Sometimes he or she will turn the volume of the TV so high that other family members will be driven out of the room. If the loss progresses to a severe stage, the victim will no longer be able to communicate with family or friends without great difficulty. Hearing aids may help in some cases, but clarity will never be restored, as it is so easily with eyeglasses. This unfortunate situation can be prevented and is being prevented more and more effectively in industrial health and hygiene programs throughout the country.

Summary of the Effects of Noise

HEARING LOSS

Hearing loss is the most obvious and the best quantified of all the effects. Its progression, however, is not always obvious, as mentioned above. In fact, most losses probably begin by being temporary. During the course of a noisy day, the ear becomes fatigued and the worker will experience a temporary reduction in hearing sensitivity known as temporary threshold shift (TTS). If no noisy activities are pursued after work, the loss will often recover by the next morning. Sometimes a worker will report turning off the car radio with the ignition switch at night, and on starting the car the next morning, the radio comes on blaring. If this pattern of daily TTS is repeated over a period of months or years the loss will become what is known as a permanent threshold shift (PTS). The process will be hastened if the exposure is intense or long enough that recovery from TTS is incomplete before the start of the next exposure.

Occupational noise is not, of course, the only cause of hearing loss in a noise-exposed population. There are three types of hearing loss that occur commonly in addition to occupational hearing loss. One is called presbycusis, the loss that occurs naturally from aging. Another is sociocusis, a term coined by Aram Glorig, which is usually taken to mean the loss caused by nonoccupational sources, so prevalent in the modern "civilized" world (Glorig, 1958). Examples of sociocusic sources could be power tools, lawnmowers, amplified music, etc. Military noise exposure, when it precedes industrial exposure, is considered sociocusis. The third category of hearing loss is a very broad one, and includes all kinds of medical abnormalities. Burns refers to these losses as "pathology" (Burns *et al.*, 1977); Ward calls them "nosoacusis" (Ward, 1977). Examples would be anything from impacted ear wax, which is very amenable to treatment, to a severe deafness caused by rubella during gestation. Diseases of the outer and middle ear produce conductive hearing losses, which can be easily distinguished from noise-induced hearing losses by clinical tests. Diseases of the inner ear can be more difficult to differentiate.

It is virtually impossible to separate presbycusis or sociocusis from noise-induced hearing loss by looking at the audiogram. Even after taking a detailed history of an individual's nonoccupational activities, the contribution of presbycusis and sociocusis can only be determined by conjecture. Therefore, to subtract a presbycusis value for nonoccupational hearing loss from the hearing levels of a noise-exposed worker is sometimes not equitable. However, when noise-exposed populations are considered, standard values can be subtracted for presbycusis on the basis of data from large samples of non-noise-exposed people. The size of this subtraction depends on the extent to which both populations have been screened to remove subjects with unusually great amounts of sociocusis and pathology. Some presbycusis corrections may appear quite large because they actually include sociocusis and nosoacusis as well as hearing loss solely from the aging process. These kinds of corrections are quite valid if the experimental population is similarly unscreened (see Baughn, 1973, p. 4).

COMMUNICATION INTERFERENCE

Hearing loss is not the only adverse effect of occupational noise. There are a number of nonauditory effects, which may be classified as communication interference, safety, performance, annoyance, and physiological effects.

The developers and users of communication systems have known for a long time that noise masks, or interferes with, speech. In fact, speech communication criteria have become quite sophisticated. Predictive curves exist for the amount and kind of speech that will be intelligible in various kinds of spaces as a function of the level, frequency composition, and temporal pattern of the background noise, as well as the distance between talker and listener (Webster, 1979). The average sound levels at 500, 1000, 2000, and 4000 Hz are called speech interference levels (SILs). Webster has developed curves showing SIL, as a function of distance between talkers and the vocal effort they use, which will permit 95 percent sentence intelligibility (Webster, 1979). The SIL curves are appropriate for noisy industry but they probably are not used very often. In most normal industrial operations the noise has, from the onset, determined the kind and amount of communication that can take place. People have learned from experience that in noise levels above 80 dB they have to speak very loudly and in 85 to 90 dB they have to shout. In anything much above 95 dB they have to move very close together to communicate at all.

Industrial noise does become a communication problem when it masks speech or signals that are necessary to carry out the job or to ensure employee safety. In the first instance, the industrial hygienist or acoustical specialist will need to determine the amount of noise reduction necessary to achieve the requisite communication. The safety problem is much more elusive. Sometimes it becomes apparent only after a fatality or serious accident. There have been many reports of workers who have gotten clothing or hands caught in machines and have been seriously injured while coworkers were oblivious to cries for help. There are other stories about workers who, because they were unable to hear a warning signal, remained on the job while others were evacuated. Because of these types of incidents, some employers have installed visual warning devices. Unfortunately, in most cases the cause-and-effect relationship between noise and safety is never realized.

EFFECTS ON PERFORMANCE

The effects of noise on job performance have been studied extensively in the laboratory and somewhat less so in the field. Studies have shown that noise can increase, decrease, or have no effect on job performance, depending on the circumstances. A thorough yet succinct analysis of these studies may be found in a discussion by D. E. Broadbent (1979). In general, low to moderate levels of noise may increase job performance in monotonous tasks. Even high noise levels may increase output in terms of numbers, but errors are more likely to occur and quality will often be reduced. Tasks involving concentration are more vulnerable to noise disruption than are routine tasks, and intermittent noise tends to be more disruptive than continuous noise (Broadbent, 1979). Interesting studies have shown that people perform more poorly on tasks *after* being exposed to noise that was unpredictable and uncontrollable. The investigators attributed the poor performance to a sense of apathy or helplessness that resulted from the unpredictability and uncontrollability (Glass and Singer, 1973). Broadbent also cites some research indicating that people exhibit less helpful behavior during and after noise exposure than they do in quiet (see Broadbent, 1979, pp. 16-17).

ANNOYANCE

Perhaps the term "annoyance" is more appropriately applied to community noise, which usually is intermittent, with relatively long intervals of quiet. Because occupational noise, at least in the production industries, tends to be constant, aversion might be a better term for the effect. Occasionally the initial aversion is so strong that young workers will seek employment elsewhere after the first few days (if they have the opportunity to do so). Most will become adjusted to the situation in time, and the adjustment will be more successful if hearing protectors are properly fitted and worn from the start. However, without adequate attenuation, workers will often report fatigue, irritability, and sleeplessness, which is attributed to the noise. This kind of information sometimes appears *after* a company initiates a hearing conservation program because workers are suddenly aware of the contrast.

NOISE AND STRESS DISEASES

Some scientists view the "adjustments" that we make to modern, stressful society as harmful in the long run. They are the root of the so-called stress diseases: heart attack, stroke, ulcers, even cancer. Stress diseases occur after the body has been chronically mobilized to fight off an enemy, either real or perceived. In primitive times high levels of noise were rare and when they occurred, they signified danger. The body would go through a series of biological changes preparing either to fight or to run away. There is evidence that these changes persist, even though the stimulus is no longer viewed as dangerous. High levels of noise have been implicated particularly in cardiovascular disorders (Hattis and Richardson, 1980). Hundreds of studies have been conducted on the effects of occupational stress, many of which involved noise. Many of these studies show an increased incidence of cardiovascular disease with increasing exposure level and some show no effect. Many of the field studies are plagued by methodological problems because other stressors (in addition to noise) are extremely difficult to control and would contaminate the results. Laboratory studies on animals are difficult to generalize to humans, and studies on humans show acute effects which may or may not become chronic. There is a significant set of laboratory studies by E. A. Peterson and his colleagues investigating the effects of protracted noise exposure on monkeys (Peterson *et al.*, 1978 and 1983). Peterson found that noise around 85 to 90 dB caused chronic elevated blood pressure levels, which did not return to baseline after cessation of the exposure. The extent to which these experiments resemble the human condition is not known, but they certainly would imply a need for caution.

The Solution: Hearing Conservation Programs

An effective hearing conservation program can do more than prevent hearing loss. It can improve employee morale and general feeling of well being, it can improve the quality of production, and it may reduce the incidence of stress-related disease. Over and above the benefits of hearing saved, the hearing conservation program seems like a chance worth taking. With the right approach it can also be a vehicle for improving labor-management relations.

The first step is to enlist the cooperation of management. Without the support of the whole management team, the program has little chance for success. Managers and supervisors must understand the effects of noise, the particulars of applicable regulations, and the need for participation in the program. They must be committed to wearing hearing protectors in posted zones. Otherwise they will create the feeling that hearing protectors are a burden imposed on hourly employees, but not to be bothered with by salaried personnel.

The next step, of course, is to enlist the support of the noise-exposed employees. If possible, they should be included in the planning of the program. They will have firsthand knowledge of the noisiest machines, and sometimes will have valuable suggestions as to how to deal with them. If the plant is organized, it is a good idea to discuss program plans with the appropriate union officers to enlist their cooperation in the earliest stages. If hearing protection is necessary, it is extremely important for union officials to agree to it, to provide the necessary leadership for the rank and file. As representatives of their constituents, union leaders will be more likely to accept hearing protectors if there is a viable program for engineering controls to be achieved in the future. Any discomforts of hearing protectors are more palatable if a person is reasonably sure that he will not have to wear them forever.

Later chapters in the manual will discuss extensively the various components of hearing conservation programs. However, it would be useful to summarize them very briefly at this stage.

NOISE MEASUREMENT

Noise measurements are necessary to identify overexposed employees and the machines contributing to their overexposures. These two processes require somewhat different measurement techniques. If the noise is continuous and employees are fairly stationary, area measurements might be sufficient for both purposes, provided that measurements are taken at or near employees' stations. However, most work environments consist of noise which varies with time to some extent, and most workers are somewhat mobile, so noise exposures are not often measured so easily. Separate procedures are usually necessary.

Actual exposure measurements are needed so that workers will be identified for audiometric monitoring purposes and so that proper hearing protectors can be selected. Exposure measurements are also needed to comply with OSHA's hearing conservation amendment, except in cases where the noise is continuous, as described above. A Type II sound level meter (SLM) is appropriate for measuring exposure whenever the noise is fairly steady. In fluctuating noise, however, dosimeters or integrating SLMs are preferable, both because of limitations of the Type II meter and because of the difficulties involved in temporal sampling procedures. Impulsive noise environments require a modern dosimeter or integrating SLM. An SLM with peak-hold capability is useful for ascertaining impulse noise peaks, both for OSHA compliance and for engineering purposes.

Diagnosing noise problems for engineering solutions requires more sophisticated instruments, and in some cases, the services of an acoustical engineer. The equipment battery may include an octave band analyzer, precision SLM, tape recorder, graphic level recorder, etc. Proper diagnosis and treatment is of the utmost importance so that funds will not be squandered on ineffective control measures.

It is always a good idea to let employees know about the initiation of a noise measurement program and to enlist their cooperation. This is especially important when dosimeters are used, to prevent any mishandling that might lead to spurious measurements. The OSHA standard requires that workers exposed to daily average levels of 85 dBA or above be notified of the results of the monitoring. Employers must also allow employees or their representatives to observe the noise monitoring procedures.

ENGINEERING AND ADMINISTRATIVE CONTROLS

Engineering controls are the best long-term solution to the occupational noise problem. In some cases they can be extremely expensive, but in others they can be surprisingly cheap. Despite the promulgation of the hearing conservation amendment, they are still (at the time of this writing) required by OSHA whenever they are feasible, as the first order of compliance.

Noise problems can be divided according to the traditional three methods of control: the source, the path, and the receiver. Controlling the source is usually the most satisfactory approach, but the other approaches can be very useful as well.

Methods to control the source include isolating a machine to prevent the radiation of vibrations, putting on a muffler, and enclosing it. While enclosures can be very effective, they may lead to problems when workers are unable to manipulate parts of their machines, or the process is slowed down. One of the most effective means of source control is to redesign the machine or work process. Design for quiet can be built into a company's specifications for new machinery, avoiding the trouble and expense of retrofitting old equipment.

One method of controlling the transmission (or path) of sound is to place absorptive material on the ceilings and walls to reduce the reflection of sound waves. Another is to erect barriers or to hang lead curtains around a noisy area. These methods are particularly effective against high-frequency noise. The third approach is to treat the receiver's immediate environment. Methods include erecting an enclosure or control booth around the employee, and, of course, hearing protection. The control booth can be a very effective method of reducing exposure, but in some cases it can lead to feelings of isolation. Booths can be quite successful in cases where all the controls are located in a central area, and the room is big enough to house a number of employees.

Another approach to noise exposure control, one that OSHA gives equal status with engineering controls, is the use of administrative controls. People usually think of administrative controls as the rotation of workers between noisy and quiet jobs (or more often noisy and less noisy jobs). This practice is not widely used. Workers who are trained in a particular skill generally do not like being rotated to other jobs, particularly when the quieter jobs are often less skilled. Unions usually object to this practice. Also, rotation is not very good health practice because, while it may reduce the amount of hearing loss individuals incur, it spreads the risk among other workers. The final result tends to be that many workers develop small hearing losses rather than a few workers developing big ones (Ashford *et al.,* 1976). One thing that employers can do to mitigate the problem (although it is not a complete solution) is to provide workers with rest and lunch areas that are sufficiently acoustically treated to provide the ears with enough quiet to effect some recovery from TTS. The levels of these places when occupied should not exceed 70 dB. The only completely satisfactory solution is to reduce exposures to safe levels (8-hour time-weighted average levels below 80 dBA).

AUDIOMETRIC TESTING

Audiometric testing is an integral part of a hearing conservation program whenever employees are exposed to daily average noise levels exceeding 85 dBA. It does not actually protect workers, but it is the only way to tell if the hearing conservation program is working. Because audiometric tests depend upon a subjective human response, there is a certain amount of normal variability to be expected. Thresholds that vary 5 dB are usual in the clinic and variability can even exceed 10 dB in the field without cause for alarm. Well-calibrated equipment, well-trained testing personnel, well-instructed subjects, and quiet audiometer rooms can minimize spurious variability. Employers need to worry about noise exposure when threshold shifts are large at one frequency, when they are moderate at two or more adjacent frequencies, or when even small successive shifts occur consistently over a period of years. OSHA's current definition of a shift to worry about, a significant or "standard" threshold shift, is an average shift of 10 dB or greater at 2000, 3000, and 4000 Hz in either ear. (Presbycusis values may be subtracted from current hearing threshold levels.)

The object of the audiometric test program is to identify workers who are beginning to lose their hearing and to intervene before the hearing loss becomes worse, particularly before it gets to the handicapping stage. Intervention may be through engineering controls or through hearing protection. If hearing protection is already in use, the appearance of a significant threshold shift means that the industrial hygienist or other health professional should carefully reinstruct the worker on the proper fitting of the protector, or perhaps select a different protector. It is also a signal to the worker that more diligence is needed on his or her part.

One of the best ways to identify noise-induced hearing loss *before* it becomes permanent is to test workers at the end of the workshift, assuming that the baseline test has been conducted after an adequate period of effective quiet. If, at the end of the shift, a worker shows a threshold shift with respect to the baseline, the chances are that it is a TTS. To confirm this fact, the worker can be tested again the next morning. Then intervention can take place before the loss becomes permanent.

HEARING PROTECTION

Along with engineering controls, hearing protection is the other principal method of preventing noise-induced hearing loss. But selection, fitting, and wearing procedures are as important as the protectors themselves. Many employers were under the impression that all they needed to do was to hand out hearing protectors and the noise problem would be solved. People have begun to realize, however, that without sufficient attention to fitting, employees will experience considerable discomfort and the protectors simply will not be worn. Improperly fitting protectors, when they are worn, will allow the sound to leak past them into the ear canal and attenuation will be greatly reduced. Over recent years a number of researchers have investigated the attenuation of hearing protectors in the field, *as they are worn*, rather than in the laboratory. In general, they have found that field attenuation is about one-half of the number of decibels of attenuation that is achieved in the laboratory, and the standard deviation is about three times as large (Berger, 1983a).

The OSHA hearing conservation amendment requires employers to give employees a choice of hearing protectors. This means at least two or three different protectors (with sufficient attenuation) and preferably a choice between plugs and muffs. Protectors must be properly fitted initially, and employees must be trained in their care and use. Protector attenuation must be sufficient for an individual's noise environment. OSHA requires that protectors attenuate to time-weighted average levels of 90 dBA or below, except for people who have suffered significant threshold shifts, in which case the protector must attenuate to 85 dBA. Because hearing loss may occur in people chronically exposed to levels of 85 dBA and above, it is wise to use protectors that attenuate to 85 dBA in all cases.

Even with careful training, some individuals still have trouble wearing their hearing protectors correctly, and therefore experience reduced attenuation (Fleming, 1980). There are some good ways, however, to check protector attenuation in the plant, without having to use special semi-reverberant rooms. One method is to bring a worker into the audiometric test room, put the earphones on over the worker's plugs, and run a hearing test with and without the plugs. The greater the difference in thresholds, the better the attenuation. Some companies test all of the standard frequencies this way (500 - 6000 Hz). Other methods using various psychoacoustic techniques necessitate some modifications to the standard industrial audiometric equipment, but have the advantage of being able to be performed in moderately noisy environments (Fleming, 1980; Berger, 1983b). Although none of these methods have been standardized, they can be very useful as teaching and evaluation techniques.

EMPLOYEE TRAINING AND EDUCATION

The success of the hearing protection program will be greatly enhanced if employees are well trained and educated in hearing conservation. Besides the fitting, care, and use of hearing protectors, this includes teaching workers about the effects of noise, the advantages and limitations of hearing protectors, and the purpose and particulars of audiometric testing. It also should include an explanation of the company's noise measurement procedures, and if possible, a discussion of the company's plans to reduce noise exposure by engineering controls. Workers who understand the mechanism of hearing and how it is lost will be more motivated to protect themselves. Also, workers who are actively involved in the planning and administration of the program, particularly if they see some noise reduction goals for the future, will be more inclined to accept the inconvenience of hearing protection. In a recent study by Zohar *et al.*, the authors investigated ways to improve the acceptance of hearing protectors by a particularly resistant group of workers (Zohar *et al.,* 1980). They found that by testing them before and after the workshift, by educating them in the identification and consequences of TTS, and by posting the audiograms for all to see, the acceptance of hearing protectors improved dramatically, from 35% to 80% use.

RECORD KEEPING

The final element of the hearing conservation program is record keeping. While this may not be the most exciting element, it is critical to the program's successful functioning. Exposure records need to be kept so that the hygienist or nurse can check to make sure that hearing protectors are adequate for employees' noise conditions. Records of hearing thresholds must, of course, be kept so that audiometric testers or audiogram reviewers will know whether a worker has incurred a significant threshold shift, or whether thresholds are decreasing gradually over time. Audiologists or physicians who review audiograms or supervise audiometric testing programs will need to consult records to make sure that calibrations have been carried out properly and that background levels in audiometer rooms are low enough to permit accurate tests. Finally, all of the above records need to be kept in the eventuality of worker compensation claims. With the advent of computerization, and particularly with the help of certain microprocessor audiometers, record keeping need not be as tedious or as space-consuming as it used to be.

PROGRAM EVALUATION

One further use of records is in the evaluation of program effectiveness. Certainly audiograms are not very useful if they are filed away after the test and never evaluated. Even when significant threshold shifts are documented, and workers are notified, refitted with hearing protectors and retrained, employers' responsibilities are not completely fulfilled. They need to be constantly vigilant to make sure that the hearing conservation program is working, and not just an exercise in futility. What good does all of the measuring and testing do if 30% of the workers experience significant threshold shift each year? Then the question arises: how many significant threshold shifts are acceptable? There is no official guidance on this matter, either from OSHA or from consensus groups. Audiologists Morrill and Sterrett suggest that three to six percent would be acceptable, using a definition that is very similar to the one OSHA promulgated in 1983 (Morrill and Sterrett, 1981). This figure presumably excludes hearing loss from external and middle ear pathologies.

Some very creative and useful ways of evaluating hearing conservation program effectiveness have been developed by Larry and Julia Royster (Royster *et al.,* 1980; Royster and Royster, 1982; See Chapter 9). Their method consists of examining computerized audiograms taken over a period of five or six years. They look for a learning effect, where mean hearing threshold levels should show a slight improvement over the first four years. This signifies an effective program. Other criteria include test-retest comparisons, which is the proportion of thresholds improving compared to those decreasing, sequential test-retest comparisons, in which each audiogram is compared to the results of the preceding year, and comparisons to the thresholds of a non-noise-exposed population. By looking at the combined results of these comparisons, program managers and consultants can gain insight into the overall effectiveness of the program. The most effective program is the one where the changes are fewest and smallest, except, of course, changes for the better.

Summary

Public awareness of the occupational noise problem has increased tremendously in recent decades. Although much research remains to be done, the effects of noise are increasingly well quantified, especially the effects of noise on hearing and on communication. The solutions, too, have become well known in most instances, and all that remains is for managers, health and engineering personnel, and employees to mobilize their efforts and set about the task. The task will only be successful, however, if these participants work together in a knowledgeable and cooperative manner.

References

Air Force (1956). Hazardous Noise Exposure. U.S. Air Force Regulation No. 160-3, Washington, DC.

American Standards Association (1954). The Relationship of Hearing Loss to Noise Exposure. Report of the Z24-X-2 Exploratory Subcommittee of the American Standards Association Z24 Sectional Committee on Acoustics, Vibration and Mechanical Shock. New York, NY.

Ashford, N.A., Hattis, D., Zolt, E.M., Katz, J.I. and Heaton, G.R. (1976). Economic and Social Impact of Occupational Noise Exposure Regulations. EPA Report 550/9-77-352.

Baughn, W.L. (1973). Relation Between Daily Noise Exposure and Hearing Loss Based on the Evaluation of 6,835 Industrial Noise Exposure Cases. AMRL-TR-73-53. Aerospace Medical Research Laboratory, Wright-Patterson AFB, OH.

Berger, E.H. (1983a). "Using the NRR to Estimate the Real World Performance of Hearing Protectors," *Sound and Vibration 17,* 12-18.

Berger, E.H. (1983b). "Assessment of the Performance of Hearing Protectors for Hearing Conservation Purposes," *J. Acoust. Soc. Am. Suppl. 1, 74,* S94.

Broadbent, D.E. (1979). "Human Performance and Noise." In *Handbook of Noise Control, 2nd. Ed.,* edited by C.M. Harris, McGraw-Hill, New York, NY.

Burns, W., and Robinson, D.W. (1970). *Hearing and Noise in Industry.* Her Majesty's Stationery Office, London, England.

Burns, W. (1973). *Noise and Man,* John Murray, London, England.

Burns, W., Robinson, D.W., Shipton, M.S. and Sinclair, A. (1977). Hearing Hazard from Occupational Noise: Observations on a Population from Heavy Industry. NPL Acoustics Report Ac80. National Physical Laboratory, Teddington, England.

EPA (1981). Noise in America: The Extent of the Noise Problem. EPA Report No. 550/9-81-101. U.S. Environmental Protection Agency, Washington, DC.

Fleming, R.M. (1980). "A New Procedure for Field Testing of Earplugs for Occupational Noise Reduction." Doctoral thesis, Harvard School of Public Health, Boston, MA.

Glass, D.C. and Singer, J.E. (1973). "Behavioral Effects and Aftereffects of Noise," in *Proceedings of the International Congress on Noise as a Public Health Problem,* edited by W.D. Ward, U.S. Environmental Protection Agency, Washington, DC., 409-416.

Glorig, A. (1958). *Noise and Your Ear,* Grune & Stratton, New York, NY.

Hattis, D. and Richardson, B. (1980). Noise, General Stress Responses, and Cardiovascular Disease Processes: Review and Reassessment of Hypothesized Relationships. EPA report 550/9-80-101. U.S. Environmental Protection Agency.

Kryter, K.D. (1950). "The Effects of Noise on Man," *J. Speech Hearing Disord.* Monograph Suppl. 1.

Morrill, J.C. and Sterrett, M.L. (1981). "Quality Controls for Audiometric Testing," *Occupational Health and Safety 50.*

Newby, H.A. (1964). *Audiology, 2nd Ed.* Appleton-Century-Crofts, New York, NY.

NIOSH (1972). Occupational Exposure to Noise, HSM 73-11001. U.S. DHEW, National Institute for Occupational Safety and Health, Cincinnati, OH.

NIOSH (1973). Occupational Noise and Hearing, NIOSH Pub. 74-116. U.S. DHEW, National Institute for Occupational Safety and Health, Cincinnati, OH.

NIOSH (1975). Survey of Hearing Conservation Programs in Industry. NIOSH Pub. 75-178. U.S. DHEW, National Institute for Occupational Safety and Health. Cincinnati, OH.

NIOSH (1978). National Occupational Hazard Survey, Vol. III. NIOSH Pub. 78-114. U.S. DHEW, National Institute for Occupational Safety and Health, Cincinnati, OH.

OSHA (1974). "Occupational Noise Exposure; Proposed Requirements and Procedures." *Fed. Reg. 39:* 37773-37778. U.S. DOL, Occupational Safety and Health Administration, Washington, DC.

OSHA (1981a). "Occupational Noise Exposure; Hearing Conservation Amendment." *Fed. Reg. 46:* 4078-4179. U.S. DOL, Occupational Safety and Health Administration, Washington, DC.

OSHA (1981b), "Occupational Noise Exposure; Rule and Proposed Rule," *Fed. Reg. 46*: 42622-42639. U.S. DOL. Occupational Safety and Health Administration, Washington, DC.

OSHA (1983). "Occupational Noise Exposure; Hearing Conservation Amendment; Final Rule." *Fed. Reg. 48:*9738-9785. U.S. DOL, Occupational Safety and Health Administration, Washington, DC.

Passchier-Vermeer, W. (1968). Hearing Loss Due to Exposure to Steady-State Broadband Noise. Report No. 35. IG-TNO Research Institute for Public Health Engineering, Delft, Netherlands.

Pell, S. (1972). "An Evaluation of a Hearing Conservation Program," *Am. Ind. Hyg. Assoc. J. 33*:60-70.

Peterson, E.A., Augenstein, J.S. and Tanis D.C. (1978). "Continuing Studies of Noise and Cardiovascular Function," *J. Sound Vibration 59*, 123.

Peterson, E.A., Augenstein, J.S., Tanis, D.C., Warner, R. and Heal, A. (1983). "Some Cardiovascular and Behavioral Effects of Noise in Monkeys," in *Noise as a Public Health Problem--Fourth International Congress.* Centro Ricerche e Studi Amplifon, Milan, Italy.

Rosenblith, W.A. and Stevens, K.N. (1953). Handbook of Acoustic Noise Control, Vol. II, Noise and Man. Wright Air Development Center Tech. Report No. 52-204. Wright-Patterson AFB, OH.

Royster, J.D. and Royster, L.H. (1982). "Evaluating the Effectiveness of Hearing Conservation Programs by Analyzing Group Audiometric Data," Paper presented at the annual convention of the American Speech-Language-Hearing Assoc., Toronto, Ontario.

Royster, L.H., Lilley, D.T., and Thomas, W.G. (1980). " Recommended Criteria for Evaluating the Effectiveness of Hearing Conservation Programs," *Am. Ind. Hyg. Assoc. J. 41*:40-48.

U.S. DOL (1969). "Occupational Noise Exposure." *Fed. Reg. 34:*7946-7949. U.S. Dept. of Labor, Bureau of Labor Standards, Washington, DC.

Ward, W.D. (1977). "Effects of Noise Exposure on Auditory Sensitivity." In *Handbook of Physiology, Vol 9: Reaction to Environmental Agents,* edited by D.H.K. Lee, American Physiological Society, Bethesda, MD. 1-15.

Webster, J.C. (1979). "Effects of Noise on Speech," in *Handbook of Noise Control, 2nd. Ed.* edited by C.M. Harris, McGraw-Hill, New York, NY.

Zohar, D., Cohen, A. and Azar, N. (1980). "Promoting Increased Use of Ear Protectors in Noise Through Information Feedback," *Human Factors 22,* 69-79.

Noise and Hearing Conservation Manual, edited by
E.H. Berger, W.D. Ward, J.C. Morrill and L.H. Royster

2 Physics of Sound

Paul B. Ostergaard

Contents

	Page
Basic Quantities	19
Physical Units and Sound Radiation	22
Noise Reduction and Isolation	26
Decibel Addition	28
Example 2.1	29
Sound Power and Sound Pressure	30
Example 2.2	34
Example 2.3	35
References	36

Basic Quantities

In air, sound is usually described in terms of variations of pressure above and below the ambient atmospheric pressure. These pressure oscillations (p), commonly known as sound pressure, are generated by a vibrating surface or turbulent fluid flow causing high and low pressure areas to be formed which propagate from the source as sound. *Sound* is then defined as oscillations in pressure in a medium with elasticity and viscosity. Sound may also be defined as the auditory sensation evoked by the oscillations in pressure described above (ANSI, 1960).

The rate at which the pressure oscillations are produced is the *frequency* (f) which has units of hertz (Hz). One hertz is equivalent to one cycle per second (cps). A subjective characteristic of a sound related to the frequency is *pitch,* which depends mainly on the frequency of the sound stimulus incident on the human auditory system and to some extent on its sound pressure.

The frequency range of the human ear varies considerably among individuals. A young person with normal hearing will be able to perceive sounds with frequencies between approximately 20 and 20,000 Hz at moderate sound pressures. With increasing age the upper frequency limit tends to decrease.

Sound may consist of a *pure tone* (a single frequency) where the pressure is a sinusoidal function of time, or it may consist of a combination of many frequencies. In industry, the latter is usually the case.

Noise is defined as unwanted sound (ANSI, 1960), a definition that includes both the psychological and physical nature of the sound. Under certain conditions, noise may serve a useful purpose such as masking more distracting sounds. The terms "noise" and "sound" are often used interchangeably.

The *speed* (c) with which sound travels through a medium is determined primarily by the density and the pressure of the medium, characteristics that are in turn dependent on temperature. For practical purposes, the speed of sound is essentially constant. In air, the speed of sound is approximately 344 m/sec (1130 feet/sec). In liquids and solids the speed of sound is higher: in water about 1500 m/sec (4900 ft/sec), and in steel about 6100 m/sec (20,000 ft/sec).

The distance travelled by the sound wave during one sound pressure cycle is called the *wavelength* (λ), which is measured in meters (m) or feet (ft). The speed of sound is related to the wavelength and the frequency as shown in Equation (2.1):

$$c = f\,\lambda \tag{2.1}$$

where c is the speed of sound in m/sec (ft/sec), f is the frequency in hertz and λ is the wavelength in meters (ft).

The amplitude of the sound pressure disturbance is directly related to the displacement amplitude of the vibrating sound source. The pressure is expressed in units of force per unit area. The unit of *pressure* is pascals, Pa, which is the same as newtons per square meter (N/m^2). Pressure can also be expressed in dynes per square centimeter ($dynes/cm^2$) and microbars (μbar).

Almost all phenomena in acoustics are frequency dependent. Because the frequency range in acoustics is too broad to handle as a single unit, it is often broken into smaller ranges. The most common frequency bandwidth (or range of frequencies) used for noise measurements is the *octave band.* A frequency band is said to be an octave in width when its upper band-edge frequency f_2 is twice the lower band-edge frequency f_1:

$$f_2 = 2 f_1 \tag{2.2}$$

where f_1 and f_2 are the lower and upper frequency band limits. Octave-band measurements are used for noise-control work because they provide a useful amount of information with a reasonable number of measurements.

When more detailed characteristics of a noise are required, as might be the case for pinpointing a particular noise source in a background of other sources, it may be necessary to use frequency bandwidths other than octave bands. One-third-octave and narrower bands are used for these purposes. A *one-third octave band* is defined as a frequency band whose upper band-edge frequency, f_2, is the cube root of 2 times the lower band-edge frequency f_1:

$$f_2 = \sqrt[3]{2}\, f_1 \tag{2.3}$$

where f_1 and f_2 are the lower and upper frequency band limits, respectively. There are three one-third octave bands in each octave band.

The *center frequency,* f_c, of any of these bands is the geometric mean of the high and low band-edge frequencies:

$$f_c = \sqrt{f_1 f_2} \tag{2.4}$$

where f_c is the band center frequency, and f_1 and f_2 are the lower and upper frequency band limits, respectively. The center frequency of each octave and one-third octave band is used to designate the band. The internationally standardized octave bands are centered at 31.5, 63, 125, 250, 500, 1000, 2000, 4000 and 8000 Hz (ANSI, 1984). Octave and one-third octave band frequencies and upper and lower band-edge frequencies may be found in Table 3.3.

It should be noted that the upper and lower band-edge frequencies used to describe a frequency band do not mean that all sound outside the designated frequency band is excluded. Since commonly available filters do not provide total rejection of out-of-band signals, the band-edge frequencies actually represent the 3-dB-down points of the filter's frequency response. This response should meet one of the types specified in ANSI Specifications for Octave, Half-octave, and Third-octave Band Filter Sets, S1.11-R1976. Two adjacent filters have an equal response to a pure tone at the crossover frequency, which is the upper 3-dB-down frequency on the lower frequency band and the lower 3-dB-down point on the upper frequency band. Digital filters utilizing modern electronics have sharp-edged filters so that virtually no sound outside the frequency band is included in the measurements.

In acoustics, decibel notation is utilized for most quantities. The *decibel* (dB) is a dimensionless unit based on the logarithm of the ratio of a measured quantity to a reference quantity. Thus, decibels are defined as follows:

$$L = k \log_{10} \frac{A}{B} \quad \textbf{(2.5)}$$

where L is the level in decibels, A and B are quantities having the same units, and k is a multiplier, either 10 or 20 depending on whether A and B are measures of energy or pressure, respectively. Any time a "level" is referred to in acoustics, decibel notation is implied. In acoustics all *levels* are referred to some reference quantity which is the denominator, B, of the equation.

Physical Units and Sound Radiation

Sound Power (W) is the total acoustic output of a sound source measured in acoustic watts. For convenience, the sound power output of a source is expressed in terms of its *sound power level,* L_W. By definition, sound power level, L_W, in decibels, is:

$$L_W = 10 \log_{10} \frac{W}{W_0} \quad \textbf{(2.6)}$$

where W is the sound power of the source, and W_0 is the reference sound power. The reference sound power is 10^{-12} watt. Figure 2.1 shows the relationship between sound power in watts and sound power level in dB re (referred to) 10^{-12} watt.

As sound power is radiated from a point source in free space, the power is distributed over a spherical surface, so that at any given point there exists a certain sound power per unit area. This is designated as *intensity* and has the units of watts per sq m. Although intensity diminishes as distance from the source increases, the power that is radiated, being the product of the intensity and the area over which it is spread, remains constant.

There is no way to measure sound power directly. Modern instruments make it possible to measure intensity, but the instruments are expensive and must be used carefully. Under most conditions of sound radiation, sound intensity is proportional to the square of the sound pressure. Sound

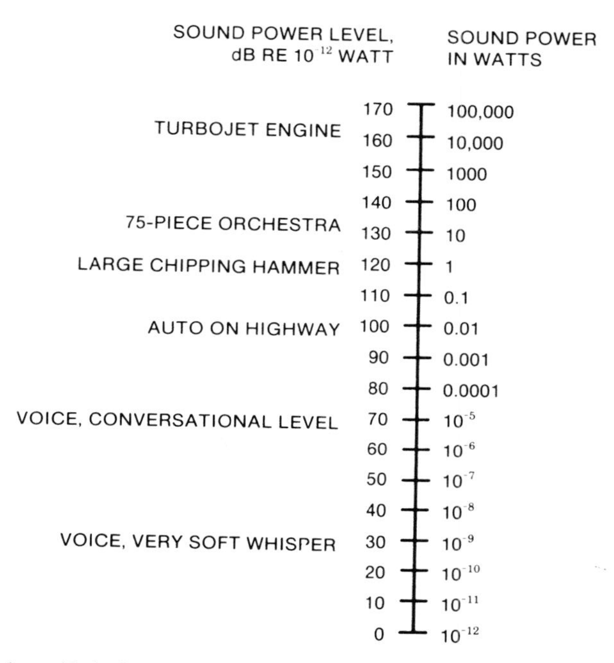

Figure 2.1 — Relationship between sound power level (L_w) and sound power (W).

pressure can be measured more easily, so sound measuring instruments are built to measure the *sound pressure level* (SPL) in decibels (ANSI, 1983).

The sound level read on a sound level meter is referred to a reference sound pressure, which for measurements in air is 20 micropascals (μPa) (also expressed as 20 $\mu N/m^2$). This reference is arbitrary; it approximates the normal threshold of human hearing at 1000 Hz. The equation for the sound pressure level L_p, in dB, is:

$$L_p = 20 \log_{10} \frac{p}{p_0} \qquad \textbf{(2.7)}$$

where p is the measured root-mean-square (rms) sound pressure and p_o is the reference rms sound pressure.

Note that the multiplier is 20 and not 10 as in the case of the sound power level equation. This is because sound power is proportional to the square of the pressure and because $10 \log p^2 = 20 \log p$. The rms value is used since it is the effective value of the varying sound pressure.

Figure 2.2 shows the relationship between sound pressure in pascals and the sound pressure level in dB re 20 μPa. It, as well as Figure 2.1, illustrates the advantage of using the decibel notation rather than the wide range of pressure or power. Note that any range over which the sound pressure is doubled is equivalent to 6 dB whether at low or high levels. For example, a change in pressure from 20 to 40 μPa (which might be of interest in hearing measurement) or from 10 to 20 Pa both represent a change of 6 dB in SPL.

The *A-weighted sound level* (dBA) is used in many noise regulations. It is the single number reading obtained from a sound level meter utilizing the A-weighting network (see Chapter 3). The A-weighted sound level in decibels can be measured directly or may be computed from octave or one-third-octave band SPL measurements using the weightings given in Table 3.1. (See below in Example 2.2.) These weightings represent adjustments applied to the measured SPLs in each frequency band. Note that the low frequencies are markedly reduced in importance.

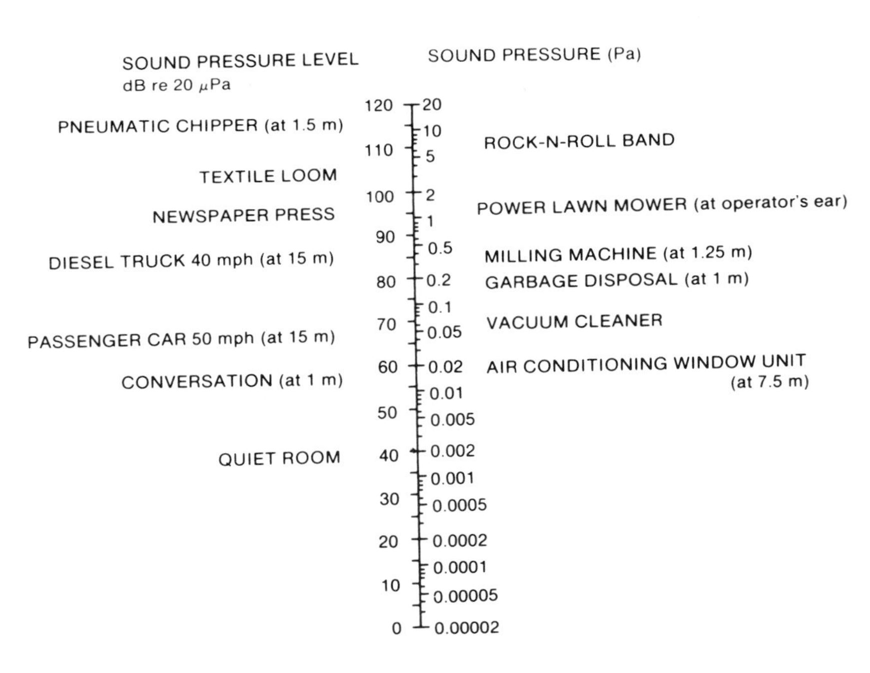

Figure 2.2 — Relationship between sound pressure level in decibels (dB) and sound pressure in Pa (N/m^2).

Directivity Factor (Q) is a dimensionless quantity that is a measure of the degree to which sound emitted by a source is concentrated in a certain direction rather than radiated uniformly in a spherical pattern. Directivity factors for radiation patterns that are associated with various surfaces surrounding a sound source are shown in Figure 2.3. Each sound radiation pattern is actually a portion of a spherical radiation pattern, *i.e.*, a fraction of the area of a sphere ($4\pi r^2$).

For free-field radiation (see below), the radiation pattern is illustrated in Figure 2.3A. This type of radiation is non-directional and Q = 1. In the following examples of directivity, the sound source and distance of an observer, r, remain the same. For hemispherical radiation of sound (Figure 2.3B), where the sound source is on a floor in the center of a large room or at ground level out-of-doors, the sound intensity at any particular distance r would be twice as great as that in free-field radiation, and hence the SPL will be 3 dB greater, because the area into which the sound radiates has been reduced by a factor of 2, *i.e.* ($4\pi r^2/2$). If the sound source is near the intersection of the floor and a wall of a room such that the area which the source radiates is one-quarter of a sphere (Figure 2.3C), Q would be 4 since the area has been reduced by a factor of 4: ($4\pi r^2/4$).

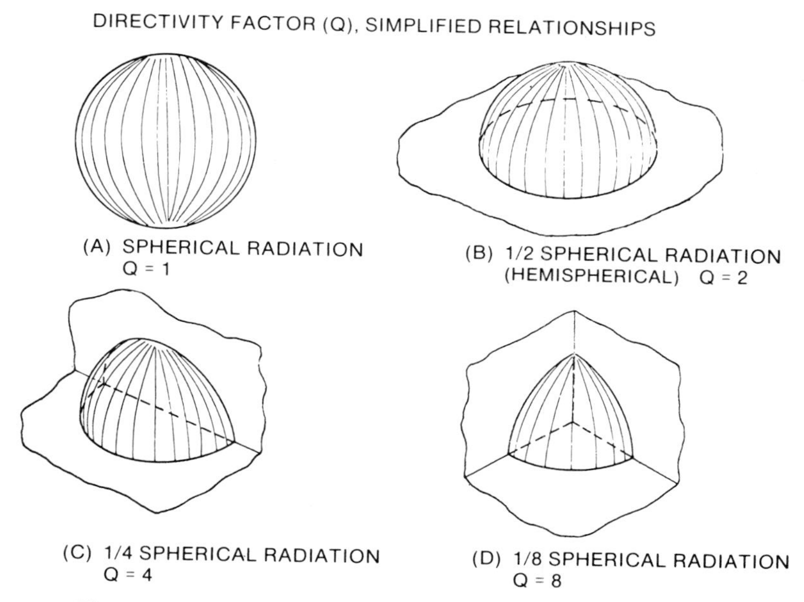

Figure 2.3 — Directivity factor (Q), simplified relationships.

In like manner, if the sound source is near the intersection of the floor and two walls (Figure 2.3D), the sound source radiates into one-eighth of a sphere so that Q would equal 8. Note that the area in which the sound radiates is the area of the sphere divided by the directivity factor.

In the discussion above, the directivity of the noise source is only related to the source location in relation to the reflecting surfaces. If the noise source itself is directional, both influences must be considered, because the sound pressure will not be uniform over the spherical area.

For a given sound source, Q is the ratio of the sound pressure, measured at some distance r, to the sound pressure expected at the same distance r, but under conditions in which the source radiates the same total acoustic power in a spherical pattern. Knowing the directivity factor of the sound source, one can modify the directivity factor of the source position in the room to obtain a combined directivity factor accurate enough for most applications of industrial noise control (Beranek, 1971; Harris, 1979).

A term related to the directivity factor is the *directivity index* (DI). The directivity index is related to the directivity factor Q as follows:

$$DI = 10 \log_{10} Q \tag{2.8}$$

From the discussion above on directivity and Equation (2.8) it can be seen that for hemispherical radiation (Q = 2), DI = 3 dB, and hence the SPL measured on a hemisphere for a source of given sound power output is 3 dB higher than for spherical radiation.

Noise Reduction and Isolation

Several additional terms need to be defined for use in noise control. A useful term is *noise reduction* (NR). This is the difference between the sound pressure levels measured at two locations, one on either side of a noise control device (*i.e.*, an enclosure or a barrier):

$$NR = L_{p1} - L_{p2} \tag{2.9}$$

where NR is the noise reduction in decibels, L_{p1} is the sound pressure level at location 1 and L_{p2} is the sound pressure level at location 2.

Insertion loss (IL) is the difference in sound pressure level at a fixed measuring location before and after a noise control method has been applied to the noise source:

$$IL = L_{p1} - L_{p2} \tag{2.10}$$

where IL is the insertion loss in decibels, L_{p1} is the sound pressure level before noise control and L_{p2} is the sound pressure level at the same location after noise control.

Attenuation is the reduction in sound pressure level in decibels as one moves further and further from a noise source (*i.e.*, out-of-doors or down an air-conditioning duct system).

In working within spaces with noise sources an important consideration is the *sound absorption coefficient* (α) of the materials in the space. All materials absorb some sound but to be "sound absorbing" a material should absorb a good fraction (*i.e.*, at least half) of the sound energy incident upon it. The sound absorption coefficient is dimensionless. It represents the fraction of the sound incident upon the material that is absorbed; *i.e.*, if α is 0.75, 75% of the incident sound is absorbed. The sound absorption of architectural materials is normally measured in the laboratory using a diffuse sound field. Sound absorption is useful in controlling noise in a reverberant sound field (see below and Chapter 12 for a fuller discussion).

In evaluating the sound isolation provided by an impervious barrier between two spaces, the *sound transmission loss* (STL) must be known. The sound transmission loss, in decibels, of a panel is 10 times the logarithm to the base 10 of the ratio of the sound power incident upon the panel to the sound power transmitted through the panel to the adjoining space (ISO, 1978; ASTM, 1981). The STL of a material or structure is governed by the physical properties of the material and type of construction used. The STL of a wall generally increases 5 to 6 dB for each doubling of wall weight per unit of surface area or for each doubling of frequency.

Multiple-wall construction with air spaces provides more attenuation than a single wall of the same weight would provide. However, considerable care must be taken to minimize the rigidity of connections between multiple walls when they are constructed, or any advantages in sound isolation can be nullified (Beranek, 1971; Harris, 1979).

Sound leaks which result from cracks or holes in a wall, or from windows or doors in a wall, can severely limit the sound isolating characteristics of the wall. Care must be exercised throughout construction to prevent leaks that may be caused by electrical outlets, plumbing connections, telephone lines, etc., in otherwise effective walls. (See Chapter 12).

Figure 2.4 shows theoretical sound transmission loss characteristics of solid walls as a function of weight and frequency. However, it is recommended that measured STLs be used where available (see Table 12.1). For discussion of measurement techniques for determining STL the reader is directed to Beranek, 1971; ISO, 1978; Harris, 1979; and ASTM, 1981.

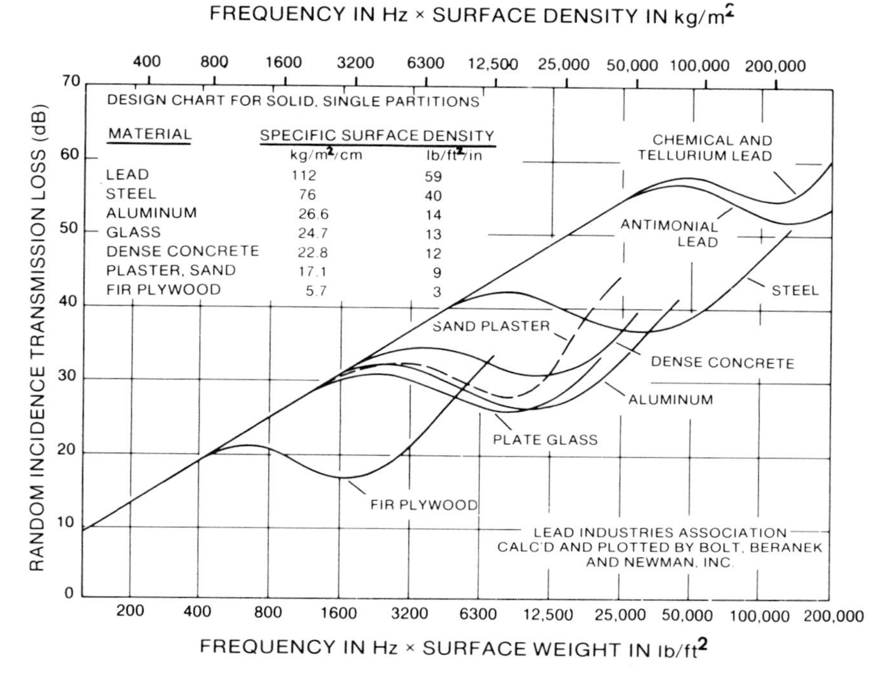

Figure 2.4 — Theoretical sound transmission loss as a function of surface weight and frequency for various homogeneous materials.

Decibel Addition

Often, it is necessary to *add levels* in decibels. An example is the combination of frequency-band sound levels to obtain the overall or total sound level. Another example is the estimation of the final sound pressure level as a result of adding a new machine of known sound output to an existing noise environment of known characteristics.

For addition of SPLs of two separate, random noise sources, use Table 2.1. To add one random noise level, $L_{p(1)}$, measured by itself at some point, to another, $L_{p(2)}$, measured by itself at the same point, the numerical difference between the levels is used to find the amount to be added to the larger of $L_{p(1)}$ or $L_{p(2)}$ in order to obtain the resultant of the two levels. If more than two are to be added, the resultant of the first two must be added to the third, the resultant of the three sources to the fourth, *etc.,* until all levels have been added, or until the addition of smaller values does not add significantly to the total (see additional discussion, Chapter 4).

TABLE 2.1
Table for Combining Decibel Levels of Noises with Random Frequency Characteristics

Numerical Difference Between Levels L_1 and L_2	L_3: Amount to Be Added to the Higher of L_1 or L_2
0.0 to 0.6	3.0
0.7 to 1.6	2.5
1.7 to 3.1	2.0
3.2 to 4.7	1.5
4.8 to 7.2	1.0
7.3 to 13.0	0.5
13.0 to ∞	0.0

Step 1: Determine the difference between the two levels to be added (L_1 and L_2).

Step 2: Find the number (L_3) corresponding to this difference in the table.

Step 3: Add the number (L_3) to the higher of L_1 and L_2 to obtain the resultant level (L_{pt}).

The equation which forms the basis for Table 2.1 is

$$L_{pt} = 10 \log_{10} [\sum_{i=1}^{N} 10^{L_{pi}/10}] \quad \textbf{(2.11)}$$

where L_{pt} is the total SPL in decibels generated by the N sources and L_{pi} represents the individual SPLs to be added.

This equation can be used with a hand-held calculator to add decibels.

EXAMPLE 2.1

The overall SPL produced by a random noise can be calculated by adding the SPLs measured in individual octave bands. As an example, consider the data presented in Table 2.2.

When adding a series of levels, begin with the highest level so that calculations may be stopped when lower values are reached which, when added in, do not contribute significantly to the total. In this example, the levels of 100 dB and 97 dB have a difference of 3 dB, which corresponds to 2.0 dB in Table 2.1. Thus, the combination of 100 dB and 97 dB will result in 100 + 2 = 102 dB. Combining 102 with 95 dB, the next higher level, gives

TABLE 2.2
Octave-band Sound Pressure Levels Utilized in Example 2.1

Octave Band Center Frequency (Hz)	31.5	63	125	250	500	1000	2000	4000	8000
Sound Pressure Level in dB re 20 μPa	85	88	94	94	95	100	97	90	88

102 + 1 = 103 dB, which is the total of the highest three bands. This procedure is continued, adding one band at a time, and the overall SPL will be found to be about 104 dB.

To use Equation (2.11) for addition of the octave-band SPLs in Table 2.2 with a hand-held calculator or computer, begin by keying in the first SPL; divide by 10 and raise 10 to the result, *i.e.*, take the antilog of $0.1 \times L_{p1}$. Key in the second SPL, divide by 10 and raise 10 to the result. Add the two numbers. This is now the sum of the antilogs of one-tenth of the two SPLs.

Continue to add the antilogs of one-tenth of the remaining SPLs until all have been added. When all of the additions have been completed, take the logarithm (to the base 10) of the result and multiply it by 10. This will be the total of all of the levels added.

In using a calculator, it is easier to add all of the levels and not stop to check if the lower levels contribute only a small amount to the sum. To check would mean that it would be necessary to take 10 times the logarithm of each sum and then reconvert the the sum back into the antilog form again before adding a new level.

Sound Power and Sound Pressure

Many noise-control problems require a practical knowledge of the relationship between sound power levels and sound pressure levels. These relationships allow calculations of the sound pressure levels that a particular machine will produce in a specified environment, based upon a known sound power level of the machine.

Consider a sound source which is small. This source is radiating sound equally in all directions. In a *free field* there are no reflections of sound, and hence the sound pressure level will be the same at any point equidistant from the source, that is, any point on the surface of a sphere centered on the source. The sound intensity will diminish inversely as the square of the distance, r, from the source since the sound energy is spread over the area

($4\pi r^2$) of the sphere. Thus in a true free field, the SPL decreases 6 dB with each doubling of distance from the source. However, under most real-life conditions, particularly indoors, reflection of sound (for example, from the floor or ground) will reduce the attenuation due to distance. Doubling the distance will produce, instead of a decrease of 6 dB, a drop of only 4-5 dB.

The sound power level of a source is generally independent of environment. However, the sound pressure level at some distance, r, from the source, is dependent on that distance and on the sound-absorbing characteristics of the environment. Therefore, these factors must be stated when expressing the SPL of a sound source at a particular place.

The simplest relation between sound power level and sound pressure level is found for a free-field non-directional sound source. The relation is given by the following equation:

$$L_p = L_w - 20 \log_{10} r - k + T \quad \textbf{(2.12)}$$

where L_p is the sound pressure level in decibels re 20 μPa, L_w is the sound power level in decibels re 10^{-12} watts, r is the distance from the source in meters or feet, k is 11.0 dB for metric units and 0.5 dB for English units, and T is a correction factor for atmospheric temperature and pressure in decibels. Since most industrial noise problems are concerned with air at or near standard conditions, that is, 20° C (68° F) and 1.013×10^5 pascals (30 inches of mercury or 14.7 psi), T is usually negligible. For extreme conditions see correction factors shown by Figure 2.5.

If the sound is directional in a free field, the relationship between sound power level and sound pressure level becomes

$$L_p = L_w - 20 \log_{10} r + DI - k + T \quad \textbf{(2.13)}$$

where DI is the directivity index in decibels.

Many industrial noise problems are complicated by the fact that the noise is confined in a room. Reflections from the walls, floor, ceiling, and equipment in the room create a *reverberant sound field* which exists in conjunction with the sound radiated directly from the source to the receiver (the *direct sound field*). The reverberant field alters the sound wave characteristics from those described above for free-field radiation. If reverberant-field SPLs are uniform throughout a room and sound waves travel in all directions with equal probability, the sound is said to be *diffuse*. In actual practice, perfectly directional, reverberant, or diffuse sound fields rarely exist; rather, in most cases, the sound fields are usually something in between.

A variable which relates the sound power level to the SPL in a room is the *room constant (R)* in square meters (feet). The room constant is an

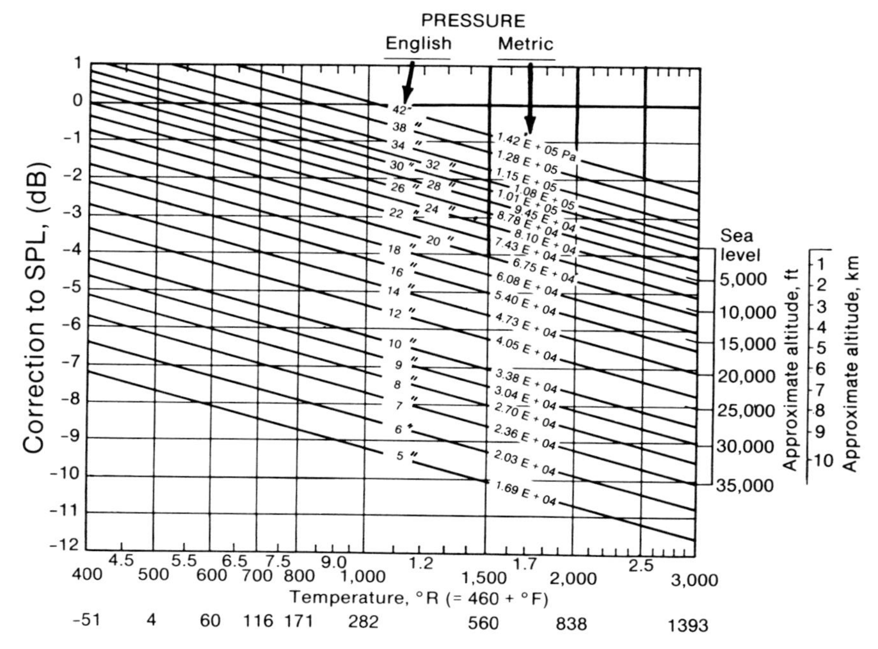

Figure 2.5 — Corrections to sound pressure level for temperature and barometric pressure. The temperature is in °R or °C, and the atmospheric pressure is in both inches of mercury and in Pa. Zero correction is for 68°F and 30 in. Hg. After Beranek (1954).

indication of the amount of sound absorption in a room; the smaller the room constant, the more reverberant the room. The room constant R can be calculated at any frequency from the following equation:

$$R = \frac{S_t \bar{\alpha}}{1 - \bar{\alpha}} \tag{2.14}$$

where R is the room constant in square meters (feet), S_t is the total surface area of the room in square meters (feet) and $\bar{\alpha}$ is the average sound absorption coefficient of the room surfaces computed using Equation (2.15)

$$\bar{\alpha} = \frac{\sum_{i=1}^{N} S_i \alpha_i}{\sum_{i=1}^{N} S_i} \tag{2.15}$$

where S_i is the area of individual room surfaces in square meters (feet) and α_i is the corresponding sound absorption coefficient for the *i*th surface.

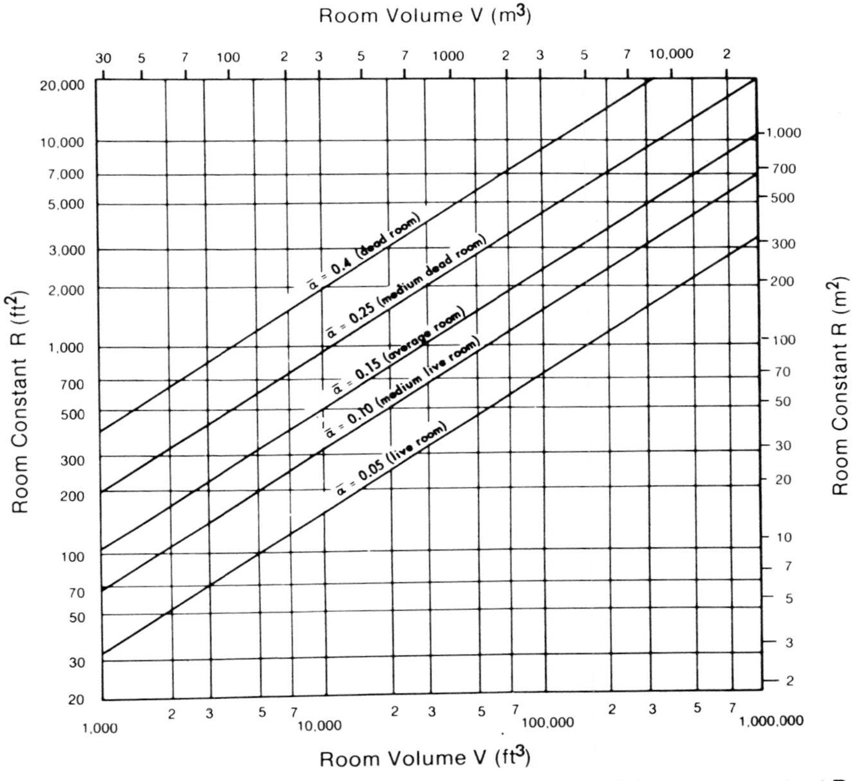

Figure 2.6 — Room constant for typical rooms. Value of the room constant R as a function of room volume for rooms with proportions of about 1:1.5:2. The parameter is the average sound-absorption coefficient for the room. After Beranek (1954).

The room constant R can be estimated from Figure 2.6.

In an enclosed space, the relation between sound power level and sound pressure level is as follows:

$$L_p = L_w + 10 \log_{10} \left[\frac{Q}{4 \pi r^2} + \frac{4}{R} \right] + k + T \qquad \textbf{(2.18)}$$

where Q is the directivity factor, the other variables are as defined in Equation (2.12), and k is 0 dB for metric units and 10.5 dB for English units. This relationship is shown graphically in Figure 2.7 without temperature and pressure corrections.

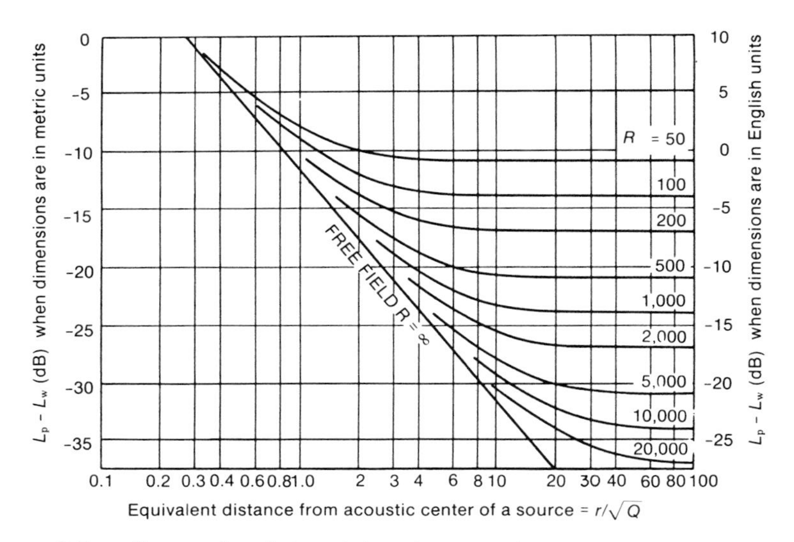

Figure 2.7 — Curves for determining the sound pressure level in a large room relative to the sound power level as a function of the directivity factor Q, distance r and room constant R. The ordinate is calculated from Equation (2.18). When all dimensions are in English units with length in feet, use left hand ordinate; for metric dimensions with length in meters, use right-hand ordinate. From Beranek (1971).

Figure 2.7 is very useful in noise control work because it combines the basic characteristics of sound fields in a manner that is easy to use. The horizontal portions of the room constant curves illustrate the decreasing sound level for a given sound source as the amount of sound absorption is increased. The downward slanting line illustrates the free-field decrease in sound level with increasing distance. The curved portions of the room constant curves show the transition from a free field to a reverberant field.

EXAMPLE 2.2

A vendor is proposing a machine for installation in a room where it would be the only noise source of consequence. The sound power level spectrum of this machine is shown by line 1 of Table 2.3. The machine is to be installed on the floor in the center of a live room having dimensions of 18.3 × 42.7 × 6.1 m (4770 m^3). If the operator is positioned 1 m from the machine during his work shift, would the noise level at the operator's position exceed 90 dBA?

From Figure 2.6 for a live room of 4770 m^3, R is approximately 90 sq m. The operator position dictates that r = 1m. Since the machine is to be

TABLE 2.3
Computations Described in Example 2.2

Line No.	Description	Frequency - Hz Center Frequency of Octave Band 63	125	250	500	1000	2000	4000	8000
1	L_w (given) (dB) re 10^{-12} watt	81	95	90	89	88	84	79	65
2	L_p (L_w - 7 dB) re 20 μPa	74	88	83	82	81	77	72	58
3	Correction for A scale From Table 3.1	-26	-16	- 9	- 3	0	+ 1	+ 1	- 1
4	L_p (dBA)	48	72	74	79	81	78	73	57

Logarithmic sum of line 4 per Table 2.1 = 85 dBA.

installed on the floor near the center of this large room, hemispherical radiation would indicate Q = 2. Referring to Figure 2.7, compute $r/\sqrt{Q}$ which equals 0.71, and proceed vertically from 0.7 to R = 90 sq m. and horizontally to the left to find that the sound pressure levels are 7 dB below the sound power levels of line 1 of Table 2.3. Hence, the sound pressure levels shown on line 2 are obtained. To obtain the A-weighted sound levels from the octave-band levels, determine the appropriate correction factors from Table 3.1 as shown on line 3. By subtracting *arithmetically* the correction factors of line 3 from the sound pressure levels of line 2, obtain the octave-band A-weighted sound levels shown on line 4. Now, by adding the A-weighted levels of line 4 using Table 2.1, obtain an A-weighted sound level of 85 dBA. Thus the noise level at the operator's position does not exceed 90 dBA.

EXAMPLE 2.3

A machine is to be installed in the corner of a reverberant room. The dimensions of the room are 6.1 × 15.2 × 30.5 m (2830 m^3). From Figure 2.6, R is 65 sq m. Since the machine is to be installed in one corner of the room, Q = 8. The operator has to be as close as 1 m to the machine, so r = 1m. The octave-band SPL design criteria are shown by line 1 of Table 2.4. Line 2 shows a 6-dB reduction to allow for background noise from other noise sources in the room. What would the maximum allowable octave-band sound power levels be for this new machine?

By subtracting 6 dB from the environmental criteria, the single machine criterion is obtained as shown by line 3. But this is the sound pressure level and the requirement is for sound power level. Referring to Figure 2.7 the relationship between SPL and sound power level can be found. Starting

TABLE 2.4
Computations Described in Example 2.3

Line No.	Description	Frequency - Hz Center Frequency of Octave Band							
		63	125	250	500	1000	2000	4000	8000
1	Environmental Criteria dB re 20 μPa	106	100	94	90	90	90	90	91
2	Allowance for Multiple Noise Sources, dB	6	6	6	6	6	6	6	6
3	Single Machine Criteria (L_p) dB	100	94	88	84	84	84	84	85
4	Single Machine Criteria (L_w), dB re 10^{-12} watt	101	95	89	85	85	85	85	86

from $r/\sqrt{Q} = 0.35$ and proceeding upward to R = 65 sq m, and horizontally to the left, the graph indicates that the SPL is 1 dB less than the sound power level. Hence, add 1 dB to line 3 to obtain the allowable maximum octave-band sound power levels as shown on line 4. This would then be used as a machine purchase specification.

References

American National Standards Institute (1960). "American Standard Acoustical Terminology," S1.1-1960, New York, NY.

American National Standards Institute (1976). "American Standard Specification for Octave, Half-Octave, and Third-Octave Filter Sets," S1.11-1976, New York, NY.

American National Standards Institute (1983). "American Standard Specification for General Purpose Sound Level Meters," S1.4-1983, New York, NY.

American National Standards Institute (1984). "American National Standard, Preferred Frequencies, Frequency Levels, and Band Numbers for Acoustical Measurements," S1.6-1984, New York, NY.

American Society for Testing Materials (1981). "Recommended Practice for Laboratory Measurement of Airborne Sound Transmission Loss of Building Floors and Walls," Designation E-90-81, ASTM, Philadelphia, PA.

Beranek, L.L. (1954). *Acoustics*, McGraw-Hill Book Company, New York, NY.

Beranek, L.L. (Ed.) (1971). *Noise and Vibration Control*, McGraw-Hill Book Company, New York, NY.

Harris, C.M. (Ed.) (1979). *Handbook of Noise Control*, McGraw-Hill Book Company, New York, NY.

International Organization for Standardization (1978). "Acoustics — Measurement of Sound Insulation in Buildings and of Building Elements — Part 4: Field Measurements of Airborne Sound Insulation Between Rooms," ISO/R 140-1978, ISO, Geneva, Switzerland.

Noise and Hearing Conservation Manual, edited by
E.H. Berger, W.D. Ward, J.C. Morrill and L.H. Royster

3 Sound Measurement: Instrumentation and Noise Descriptors

John J. Earshen

Contents

Page

Introduction 38

Generic Sound Level Meter Instruments 41
- Basic SLM 42
- Graphic Level Recorder 44
- Dosimeter 44
- Integrating/Averaging SLM 45
- Community Noise Analyzer 45

Sound Level Meters 45
- Types 45
- Impulse Dynamics 46
- True Peak SLM 47
- Weighting Filters 47
- SLM Measurement Tolerances 50
- Root Mean Square Value (RMS) 50
- Crest Factor and Pulse Range 53

Accessories and Other Instruments 54
- Microphones 54
  - Condenser 55
  - Electret 55
  - Ceramic 55
  - Piezoelectric 56
  - Electrodynamic 56
- Acoustical Calibrators 56
- Frequency or Spectrum Analyzers 58
- Magnetic Tape Recorders 61

Noise Measures 62
- Average Levels 63
  - Equivalent Continuous Sound Level — L_{eq} 63
  - Other Equivalent Continuous Sound Levels—L_{DOD}, L_{OSHA}, TWA 65

Dose Measures 67
Noise Dose 67
Sound Exposure Level—SEL 68

Instrument Selection and Utilization 70
Use of Microphones 70
Limits on Measurement Imposed by System Noise Floor 73
Addition of Decibels and Corrections for Background Noise 76
Calibration 76
Application and Operation of SLMs 79
Use of FAST or SLOW Dynamics 79
Averaging by Mental Estimation 80
Dosimeter Use 81
Verification of Setting 82
Cutoff Characteristics 83
Microphone Placement 83
Measurement Artifact 84
Impulse Noise 85
Wet Environments 85
Personal *vs.* Area Monitoring 86
Potential Discrepancies between SLMs and Dosimeters 87
Microphone Placement 88
Personal *vs.* Area Monitoring 88
Differential Susceptibility to Noise Artifact 88
Response to Impulsive Noise 88
Earphones and Auditory Perception 89
Susceptibility to Electromagnetic and Radio Frequency Interference ... 90
Scope of Frequency Analyses 91

Concluding Remarks 92

References 94

Introduction

In industrial hygiene practice, sound is measured to assess its detrimental effects on humans. These include hearing damage, speech interference and annoyance, with emphasis on the first. Should mitigation of these effects be necessary, additional collection and processing of data may be required to implement engineering control measures and institute hearing conservation programs.

On initial examination, there seems to be an overwhelming number of instruments applied to sound measurement. Recent incorporation of microprocessors in instruments and increased use of computers for data acquisition, problem solution, recordkeeping, and retrieval, all tend to intimidate the novice. In fact, however, modern developments have considerably simplified solution of problems formerly requiring tedious effort. It is now possible to address measurement and control problems which only a decade ago had no feasible solutions. A few basic facts viewed in proper perspective greatly simplify and speed understanding of the characteristics, functions, and applications of the family of acoustical instruments.

Sound is a dynamic pressure fluctuation superposed on atmospheric pressure. It varies in both time and space and is propagated as compressional waves. If simultaneous pressure measurements could be made at all points of a defined space and recorded as continuous functions of time, the sound field would be uniquely determined. Clearly, this is impossible. Practical measurements are constrained to observation of a finite number of points, each of which may exhibit unique time variability. The basic measurement problem then involves choosing the number and location of observation points and deciding whether to record detailed time histories of pressure variations or to perform on-site averaging or other data-compression operations (for example, the basic sound level meter measures a short-duration moving average of the square of observed pressure).

From the fundamental instrument observations, descriptors characterizing the physical properties of the sound field are derived. Rarely are the descriptors stated in units of pressure or other physical units. Instead, the decibel notation is used. This stems from historical precedents in psychoacoustics and circuit design.

For application in the physical sciences and in engineering practice, the basic measurements are typically processed to produce statistical and averaged descriptors based on functions of time, and power spectrum descriptors based on functions of frequency. Such descriptors are generally inadequate to predict and characterize the physiological and subjective impact of noise on humans. Therefore, many psychophysical measures have been established (and continue to be established). Though they are derived from physical measurements, they are empirical and are used to quantify physiological and psychological effects on humans of specified sound exposures.

This chapter describes the generic instruments used for sound measurement and presents guidelines for computation of noise measures and for instrument selection and utilization. A guide to commercially available instruments and their manufacturers is published annually in *Sound and Vibration* magazine (Anon., 1984).

Nearly all instruments used for noise analysis have evolved from the basic sound level meter (SLM) which senses acoustic pressure and indicates sound levels. Specialized instruments have been developed to overcome SLM limitations and to automatically compute objective and subjective noise measures derived from sound level measurements.

Noise measures are essential for quantifying hazard to hearing and they are pivotal in establishing regulatory protection against such hazards and controlling environmental quality. In such applications, validity of the measures must be supported by statistical evidence. Even when mathematical confidence levels associated with statistical inferences are high (which is not always the case), many people feel uneasy about having their physical and economic welfare so governed. Matters are further complicated when several different procedures can be used to generate a measure. Some are fully automatic; others involve human participation and interpretation of dynamic readings of instruments. The results may be identical for slowly varying noise but may differ significantly when the noise is highly variable and composed of transient sound waveforms of short duration (*e.g.* punching, forging, and variable-duty-cycle noise sources). It is no surprise then that disputes arise about the comparative accuracy and validity of measures obtained with alternative measurement procedures and instruments. Resolution of conflicts by rigorous mathematical analysis can be difficult because analytic representation of human operator functions and judgments, and correct representation of instrument dynamics, are elusive.

Even when identical instruments and procedures are employed to make successive measurements of exposure to noise presumed to have constant properties, the end results may be highly variable. Such variability may be caused by subtle but significant changes in the noise sources, the position and orientation of human subjects, and an inability to replicate instrument performance and measurement procedures. Fortunately, many of the measurements made in industry can be replicated with a high degree of confidence. Nevertheless, when initiating a measurement program and evaluating results, careful attention should be paid to observing and assessing causes and limits of variability.

The industrial hygienist must be aware of the types of problem that may be encountered in a measurement program. To that end, an appreciation must be obtained of the characteristics and limitations of noise measures, the properties and applications of measuring instruments, and basic measurement procedures. Above all, the goal should be to acquire sufficient proficiency to recognize when questionable results are being obtained and thus conclude when the aid of a specialist should be sought.

Generic Sound Level Meter Instruments

Figure 3.1 illustrates functional relationships among several generic instruments. At the input, pressure is sensed by the microphone whose electrical output is connected to an amplifier which normally includes a frequency-selective filter. The most common filters include the A, B, and C response curves and octave or fractional octave types. Their characteristics are discussed later in this chapter. Other response curves, may be included for special noise measures. The next block indicates a squaring operation required because sound pressure level (SPL) is a function of pressure squared. Next is an exponential averaging filter; this produces a form of moving average. Functionally, this block also characterizes the instrument's dynamic response characteristics (SLOW or FAST). Next is a switch included to illustrate how changes can be made to select one instrument configuration or another.

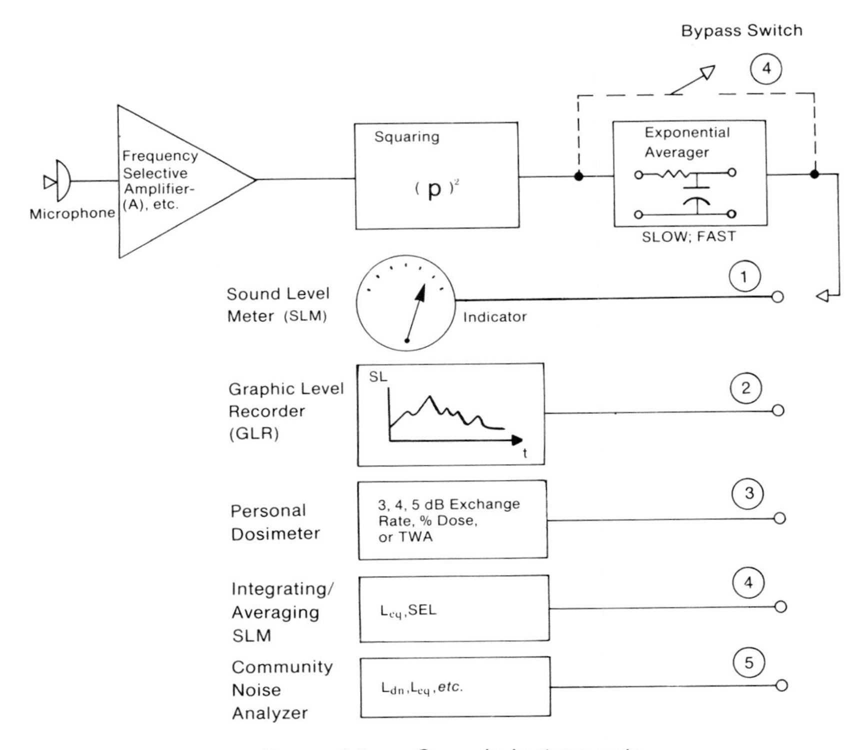

Figure 3.1. — Generic instruments.

BASIC SLM

If the switch is placed in position 1, an indicator is connected and the resulting configuration is a basic SLM. The meter's characteristics influence the definition and computation of derived noise measures.

Tracing the circuit from the input microphone, the signal at the output of the microphone will show a waveform having frequency components up to approximately 20 kHz. Should a sensitive set of high quality earphones be connected here, the sound heard would be essentially that observed by the microphone (actual connection would require a matching and isolating amplifier to drive the earphones).

After amplification and filtering, the noise signal has been frequency weighted and if auditioned would appear to have been modified by a tone control. For example, if an A-weighting filter is used, the sound in the earphones will have its low frequency components diminished and the higher pitched tones will dominate. (If the sound in the earphones is distorted or otherwise grossly different from that perceived by the operator's hearing, instrument failure or overloading should be suspected and investigated.)

The next block introduces a drastic change to the signal by the squaring process, which generates a waveform, in principle, proportional to the instantaneous power of the acoustic signal. Through the earphones it would no longer have a recognizable association with the microphone input. If this waveform drove a moving needle indicator capable of following perfectly its instantaneous variations, the operator would only see a blur. For the general case, the waveform contains frequencies up to 20 kHz. Due to limitations of human vision as well as the inability of a practical meter to respond that rapidly, such operation is clearly impractical.

To make the output usable, some means of reducing the speed and extent of indicator variations has to be found. One logical approach would be to divide the pressure squared waveform into segments and process them to produce sequential average readings. Such an instrument now exists and is a version of an L_{eq} meter. Its development came about only recently. However, the concept of the SLM was introduced over 50 years ago and the available technology of the time undoubtedly influenced the choice of a dynamic averaging process based upon an exponential averager. Similar problems existed at the time in other areas in the monitoring of rapidly varying waveforms with moving needle meters (*e.g.* modulation monitors in transmitters, recording level indicators, *etc.*). Similar solutions were applied to these problems. From a pragmatic viewpoint, a means to "slow down the meter" was absolutely mandatory to make it readable. Through internal and external controls, meter dynamics were experimentally implemented to provide acceptable performance.

Two meter dynamic characteristics were established in the early days of development and have persisted to our time. They are designated SLOW and FAST. These dynamics are described by one of two time constants. To illustrate the properties of a time constant, consider what the hypothetical response of the meter movement of an SLM would be for a suddenly applied constant sound pressure. If the meter has instantaneous response, the needle would instantly rise to conform to the input. Remember, however, that audible sound has pressure fluctuations ranging up to 20 kHz in frequency which would be invisible to the eye. By control of the physical parameters determining the needle response, it is possible to obtain a sluggish response that tends to follow the short-time average value of the applied waveform. Under such conditions, the meter needle rises exponentially toward the value of a suddenly applied constant sound amplitude. Because such a response curve theoretically only reaches the maximum value at infinite time, a convention has been adopted to characterize the rising response curve in terms of the time it takes to reach 63 percent of its maximum value. The averaging obtained with such dynamics is called exponential averaging. When a constant input is applied suddenly (step input) to such a filter, the output rises exponentially. For the SLM, the meter reading will reach 63 percent of its final steady-state reading within one time constant.

The two SLM time constants are respectively 1 second (SLOW), and .125 second (FAST). Why these specific numerical time constants were chosen rather than others of similar magnitude is difficult to ascertain. In practice, the SLOW time constant is used when one attempts to determine the average or slowly changing average value of observed sound. In contrast, the FAST time constant is used to estimate the variability of observed sound.

The literature contains a variety of papers that develop rationales for the given time constants based on psychoacoustic and physiological models of the hearing process. None present evidence to support the fundamental uniqueness of the constants in the manner of an Avogadro's number. One can reasonably speculate that instrument designers originally chose time constants and dynamics attainable with the available technology which were "comfortable" to the instrument user. Subsequent investigations in the life sciences further solidified the acceptance of the FAST and SLOW and matched them empirically to physiological and psychoacoustic parameters.

An appreciation of meter dynamics associated with basic SLMs is of more than academic interest to the industrial hygienist, since they significantly affect how noise measures are defined and computed. Though modern computing instruments can achieve the desired results in a swifter and more accurate manner, older data gathered on effects of sound on humans are based on the traditional dynamics. Accordingly, various noise measures and regulations require that computing instruments incorporate SLOW or FAST. For example, currently there is serious debate over SLM dynamics to be included in measuring noise dose or time-weighted average sound level for impulsive noise as mandated by the Occupational Health and Safety Administration (OSHA) (Rockwell, 1981; Kundert, 1982; OSHA, 1983).

GRAPHIC LEVEL RECORDER

An augmented form of the SLM, the Graphic Level Recorder (GLR), surmounts limitations of meter interpretation by an operator. Its function is shown in the block diagram when the switch is transferred to position 2, connecting the output to a chart recorder. The configuration produces a record of sound level *vs.* time, storing details that cannot be assimilated by visual observation of a meter. These data can then be analyzed and processed to generate desired noise measures.

DOSIMETER

Various measures exist which purport to quantify the threat to hearing of humans exposed to excessive sound. As discussed later, many measures are derived by processing sound level measurements obtained in the immediate vicinity of a person's ear (hearing zone) with an SLM. A dosimeter is an instrument which performs two functions. Its microphone and the instrument are placed on a person being monitored. The microphone monitors the "hearing zone" while the remainder of the instrument automatically computes the desired noise measures. If the switch in Figure 3.1 is in position 3, a basic dosimeter is configured.

These instruments are battery-powered and have evolved from simple devices computing single-number exposure measures to highly sophisticated monitors which compute and store comprehensive data on the sound field encountered by the subject. It is important to recognize that a dosimeter derives directly from an SLM and was developed to simplify measurement and computational procedures. If measures derived from SLMs and dosimeters are to be commensurate, the dosimeter must correctly duplicate the dynamic characteristics of the SLM. Specific requirements prescribing such characteristics are set forth in the pertinent ANSI standard on Personal Dosimeters (ANSI S1.25, 1978) and in various regulatory statutes (OSHA, 1983).

INTEGRATING/AVERAGING SLM

The integrating/averaging SLM is an instrument having properties similar to those of a dosimeter. However, there is an important difference. Once again referring to Figure 3.1, the described instrument configuration is attained by placing the selector switch at position 4. Note that a symbolic shorting switch is also closed around the exponential averaging block. This is the significant difference between the dosimeter and the integrating/averaging SLM. The meter dynamics are explicitly removed from the computation of measures based on integration or averaging of the frequency-weighted sound level. There is no ANSI standard as yet adopted for such an instrument although one is in preparation. At this time the only formal standard generally employed is IEC 804 (1985). Though no formal recognition is given to any but 3-dB averaging, the standard can be extrapolated to deal with 4- and 5-dB exchange averaging.

COMMUNITY NOISE ANALYZER

Yet another member of the generic group is the Community Noise Analyzer. This instrument, represented in Fig. 3.1 by position 5 of the selector switch, automates the computation of special noise measures (discussed later) and collection of data to characterize and regulate generation of noise in the community. The most prominent measures, at present, are based on SLM readings and thus require inclusion of instrument dynamics. The shorting switch is thus left open for this instrument. The measures include integration and averaging of SLM-derived sound levels as well as statistical processing of such levels. Typically such instruments are enclosed in weatherproof containers and used with special all-weather outdoor microphones. This makes it possible to conduct continuous observations outdoors over many days. In addition, these capabilities permit measurement of industrial noise for a hearing conservation program under environmental conditions so adverse as to preclude use of conventional SLMs or dosimeters.

Sound Level Meters

TYPES

Developments in electronics over the past decade have produced large-scale integrated solid-state devices which include both analog and digital circuits. In addition, the microprocessor "computer on a chip" has experienced explosive growth. As a consequence, SLMs have become smaller, more rugged, more reliable and accurate, less demanding of power, and, most significant of all, capable of being integrated into comprehensive instruments which overcome operator limitations in making readings and processing and recording data. Increasingly, SLMs are being absorbed into instruments performing multiple functions. Standards IEC 651 and ANSI S1.4 (1983) have been upgraded so that conforming instruments will yield equivalent readings under identical circumstances. The current version of ANSI S1.4 (1983) specifies four types of SLMs which differ from former designations. Specifically it provides for three grades of instruments, types 0, 1, and 2, and for a special-purpose type S, each designed with accuracies necessary for a particular use.

Type 0: Laboratory Standard — This was not described in ANSI S1.4-1971 and is intended for use in the laboratory as a high-precision reference standard and is not required to satisfy environmental requirements for a field instrument.

Type 1: Precision — An instrument intended for accurate measurements in the field and laboratory.

Type 2: General Purpose — An instrument with more lenient design tolerances than type 1, intended for general field use, particularly in applications where high-frequency (over 10 kHz) sound components do not dominate. (Use of this type of instrument is specified by OSHA and MSHA for determining compliance.)

Type S: Special Purpose — May have design tolerances associated with any of the three grades but is not required to contain all of the functions stipulated for a numbered type. (The old type 3 survey instrument included in ANSI S1.4-1971 is no longer included.)

IMPULSE DYNAMICS

For types 0, 1, and 2 the weighting filters A, B, and C and two exponential-time-averaging characteristics, SLOW and FAST, are required. An additional meter response characteristic, impulse, is defined but not required. This characteristic specifies an exponential rise time constant of 35 milliseconds and an asymmetric decay time constant of 1.5 seconds. It is mainly used at present in certain community noise measures and in rating the emissions of business machines. In North America, impulse response is not commonly used to rate industrial noise. Historically it was defined to enable SLMs to respond to short burst of sound and to hold the maximum value long enough for a human operator to perceive it before it decays. *This characteristic is not suitable* for measuring true unweighted peak sound level.

TRUE PEAK SLM

Various regulatory agencies require that accurate measurements be made of true peak SPL (*e.g.* OSHA requires that peak exposures be kept below 140 dB for impulsive sound). For such applications an optional peak measuring mode is defined. For type 1 and 2 instruments the true peak reading SLM must adequately measure a pulse of 100 microseconds duration. A type 0 true peak SLM must adequately measure a 50-microsecond pulse (see the cited standard for test procedures and tolerances).

A little clarification of terminology is in order here. Impact and impulse noise have been used interchangeably as descriptors of short transient sounds. In industrial hygiene practice an impulse (impact) sound is often defined as a transient having less than one second duration which may be repeated after a delay of more than one second.

The reader is again cautioned *not to use* "impulse" response in measuring true peak SPL.

WEIGHTING FILTERS

Various acoustical measuring instruments employ frequency selective weighting filters. In SLMs the most common are designated A, B, and C. These derive their characteristics from certain properties of human hearing. Others have bandpass filters (such as octave-band) applied to analyzing the spectral content of sound waveforms. It is informative to examine what these filters are, what they do, and why they are used.

Functionally such filters can be thought of as "tone controls." This can be demonstrated with SLMs which have provision for connecting earphones in the amplifier chain following the filters. For a sound source having broad spectral content, a distinctive change in perceived tonal quality will be observed as various weighting filters are switched in. The effect is particularly striking when octave-band or fractional octave-band filters are inserted.

The rationale for using band-selective filters (octave, fractional octave, narrow, *etc.*) to analyze spectral content of sound waveforms is self-evident. In contrast, the origin and basis for use of the A, B, and C filters is rather obscure. The A, B, and C filters were historically derived from the Fletcher-Munson curves which characterize equal loudness perception by humans for pure tones of variable frequency.

Figure 3.2 shows a set of such curves. They are interpreted as follows: Given a starting reference pure tone at a fixed SPL at 1 kHz, each one of the family of solid line curves shows how the SPL of a pure tone must be modified, when its frequency is different from the reference frequency, in order to produce a sensation of loudness equal to that produced by the reference. Observe that the curves change substantially as the reference level is varied from 100 dB down to 0 dB. This demonstrates that human hearing frequency response to pure tones is very much a function of the absolute level of the sound.

After the Fletcher-Munson curves were developed and published, proposals were set forth for design of instruments to measure the loudness of complex sounds. Agreement was reached to standardize three frequency weighting curves which would attempt to approximate the equal loudness response of human hearing at low, medium, and high SPLs.

The first was characterized by the 40-phon equal-loudness contour; its approximation by electrically achievable filter was designated the A filter. Similarly, the 70-phon curve was designated for approximation by the B filter, and finally the 100-phon contour was designated for representation by the C filter. The A, B, and C filter curves are shown in Figure 3.3, and

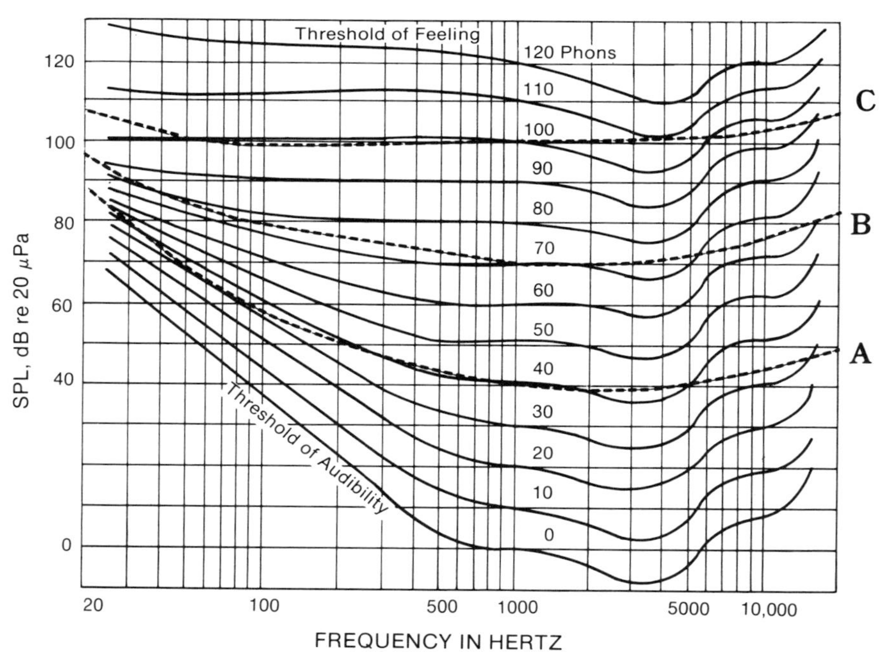

Figure 3.2. — Fletcher-Munson curves with weighting filter response overlay.

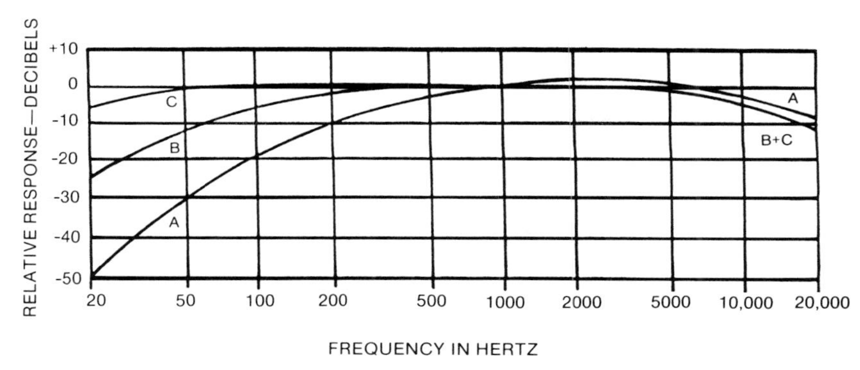

Figure 3.3. — SLM weighting curves — ANSI S1.4-1983.

have been superposed as dashed lines over the 40-, 70-, and 100-phon contours in Figure 3.2.

The curvature of the filters seems to slope the wrong way in the superposed sketches. However, this is a perfectly valid representation. Note that in actual practice the filter curves are "turned upside down" in order to modify the otherwise flat (ideal) frequency response of an unweighted instrument in order to make it approximate the chosen properties of hearing.

Various papers and instruction books tend to mislead the reader with statements such as, "the A-weighting curve approximates frequency response of human hearing and thus is incorporated in measuring instruments". A more informative and correct statement is that the A-weighting curve is an approximation of equal loudness perception characteristics of human hearing for pure tones relative to a reference of 40 dB SPL at 1 kHz. Its application to measurement of noise for hearing protection and other purposes is only vaguely, if at all, related to equal loudness perception. As a result of investigations in which a variety of weighting filters have been compared, it has been concluded that empirically derived measures using A-weighting give a better estimation of threat to hearing by given noise waveforms than do the other weightings. Other weighting schemes have been used and continue to be investigated (such as multiple octave-band filtering). Because of simplicity and acceptable results, A-weighting has received very wide acceptance. Its validity is periodically reexamined but as of the time of publication it still maintains wide acceptance. In current applications, the B filter is rarely used.

A-weighting is also used in determination of annoyance measures. Again, this use is justified by empirical data derived from studies with human subjects. A-weighting is not used to the exclusion of other descriptors in annoyance determination as extensively as it is in the area of hearing risk due to noise exposure. Nonetheless, it has very wide acceptance.

The C filter has gained some degree of acceptance in applications where impulsive or explosive blast type waveforms are encountered. Some additional use is made of C-weighting by comparing successive readings made with A and C weightings in order to determine approximately whether or not a sound has significant low-frequency components. If it does, the C reading will be higher than the A; if it does not, the two readings will be similar. Such C-A techniques are useful for example in evaluation of the adequacy of hearing protection devices (See Chapter 10). For more accurate analyses, instruments such as octave, fractional octave, and narrow-band analyzers are commonly available.

Table 3.1 presents in detail the relative response levels for the A, B, and C curves.

SLM MEASUREMENT TOLERANCES

Table 3.2 shows the allowed tolerance for relative response levels for sound at random incidence for the three types of SLMs. Careful examination of the Table shows why it is not possible to give a simple answer to the question of the accuracy of a given type of SLM. It is evident that at low and again at high frequencies the tolerance limit "opens up". Thus the accuracies must be stated in terms of envelope limits that are functions of frequency. These characteristics are imposed by the limitation of practical microphones.

To illustrate the point, consider two simple measurements made with a type 2 instrument. If a measurement is made of machine noise having frequencies ranging between 500 and 2000 Hz, Table 3.2 indicates that deviations of up to ± 2 dB will be permitted. In contrast, if measurements are made of noise having frequencies ranging between 4000 and 8000 Hz, deviations of up to ± 5 dB may exist. Attainable measurement accuracies are obviously substantially affected by the predominant frequency content of the noise.

ROOT MEAN SQUARE VALUE (RMS)

Throughout this text reference is made to RMS values of sound pressure. The use of this measure stems from the fact that sound is a dynamic pressure fluctuation whose arithmetic average value is zero; thus, the ordinary averaging process cannot be used to develop useful measures. Furthermore, decibel notation and many other acoustical measures are

TABLE 3.1
Relative Response for A, B, and C Weighting

Nominal Frequency Hz	A Weighting dB	B Weighting dB	C Weighting dB
10	-70.4	-38.2	-14.3
12.5	-63.4	-33.2	-11.2
16	-56.7	-28.5	- 8.5
20	-50.5	-24.2	- 6.2
25	-44.7	-20.4	- 4.4
31.5	-39.4	-17.1	- 3.0
40	-34.6	-14.2	- 2.0
50	-30.2	-11.6	- 1.3
63	-26.2	- 9.3	- 0.8
80	-22.5	- 7.4	- 0.5
100	-19.1	- 5.6	- 0.3
125	-16.1	- 4.2	- 0.2
160	-13.4	- 3.0	- 0.1
200	-10.9	- 2.0	0
250	- 8.6	- 1.3	0
315	- 6.6	- 0.8	0
400	- 4.8	- 0.5	0
500	- 3.2	- 0.3	0
630	- 1.9	- 0.1	0
800	- 0.8	0	0
1000	0	0	0
1250	+ 0.6	0	0
1600	+ 1.0	0	- 0.1
2000	+ 1.2	- 0.1	- 0.2
2500	+ 1.3	- 0.2	- 0.3
3150	+ 1.2	- 0.4	- 0.5
4000	+ 1.0	- 0.7	- 0.8
5000	+ 0.5	- 1.2	- 1.3
6300	- 0.1	- 1.9	- 2.0
8000	- 1.1	- 2.9	- 3.0
10 000	- 2.5	- 4.3	- 4.4
12 500	- 4.3	- 6.1	- 6.2
16 000	- 6.6	- 8.4	- 8.5
20 000	- 9.3	-11.1	-11.2

TABLE 3.2
Tolerances for Random - Incidence Response of SLMs
ANSI S1.4 (1983)

Nominal Frequency Hz	Type 0 dB	Type 1 dB	Type 2 dB
10	+2, -5	±4	+5, -∞
12.5	+2, -4	±3.5	+5, -∞
16	+2, -3	±3	+5,-∞
20	±2	±2.5	±3
25	±1.5	±2	±3
31.5	±1	±1.5	±3
40	±1	±1.5	±2
50	±1	±1	±2
63	±1	±1	±2
80	±1	±1	±2
100	±0.7	±1	±1.5
125	±0.7	±1	±1.5
160	±0.7	±1	±1.5
200	±0.7	±1	±1.5
250	±0.7	±1	±1.5
315	±0.7	±1	±1.5
400	±0.7	±1	±1.5
500	±0.7	±1	±1.5
630	±0.7	±1	±1.5
800	±0.7	±1	±1.5
1000	±0.7	±1	±1.5
1250	±0.7	±1	±1.5
1600	±0.7	±1	±2
2000	±0.7	±1	±2
2500	±0.7	±1	±2.5
3150	±0.7	±1	±2.5
4000	±0.7	±1	±3
5000	±1	±1.5	±3.5
6300	+1, -1.5	+1.5, -2	±4.5
8000	+1, -2	+1.5, -3	±5
10 000	+2, -3	+2, -4	+5,-∞
12 500	+2, -3	+3, -6	+5,-∞
16 000	+2, -3	+3, -∞	+5, -∞
20 000	+2, -3	+3, -∞	+5, -∞

For type 2 a sharp cutoff at 10 kHz is allowed and used in most instruments.

The given tolerances include both microphone and circuit performances.

based on the power associated with a particular sound which in turn is proportional to the square of sound pressure. It is possible to define an effective sound pressure, p_{RMS}, for a given averaging time, T, which produces the same average power as the actual time varying sound pressure. Thus,

$$p_{RMS} = \left[\frac{1}{T} \int_0^T p^2 \, dt \right]^{1/2} \quad \textbf{(3.1)}$$

The term inside the brackets simply denotes the averaging of p^2.

CREST FACTOR AND PULSE RANGE

A performance measure, crest factor, is used to specify the ability of a sound measuring instrument to process correctly waveforms which have peak values substantially higher than their RMS average. Crest factor is rigorously defined only for steady-state repetitive waveforms and is equal to the ratio of peak acoustic pressure to the RMS value of the waveform. Alternatively, crest factor, stated in decibels, is equal to 20 times the logarithm of the ratio of peak to RMS pressures of the waveform. Historically the concept of crest factor had its origin in electrical power distribution system measurements. In such an environment it is common to encounter periodic waveforms that depart substantially from a sinusoidal form and have high peak excursions. The crest-factor metric was appropriately extended to rating sound level meters for their capability to measure repetitive waveforms having high peaks.

To gain an appreciation for the magnitudes encountered, consider the crest factor of a steady-state (*i.e.* invariant for a long time) sinusoidal acoustical signal. For this waveform, the ratio of peak to RMS pressure is 1.41, or 3 dB. A waveform with moderately greater excursion of peak values from its RMS value could have a crest factor of 3.16: in decibels, 10 dB.

For quite some time, SLMs and dosimeters used to determine worker exposure to noise were simply required to have a minimum crest factor of 10 dB (ANSI S1.25-1978). Much of the noise encountered in industry can be successfully measured with such instruments. There are, however, waveforms such as those produced by power nailers, continuous riveters, chippers, *etc.* which have a much higher crest factor (20 to 30 dB or more). Such waveforms typically consist of repetitive short bursts of high-amplitude sound separated by periods of low-level sound. The peaks of such waveforms are much higher in value than the RMS effective average value and thus they have high crest factors. For such waveforms, instruments having insufficient crest factor capabilities will produce readings that are too low.

In recent years, attention has been focused on making reliable measurement of impulsive noise to determine its potential for hearing damage. An initial but insufficient step in that direction has been to specify SLMs and dosimeters having increased crest factors. A value that has been proposed is 30 dB (a ratio of peak to RMS pressure of approximately 31.6). Such an approach has utility for describing capability of instruments to deal with quasi-steady high-crest-factor noise waveforms such as prolonged bursts of riveting noise. A more severe problem is encountered when the noise of interest consists of short-duration high-amplitude transients occurring at random. Examples of such noises include hammering, forging, and stamping. For technical reasons that are beyond the scope of this discussion, instruments that have high steady-state crest factor ratings may perform extremely poorly when required to measure isolated impulses having the same pressure excursion as the periodic waveform.

A new measure has been adopted to rate the ability of an instrument to measure either the area under the curve generated by a pulsed transient or its average value. This measure is called Pulse Range (IEC 804-1985), and stipulates tolerances to be met in determining average pulse value for isolated pulses of given duration and amplitude. The referenced document stipulates measurement tolerances for level excursions up to 73 dB for pulses as short as one millisecond.

At present there is no corresponding approved ANSI standard that addresses the issue; however, several pertinent standards are now in preparation. Ones having direct bearing on measuring instruments include an amended version of ANSI S1.25 "Personal Dosimeters" and a new standard being prepared by ANSI Working Group S1-3, "Specification for Integrating - Averaging Sound Level Meters."

Although crest factor and pulse range specify the ability of an instrument to average or integrate sound levels to provide for example the TWA of a "spiky" waveform (*i.e.*, one containing isolated and/or numerous impulses), neither describes the capacity of the instrument to be used to measure the actual instantaneous peak values themselves.

Accessories and Other Instruments

MICROPHONES

Microphones are transducers that convert dynamic pressure variations into electrical signals that can be processed by a wide variety of measuring instruments. To select and use microphones properly for particular applications, it is necessary to understand their basic characteristics and proper method of employment (Peterson, 1980). Microphones can be categorized in terms of their basic transduction mechanisms. The primary types of microphones currently used include:

Condenser

These have a thin, stretched diaphragm which forms one plate of an electrical capacitor. The diaphragm is displaced by pressure fluctuation of incident sound waves. The second capacitor plate is stationary and is mounted in the body of the microphone behind the diaphragm. The capacitance of the two-plate configuration varies proportionally to incident sound pressure. A pressure-dependent output voltage is generated when the microphone is polarized by a well-filtered power source generating a potential of up to several hundred volts DC. Such microphones can have excellent linearity and frequency response and are available in all quality grades. They can be damaged by rough handling and are susceptible to degradation by moisture.

Electret

These are similar to condenser types in construction and performance. However, they do not require an external polarization voltage, using instead an electret built either into the diaphragm or the stationary condenser plate. An electret is an electric field analog of a permanent magnet. It is typically made of special waxes or plastics which have the ability to retain fixed electrical charge distribution (*i.e.* polarization). Such mikes are somewhat less stable and accurate than condenser types, though significant improvements have been and continue to be made. Like the condenser microphone, these also can be damaged by rough handling and are susceptible to moisture.

Ceramic

The transducer mechanism in these uses a special polarized ceramic such as barium titanate which generates an electrical potential when it is stressed. The stressing force is derived by linkage to a diaphragm that moves in response to sound pressure. Such microphones have higher output voltage for a given sound pressure than do the condenser or electret mikes. They are also more rugged than the latter types; however, they have greater susceptibility to excitation by vibration, higher noise floors, and more limited high-frequency response. Though they can be made in higher grades, they commonly are only compatible with type 2 instruments. These limitations are acceptable for dosimetry and thus they are used in such applications because of their other desirable properties. Ceramic microphones are sometimes inaccurately called piezoelectric types.

Piezoelectric

Even though this type, like the ceramic type, has the property of producing an output voltage that is a function of stress, the specific mechanisms of transduction are different. True piezoelectric microphones employ quartz or tourmaline crystals and are sometimes called blast microphones. They are highly linear and have superior ability to handle high sound pressure levels (such as those generated by sonic booms or explosive blasting) without distortion over extremely wide frequency ranges. They are very rugged but also respond to excitation by vibration and have very high noise thresholds (50 dB and above).

Electrodynamic

These microphones rely on the principle that a conductor moving through a fixed magnetic field will generate a voltage. The motion consists of movement of a diaphragm responding to sound pressure. At one time such microphones were very popular but currently have been almost completely removed from use in acoustical measurements. They are somewhat delicate and susceptible to hum pickup from electrical power lines and thus lack redeeming features to prompt their continued use.

ACOUSTICAL CALIBRATORS

Sound measuring equipment couples microphone outputs with many forms of electronic processing circuits and recording devices. Verification of accuracy and calibration of equipment functions other than the microphone are best done by insertion of appropriate electrical signals corresponding to all sound waveforms that the instruments are designed to measure. Modern equipment is quite stable unless physically mistreated, so comprehensive verification of specifications and calibration need only be done infrequently, either following repairs or to comply with company or regulating agency requirements for periodic recalibration. The user, however, must not become complacent and should learn how to detect equipment malfunction. Some older equipment has built-in electrical calibration signals that can be switched on by the operator. Such capabilities are of limited utility.

The weakest link in the measurement chain is the microphone. Sensitivity and frequency response of this component can change either as a result of varying ambient conditions or physical damage. Therefore, it is good industrial practice (often mandated by regulating agencies) to employ an acoustical calibrator providing direct excitation of microphones to verify or adjust measuring instruments before and after a test run.

A recently adopted ANSI standard for acoustical calibrators (ANSI S1.40-1984) stipulates "[Calibrators] . . . normally include a sound source which generates a known sound pressure level in a coupler into which a microphone is inserted. A diaphragm or piston inside the coupler is driven sinusoidally and generates a specified SPL and frequency within the coupler. The calibrator presents to the inserted microphone of a sound level meter or other sound measuring system a reference or known acoustic signal so one can verify the system sensitivity or set the system to indicate the correct SPL at some frequency."

Many acoustical calibrators provide two or more nominal SPLs and operate at two or more frequencies. Multiple levels and multiple frequencies are respectively useful for checking the linearity and, in a limited way, the frequency response of a measuring system. The latter is useful for gross checks of mikes and weighting filters for failure. An acoustical calibrator that produces multiple frequencies may also be used to determine a single-number composite calibration for a broadband sound through calibration at several frequencies. Additional signals such as tone bursts may be provided for use in checking some important electroacoustic characteristics of SLMs and other acoustical instruments. Such signals may also be useful in checking performance characteristics of instruments to measure sound exposure level or time-period average sound level. Adaptors may be provided to accommodate microphones having diameters different from that of the microphone for which the basic calibrator was designed.

The ANSI standard specifies tolerances for SPLs produced by calibrators. These range from ± 0.3 dB for calibration of microphones expected to be used with types 0 and 1 SLM instruments to ± 0.4 dB for calibration of type 2 instruments including dosimeters.

The frequency range over which acoustical calibrators can be designed to operate is limited by certain physical constraints. Thus, it is not practical to develop coupler type calibrators to span the entire operating frequency range of acoustical instruments. Limitations are encountered when the wavelengths of sound frequencies generated become short enough to be comparable to the dimensions of the microphone and calibrator cavity.

Commonly used calibrators generate a single level output at 1000 Hz (a frequency at which the A, B, and C weighting filters introduce no attenuation). Calibration should not be affected by switching filters. Other frequencies are sometimes provided in octave steps down to 125 Hz and up to 2000 Hz (above this frequency the constraints cited above come into play). Such calibrators provide a limited capacity for calibrating octave-band analyzers and spot-checking weighting filter frequency response.

Still more sophisticated calibrators are available which produce:

a) controlled outputs — step-attenuated for testing instrument linearity
b) repeated tone burst to test crest factor (limited to low crest factor)
c) transient signals to test instrument dynamic response (*i.e.* SLOW, FAST).

One should not place blind trust in any calibrator or instrument. A simple procedure, when several calibrators and/or SLMs are available, is to perform a cross check among them. This should produce results within the published tolerance limits.

Coupler-type calibrators should only be used with microphones for which they are intended. Instructions supplied by the manufacturer on instrument use and corrections for barometric pressure and temperature should be carefully followed. Use of single-frequency calibrators may result in overlooking damage to microphones that is manifested at frequencies other than the calibrator frequency.

It is possible that acoustic calibrators having additional capabilities will eventually be devised. It is unlikely, however, that coupler-type calibrators will be developed to generate very short impulsive waveforms. In the foreseeable future, verification of an instrument's capability to measure short impulsive transients will depend upon electrical testing combined with separate testing of microphone response under laboratory conditions.

FREQUENCY OR SPECTRUM ANALYZERS

Generally, determination of individual noise exposure for assessment of potential hearing damage is based on measurement of sound level with A-weighting, SLOW response. Some criteria exist for assessing potential hearing damage based on measurements made with instruments having octave-band filters replacing the A-weighting. Such information is more commonly used for analysis pertaining to implementation of engineering controls and, in certain instances, for the evaluation of performance of hearing protectors.

Accurate measurement of the spectral characteristics of noise and correct interpretation and application of results are complex undertakings, in most instances requiring substantial training and experience. The following discussion is primarily presented to provide background information.

To analyze the spectral content of sound, instruments having frequency selective filtering capabilities must be used. Some of these are generically related to SLMs; others are similar but have distinctive properties. For the generic group, the frequency-selective filters are substituted for the A, B, and C filters (see Figure 3.1). The final output may be a meter indication or a graphic plot. The significant similarity to the SLM is the inclusion of an exponential averaging function (typically SLOW). The output of the analyzers provides a measurement of the sound level in each filter frequency passband, thus representing the components of the observed sound as a function of frequency.

The most common spectrum analyzer is the octave-band analyzer which uses selectable filters having frequencies related by multiples of 2. The passband of these filters is bounded by f_2, the upper band-edge frequency, and f_1, the lower band-edge frequency. The center frequency f_c for the bands is the geometric mean of the band-edge frequencies.

$$f_c = (f_1 f_2)^{1/2} \quad \textbf{(3.2)}$$

Table 3.3 shows the mean frequencies which are defined by ANSI S1.11 (1971). A simple way to remember the mean frequencies of octave filters is to start with the commonly used calibrator frequency of 1000 Hz. The higher frequency filters can then be identified by repeatedly multiplying by 2 (*e.g.* 1000, 2000, 4000, *etc.*), and the lower frequency filters can be obtained by successively dividing by 2 (*e.g.* 1000, 500, 250, *etc.*).

Octave-band filters have progressively wider bandwidths as the mean frequency is increased. Each successive octave filter has double the bandwidth of the lower adjacent filter. Should narrower band filtering be desired, 1/3 octave filters are available (See Chapter 2) and their properties are listed in Table 3.3. (A common nomenclature is to call the octave-band types "1/1 octave" and one third octave types "1/3 octave.") A still narrower set of 1/10 octave filters are available from some manufacturers but these have not as yet been listed in the ANSI standard.

The previously described filters are available either as built-in features of certain SLMs (including integrating/averaging SLMs) or can be provided in the form of external plug-ins. In such configurations, the filters can only be used one at a time. If the sound waveform being analyzed is constant, analysis is performed by sequential recording of each filter output. This process may be operator performed or automated with a graphic level recorder which identifies the filter in use with a corresponding portion of record.

Many noise sources generate sound waveforms that are highly variable in time and frequency. Spectral measurements using sequential filters are difficult or impossible under such conditions. Real time analyzers can be used to advantage in such applications. This type of analyzer uses a bank of parallel filters to observe the desired noise source continuously. The output of such an analyzer produces a record of spectral content which varies with

TABLE 3.3
Center and Approximate Cutoff Frequencies for Contiguous-Octave and One-Third Octave Bands ANSI S1.11 (1971) and ASHRAE (1985)

	1/1 Octave Bands			1/3 Octave Bands		
Band	Lower	Center	Upper	Lower	Center	Upper
14				22.4	25	28
15	22.4	31.5	45	28	31.5	35.5
16				35.5	40	45
17				45	50	56
18	45	63	90	56	63	71
19				71	80	90
20				90	100	112
21	90	125	180	112	125	140
22				140	160	180
23				180	200	224
24	180	250	355	224	250	280
25				280	315	355
26				355	400	450
27	355	500	710	450	500	560
28				560	630	710
29				710	800	900
30	710	1000	1400	900	1000	1120
31				1120	1250	1400
32				1400	1600	1800
33	1400	2000	2800	1800	2000	2240
34				2240	2500	2800
35				2800	3150	3550
36	2800	4000	5600	3550	4000	4500
37				4500	5000	5600
38				5600	6300	7100
39	5600	8000	11 200	7100	8000	9000
40				9000	10 000	11 200
41				11 200	12 500	14 000
42	11 200	16 000	22 400	14 000	16 000	18 000
43				18 000	20 000	22 400

time. The term "real time" implies that the spectral content is being observed continuously as it varies in time. The output of each filter channel may be exponentially averaged or it may have some other smoothing. Depending on the method used, the results may duplicate SLM-type measurements or they may be different. Measurement of time-varying spectra is a complex undertaking which requires understanding of analytic methods that are beyond the scope of this text. A basic assessment can be made by observing the output of individual filter channels. If the outputs change very slowly or remain constant, then the measurement results can be viewed with confidence. If high variability is observed, careful application of the instrument manufacturer's guidelines must be made.

Many other analyzers are available which have greater resolution in frequency. One is a constant percentage bandwidth type. The filter bandwidth is a constant percent of the center frequency. Like the octave-band and fractional octave types, the bandwidth in hertz increases with the center frequency chosen. Representative values are one to two percent. These are available as parallel banks of filters, or as sequentially switched filters. Another common type uses a scanning filter swept across the frequencies of interest. Even narrower resolution in frequency is possible with digital computer-based analyzers using fast Fourier transform (FFT) algorithms. Such analyzers have fixed bandwidth resolution. The cautionary remarks about care in analyzing spectra varying in time apply even more gravely when extremely narrow band analyses are attempted.

There is one type of instrument that can be used with minimal precautions for analyzing time-varying spectra, one which is capable of doing averaging over periods of time that are long, relative to the temporal variability of the spectra. An integrating/averaging SLM with selectable 1/1 or 1/3 octave filters is such an instrument. Another, more easily applied instrument has parallel bands of filters whose outputs are averaged over long periods. One useful application is to determine octave-band sound levels averaged over a work shift. Should such a measurement be undertaken, the conservative choice would be to use true equal energy averaging (3-dB-per-doubling-duration exchange rate). If circumstances mandate a less conservative choice, instruments are available which will perform the averaging on a 4-dB or 5-dB exchange basis, respectively, compatible with the regulatory practices of the Department of Defense (DOD) and Federal OSHA.

MAGNETIC TAPE RECORDERS

Magnetic tape recorders can be used to store noise waveforms for subsequent analysis and to establish a permanent record. Recorders can be extremely useful instruments but they should be chosen and used with great care. They must have performance capabilities that are equal to or better than those of the instruments which will be ultimately used for playback analysis (within the measurement range of interest). This is not easily accomplished. There are many traps for the novice, so use of tape recorders should not be entered upon lightly and should be resorted to only when no other solution is possible. The discussion that follows does, however, provide useful references for interpreting results reported in contemporary publications.

Instrumentation grade recorders are available in many forms. Some have the capacity to record from 2 to 16 channels simultaneously. Others have enhanced low-frequency capabilities through use of special modulation schemes. If measurement of vibratory motion is needed in conjunction with sound measurement, multichannel recorders are particularly useful.

An important feature of an instrumentation type recorder is the ability to set input levels with a calibrated attenuator. After initial calibration, it is frequently necessary to change internal amplification. An uncalibrated volume control results in loss of calibration.

Even an instrumentation grade tape recorder has to be used with great care, especially if impulsive noise is present. It should be recognized that even under ideal circumstances, only the top 50 dB of signal excursion can be transcribed. In addition, it is necessary to ensure that the tape recorder used has a high degree of phase linearity in order to preserve the original impulse shape. If this precaution is not observed, it is quite possible to use a tape recorder which may have a very flat frequency response but because of poor phase linearity will produce considerable distortion in the pulse waveforms. The net result may manifest itself in playback pulses that are smeared out in time in contrast to the original pulses which possessed a very short duration envelope.

Currently research is being conducted at the National Institute of Occupational Safety and Health (Erdreich, 1984), and at several other laboratories where efforts have been made to obtain reference recordings of impulsive sounds. The tape recorders used have frequency bandwidths of 40 kHz and are used with carefully matched phase-linear microphones.

Present day analytic instruments have capabilities that are generally far superior to those obtainable with tape recorders and it is recommended that measurements be made directly on the noise of interest.

The future will likely produce great improvements in recording. Research and development is currently proceeding to produce various new media and techniques for digital recording of sound.

Noise Measures

A small number of noise measures are extensively employed for evaluating and regulating noise hazards and annoyance. They are derived from basic sound level measurements. Such data can be obtained with an SLM and subsequently processed into required noise measures, or the entire process can be automated with instruments having enhanced capabilities.

The measures discussed here are derived from time-function variables. Of the almost limitless number of noise measures which have been defined (Schultz, 1972; Lipscomb, 1978; Harris, 1979), the ones most accepted in industrial hygiene practice are based on A-weighted sound levels. Subsequent discussion defines and explains application of two groups of such measures: average levels and dose measures.

AVERAGE LEVELS

Equivalent Continuous Sound Level — L_{eq}

The average is perhaps the most obvious measure for classifying a time-varying sound level for a given observation period. Since sound levels are expressed in decibels which are logarithmic, they should not be averaged arithmetically. Because of the decibel notation, the definition and calculation of this intuitive measure may be intimidating on initial consideration. Recall that sound level is defined as ten times the logarithm of the ratio of sound pressure squared, normalized with respect to the reference pressure (at the threshold of hearing) squared. Pressure squared is proportional to power. To obtain the desired result, it is the underlying time-varying power function that must be averaged arithmetically. Finally the result is transformed back into decibels to become the equivalent continuous sound level, L_{eq} (commonly called equivalent sound level or simply, equivalent level). Among professional acousticians the term average sound level (L_{av}) is frequently used. The equivalent sound level terminology was probably adopted to caution against indiscriminate arithmetic averaging of decibel readings. Another way of defining L_{eq} is that it equals the continuous sound level which, integrated over a specific time, would result in the same energy as a variable sound level integrated over the same time. The relationship is:

$$L_{eq} = 10 \log \left[\frac{1}{T} \int_0^T 10^{L_A/10} \, dt \right] \quad \textbf{(3.3)}$$

where t is the time in seconds, T is the observation time and L_A is the A-weighted instantaneous sound level. The meter dynamics are normally not explicitly contained in the defining relation and in most applications are not relevant. If particular dynamics are required, they are stated separately.

For sound levels obtained over discrete increments of time (from SLMs or GLRs) the relationship becomes:

$$L_{eq} = 10 \log \left[\frac{1}{T} \sum_{i=1}^{N} (10^{L_{Ai}/10}\ t_i) \right] \quad \textbf{(3.4)}$$

where i is the i_{th} interval, N is the total number of intervals, t_i is the duration of the i_{th} observation interval, and L_{Ai} is the A-weighted sound level during t_i. The comments concerning meter dynamics that follow equation (3.3) apply here as well.

Equation (3.4) can also be used to combine equivalent sound levels obtained at a common point during contiguous time periods. Simply substitute L_{eq} obtained during an observation period t_i, for L_{Ai}.

The equivalent sound level, like other measures, has many possible applications. Instances where the industrial hygienist may encounter it include measurements of the average noise level emitted by a source at a specified location, community noise regulations, and regulations governing hearing protection.

In community noise regulations (ANSI S3.23-1980) equivalent sound level is commonly specified to control maximum allowed average sound level as measured at the boundary or property line of a source. The averaging period may be 24 hours or it may consist of a number of contiguous segments during each of which a different maximum L_{eq} may prevail.

A special version of L_{eq} is the day-night average level, L_{dn}. This is only valid for a 24-hour period and is computed the same way as would be a 24-hour L_{eq}. The actual prevailing sound level in the calculation has a 10-dB penalty added between 2200 to 2400 and 0000 to 0700 hours. The concept is based on the premise that people are more annoyed by a given level of noise during nominal sleeping hours.

Though at present, equivalent sound level is used only by a limited number of regulatory bodies for hearing protection in North America, it is very common throughout the rest of the world. Much debate is in progress over possible adoption in North American jurisdictions, *e.g.* Shaw (1985). In addition, recent damage risk standards (ISO 1999-1985) and recommendations (Von Gierke *et al.,* 1981) use L_{eq}.

To gain insight into the development of other related measures, consider certain aspects of L_{eq}. Many jurisdictions limit the exposure of unprotected humans to noise by stipulating that a given L_{eq} may not be exceeded for a baseline work shift (typically 8 hours) during a calendar day. It is not correct to conclude that the momentary SPL may never exceed the numerical value of that equivalent level. The regulatory agency is really stipulating a maximum daily accumulation of energy not to be exceeded. The same total energy can be accumulated for an infinite set of combinations of power amplitude and exposure duration. The relationship simply is that the product of the two must remain constant under the regulatory limit (*i.e.* equal energy must be maintained). It further follows that doubling the power amplitude requires a halving of exposure duration. If equal energy is maintained, it then follows that a 3-dB exchange rule applies to sound level and exposure duration. (A doubling of power corresponds to a 3-dB increase in level; a halving corresponds to a 3-dB decrease. The exposure durations are respectively halved and doubled.)

Next consider the concept of noise dose. A daily limit on accumulated sound energy was introduced in the preceding discussion. Accordingly, the limit could be stated in joules. More commonly, regulatory agencies normalize the allowed limit by stating it as percentage dose, where a dose of 1.00 (100%) corresponds to 8 h of work in some specified level of noise.

Other Equivalent Continuous Levels—L_{DOD}, L_{OSHA}, TWA

Additional functions have been defined to serve the needs of other empirically-derived models relating hearing damage to noise exposure. Regulations based on such models in effect require that the product of pressure squared, raised to a fractional exponent, and time be constant. Unless the reader can easily visualize the functional relationships, the consequences of the statement may not be immediately apparent.

For the modified product, an analogous trade-off relationship exists between change in sound level in decibels and change in exposure time. To fit particular damage models, two exchange rates have been chosen, 4 and 5 dB (Botsford, 1967), the first by DOD and the second by OSHA.

There is a mistaken tendency to seek justification for applying one trading ratio in preference to another on the basis of intrinsic properties of the constant product relations used. There simply is no basis for so doing (at least at this stage of knowledge about hearing damage). The chosen damage model is the absolute governing factor. Similarly, there is a mistaken tendency to work backward from instrument dynamics to select or challenge a damage model. The reader will likely encounter such disputes in practice. Perceiving the principles involved is not particularly difficult; however, comprehending and critically assessing the complex interaction between legal, physiological, physical, economic and political factors pose a challenge of the first magnitude. [One case in point is illustrated by the current multiplicity of widely diverging views on how to measure, regulate, protect, and thereby conserve hearing in the presence of impulsive noise (Von Gierke *et al.*, 1981; Suter, 1983; Erdreich, 1984; Shaw, 1985; Hess, 1985)].

In ANSI S1.25 (1978) a normalized dose based on the concept discussed above is defined and presented in a parametric equation. By appropriate choice of parameters it is possible to define dose and to derive functions for computing equivalent continuous sound levels based on 3-, 4-, or 5-dB exchange rates. Though there is much effort to standardize, the terms $L_{av}(4)$, $L_{eq}(4)$, and L_{DOD} are used interchangeably in North America as are $L_{eq}(5)$, $L_{av}(5)$, and L_{OSHA}. The function given below, used in ANSI S1.25 (1978), is identical to L_{eq} (Equation 3.3) except that the leading coefficient has been replaced by q, and the meter response is now explicitly stipulated as SLOW.

$$L_{av}(Q) = q \log\left[\frac{1}{T}\int_0^T 10^{L_{AS}/q}\,dt\right] \tag{3.5}$$

where Q is the exchange rate in decibels and L_{AS} is sound level that is A-weighted and exponentially averaged with SLOW dynamic response. The use of L_{AS} is mandated by current OSHA practice for both SLMs and dosimeters. (See section on "Potential discrepancies between SLMs and dosimeters" in this chapter for discussion of the effect of using L_{AS} on measurement of impulse noise.)

The variable q is defined as

$$q = Q/\log 2 \tag{3.6}$$

For a 3-dB exchange rate, q equals 10, for a 4-dB rate it is 13.3, and for a 5-dB exchange rate q equals 16.6.

Time Weighted Average (TWA), a measure used by OSHA, is similar to L_{OSHA} although they cannot be used interchangeably. To compute TWA using Equation 3.5, the coefficient 1/T must be replaced by 1/8 and time must be expressed in hours. Regardless of the actual duration of noise exposure, the averaging period for TWA is always 8 hours, corresponding to a reference workshift duration.

There is an additional factor that must be considered in computing TWA and the related measure of noise dose (see following section). Some regulatory agencies define a threshold (or cutoff) sound level, L_{co}. Any sound that is observed to be below that level is discarded. Thus for:

$$L_{AS} \geq L_{co} \tag{3.7}$$

L_{AS} is not affected. However for:

$$L_{AS} < L_{co} \tag{3.8}$$

L_{AS} must be replaced with minus infinity.

For discrete sound level inputs,

$$L_{av}(Q) = q \log \left[\frac{1}{T} \sum_{i=1}^{N} (10^{L_{ASi}/q}\ t_i) \right] \qquad \textbf{(3.9)}$$

L_{ASi} is the level based on A-weighted SLOW response during interval i. It is compared to the threshold level as in Equations (3.7) and (3.8), and an appropriate value for L_{ASi} chosen.

The concept of a threshold level rests on tenuous grounds but is well entrenched and its implications should be understood. Use of a threshold implies that exposure to sound above the threshold level is potentially damaging but sound below the threshold abruptly becomes nonhazardous. No published evidence of such an effect exists. The use of threshold levels has its origins in administrative interpretation of regulatory practices.

DOSE MEASURES

Noise Dose

As given in ANSI S1.25, normalized noise dose in percent (the previous threshold constraints on L_{AS} still apply) is:

$$D = \frac{100}{T_c} \int_0^T 10^{(L_{AS} - L_c)/q}\ dt \qquad \textbf{(3.10)}$$

where T_c is the criterion sound duration and L_c is the criterion sound level; the two are chosen in conjunction to establish what constitutes 100% dose. For OSHA at present, T_c equals 8 hours and L_c is 90 dBA. T is the measurement duration, in hours.

If the sound level readings are discrete, dose in percent is:

$$D = \frac{100}{T_c} \sum_{i=1}^{N} 10^{(L_{ASi} - L_c)/q}\ t_i \qquad \textbf{(3.11)}$$

where L_{ASi} is the A-weighted exponentially averaged SLOW sound level in the ith interval.

Some readers will be more familiar with the alternative definition of noise dose:

$$D = 100 \left[\frac{C_1}{T_1} + \frac{C_2}{T_2} + \ldots + \frac{C_i}{T_i} + \ldots + \frac{C_N}{T_N} \right] \qquad \textbf{(3.12)}$$

where C_i is the actual time duration of the i_{th} interval during which a worker is exposed to a constant sound level L_{ASi}. The values for T_i are taken from Table G-16a in Appendix A of the Hearing Conservation Amendment (found in Appendix I of this manual). These values represent the maximum permissible times during which a worker can be exposed to a corresponding sound level, L_{ASi} without exceeding a dose of 100%.

A noise dose can be determined by the above methods for any observed exposure time. There is a corresponding TWA which will generate the same dose after an exposure of 8 hours.

The published table for T_i (Table G-16a) is computed from the relationship

$$T_p = T_c / 2^{(L_{AS} - L_c)/Q} \tag{3.13}$$

where T_p is the maximum permissible time (in hours) of exposure at a given level of L_{AS}. Based on Equation (3.13), Figure 3.4 shows a series of curves relating maximum permissible exposure time to A-weighted sound level for particular durations and criterion levels. Another nomenclature equivalent to criterion level is Permissible Exposure Level (PEL) expressed in dBA.

The reader may sketch in curves for any combination of PEL, criterion sound duration time, and exchange rate. Three curves are actually shown, all based on an 8-hour criterion sound duration. The 5-dB exchange, 90 dB PEL curve represents OSHA practice. The 4-dB exchange, 84 dB PEL represents DOD practice. Finally, the 3-dB exchange, 90 dB PEL curve represents the practice in some European countries. The warning that special limitations may apply is based on the fact that permissible exposure time and sound level may not be exchanged without restrictions. For example, currently OSHA does not allow *any* duration of exposure above 115 dBA, except for impulsive sound..

Sound Exposure Level—SEL

An alternative that has been advocated for use in regulating hearing protection is SEL. It is used only with a 3-dB exchange rate; no meter dynamics are specified. It is defined as:

$$SEL = L_{eq} + 10 \log \left(\frac{T}{T_0} \right) \tag{3.14}$$

In the above formula, L_{eq} is computed for the observation time T given in seconds. T_0 is a reference duration of one second. The interpretation of SEL is that it represents a sound of one second duration which contains the same amount of acoustical energy as would be obtained by integrating a sound level that is varying in time over the time period T.

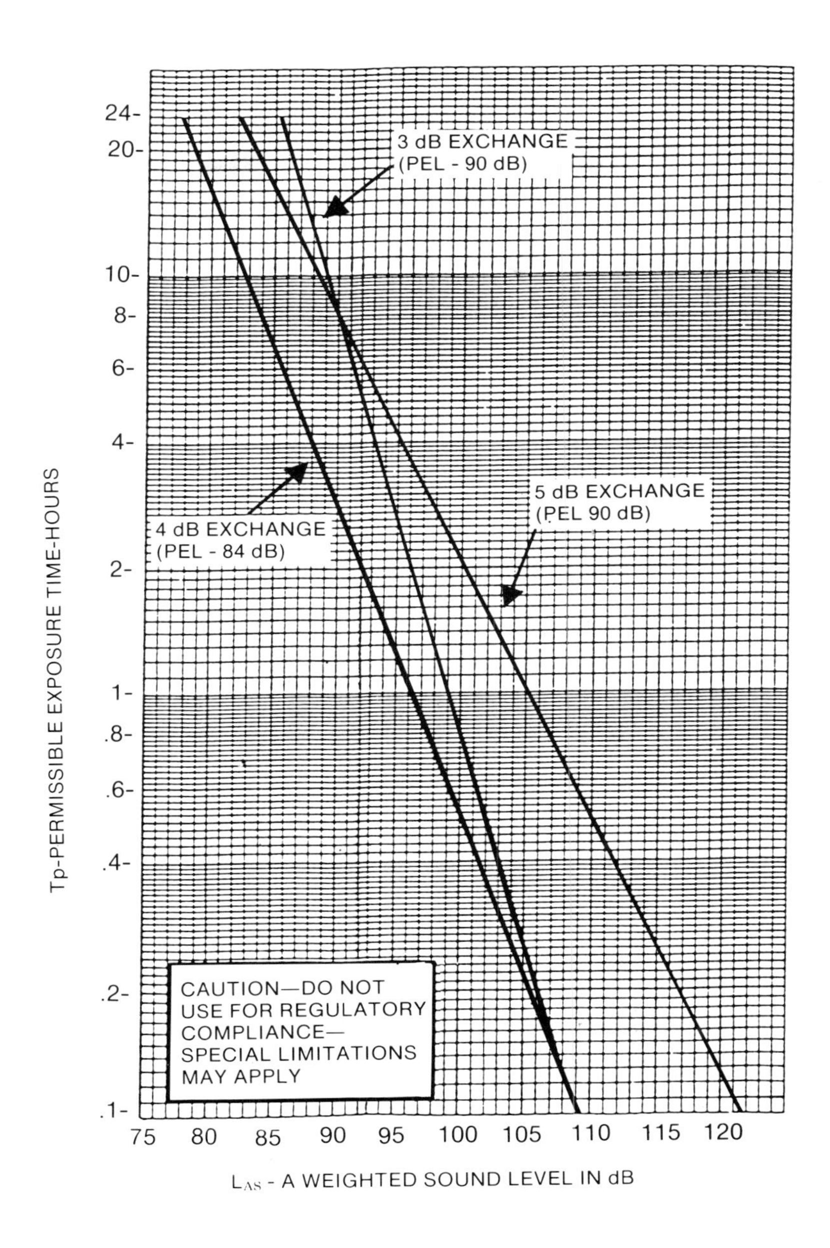

Figure 3.4. — Examples of permissible exposure time *vs.* A-weighted sound levels.

Instrument Selection and Utilization

When noise measurements are to be made, it is important to learn the capabilities and limitations of the measuring instruments and to acquire sufficient knowledge of acoustic field properties so as to know when a reliable measurement can be made. However, when variability is high and reliability of measurements is low, the knowledge and experience of a professional acoustical consultant may be necessary. The importance of enlistment of such assistance cannot be overemphasized if noise reduction is contemplated by application of engineering controls. Such controls can be quite expensive and can be rendered ineffective if small but crucial details are overlooked.

Subsequent discussion emphasizes the use and salient operating characteristics of instruments; the objectives and procedures for conducting surveys are addressed in the next chapter.

USE OF MICROPHONES

The primary sensor for noise measurement is the microphone. Proper application requires selection of a type matched to the properties of the field being measured. This information can be found in instrument manufacturers' instructions and application notes. Extensive discussions are presented in several noise handbooks (Beranek, 1971; Harris, 1979; Peterson, 1980). In addition, ANSI S1.13 (1971) and S1.30 (1979) describe standardized procedures for use of microphone and instruments in making measurements. Also, Chapter 2 of this text contains definitions and properties of free and diffuse sound fields.

Regardless of the microphone type used, some basic facts and procedures pertain to all measurements. Whether attached to an instrument or handheld, a microphone is susceptible to errors in measurement produced by reflections. Figure 3.5 (Peterson, 1980) illustrates the variations in measurement that can occur due to the presence of the operator and an instrument case. Such effects can be minimized by the simple expedient of orienting the microphone and operator in a plane perpendicular to the propagation path from the source. The worst possible configuration is to stand either in front of the microphone and thus risk casting an acoustic shadow, or behind the microphone and risk reflective interference. If a diffuse field is being measured, there is no identifiable propagation path, so the microphone should be held as far from the operator as is practical. An extension cable may be used to isolate the microphone. It is best to insert a preamplifier between the microphone and cable to counteract effects caused by electrical losses in the cable.

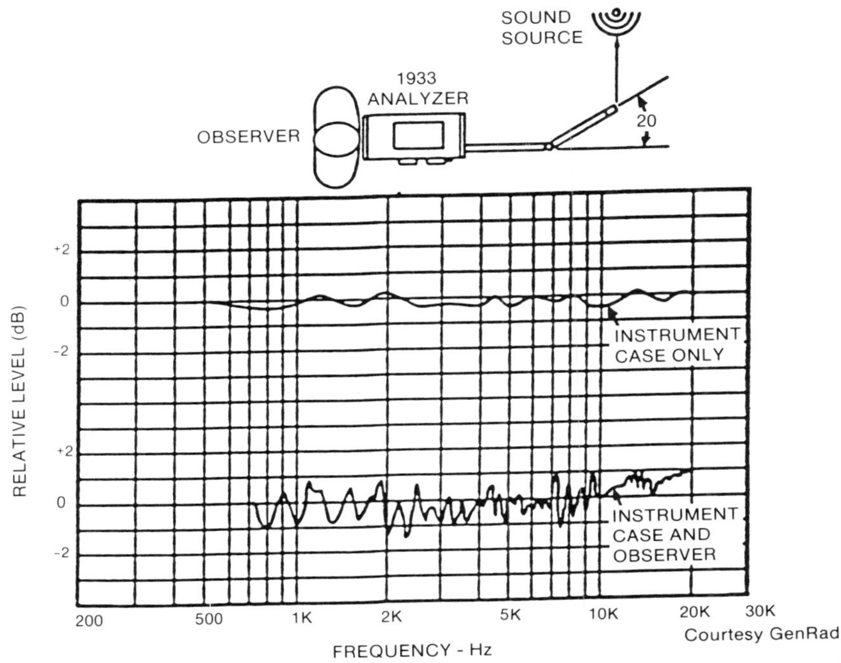

Figure 3.5. — Errors introduced by presence of SLM instrument case and operator with microphone extended on boom in a directional sound field.

Similarly, care should be exercised in measurements near reflecting surfaces, such as walls or large objects. "Near" is a relative term; the governing parameter is the wavelength of the sound being measured. Good practice dictates staying at least 1/4 wavelength away from a reflector. (Such a restriction cannot be observed, of course, when the noise exposure of a person is being measured.) In measurement of noise containing many frequency components, it is the wavelength of the lowest frequency component that governs. For 100 Hz, a 1/4 wavelength is approximately 76 cm. At 200 Hz, the corresponding distance is 38 cm. Distances can be scaled up and down to match frequencies of interest.

The guideline given will fail in the special situation where a standing wave exists. Such an effect occurs when a noise source radiates sound toward a reflecting surface which is uniform and does not scatter the sound. If the sound is reflected toward the source, the incident and reflected waves superpose and produce a "standing wave" by constructive and destructive interference. This condition is detected by moving the microphone along the principal propagation path and noting that a succession of maxima and minima of sound level occur. To obtain an estimate of the average sound field note the maximum and minimum levels. If the difference is less than 6 dB, use the arithmetic average; if greater than 6 dB, use a value 3 dB below the maximum.

The admonition that the operator taking measurements should strive to minimize disturbance of the ambient sound field by appropriate mike positioning or use of extension cables on the microphones is very appropriate and correct. It should be observed in investigating sound fields for engineering control purposes. The related question, whether to remove workers from a measuring zone when the objective is to determine their personal exposure, is not simple to answer. It is the noise actually impacting the worker's ears that matters. Unfortunately, there is no clear agreement between the professional community and regulatory agencies as to exactly what field values must be observed under specified conditions. For example, if a sound level measurement is made at the very opening of the auditory canal, should the value obtained be corrected to the value that would have existed in an undisturbed ambient field, at the center of the employee's head? This is one of a number of perplexing issues that affect choice and implementation of noise measures used to predict hearing damage. Current OSHA practice stipulates locating the microphone in the worker's hearing zone (a hypothetical sphere extending approximately 30 cm away from the head.)

A fundamental split exists over the types of microphones to be used in measurements defined by standards of European origin *vs.* those of North American origin (Bruel, 1981). The European approach is based on free-field microphones which have directional properties. In use they are pointed at the source of the noise. In North America, random-incidence microphones are preferred and have an average response that is independent of the angle of arrival of a sound wave. The user of microphones and associated instruments should verify that the appropriate type is selected for measurements complying with applicable technical and regulatory standards. The manufacturer's instructions and applications engineers should be consulted for clarification.

Basic commonly used SLMs and dosimeters encountered in North America have random incidence microphones. Choice of other, specialized microphones becomes necessary only when very versatile instruments which permit microphone interchangeability are considered for complicated measurements.

Long exposure of any microphone to very high humidity should be avoided. This is especially true when condensation of water on the microphone may take place. Although ceramic and condenser microphones are not damaged by exposure to high humidity, their operation can be seriously affected unless proper precautions are taken. Some microphones have extremely small ports which can be blocked by water from condensation. This materially affects performance.

For proper operation, it is essential that very little electrical leakage occurs across the microphone circuit. The exposed insulating surface of the microphone is specially treated to maintain this low leakage, even under conditions of high humidity. Despite this precaution, the leakage can be excessive under extreme conditions. It is then advisable to keep the microphone at a temperature higher than ambient to reduce the leakage. The preamplifiers supplied by some manufacturers have an electrical resistance heater to achieve this goal.

In climates where the humidity is normally high, it is recommended that the microphone be stored at a temperature above ambient to avoid condensation. An instrument malfunctioning due to moisture will emit a crackling sound similar to static on an AM radio, which can be heard if the output signal is monitored with earphones.

An electret microphone can stand normal variations in temperature and humidity for long periods without a significant change in sensitivity. However, if the humidity is normally high, an electret microphone should be stored in a small jar containing silica gel. It is recommended that the electret microphone not be exposed to relative humidities in excess of 90% (NBS, 1976).

Although most noise measurements are made indoors at average room temperatures, some measurements must be made at higher or lower temperatures. Under extreme conditions, it is essential to know the limitations of the equipment as specified by the manufacturer. Most microphones will withstand temperatures of −25 °C to +55 °C without damage. Preamplifiers for condenser microphones are limited to about 80 °C. SLMs usually cannot be operated below −10 °C without special low temperature batteries. Microphones are usually calibrated at normal room temperature. If a microphone is operated at other temperatures, its sensitivity will be somewhat different and a correction may have to be applied if specified in the manufacturer's recommendations.

LIMITS ON MEASUREMENT IMPOSED BY SYSTEM NOISE FLOOR

An ideal microphone and associated amplifiers should have exactly zero output if no sound is present. Real instruments have self-generated internal noise sources; therefore, the output can never be exactly zero for a total absence of sound.

Modern instruments have self-generated noise levels that are much below commonly encountered industrial levels, so they impose no constraints on such measurements. There are circumstances which necessitate examination of the limits imposed by system noise floors. One example involves measurement of community background noise when intrusive industrial sources are shut down. Another, of particular current significance, pertains to octave-band measurement of background noise in audiometry booths and rooms.

The principal sources of noise which establish a system measurement floor are the electrical circuit noise and noise generated by air turbulence at the microphone. Both noises have individual spectral characteristics. It is therefore necessary to obtain data stipulating the noise floor levels for every weighting filter and microphone that might be selected for use with the SLM or other instrument. Such data are usually available from the instrument or microphone manufacturer.

A possible point of confusion to be avoided results from the specific manner in which the threshold data are identified. No universal nomenclature is recognized, so descriptors may vary among manufacturers. Conceptually, the floor levels may be identified as the "self noise level" or the "minimum measurable level". The two are not synonymous. Consider hypothetically a self noise level (electrical) of 25 dB being stipulated for a given filter (for example, A-weighting). In the complete absence of acoustic noise at the microphone, an SLM can be expected to have an average reading of 25 dB. If a broadband noise, having an A-weighted envelope and a level of 25 dB is presented to the microphone, the SLM will have an average reading of 28 dB (recall that summing two random noise signals having equal levels results in a combination having a level 3 dB higher). It is evident that this SLM cannot be used simply to measure a true noise having a level of 25 dB. Standards of good practice dictate that the self noise floor be at least 10 dB below the level at which a minimum reliable measurement can be made. Table 3.4 shows data provided by one manufacturer for a type 1 SLM having optional octave-band filters. Note that the particular microphone used affects the minimum levels.

A second limit is imposed by turbulence-generated noise produced by air flow past the microphone. Such noise increases with the magnitude of the velocity of the flow. A manufacturer may supply data for wind-generated noise as a function of specified filters. For a given velocity, noise level is greater in amplitude at lower frequencies. It is commonly observed that an

TABLE 3.4
Nominal Minimum Measurable Noise Levels (dB re 20 μ Pa)

Microphone Type	Wide Band				Octave Band Center Frequency (Hz)									
	A	B	C	Flat	31.5	63	125	250	500	1K	2K	4K	8K	16K
1″ Electret condenser	21	23	31	38	24	22	19	17	15	13	12	13	15	16
1/2″ Electret condenser	32	35	39	48	37	34	31	32	26	24	23	22	22	22
1″ Ceramic	19	18	20	28	13	11	10	10	10	9	9	12	15	16

A-weighted reading of turbulence noise will be substantially lower than that taken with C or flat weighting because the A-weighting filter discriminates against low frequencies.

Airflow noise can be greatly reduced by enclosing the microphone in a windscreen. The screen may range from a small plastic foam ball to a wire-framed sphere several feet in diameter and covered with a fine nylon mesh. Such screens must be nearly transparent acoustically, and they must smooth out the ambient airflow to minimize turbulence and thus noise.

Data reported for wind screens are typically stated as equivalent self noise for a given wind speed. It is the responsibility of the user to decide how much margin to use in deciding the minimum measurable level of ambient noise when a given wind speed is encountered. For example, one manufacturer stipulates a self noise of 56 dBA for a screened microphone exposed to 25-mph wind. If one chooses to use a 10-dB margin, the minimum measurable level is 66 dBA. Clearly this may be an unacceptable level for some community noise measurements. One solution is to wait for a calm day.

A simple test to explore whether turbulence noise exists is to make the measurement with and without a wind screen. If use of the screen lowers the reading, turbulence is present. If the reading is lowered by 10 dB or more, the reading is dominated by turbulence. This can also often be detected by listening to the output of the SLM with high quality headphones. Turbulence will add a distorted characteristic to the sound that is not present in the actual sound field.

Turbulence noise is typically highest at low frequencies. If there is a suspicion that it exists, care must be exercised to assure that high-level low-frequency components do not overload the measuring instrument. Refer to instrument operating instructions and microphone manufacturer instructions for guidance on specific precautions.

In industrial environments with high sound levels, turbulence-generated noise is not very often of concern. Should significantly high air velocity flows be encountered, the possibility of interference with the sound measurement should be evaluated (*e.g.*, at a work position near the opening to a glass melting furnace, high-velocity cooling air jets may be encountered.)

ADDITION OF DECIBELS AND CORRECTIONS FOR BACKGROUND NOISE

In performing noise measurements, it is often necessary to add or subtract sound level readings obtained at a common point. Simple arithmetic addition or subtraction of the levels in decibels will give an incorrect result. Chapter 2 discusses the proper methods to use. Frequently a quick estimate of summations must be obtained without resorting to tables or formulas, and is achieved by remembering the following: For equal values of the levels to be added, add 3 dB to the common level. For levels differing by 10 dB, add 0.4 dB to the higher one. It is evident that for a difference of 10 dB or more, the higher level is dominant.

In using the above it is important that the sound levels involve the same frequency weightings. For example, if one of the levels is A-weighted, then the other one must also be A-weighted. It is not correct to use the given method to add and subtract an A-weighted and a C-weighted level measurement.

When measurements are made on an individual noise source, the possibility exists that errors will be introduced by the added contribution of background noise. To eliminate such errors, two identical sets of measurements should be made, one with the noise source of interest operating and a second with it inoperative in order to obtain the level of background noise. From such data, the noise from the individual source can be determined through use of the nomograph of Figure 3.6. The horizontal scale represents the difference between the total noise and background alone in decibels. The vertical scale indicates the correction to be subtracted from the total to obtain the desired source noise level. It is evident that for a difference of 10 dB between total and background alone, the correction is only 0.4 dB. This justifies the conclusion that background contributions may be ignored when the sum is 10 dB or more above background.

CALIBRATION

The electronic circuits of modern instruments exhibit a high degree of stability and typically do not have to be adjusted on a day-to-day basis. Microphones used with contemporary instruments are also quite stable but have substantially greater susceptibility to damage by environmental conditions. It is good practice to calibrate measuring instruments with an

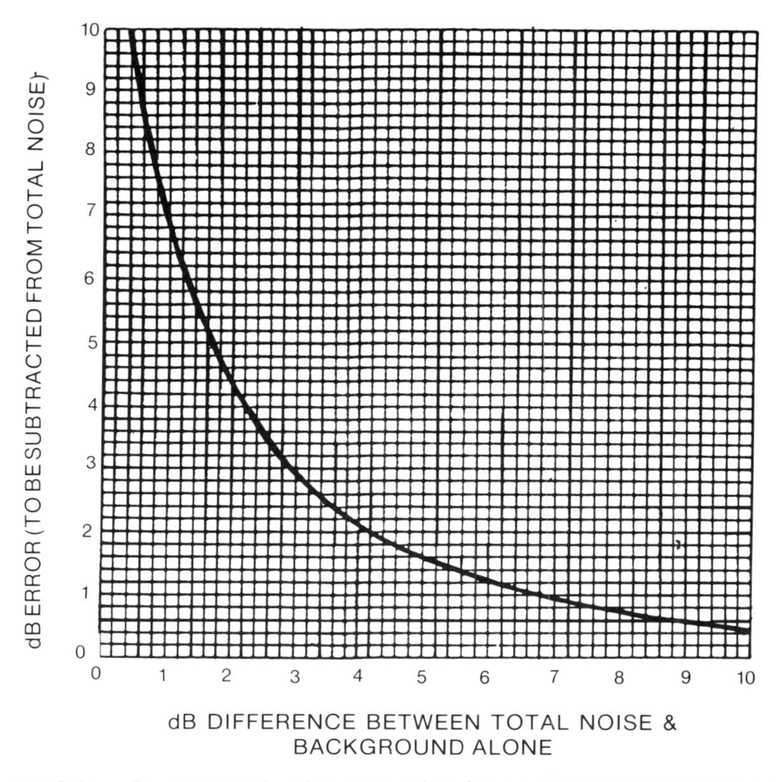

Figure 3.6. — Background noise correction for sound level measurements.

acoustical calibrator at least at the beginning and end of the work day. If the statements about stability are given credence, why should effort be devoted to frequent calibration? The primary purpose is to provide verification of continued reliable operation of the instrument rather than to insert minor corrections repeatedly. Corrections of a fraction of a decibel are usually counterproductive (also see Chapter 4).

Two guidelines determine when and how to calibrate an instrument acoustically. Regulatory agencies have individual requirements that may mandate particular calibration procedures. These should be studied and complied with. In addition, instrument manufacturers provide recommendations for calibration procedures. Some manufacturers of high-accuracy instruments such as type 0 and 1 actually recommend against fine-tuning the instruments to conform to the output of a calibrator. Instead it is recommended that a record of the difference between the indicated calibrator output and the instrument reading be kept for possible subsequent analysis. Such a procedure is motivated in part by the limited accuracy that can be maintained in an acoustical calibrator (ANSI S1.40-1984).

Probably the most important function to be performed by the calibrator is to detect gross changes in instrument performance, in particular the microphone. Calibrators that have multiple frequency outputs are particularly useful for this. In condenser and electret microphones there is always the possibility of puncturing the diaphragm, particularly because such microphones are easily disassembled. Should this happen, the low-frequency response of the microphone tends to be affected more than the high. With a multiple frequency calibrator, this effect can be detected.

Calibrators should be selected with care. It is not safe to assume that because a microphone fits snugly into a calibrator cavity that the internally generated pressure is necessarily at its nominal value. The presence of the microphone itself can affect the calibration signal. The proper approach is to consult the operating handbook of the calibrator and/or instrument to seek assurance that a given microphone and calibrator are compatible.

Ambient atmospheric pressure has an effect on the output level of most calibrators. The effect is instrument-specific. Again, the appropriate manufacturer's information should be consulted to apply necessary corrections.

When possible, perform a calibration with two or more calibrators to obtain comparisons. If the variations are outside published tolerances, send all in for recalibration. (In the short run, use majority vote selection if possible.)

The important question "How often must calibrators be recalibrated?" has no simple answer. Certainly any time a calibrator has been dropped or is otherwise suspected to have been damaged, it should be recalibrated. Regulatory agencies sometime stipulate that calibration be performed at specified intervals. To retain confidence, establish a regular schedule for recalibration.

As previously discussed, there are sophisticated calibrators which permit testing of instrument time constants, linearity, crest factor, *etc.* These are useful instruments but should be used in the same spirit as the basic calibrator: namely, monitoring verification of continued satisfactory performance rather than fundamental calibration. An important additional step in using calibrators is to verify that the internal batteries are in good condition, sufficient to permit acceptable output by the calibrator. Some instruments have automatic shutdown when battery voltage falls below an acceptable level; others require that the user monitor a battery-condition indicator. Again it is good practice to be familiar with all the manufacturer's recommendations.

APPLICATION AND OPERATION OF SLMS

Reading a sound level meter is not difficult when the indicator needle is constant; problems arise only when it is fluctuating. Instructions of instrument manufacturers and those incorporated in some standard reference texts for interpreting variable readings (*e.g.*, Olishifski, 1975) have to be very carefully assessed to determine when they are appropriate to the required objectives of the measurement. To appreciate the significance of what follows, one must have a basic understanding of the workings and dynamics of an SLM as well as the definitions of the noise measures used to regulate hazards to hearing. Difficulties arise because guidelines presented in many references strictly apply only to measurements pertaining to engineering controls and regulatory compliance based on 3-dB exchange averaging. When measurements are used for predicting or verifying compliance with regulations based on other than 3-dB exchange, subtle but vital constraints arise. The following paragraph is quoted from the third edition of the AIHA Industrial Noise Manual, (1975), p. 18.

"Meter Response. Since most industrial noises are fluctuating in nature, sound-level meters are provided with FAST and SLOW ballistic (response) characteristics. If the noise is steady enough and the meter pointer fluctuates over only a few decibels, FAST response is used and the measurement might be recorded as, say 85-91 or 88 ± 3 dB. When an average sound pressure level is desired and fluctuations are less than 6 dB, a simple average of the maximum and minimum levels is usually taken. If the range of fluctuation is greater than 6 dB, the average sound pressure level is usually taken to be 3 dB below the maximum level. *In selecting this maximum level, it is also customary to ignore any unusually high levels that occur infrequently.* The SLOW meter response should be used to obtain an average reading when the fluctuations on FAST response are more than 3 or 4 dB. For steady sound, the reading of the meter will be the same on either the SLOW or FAST positions. For fluctuating sounds, the SLOW response provides a long time average reading."

The foregoing is representative of what has been accepted practice in acoustical engineering and in industrial hygiene. The principles are correct and adequate when averaging is based on a 3-dB exchange (equal energy) basis. In contrast, these guidelines can lead to incorrect conclusions when 4- and 5-dB exchange averaging must be used for regulatory compliance.

Use of FAST or SLOW Dynamics

The choice cannot be made arbitrarily based on the character of the noise. Current OSHA regulations and the ACGIH guidelines (Jones, 1984) explicitly state that determination of worker noise dose must be made with SLOW meter dynamics. When a 4- or 5-dB exchange rate is used, the substitution of FAST or even elimination of the meter dynamics as suggested by some practitioners (Erdreich, 1983) can have a significant impact on the measured dose. This effect is marked when the noise contains short transients of less than 1 second duration (Earshen, 1980; Kamperman, 1980).

The mathematical rationale for this is somewhat involved and beyond the scope of this discussion. However, a simple explanation is that the presence of the meter dynamics in the signal processing chain modifies the waveform that is ultimately averaged and used to generate noise dose. It is a basic fact that the averaging, on a 3-dB exchange basis, of a waveform modified by SLOW or FAST meter dynamics produces final results that are essentially independent of the meter dynamics chosen. Unfortunately, this is not the case for 4- and 5-dB exchange rates.

There is no rigorous mathematical model for predicting hearing damage that one can use to justify use of a given exchange rate or given meter dynamics. As previously cited, such choices are made on the basis of empirical models. Thus, choice of instrumentation must comply with the constraints associated with the model used by a given regulatory agency. While challenges to the appropriateness of a given model may be valid on physiological grounds, resolution cannot be attained by advocating departure from specified procedures on mathematical grounds.

Averaging by Mental Estimation

One of the prominent places where the inadequacy of decibel notation becomes evident occurs when an SLM is used to measure a varying noise level. The human mind is simply not equipped to perform biased averaging of readings presented on a logarithmic scale.

To illustrate the point, consider some very simple cases. Assume that a measurement is being made of a noise source that alternates between two sound levels having identical spectral properties and separated by 10 dB. Each level is on sequentially for exactly 10 seconds. A realistic example of such a source would be a compressor with a cyclic load. With an SLM set on SLOW, both levels can be easily observed and so would produce readings 10 dB apart. If the two levels were 100 and 90 dB respectively, customary methods of reading linear instrument meters would lead to the conclusion that the "average value" should be 95 dB. The guidelines given in the above cited reference, however, tell us that the average level should be approximately 97 dB, which indeed it is provided that equal-energy averaging is required. This conclusion would be much more obvious if the decibel notation were not used.

If we transform the sound level readings into their underlying power ratios, we would find that the 100-dB reading has a magnitude that is 10 times as large as that corresponding to the 90-dB reading. The arithmetic average of 1 and 10 is 5.5, which is very nearly just half the value of the larger quantity, which is what the 3-dB reduction, from 100 to 97 dB, implies. However, consider what the average level will be when 4-dB exchange and 5-dB exchange averaging is required. The average level for the 4-dB case will be 96 dB and for the 5-dB case will be 95 dB. Clearly the results are not identical for the 3 cases cited. The user must therefore exercise caution when using an SLM. If the fluctuations are small, say ± 3 dB, averaging by eye is acceptable. However, it is not that the small size of the fluctuation makes such visual averaging mathematically correct; it simply produces results that are within the bounds of accuracy achieved for the type of measurement discussed.

A further complication can be encountered. In the example, equal durations were stated for the levels to be averaged. This is an unrealistic assumption for most fluctuating noises. Accordingly, the operator must account for the time variability and enter appropriate time weightings if mathematical rigor is to be observed. Clearly this is ordinarily beyond the visual and mental capabilities of the operator. It is evident that fluctuations exceeding 6 dB lead to a complicated situation and require careful attention to the details of the time waveform. This can be achieved by producing a hard-copy record of the sound level by means of a graphic level recorder. Alternatively the problem of recording and data reduction can more easily be solved by using a dosimeter or integrating/averaging SLM having proper computational parameters to achieve the objective automatically.

DOSIMETER USE

Before use, a dosimeter must be calibrated (some regulatory agencies also require calibration verification after use.) Such calibration may be a simple task similar to calibrating an SLM or it may be somewhat more complicated. Some dosimeters can operate as conventional SLMs as well as dosimeters. Calibrating such an instrument is done with an acoustical calibrator by setting the indicated sound level to correspond to the known level of the calibrator. Other types of dosimeters are calibrated by supplying a calibrator signal for a specific time. This combination produces a predictable noise dose which should be indicated by the dosimeter. If the reading is incorrect, then the dosimeter must be adjusted in the manner prescribed for the particular instrument.

Modern dosimeters are capable of computing noise dose or time weighted average using 3-, 4-, 5- and sometimes 6-dB exchange rates, one of several 8-hr criterion levels such as 80, 84, 85, and 90 dBA, and different cutoff levels such as 80 and 90 dB. Other combinations of parameters are also available. The selection of these parameters is performed by one of the following means:

1) Factory setting. Some dosimeters cannot be reset. Fixed values of the exchange rate, criterion level and cutoff level dictate that different applications can be achieved only by choosing different model numbers.
2) Internal wiring or switch changes. By resoldering internal connections or by repositioning internal switches, changes can be effected.
3) By exchange of PROMs. Microprocessor-based instruments use program instructions contained in plug-in components called Permanent Read Only Memories (PROMs).
4) Front panel keyboard selection. Some dosimeters are in effect special purpose microprocessors which can be reprogrammed or have their parameter choices changed through a keypad or keyboard. (Another version of this type downloads instructions from an external data source into the operating memory of the dosimeter processor.)

The above selections are only required when a dosimeter is first employed or when changes are made in the type of measurement desired. An example of the latter is preparing to make a series of measurements using a 90-dBA cutoff for the purposes of determining requirements for engineering controls and subsequently making a measurement with an 80-dBA cutoff for determining the need for a hearing conservation program. (These modes of measurement are dictated by current OSHA regulations.)

Measurement of noise and worker exposure to noise is far from an exact science. Many variables other than limitations of instrument accuracy affect the results. Reasonable effort should be devoted to careful use of the best possible instruments, but it is vital to recognize that measurement results having resolution finer than 1 dB really do not lead to better, more effective protection of hearing.

Verification of Setting

It is very useful to have simple methods for verifying the operating parameter settings of a dosimeter. In the case of exchange rate, if the instrument can indicate average sound level, the following procedure can be followed. Supply an acoustical input with a calibrator for a period of time long enough to obtain stable average level readings. This could be as little as 30 seconds; however, the manufacturer's instructions should be followed. The average level read at the end of the fixed period should correspond exactly to the output level of the calibrator. Next, the calibrator is turned off but kept in place over the microphone to provide an acoustic shield. Finally, the dosimeter is again placed in operation for a period of time identical to the first. At the end of the second period, the average level should be read. Since no additional acoustical input was supplied, but the averaging time was doubled, the value in decibels should drop by an amount corresponding to the exchange rate. Thus for an initial calibrator input of 94 dB, the final average level should be 89 dB for a 5-dB exchange rate, 90 dB for a 4-dB exchange, and 91 dB for a 3-dB exchange rate.

Cutoff Characteristics

A test of the cutoff characteristics of the instrument can be performed with a calibrator having a variable output. The current dosimeter standard (ANSI S1.25-1978) stipulates that a signal supplied to the dosimeter at the cutoff level in dB should be completely integrated. As the level is lowered, there should be a progressively smaller effect on the computed output until there is an absolute cutoff 3 dB below the nominal cutoff level. The standard cited is under revision and it is likely that this tolerance will be tightened up. Refer to the most up-to-date standard for guidance.

Microphone Placement

Kuhn and Guernsey (1983) have analyzed published and unpublished data on sound distributions about the human head and torso. They conclude that differences as large as 5-7 dB relative to the undisturbed field can exist. Seiler (1982) reports an experimental evaluation of positioning one-half inch microphones on the human body and indicates that the practice prescribed by the Mine Safety and Health Administration (MSHA) is an acceptable compromise. MSHA requires that the microphone be placed on the top middle of the shoulder with the microphone pointing upward if it has enhanced directional characteristics along its axis. However, with modern small microphones, 1/2 inch or less in diameter, orientation is generally not a problem since they are nondirectional in the frequency range of interest.

At present, OSHA does not stipulate where a microphone has to be positioned nor does ANSI S1.25. The individual conducting dosimetry measurements has broad latitude in choosing microphone position. The previously described position on the shoulder is strongly recommended unless some specific reason for an alternative position exists.

The user should be aware of possible field disturbances due to microphone positioning but should not be prejudiced against using dosimeters (it is the preferred method for OSHA.) In a very large percentage of measurement situations, the noise field encountered is diffuse and generally has frequency content that will not be associated with significant field distortion. In addition, use of a measurement method which does not employ a person-mounted microphone, *e.g.* an SLM, still requires that the microphone be located within "the hearing zone", a sphere whose surface is about 30 cm distant from the head, so that there still exists the possibility of field distortion.

Measurement Artifact

Some dosimeters can record true unweighted peak SPL. Under current OSHA regulations, no worker may be exposed to peaks in excess of 140 dB. With instruments having peak-reading capabilities, care must be exercised in interpreting readings because microphones do respond to tapping and brushing. It is perfectly possible in an ambient noise field of 90 to 100 dB to produce readings that may exceed 140 dB by intentionally or accidentally tapping the microphone. In a high ambient noise level, neither the test conductor nor the worker wearing the mike may be aware of such rubbing and tapping. To resolve the problem, the subject should be observed very carefully during the period of time that exposure to impulsive sound is being evaluated. Should any contact be made with the microphone, the readings should be questioned. There is nothing, of course, to prevent the dosimeter from being placed on a stand for determining the possible peak values.

Another artifact that can be minimized by observation of the workers is deliberate blowing, shouting, or whistling into the microphone. In fact, procedures used by OSHA inspectors specify that a worker wearing a dosimeter must be kept under surveillance.

Susceptibility to noise artifact generation from shock and vibration of the microphone can be minimized by protecting it with windscreen foam caps provided by most manufacturers. Use of windscreens will also provide possible protection against turbulence-generated noise when the worker-mounted microphone is subjected to air blasts. If the manufacturer supplies data relating air velocity past the windscreen to turbulence-induced noise artifact, the velocity of the air flow should be measured and appropriate judgment made about the possible contribution of the artifact to the overall measurement. A separation of 10 dB or more between the artifact and the actual ambient noise will assure minimal impact on measurement accuracy.

A profiling dosimeter provides additional information which facilitates exercise of judgment about the validity of individual portions of the exposure. Consider the case of a profile (temporal record) consisting of a series of 1-minute contiguous sound level averages. If it is known that the work areas where the subject being monitored is authorized to enter have a maximum sound level of 112 dBA and a minimum sound level of 80 dBA, then finding any 1-minute average that falls below the minimum or above the maximum will indicate that the exposure measurement should be questioned.

Impulse Noise

Dosimeters covered by the basic standard (ANSI S1.25) are not required to measure noise that is predominantly impulsive. Such dosimeters have a nominal crest factor capability of 10 dB. (See previous definitions of crest factor and pulse range.) However, an instrument that has a crest factor of 10 dB cannot safely be assumed to have a pulse range that is also 10 dB, and even if it does, this is woefully inadequate for observing impulses that can be encountered in industry. (Even high-quality SLMs are not much better in this respect in that they have a capability of not much better than 20 dB.)

What pulse range capability should be required and how short an impulse should be measured are much in debate but there are several benchmarks to be examined. OSHA, in its initial introduction of the Hearing Conservation Amendment of 1981, in effect stipulated a 30-dB pulse range. Modern instruments now appearing on the market are capable of pulse ranges in excess of 53 dB. What should the maximum requirement be? The total excursion of impulses is difficult to categorize, but a meaningful rule of thumb can be used. If we assume that hearing damage begins to manifest itself somewhere around 80 dB, and if we accept the limitation that 140 dB must never be exceeded, we conclude that an instrument that has a pulse range of 60 dB should be adequate. This in fact is the goal at which instrument manufacturers are now aiming.

Along with the pulse range identified, it is necessary to stipulate how short the impulses to be measured can be. After reviewing data dealing with industrial impact noise, Erdreich (1983) has concluded that impulsive noise of less than 20 milliseconds duration is seldom encountered in industry.

The only instrument standard at this time which stipulates pulse duration and pulse range requirements that can be applied to dosimeters is IEC 804 (1985). It presents performance requirements for pulses as short as 1 millisecond.

Wet Environments

Another microphone and instrument problem encountered in practice but rarely addressed in manufacturer's instructions is what to do to monitor a worker in an extremely wet or rainy environment.

This is a challenging problem because any physical covering used to protect the microphone may affect its frequency response. Many enclosures have been tried, including sandwich bags, balloons, plastic wrapping material, *etc.* Under controlled laboratory conditions, each has sometimes been found to affect frequency response and therefore cannot be used without constraints. Nonetheless how can the problem be handled? An effective but limited solution is to wait until it stops raining and foul weather subsides. Another is to use an all-weather microphone and enclosure, but cost, weight, and size limit this option severely for body-mounted dosimeters. Is is possible to make area measurements with such microphones and to rely on analysis to infer worker exposure.

As an absolutely last resort, covering the microphone with a makeshift waterproof enclosure may be the solution; however, extreme care must be used to detect and eliminate errors. *Serious errors can occur.* The thinnest and limpest covering should be employed. An inexpensive sandwich bag has on occasion been used by some practitioners. It should be placed over a microphone already covered by a windscreen. The bag should be placed in position very loosely and tied securely to hold water out. It is mandatory to verify that there is minimal effect on the frequency response of the microphone. Accurate evaluation is difficult, requiring laboratory measurements. If a noise to be monitored can be observed from a dry place, the validity of the scheme can be tested by comparing successive measurements with the microphone wrapped and unwrapped. This is not easy to do, but can be accomplished with diligence. If the differences in average sound level do not vary by more than several dB, the scheme should be usable. *This makeshift solution should not be relied upon for compliance, but it can provide valuable survey data when no alternative exists.*

Personal vs. Area Monitoring

Figure 3.7 presents results of an informative test performed to compare results obtained by monitoring an individual worker with those obtained by monitoring the area in which he worked. Two identical profiling dosimeters were used: one on the worker, the other fixed. Both were set to compute sound level averaged on a 5-dB trading relation in one-minute increments. The work zone was a large bay containing machines distributed throughout the area. The worker under observation performed a multiplicity of tasks at or in the vicinity of a number of machines. The test was conducted over a period of approximately 430 minutes.

The dotted curve depicts the worker-mounted dosimeter results, while the solid one shows fixed dosimeter results. The two are different and it is evident that there is little correlation between them. At times the employee average was higher and at other times the area average was higher. The

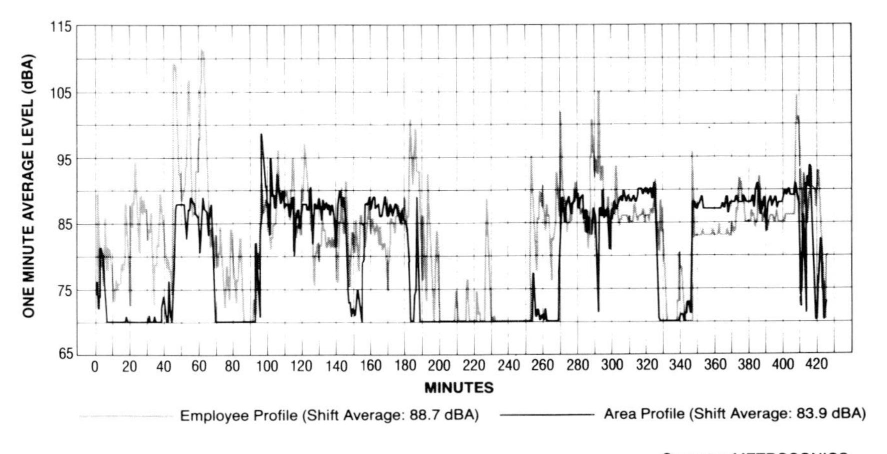

Figure 3.7. — Comparison of records of personal and area monitoring with profiling dosimeters.

reason for these differences is that the machines in the room had variable, unsynchronized operating cycles and the worker moved from one work location to another in a complex sequence difficult to predict.

The average overall sound level obtained from the personal monitoring instrument was 88.7 dBA while that from the area monitor was 83.9 dBA.

These results should be examined in the context of current OSHA requirements (OSHA, 1983) that a hearing conservation program be instituted if a worker is subjected to an 8-hour average sound level over 85 dBA. In this case, the personal data indicate that a program is required while the area data do not. The example shown does not suggest that the two forms of data gathered will always have the same relative values. It is presented to illustrate the complexity of worker exposure patterns that can occur and to caution the reader against simplistic inferences based on area measurements.

POTENTIAL DISCREPANCIES BETWEEN SLMS AND DOSIMETERS

Fundamentally, sound level meters and dosimeters should produce identical measurement results since they are generically related in a prescribed manner. When comparisons are made, it is important that the particular instruments involved have commensurate performance specifications. Discrepancies *are possible* but they are invariably caused by improper or incompatible utilization of the instruments. The manifestations of and remedies for such discrepancies are summarized in this section.

Microphone Placement

The microphone used with a dosimeter is most frequently placed on the body of the person being monitored. Sound reflection may produce spurious readings in this case. In contrast, the microphone of an SLM is commonly located away from nearby objects during measurements. In such an eventuality, differences in results may be produced. Note, however, that OSHA requires the measuring microphone to be in "the hearing zone" of the person being monitored. If results are to be consistent, the two microphones should (and certainly may) be positioned at the same spot. A dosimeter microphone is not restricted to body positioning. If it is desired to operate an SLM with a tripod-mounted microphone, the same can be done with the dosimeter microphone.

Personal vs. Area Monitoring

In addition to effects produced by positioning of microphones near or away from the person being monitored, significant discrepancies can result from comparing personal to area monitoring as was discussed previously. The basic configurations of the instruments tend to encourage personal monitoring for the dosimeter and area monitoring for the SLM. However, neither instrument is uniquely restricted to one measurement mode alone. Used comparably, the instruments should give consistent results.

Differential Susceptibility to Noise Artifact

All microphones are sensitive to shock and vibration. It is possible, however, that an unprotected dosimeter microphone might, unnoticed, be subjected to such artifact. An SLM used by a skilled operator tends to avoid this risk. For the dosimeter, care should be taken to provide shock and vibration insulation. If the measurement is likely to be critically affected, observation of the wearer to detect such occurrences may be necessary.

Response to Impulsive Noise

In the portion of the text discussing the application and operation of SLMs, reference is made to outdated recommendations for interpreting SLM meter readings: "... it is customary to ignore any unusual high levels that occur infrequently." Should this practice be followed, errors of omission can occur in the presence of impulsive sound. Claims have been made that dosimeters read "high" in the presence of impulsive noise as compared to SLMs. If "unusually high levels that occur infrequently" are ignored, it is reasonable to expect that dosimeters which do not ignore such transients will indeed give higher readings.

In its time, the cited practice was probably justified because SLMs had poor or inconsistent response to impulsive noise. Contemporary standards and instruments have remedied these shortcomings.

Even if the SLM is diligently monitored and recorded, discrepancies may still occur. The dosimeter continuously computes average sound level (or related dose) using the correct (OSHA-mandated) 5-dB exchange algorithm while following minute transient fluctuations in sound level. In contrast, the human eye-brain system lacks the ability to note and record, or to note and correctly average, slight transient changes in sound levels.

When mandated SLOW response is used with 5-dB exchange averaging, the results produced by dosimeters (or appropriately equipped averaging SLMs) are correct and accurate within allowed instrument tolerances. Results derived from human observers using SLMs tend to produce dose values that are too low. Experimental verification has been obtained (Earshen, 1985) through use of a video camera and tape recorder to monitor SLM meter readings in detail. Playback of recordings, one frame at a time, enabled accurate meter readings to be made. Analyses of readings obtained with a GenRad 1933 and a B&K 2209 SLM demonstrated that their SLOW dynamic responses are commensurate with the prescribed SLOW response of dosimeters. Furthermore, the noise dose measured by four dosimeters in current use agreed closely with that computed from the video recordings of the SLMs. (Both sets of instruments were exposed to identical impulsive transients.)

Controversy exists over the proper method to account for contributions by impulsive noise. The conservative approach (current OSHA practice) is to use an automatic instrument (dosimeter or integrating/averaging SLM) that has the ability to handle impulses. Since no official standard exists, the instrument manufacturer should be consulted for recommendations.

As previously stated, the mandatory SLOW dynamics in either SLMs or dosimeters tend to accentuate the contribution to noise dose by short (less than 1 second) sound bursts (impulses). The results are not erroneous; rather they have been confirmed by mathematical analysis (Kamperman, 1980; Earshen, 1985). The accentuated contribution can be improperly overlooked or minimized because of human observer limitations in assimilating and processing rapidly changing readings on a logarithmic scale.

It is a fact that the accentuated contributions can intentionally be minimized or eliminated in a dosimeter by substituting FAST response. Note carefully that such a change implies that an alternative hearing damage model is being applied and is contrary to current OSHA requirements.

EARPHONES AND AUDITORY PERCEPTION

Many general-purpose SLMs have an output jack in the amplifier chain placed after the weighting filter. By connecting a suitable high fidelity set of earphones, the frequency-weighted sound can be monitored aurally. If monitoring is to be accomplished in a noisy environment, a circumaural noise-excluding headset is required. With such instrumentation, a skilled operator can perceive characteristics of noise literally impossible to analyze with an instrument alone. The ear/brain system can process information and perform sophisticated discrimination and recognition tasks yet to be equalled by automated instruments or computers. The latter complement human judgment and perception through their ability to quantify, process, record, and retrieve large volumes of data at high speed.

An example where auditory perception has been applied is in signature identification. Suppose that an area previously having a noise level acceptable for regulatory compliance is found by a resurvey to have a level that has increased by 6 dB. When heard directly with no electronic aids, the increased noise shows no apparent difference from the original. On earphones following an A-weighting filter, dominant low-frequency components are removed and a distinctive sound signature may be perceived and recognized by a skilled operator to be indicative of abnormal machine operation. Despite significant advancement in signal processing, recognition of subtle changes in machine-generated noise frequently can be most easily performed by auditory perception.

Another example is detection of wind- and air-flow-generated noise by turbulence at the microphone. One way to accomplish this is to compare the sound in the earphones with the sound directly heard.

SUSCEPTIBILITY TO ELECTROMAGNETIC AND RADIO FREQUENCY INTERFERENCE

In industry, measuring instruments are sometimes employed in environments having high levels of electromagnetic fields. Although modern instruments do not exhibit strong susceptibility to such interference, it is nonetheless possible that conditions can be encountered in which care should be exercised. Performances specified in SLM and dosimeter standards primarily define resistance to interference from power lines. Other sources of interference include vehicular ignition systems, arc furnaces, induction heaters, radio frequency heaters and welders, arc welders, communications transmitters, *etc.*

In the majority of cases, electromagnetic interference will produce readings higher than those resulting from acoustical noise alone. A simple procedure for investigating suspected occurrences involves using unpowered acoustic-cavity calibrators. Such calibrators usually provide substantial acoustical attenuation when they are covering a microphone. When interference is suspected, the reading of the instrument in the suspect location should be noted, as the instrument is normally employed. The measurement should then be repeated under identical conditions but this time with the manufacturer's recommended calibrator covering the microphone. A substantial reduction should be observed (at least 10 dB); if not, interference must be suspected and further efforts to locate the source (or shield the instrument) must be made. If only a minor contribution from interference exists (more than 10 dB down), its effect on measurement accuracy will be insignificant.

Caution in applying the foregoing principle must be exercised, however, in cases where high levels of acoustic noise at low frequencies exist. For such low-frequency noise, the attenuation of the calibrator walls may not be in excess of 10 dB. To be sure that a particular calibrator has sufficient acoustic isolation, it should be tested (in a low frequency noise) in an area known to be free of electrical interference.

For some types of instruments substitution of a dummy microphone may be used to check for electrical interference. The dummy microphone has electrical properties identical to those of an active mike but has no response to sounds. In effect, this results in total elimination of an acoustical input. Therefore, the remaining responses observed must be entirely contributed by electromagnetic interference.

SCOPE OF FREQUENCY ANALYSES

In planning frequency analysis of noise, it is important to heed the prior admonitions that such analyses require skill and understanding of acoustical engineering and instrument applications. If measurements are to be made, one must first establish the extent to which analyses will be detailed. It is generally true that the analyses should be performed at the coarsest resolution in frequency compatible with the end objective. Figure 3.8 provides some instructive examples.

Consider a work station where a high noise level is produced by a nearby machine having rotating components. Initially, to determine compliance with OSHA regulations for hearing protection, a measurement can be made with an A-weighted filter. Results can be processed to produce either a 5-dB exchange averaged sound level or to generate a partial- or full-shift worker noise dose. Should noncompliance be identified, the next step might be to make octave-band measurements. The data so obtained could be used for specifying the attenuation required of an enclosure or intervening wall.

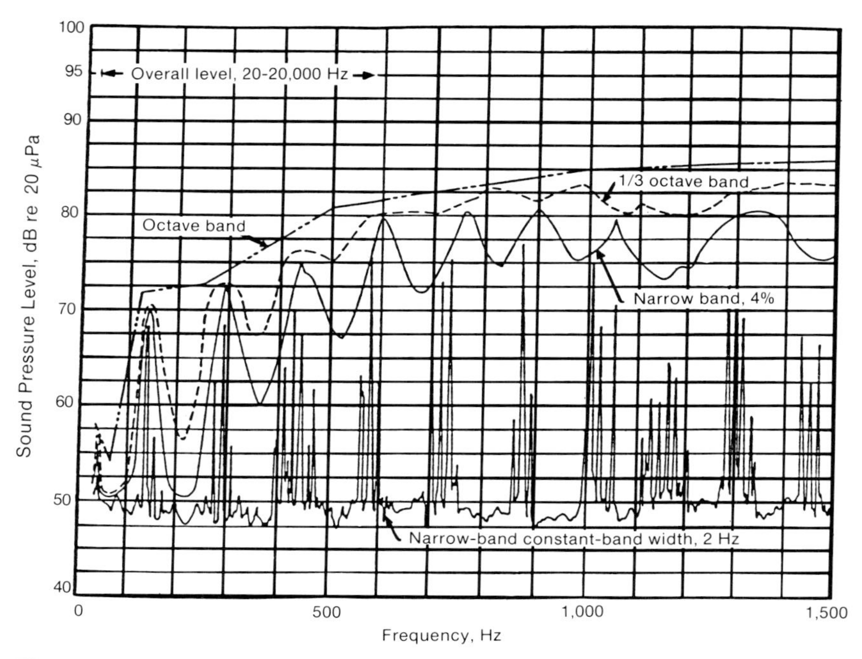

Figure 3.8. — Results of analyzing a noise source with multiple bandwidth analyzers.

To characterize the source in greater detail, measurements can be repeated with filters of successively smaller bandwidth. Figure 3.8 shows results for octave, 1/3 octave, 4%, and 2-Hz bandwidth analyses. Each successive measurement reveals greater detail about the source. It can be a mistake to go immediately to the highest possible resolution in frequency because unwieldy quantities of data will be rapidly accumulated. Also, high resolution in frequency requires a noise source with a reasonably stable noise spectrum. A good starting point is to use octave-band filters. As more detail becomes necessary, higher resolution analyses can be conducted.

Concluding Remarks

Measurement of noise, generation of integrated noise measures, and application of results to protect hearing and minimize annoyance under regulatory control, all require the exercise of judgment. The design, operation and application of instruments are inextricably tied to the metrics relating noise exposure to physiological and psychological effects and to value judgments implicit in regulatory practices. Comprehension of hearing damage processes involving exposure to noise is still limited, yet of necessity regulatory practices must be founded on limited scientific facts. In this state of affairs, great care must be exercised when advancing or interpreting advocacy positions supported solely by analyses founded on principles of physics while ignoring the underlying physiological processes. (For example, far too frequently, arguments are advanced claiming that 5-dB-exchange-based noise dose and $L_{av}(5)$ are without merit because they violate fundamental mathematical principles. That thesis stated alone is unsupported because it tacitly presumes, but does not prove, that a linear process must be pivotal in a hearing damage model.)

The reader has been besieged in this chapter with widely varying topics having crucial impact on the design and employment of noise measuring instrumentation. The objective has been to place in perspective the measurement challenge faced by the industrial hygienist. Measurement of noise, assessing its effects on humans, formulating and interpreting regulatory practices and effecting control and mitigation are each substantial tasks requiring extensive professional training and experience. Of these, analysis of spectral composition of noise and design of engineering controls are the most removed from the normal training and background of industrial hygienists. They should be addressed with caution.

It has been shown that sound level meters and their extensions, dosimeters, frequency analyzers, graphic level recorders, and integrating/averaging SLMs, are related in prescribed ways. For regulatory compliance, it is especially important that final noise measures that are generated must be based on commensurate data. To produce such data, when different types of instruments are used, each must conform to performance specifications that are stated or inferred by regulatory agencies.

Given sufficient understanding of objectives and constraints, many noise measurements can be performed with a basic SLM. It should be evident from the text that modern instruments are available which greatly improve the efficiency and versatility of measurements. Moreover, some instruments permit measurements that are beyond the limits of a human operator's ability to observe and interpret.

Computers are now well established both as components of measurement instrumentation and as means to store, retrieve, and process industrial hygiene information. Though it is possible to use basic SLMs for noise measurements and to store, retrieve, and process information without computers, the industrial hygienist must identify the true objectives for measurements and information handling before starting a program. In selecting instruments and means for utilizing and managing resulting information, the total costs associated with the overall objectives must be identified and used as a basis for selection. Focusing on individual system component costs alone can lead to erroneous conclusions. For example, the cost of an SLM, clipboard, paper, and pencil is deceptively low compared to computerized instruments and data handling systems. However, when personnel costs associated with manual monitoring and risk of damage claims are factored in, conclusions about cost *vs.* effectiveness may be drastically changed.

References

AIHA (1975). *Industrial Noise Manual,* Third Edition. American Industrial Hygiene Association, Akron, OH.

Anon. (1984). "Dynamic Measurement Instrumentation Buyer's Guide," *Sound and Vibration 18(3),* 33-35.

ANSI S1.11 (1966) (R1971). "Specification for Octave, Half-Octave, and Third-Octave Band Filter Sets," American National Standards Institute, New York.

ANSI S1.13 (R1976). "Method for Measurement of Sound Pressure Levels," American National Standards Institute, New York.

ANSI S1.25 (1978). "Specification for Personal Noise Dosimeters," American National Standards Institute, New York.

ANSI S1.30 (1979). "Guidelines for the Use of Sound Power Standards and for the Preparation of Noise Test Codes," American National Standards Institute, New York.

ANSI S1.4 (1971). "Specification for Sound Level Meters," American National Standards Institute, New York.

ANSI S1.4 (1983). "Specification for Sound Level Meters," American National Standards Institute, New York.

ANSI S1.40 (1984). "American National Standard Specification for Acoustical Calibrators," American National Standards Institute, New York.

ANSI S1.6 (1967) (R1976). "Preferred Frequencies and Band Numbers for Acoustical Measurements", American National Standards Institute, New York.

ANSI S3.23 (1980). "Sound Level Descriptors for Determination of Compatible Land Use," American National Standards Institute, New York.

ASHRAE (1985). "Sound and Vibration Fundamentals," in ASHRAE Handbook — Fundamentals, ASHRAE, Atlanta.

Beranek, L.L. (1971). *Noise and Vibration Control,* McGraw Hill, New York.

Botsford, J.H. (1967). "Simple Method for Identifying Acceptable Noise Exposures," *J. Acoust. Soc. Am. 42*, 810-819.

Bruel, P.V. (1981). "Sound Level Meters, The Atlantic Divide," *Proceedings Inter-Noise '81,* Noise Control Foundation, Poughkeepsie, NY.

Earshen, J.J. (1980). "On Overestimation of Noise Dose in the Presence of Impulsive Noise," *Proceedings Inter-Noise '80,* Noise Control Foundation, Poughkeepsie, NY.

Earshen, J.J. (1985). "Noise Dosimeter Transient Response Characteristics," *J. Acoust. Soc. Am. 78,* Suppl. 1, S4.

Erdreich, J. (1984). "Problems and Solutions in Impulse Noise Dosimetry," *Sound and Vibration 18(3),* 28-32.

Harris, C.M. (1979). *Handbook of Noise Control,* McGraw Hill, New York.

Hess, P.W. (1985). "The Audiodosimeter Problem," *Noise and Vibration Control* (in press).

IEC 651 (1979). "International Electrotechnical Commission Standard, Sound Level Meters," American National Standards Institute, New York.

IEC 804 (1985). "International Electrotechnical Commission Standard, Integrating/Averaging Sound Level Meters," American National Standards Institute, New York.

ISO 1999 (Draft, 1985). "Determination of Occupational Noise Exposure and Estimation of Noise Induced Hearing Loss," International Organization for Standardization, Switzerland.

Jones, H.H. (1984). "Historical Development of the Walsh-Healy Noise Regulation," *Ann. Am. Conf. Ind. Hyg. 9,* 187-191.

Kamperman, G.W. (1980). "Dosimeter Response to Impulsive Noise," *Proceedings Inter-Noise '80,* Noise Control Foundation, Poughkeepsie, NY.

Kuhn, G.F., and Guernsey, R.M. (1983). "Sound Pressure Distribution about the Human Head and Torso," *J. Acoust. Soc. Am. 73,* 95-105.

Kundert, W.R. (1982). "Dosimeters, Impulsive Noise, and the OSHA Hearing Conservation Amendment," *Noise Control Engineering Journal,* November-December, pp. 74-79.

Lipscomb, D.M. and Taylor, Jr., A.C. (1978). *Noise Control Handbook of Principles and Practices,* Van Nostrand Reinhold Co., New York.

NBS (1976). "Environmental Effects on Microphones and Type II Sound Level Meters," Technical Note 931 (1976). National Bureau of Standards, Washington, DC.

OSHA (1983). "Occupational Noise Exposure; Hearing Conservation Amendment," *Federal Register 48(46),* Tuesday, March 8, 1983, 9738-9785.

Olishifski, J.B. and Harford, E.R. (1975). *Industrial Noise and Hearing Conservation,* National Safety Council, Chicago.

Peterson, A.P.G. (1980). *Handbook of Noise Measurement,* GenRad, Concord, MA.

Pierce, F.D. and Parker, R.D.R. (1983). "A Field Evaluation of Noise Measuring Instruments," *Am. Indust. Hyg. Assoc. J. 44,* 665-670.

Rockwell, T.H. (1981). "Real and Imaginary OSHA Noise Violations," *Sound and Vibration 15(3),* 14-16.

Schultz, T.J. (1972). *Community Noise Ratings.* Applied Science Publishers, London.

Seiler, J.P. (1982). "Microphone Placement Factors for One-Half Inch Diameter Microphones," M.S. Thesis, U. of Pittsburgh Grad. School of Public Health.

Shaw, E.A.G. (1985). "Report of the Scientific Advisor to the Special Advisory Committee on the Ontario Noise Regulation," Ontario Ministry of Labour, Toronto.

Suter, A. (1983). "The Noise Dosimeter on Trial," *Noise/News 12(1),* 6-9.

Von Gierke, H.E., Robinson, D.W., and Karmy, S.J. (1981). "Results of the Workshop on Impulse Noise and Auditory Hazard," Univ. of Southampton, Institute for Sound and Vibration Research, Memorandum 618.

Noise and Hearing Conservation Manual, edited by
E.H. Berger, W.D. Ward, J.C. Morrill and L.H. Royster

4 Noise Surveys and Data Analysis

Larry H. Royster
Elliott H. Berger
Julia Doswell Royster

Contents

Page

Introduction 99
An Overview 99
Political Considerations 99

Why Conduct a Sound Survey? 100
Estimate Potential Noise Hazard 100
OSHA and Other Government Regulations 100
Input to Company's HCP 101
Workers' Compensation Regulations 101
Safety Considerations 101
Special Requests 102

Classifying Noise Exposures 102
Recommended Classification Scheme 102
Alternative Classification Schemes 104
The Blanket Approach 104
Individual Worker Classification 105
General Classification Guidelines 106

Presurvey Considerations 107
Types of Sound Surveys 107
Educational and Informative Efforts 108
"Smell the Roses" 109
Collect Information on Environment/Equipment 109
Prepare a Check-off List and Survey Outline 111

Instrumentation Considerations 112
SLMs and/or Noise Dosimeters 112
Expected Accuracy 114

The Basic Sound Survey 115
Purpose of the Basic Sound Survey 115
Information to Record 116
Supporting Information 116

Calibration Data 117
Survey Sound Level Data 120
Collecting the Data 125
Area Survey 125
Work Station/Job Description Survey 125
Evaluating the Data for OSHA 126
Area Survey 126
Work Station/Job Description Survey 131
The Detailed Sound Survey 132
Purpose of the Detailed Sound Survey 132
Collecting the Data 133
Sampling Methodology 133
Noise Dosimeter 133
Sound Level Meter 135
Sampling Procedure Examples 137
Sound Level Contours 140
The Engineering Noise Control Survey 141
Purpose of the Engineering Noise Control Survey 141
Determining the Dominant Noise Sources 141
Area Sources 142
Contributors to Each Dominant Noise Source 144
Information to Record 146
Special Noise Control Surveys 146
Statistical Factors in Sound Survey Methods 150
Drawing Conclusions from Sample Data 150
Sampling Considerations 152
Use of Confidence Intervals 155
Use of Tolerance Limits 157
Example Statistical Applications 157
Basic Descriptive Techniques 157
Making Inferences 163
Using Statistics 169
Report Preparation 169
Goals 169
Potential Audience 170
Consider the Political Constraints 170
Report Format 170
Record Keeping 172
Guidelines for Updating Survey Findings 172
The End 174
Acknowledgments 174
References 174

Introduction

AN OVERVIEW

The primary goal of this chapter is to present guidelines for the development and execution of the sound survey phase of a company's hearing conservation program (HCP) (L.H. Royster, Royster, and Berger, 1982). Information contained in other chapters of this manual should also be considered, especially Chapter 3 on instrumentation, as well as the general literature (Alpaugh, 1975; ANSI, 1976; Wells, 1979; Gasaway, 1985). In the opinion of the authors of this chapter there does not exist any one set of sound survey guidelines that would be appropriate for all of U.S. industry (L.H. Royster and Royster, 1984). Therefore, it is important for the sound surveyor to consider existing company constraints before developing and implementing the sound survey phase of the HCP.

The sound surveyor must also keep in perspective the importance of the sound survey efforts as they relate to the other phases of the company's HCP. To allow one phase of the HCP, such as the sound survey phase, to operate independently of all other phases could result in the sound surveyor's spending excessive funds on possibly unnecessary equipment or other items, when these monies could have been put to better use in presenting educational programs, purchasing more adequate audiometric instrumentation or implementing necessary additional training for other HCP personnel.

POLITICAL CONSIDERATIONS

The sound surveyor must recognize that the organization's political climate is determined by the sum of many individual motivations. It is important that the sound surveyor be sensitive to this climate, listening to the concerns of all affected parties but not making early commitments or expressing preliminary judgments of environmental conditions or actions that may be necessary.

If the sound surveyor tends to project an attitude of superiority toward the workers, they may be resentful and can easily find ways to significantly alter the noise environment so that the measured data end up being an inaccurate representation of actual conditions. In addition, they can manage to drop one or more of the noise dosimeters or "accidentally" get the microphone cord caught in a piece of equipment.

The sound surveyor should take the time to express an outwardly friendly attitude toward the workers and give them an opportunity to voice opinions or concerns about the noise environment. This approach will not only minimize the misuse of the equipment and distortion of the environment by the workers, but will on occasion provide significant additional input data that may result in a more accurate estimate of actual worker time-weighted average (TWA) exposures.

The sound surveyor should also consider the interests and needs of other personnel who may be affected by the results of the survey. For example, production managers and front line supervisors are often concerned about the potential impact that the sound survey findings may have on production efficiency and annual personnel evaluations. Management, supervisors and employees in work areas without a requirement to wear hearing protection devices (HPDs) will be watching with interest. Each of these groups has been observed to attempt to alter the noise environment, such as by shifting noisy work efforts to second or third shifts when sound surveys are not normally conducted, in order to project a more favorable climate.

The preceding discussion indicates the importance of considering the political implications of a sound survey. A moderate effort on the political front during the planning stage can have a very positive impact and potentially reduce the level of effort necessary to obtain reasonable results. Obviously the sound surveyor should have a strong interest in the content of the company's educational and motivational program (Chapter 11). This phase of the HCP, if properly developed and implemented, can assist the sound surveyor in preparing the workers not only for the sound survey phase but for active participation in the company's overall HCP.

Why Conduct a Sound Survey?

ESTIMATE POTENTIAL NOISE HAZARD

The principal reason for conducting a sound survey should be to establish the noise environment's potential for producing a permanent noise-induced hearing loss. To estimate this potential hazard, it is essential that a reasonably accurate sound survey data base be obtained.

OSHA AND OTHER GOVERNMENT REGULATIONS

One of the strongest incentives over the past several years for U.S. industry to conduct sound surveys has been federal regulations, the most important being the Occupational Safety and Health Act (OSHA, 1983; see Appendix I of this manual). The sound survey data base provides information that is normally required to satisfy the legal requirements of OSHA and other similar governmental regulations.

INPUT TO COMPANY'S HCP

Sound survey results will also be utilized in making important decisions with respect to all phases of the company's HCP, especially identification of the employees that will be included in the program. In addition, decisions made by management in selecting the HPDs to be offered to employees are often based on the findings of the survey, especially for work environments where the employee's TWA is above 99 dBA (L.H. Royster and Royster, 1985b).

The sound survey results assist management in selecting work areas for possible engineering noise control efforts. The information generated by sound surveys is also useful when one attempts to explain to workers why they are, or are not, included in the HCP or why a particular HPD is not acceptable for a particular work station or job classification. The results of the sound survey must be considered in making other administrative decisions such as selecting a location for the audiometric test facility.

In summary, the sound survey provides the necessary information to solve many of the typical problems that arise on a day-to-day basis in running an effective HCP.

WORKERS' COMPENSATION REGULATIONS

Workers' compensation for noise-induced hearing loss provides an additional incentive for U.S. industry to conduct sound surveys (Berger, 1985). However, the type of data required for OSHA compliance and for workers' compensation purposes may differ. As an example, when conducting a sound survey for the purpose of satisfying OSHA requirements, the surveyor is mainly attempting to determine the employees' TWAs, *i.e.* equivalent daily exposures. As a consequence, when TWAs are less than the 85-dBA action level, the sound surveyor may fail to record or maintain the data. However, in at least one state, employees can file for compensation for noise-induced hearing loss if they are exposed to a sound level of 90 dBA or greater regardless of the duration of exposure. Of course, the fact that a TWA is below 85 dBA does not ensure that it came only from exposure to sound levels below 90 dBA. Therefore, failure to maintain adequate data for environments that do not have to be included in the company's HCP could lead management to overlook potential problem areas and also result in inadequate documentation for legal purposes.

SAFETY CONSIDERATIONS

Another reason for conducting sound surveys is to investigate potential safety hazards related to employee communication and detection of warning signals. At all employee work stations, whether or not the daily TWA exceeds the OSHA action level, it is essential that acoustic warning signals be detectable above the background sound level. The data measured during the sound survey provide part of the information necessary to estimate the adequacy of the company's audio warning system.

Recently it has become very popular for employees to request to listen to personal radios. Management should decide whether the use of personal radios will create a potential hazard such as hearing damage, speech interference or masking of warning signals. Again, sound survey data will be required to estimate the potential noise hazard and provide recommendations to management (See Chapter 10, "Recreational Earphones").

SPECIAL REQUESTS

The individual responsible for conducting the company's sound surveys is frequently asked to make additional measurements that may seem a little bit out of the ordinary, but are still part of the job. Examples include requests to survey the production manager's office or conference room, the company cafeteria, a special secretarial office or even the engine room of the boss' private boat. Other examples include sound surveys of the computer room and similar areas that are not normally included in the company's HCP. Although some of these requests may seem inappropriate, the sound surveyor should keep in mind that checking the sound levels in these areas and explaining the results is in fact an extension of the company's education program. These types of requests indicate an acceptance, on the part of management and other individuals requesting the service, of the fact that noise has its effects and that hearing conservation is important.

Classifying Noise Exposures

RECOMMENDED CLASSIFICATION SCHEME

The basic goal of most sound survey efforts is the determination of TWAs for a worker or for a particular work station or job classification. There may be other goals as well, such as determining the TWAs for work areas or identifying the dominant noise source(s) for engineering control purposes, but for the majority of sound surveys conducted in general U.S. industry, the goal is employee noise exposure determination. In order to provide guidance for the sound surveyor as to the level of effort and accuracy necessary in determining employee TWAs, a practical scheme for grouping the predicted TWAs is desirable.

The classification scheme recommended is presented as Table 4.1 (L.H. Royster and Royster, 1985b). Five exposure ranges are used for grouping

TABLE 4.1
Recommended Scheme for Classifying TWAs for Sound Survey Purposes

TWA, dBA	Classification
84 or below	A
85 - 89	B
90 - 94	C
95 - 99	D
100 or above	E

the estimated TWAs. Schemes that employ a larger number of categories will not ensure better employee protection and at the same time are cumbersome and impractical, often creating unnecessary administrative burdens on the nurse, the audiometric technician and other company personnel. Additional reasons for limiting the classification scheme to the five ranges shown in Table 4.1 include the inability to estimate real-world TWAs more accurately than the defined 5-dBA ranges, the variability of real-world levels of protection provided by HPDs, and the fact that 97% of all industrial TWAs are less than 100 dBA (Bruce, Jensen, Jokel, Bolt and Kane, 1976; OSHA, 1981).

We have observed situations and reviewed reports in which individuals spent far too much time and resources in trying to determine employee TWAs to an accuracy of ± 1 dBA. The use of a classification scheme such as the one presented in Table 4.1 should minimize attempts by sound surveyors to predict TWAs to unreasonable levels of accuracy and will therefore encourage more effective utilization of the limited available resources.

The lowest exposure classification A, as shown in Table 4.1, is included for work areas such as computer rooms, and similar work environments in which the employees may request hearing conservation related information and HPDs, but where the employees may not be included in the company's HCP. As pointed out in Chapter 10 on HPDs, it may be necessary to limit use of hearing protection for some employees in low noise level environments based on safety considerations.

Exposure classification E includes all noise exposures of 100 dBA or greater because the magnitude of real-world protection provided by most HPDs (reference Chapter 10) is potentially inadequate at this level. As a consequence, HCP personnel should give special attention to the portion of the workforce classified as E.

ALTERNATIVE CLASSIFICATION SCHEMES

The Blanket Approach

An alternative to a multi-tier classification scheme is to assign one TWA to all employees in a given department or plant, based upon the highest individual employee TWA that is measured. To illustrate this type of blanket approach, assume that in a production area 70% of the employees are exposed to TWAs of 85-89 dBA, while 20% have exposures less than 85 dBA and the remaining 10% have exposures in the 90-94 dBA range. The blanket approach would require that the total work area be classified as C, with the result that all employees working in this area would be included in the HCP and be required to wear HPDs. On each employee's audiometric record a TWA of 90-94 dBA would be indicated. Now what are the potential problems and benefits from using the blanket approach?

The most obvious problem is that a large segment of the workforce would be required to wear HPDs in relatively low levels of noise. For these workers, a significant percentage could experience communication-related problems, especially employees with significant preexisting hearing loss. Workers who are forced to use HPDs in the lower noise level environments, *i.e.* TWAs less than 90 dBA, are more resistant to using HPDs, with the result that management typically will not strictly enforce HPD use (L.H. Royster & Royster, 1985b).

A second problem with the blanket approach is that the noise hazard will have been exaggerated for a high percentage of the workforce, leading the affected employees to believe that the hazard is greater than it really is. If the employees who are not wearing the HPDs properly are not being flagged by the HCP because the noise hazard was overstated, then these workers will begin to downplay the importance of the HCP. If in the future they are moved to work areas where the noise hazard is in fact serious, they may fail to participate willingly in the company's HCP as a consequence of their previous experience.

A third problem created by the blanket approach concerns the need to indicate a TWA on the employee's audiometric record. Since the blanket approach may result in an inflated TWA, the company's ability to properly judge the level of protection being provided the noise-exposed work force (by type of HPD utilized, comparison of hearing threshold level changes, *etc.*) through analysis of the company's audiometric data base may be significantly reduced. Thus, for example, when an employee exhibits a significant threshold shift (standard threshold shift if OSHA regulations are utilized), the individual who reviews employee audiometric records may be misled into assuming that the indicated shift is due to on-the-job noise exposures, when in fact the shift was due to off-the-job noise exposures.

Now, what are some of the benefits of the blanket approach, or classifying the work area as C? The benefit most often stated is avoidance of the problem encountered when an employee who works in an area classified as B is identified as having a significant threshold shift and is required to utilize HPDs. The comment is, "The supervisor can't enforce the use of hearing protection by one or two employees when the majority of the employees do not have to use them." Although there is some truth in this statement, it is also true that in many work areas different employees must use different types of safety equipment (such as eyeglasses and safety shoes) depending on the piece of machinery operated or the location of their respective work stations. Why should the use of hearing protection be different? Why penalize, in this instance, 90% of the workers because of the hesitation to enforce the use of hearing protection by 10% of the workers?

A second stated benefit is that the single-number area classification scheme reduces the administrative difficulty of establishing several different classifications for the work area and having to record them on the employee's audiometric record each year. A third benefit is that the sound survey effort is significantly reduced since the number of samples needed is reduced.

Individual Worker Classification

The opposite of blanket classification of work areas is the establishment of a TWA for each worker. This approach is often justified based on the assumption that an accurate knowledge of each worker's TWA will result in significantly greater protection for the workforce. In addition, sound surveyors have attempted to imply that an "accurate" estimation of each worker's TWA will somehow result in a lower potential compensation cost to the company for on-the-job noise-induced hearing loss. However, data in the literature and experience of the authors do not support these claims. Indeed, the models that exist for predicting hearing loss, level of protection provided by HPDs, and the hazardousness of a particular noise environment, have no greater accuracy than the 5-dBA ranges found in Table 4.1.

Another reason for limiting the effort expended in predicting individual TWAs is the significant mobility and changing noise exposure histories of employees who switch between job functions on a yearly, weekly and even daily basis.

Unless a realistic classification scheme is selected, HCP personnel will spend unnecessary effort and funds in trying to define TWAs to a much greater accuracy than is warranted.

GENERAL CLASSIFICATION GUIDELINES

To illustrate our recommendations, consider the following two examples. Assume that on a work floor 80% of the workforce is exposed to an estimated TWA of 88 dBA and is scattered throughout the total work area, while 10% have TWAs estimated in the 91-94 dBA range. The 10% of the work force with TWAs of 91-94 dBA are located at work stations at one end of the production area and their work stations are associated with specific pieces of production machinery.

For this work environment, the following classification procedure was implemented. The 80% of the population with an estimated TWA of 88 dBA was classified as B. The specific job functions where the 10% of the employees exhibited TWAs in the 91-94 dBA range were classified as C. These employees in area C were required to wear hearing protection and their area and machines were appropriately posted. In addition, all employees who entered the posted area were required to wear HPDs regardless of the time spent in the area.

Obviously the remaining problem is what to do with the 10% of the employees who exhibited TWAs less than 85 dBA. Since they were working in a potentially harmful noise environment and their work stations were not clearly isolated from other work stations in the same area where higher TWAs had been established, they could be exposed to higher TWAs. Therefore, their job functions were classified as B so that they would be a part of the company's HCP, receive annual audiometric evaluations and be required to utilize HPDs if they exhibited an OSHA standard threshold shift or a significant threshold shift as defined by company HCP policy.

As a second example consider two adjoining work areas with the following characteristics. In production area 1, 80% of the employees exhibited TWAs in the 91-94 dBA range and 20% had TWAs in the 87-89 dBA range. In production area 2, 90% of the employees exhibited TWAs in the 86-89 dBA range, while 5% exhibited TWAs in the 80-84 dBA range and 5% exhibited TWAs in the 90-94 dBA range.

Production area 1 was classified as C and all employees working in this area were required to wear HPDs, including the employees working in one end of the room who exhibited TWAs in the 87-89 dBA range. Based on the information received from the foreman in this area as to the general mobility of his work force, it was decided that the potential for exposure to the higher sound levels was significant. Therefore, requiring the use of HPDs by all employees was recommended.

Production area 2 was classified as B for all employee work areas exhibiting TWAs less than 90 dBA, including the 5% of the employees who exhibited TWAs less than 85 dBA. At those work stations where the workers' estimated daily TWAs were above 90 dBA, the stations were classified as C and posted as HPD-required work areas where the use of hearing protection was enforced.

An additional problem, not necessarily unique to this production facility, involves the frequency of employee movement between jobs within each department and between departments. This type of mobility creates questions regarding the proper classification to be placed on the employee's audiometric record. In general, it is recommended that the employee's highest noise exposure classification during the year be utilized so that the employee's highest potential exposure category is indicated.

Presurvey Considerations

TYPES OF SOUND SURVEYS

For discussion purposes, we have elected to group sound survey efforts into three categories as shown in Figure 4.1. The three categories defined are the basic-screening, detailed, and engineering noise control surveys. The type of survey selected will depend upon several factors.

When an environment is surveyed for the first time, usually a basic sound survey will be conducted. The basic sound survey includes a general screening survey of all plant areas to identify environments that do not present

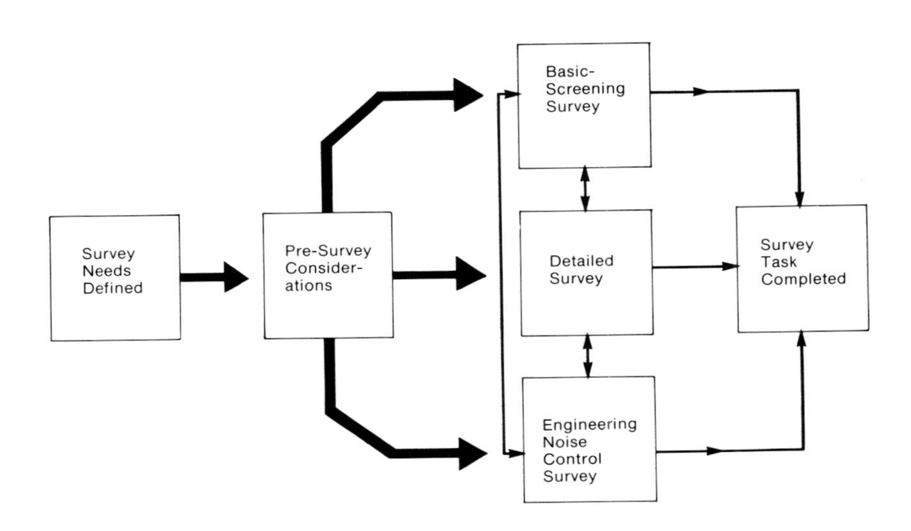

Figure 4.1 — Diagram of work efforts for the sound survey phase of the HCP.

any noise hazard to the worker. For job environments where the employees' TWAs are easily defined (noise levels and exposure periods are readily determined), the data collected during the basic survey may be sufficient to satisfy the objectives of the sound survey phase of the HCP. The basic sound survey may be thought of as a short term effort. That is, for a plant of 200-300 employees, the basic sound survey would not require more than one or two days to conduct.

If, during the basic sound survey, worker and/or job classifications are found where high variability in the sound levels requires the use of a different instrument or more detailed sampling, then a detailed sound survey would be scheduled. Likewise, if a work area is surveyed where several job classifications exist and the measured sound levels indicate that different classifications may be required (Table 4.1), then a detailed sound survey would be called for. Therefore a detailed sound survey, as the name implies, should be thought of as a survey effort that will require several days to complete. However, after the initial detailed sound survey has been conducted, then followup detailed sound surveys of worker positions or areas could be completed over shorter time periods.

If, during the basic sound survey, a noise source is identified that is a main contributor to the employee's TWA or area's noise level, or which is creating a communication or annoyance problem, then a noise control survey may be scheduled. The need for a noise control survey is often established by observations made during a sound survey or as a consequence of analyzing the resulting data. A noise control survey of a new piece of production equipment may also be requested by management based on sound data supplied by the equipment's manufacturer. In other words, the noise control survey may originate from more than one source. The main objective of the engineering noise control survey is the collection of data for noise control purposes.

Although we have elected to separate sound survey efforts into three categories for discussion purposes, practically speaking any one industrial sound survey effort will most likely include objectives that cross more than one survey classification. Also, as Figure 4.1 indicates, there exist many feedback paths among the three classifications. As an example, after the recommendations from an engineering noise control survey have been implemented, it is often necessary to conduct a new survey to establish the effect on the workers' TWAs.

EDUCATIONAL AND INFORMATIVE EFFORTS

It is important that employees, supervisors and management be informed of the purpose for and approximate dates of all sound survey efforts. The reasons for notification include the need to point out to personnel the necessity for measured sound levels to be representative of normal production activities, and to ensure that the production equipment will not be down for routine maintenance on the planned day of the survey and that appropriate personnel will be available to assist the surveyor. It is not at all uncommon for the sound surveyor to arrive at the survey site only to find out that one or more of the major noise sources are down for maintenance. Even if the surveyor properly communicates with management, some planned survey attempts may have to be rescheduled due to such factors as equipment failure, modification of production plans and personnel-related problems.

Seasoned sound surveyors know that the success of the sound survey efforts can depend significantly on the experience and competence of the company personnel who assist the surveyor in collecting the data. Often an experienced mechanic or machine operator can provide more effective assistance in identifying a noise source than a company engineer who is remote from production efforts. Failure to properly inform management can end up creating a situation where the employee(s) needed to conduct a successful sound survey are either on vacation or not available due to other scheduled duties.

"SMELL THE ROSES"

An inexperienced sound surveyor will tend to show up on the morning of the scheduled survey, calibrate the equipment and immediately proceed to start monitoring the noise environment. The result is often a need to resurvey part or all of the plant because the sound surveyor failed to note some critical fact, such as improper operation of a significant contributor to the noise environment. It is always advisable to leave the instruments in the office area that will serve as survey headquarters and, as the saying goes, "smell the roses" before actually beginning the survey. It is amazing the amount of additional information that can be gained by walking through the work area to be surveyed with pencil, paper and clipboard and recording general observations prior to conducting the actual survey. By becoming familiar with the environment, the surveyor will be able to more effectively relate the movement of the dial and/or digital indicator to the contributing noise source(s).

COLLECT INFORMATION ON ENVIRONMENT/EQUIPMENT

The surveyor should attempt to collect as much information about the work environment and equipment as possible prior to the scheduled sound survey. This includes, for example, all past sound surveys conducted either by plant personnel or outside consultive sources. Even if less than desirable survey procedures were followed in obtaining these data, they often provide useful information about the noise environment, production equipment modifications and dominant noise sources.

If practical, the surveyor should discuss the noise survey with an approachable supervisor and/or machine operator. These individuals will often remember information about equipment modifications that seemed to affect the production area noise level. It is essential that the surveyor not be critical of the information given by these non-professionals. Although their statements may not be technically correct, they can supply important information which may be useful for noise control or sound survey purposes.

An an example, one of the authors investigated a noise problem that was created by a paper-drive train at a paper production facility. The plant engineering staff had studied the problem for several weeks prior to the planned survey and had failed to establish the cause of the noise problem. Management stated that the noise problem had surfaced approximately two years earlier.

Since experience had shown the need to communicate with supervisors and machine operators, an informal meeting with the area supervisor and head machine operator was scheduled on the morning of the planned survey. During the meeting, the machine operator asked the surveyor why he was interested in talking to non-engineers. Upon being told that front line supervisors and machine operators had been very helpful in the past in identifying possible noise sources, the machine operator volunteered the information that the noise level had changed when the engineering staff had made modifications to the hydraulic lines in order to increase the production capacity of the system. Although the modifications did not succeed, the system had not been returned to its original condition.

The company engineering staff was questioned about the earlier equipment modifications and the information provided by the machine operator was confirmed. The lines were eventually restored to their original design and the noise levels returned to an acceptable level. Although for general interest purposes a few sound readings were made during the initial visit to the plant, they were not necessary to attain a solution of the noise problem.

When conducting a sound survey for noise control purposes, the surveyor should assemble general noise control information concerning the type of equipment being studied. Possible information sources include the equipment manufacturer and pertinent trade journals and publications. Sometimes it will be found either that no noise control option is available or that past noise control efforts have not been successful, making any quick fix unlikely. In this case, the surveyor should consider other solution options.

A willingness to obtain information from all potential sources is a desirable characteristic of effective sound surveyors.

PREPARE A CHECK-OFF LIST AND SURVEY OUTLINE

No matter how experienced the surveyor may be, it is important that a check-off list be developed and utilized. Otherwise it is easy to overlook items that may be needed during the survey. A representative list might include:

(1) sound level meter(s) (SLMs) and/or noise dosimeter(s) and possible backup unit(s)
(2) tape recorder for recording noise samples for later analysis or for educational purposes
(3) when necessary, a graphic level recorder or other means of obtaining sample recordings of level as a function of time
(4) calibrator and calibration forms
(5) extra batteries and connector cables
(6) clipboard, paper, survey forms, extra pencils and pens, camera and film
(7) floor and equipment location markers (tape measure, chalk, masking tape, pens and markers)
(8) calculator and reference publications or tables that might be needed
(9) personal HPDs, safety glasses, safety shoes, *etc.*
(10) belt or lightweight vest to support a noise dosimeter and safety pins or velcro strips to secure noise dosimeter microphone cable when needed
(11) windscreen for windy work areas or for microphone-damaging environments and a thin plastic cover for use in protecting the instruments from harmful environments
(12) timer and flashlight
(13) one or two foam sheets for use in isolating the equipment from vibrating surfaces or for easing the load on aged knees when conducting "low level" sound surveys
(14) a small recorder for recording pertinent comments before, during and after the survey

Even if the sound surveyor is highly experienced, it is recommended that a brief summary of the objectives of the survey be prepared and that a preliminary outline of procedures be developed. When sound surveys are conducted at locations that have not been measured previously, the presurvey-prepared outline will often have to be altered due to unforeseen circumstances. Regardless, more effective sound surveys will be conducted by using a planned outline than by using the do-it-as-you-go approach.

Instrumentation Considerations

SLMs AND/OR NOISE DOSIMETERS

In choosing equipment one should ask the question, "Why are we interested in establishing a TWA?" If the concern is for the potential of the noise environment to produce a permanent noise-induced hearing loss, the estimated TWA will be compared to some damage-risk criterion. Virtually all such data bases have been established using an SLM (A-weighted, slow response), with the resultant TWAs correlated with the magnitude of hearing damage exhibited by the noise-exposed population. Therefore, when hearing damage risk is being assessed, the most appropriate instrumentation and survey procedures to utilize would be those identical to the ones involved in conducting the original damage-risk studies.

The limited number of reports in the literature concerning the equivalence of TWAs measured using SLMs and noise dosimeters indicate that TWAs obtained with the two types of instruments differ (Erlandsson *et al.*, 1979; Walker, 1979; Jones & Howie, 1982; Shackleton & Piney, 1984). Typically the data obtained using noise dosimeters are higher by 2-3 dBA.

As an additional point of reference and also as a demonstration of the adequacy of the SLM sampling methodology presented in a following section, 79 North Carolina OSHA sound surveys conducted over a period of several weeks were analyzed. The procedure followed by these inspectors involved measuring the TWAs of three employees at each plant site with both SLMs and noise dosimeters. The mean difference between the 237 pairs of TWAs was 0.6 dBA, with the noise dosimeters indicating the higher TWAs.

Several reasons have been given for the observation that noise dosimeters tend to predict higher TWAs, including the ability of the dosimeter to more accurately follow the noise signature over time and more accurately predict the employee's time-level history for conditions such as when employees move in close to their equipment, and the effect that the employee's body has on the sound field such as when the dosimeter's microphone is placed on the shoulder. When the dosimeter's microphone is placed on a wearer's shoulder in a pink noise field, the average increase in the measured A-weighted sound level is 0.4 dBA (GenRad, 1976). However, if the employee is working in a directional sound field, such as when the employee is in close to a noise source, the effect of the body on the sound field may be as high as 2 dBA for perpendicular incidence (GenRad, 1976).

The higher estimations of employee TWAs obtained using noise dosimeters may be inappropriate to use in risk assessment since the existing damage risk data base reflects the use of SLMs of an earlier vintage and a corresponding sampling methodology. *Simply because one instrument provides higher estimated TWAs for a given work environment than another unit does not necessarily make it the instrument of choice.*

If the surveyor's main concern is strict compliance with existing OSHA regulations and if the particular OSHA enforcement office relies completely on noise dosimeter samples (as is often the case), then it may be expedient to use dosimeters in conducting the sound survey or use SLMs and attempt to account for the expected differences by adjusting the predicted TWAs accordingly.

Since the objectives of the sound survey include the measurement of sound levels as well as the determination of TWAs, it is essential that an SLM of at least type 2 capability be available along with an appropriate calibrator. For surveying jobs such as driving bulldozers, working in highly restricted areas and maintenance/repair functions, it is important that a noise dosimeter be available since the use of an SLM by a surveyor could create a safety hazard.

At least two products are presently available that provide within the same instrument both SLM and noise dosimetry capabilities, and also the capacity for later recall of the level-versus-time histories of worker exposures. The ability to review the employee's daily level-versus-time history can afford significant insight as to the noise sources that contributed to the measured TWA, and can also enhance other areas of the HCP such as educational presentations (DiBlasi, Suuronen, Horst, and Bradley, 1983).

Therefore we recommend that sound survey instrumentation include at least one SLM and one noise dosimeter (or a dual capability instrument) with appropriate calibrators. Other useful instruments that should be considered are an octave-band filter, recorders and real-time analyzers (Chapter 3). If placed in willing and capable hands, these instruments can produce additional data, not obtainable with the SLM and noise dosimeter, that can enhance the hearing conservation effort.

It should also be pointed out that sophisticated instrumentation is not always needed. Companies with very hazardous noise environments have established effective HCPs based on sound surveys conducted by company nurses or audiometric technicians who borrowed a type 2 SLM and calibrator for a limited time period each year. The critical factors are the dedication of the personnel and the HCP policies, not the instrumentation.

EXPECTED ACCURACY

The sound surveyor often asks, "What is the level of accuracy to be expected from SLMs and noise dosimeters?" Obviously the level of accuracy expected depends not only on the instruments' inherent accuracy but also on the sampling and data analysis procedures employed.

One indication of the expected error range for field-utilized sound survey instrumentation is given by OSHA's policy of issuing citations for a violation of the noise standard only when the estimated TWA exceeds the target level by 2 dBA (OSHA, 1984). That is, the stated 85-dBA action level for initiating an HCP becomes 87 dBA. This 2-dBA factor is assumed by OSHA to account for errors in calibration and equipment accuracy as well as variability in measurement techniques.

The first author routinely checks the accuracy of the SLMs, noise dosimeters and calibrators for one OSHA state agency (Royster, 1980). The type 2 SLMs and calibrators, usually in groups of 10, are given pre- and post-calibration evaluations. Initially the instruments are cross-calibrated prior to any adjustments. That is, each calibrator is utilized to check the output for all SLMs before it is compared to the laboratory reference standard and adjusted as needed. A final post-calibration check is completed by again recording the SLM outputs using each calibrator.

The standard deviation for the recorded pre-calibration SLM output sound levels across instruments and calibrators is typically 0.3-0.5 dBA. After each calibrator's output has been checked, and adjusted as necessary, and each SLM has been readjusted using each unit's calibrator, the standard deviation for the recorded SLM outputs normally decreases to 0.2-0.3 dBA. These findings, based upon single-frequency pure-tone calibration at 1000 Hz, suggest a high degree of precision.

The ability of SLMs to accurately measure sound level is usually limited by the microphone's frequency response characteristics or changes that might occur due to rough or improper handling (Delany *et al.,* 1976; Peterson, 1980; Bruel, 1983). Therefore the accuracy of the SLMs is also checked by exposing the units to a 500-4000 Hz pink noise in a semi-anechoic environment. Recall that all units are previously adjusted to the same indicated output level using their respective calibrators. When the measured output levels for the SLMs are compared to the output of the laboratory's standard, it is not uncommon to discover at least one unit where the sound level is higher or lower by 2 dBA than is indicated by the reference system. Replacement of the unit's microphone has always corrected the error.

For all SLMs with meter movement indicators that exhibited the above problem, it has also been observed that a shift in sound level of more than 0.5 dBA occurred when the SLM and attached calibrator were rotated from an upright to a horizontal position. In addition, it has often been observed that the instrument's case would exhibit indentations and/or scratches, a further indication that the unit had received rough treatment.

Experience from evaluating field-utilized noise dosimeters using a pre- and post-calibration check and semi-free-field evaluation procedures similar to those described above has yielded similar variations in the measured data. Therefore, even without consideration of the accuracy of noise dosimeter sampling procedures and the effect of the location of the dosimeter's microphone, measurement inaccuracies of ± 2 dBA across units for in-the-field-utilized type 2 noise dosimeter instrumentation could be expected.

The Basic Sound Survey

PURPOSE OF THE BASIC SOUND SURVEY

The main objective of the basic sound survey is the identification of those areas where the worker TWAs are 85 dBA or higher. The findings may suggest the need for a more detailed sound survey effort. However, for those work stations where the noise levels are steady and therefore are easily measured, the findings from the basic sound survey effort may be sufficient to adequately establish appropriate TWAs.

A second objective of the basic sound survey is the determination of typical sound level ranges for all work stations where the predicted TWAs are less than 85 dBA. Obviously one may ask why these areas should be surveyed and documented. In addition to the reasons presented earlier (see section entitled "Why Conduct a Sound Survey?"), production equipment is often moved from one location to another, or equipment may be replaced by units with different noise characteristics. The existence of sound survey data allows management to estimate the effect of these equipment modifications on existing noise levels. Also, the basic sound survey should identify work areas where the sound levels could create possible communication difficulties. Employees who are located in these typically non-HCP areas are often aware of the fact that the noise in the area is loud and as a result may question management as to the need to be furnished HPDs and/or be included in the company's HCP. The data collected during the basic sound survey will assist company personnel in providing an adequate response to these and other related questions for these non-HCP areas.

At one plant site where an employee was injured and was not able to obtain immediate assistance due to a background noise level which was slightly less than 85 dBA, the employee filed a lawsuit contending that the environmental conditions, which were under the control of the employer, were directly responsible for the additional personal injury that was incurred. The employee also claimed that the employer had not properly documented the environmental noise conditions where the accident occurred and that the employer should have been aware of the potential environmental hazard. This claim of neglect on the part of management would have been avoided if company personnel had properly documented the noise environment for all areas of the company, not just the one production area that exhibited TWAs higher than 85 dBA.

A third objective of the basic sound survey may be the establishment of a sound level data base that can be utilized to broadly estimate TWAs for workers and/or job descriptions in order to stratify the workforce for statistical sampling purposes, that is, to divide the workforce into groups of similar noise exposures, as described in the section on statistics later in this chapter.

During the basic noise survey, all areas at the plant site under study should be visited and the existing noise levels documented. Areas that are often overlooked are computer rooms (where the noise exposures can approach a TWA of 85 dBA) and isolated work areas where the sound level is 90 dBA or higher (pump and generator stations, and postal rooms with paper-shredders or mail-handling machinery).

INFORMATION TO RECORD

Supporting Information

During the time of the initial visit to the plant site and production areas, data concerning the typical work schedules for the general plant population should be collected. This information, along with the noise level samples obtained during the basic sound survey, will allow the surveyor to establish estimates of the expected TWAs for all work stations and/or work locations. During or following the initial visit to the plant facility, floor plan drawings should be obtained, if available, or general sketches of the work areas should be completed. Photographs or slides of the areas and equipment studied are also very useful in recalling information at a later time.

The sound surveyor should always keep in mind that data acquired during any visit to the plant facility not only will be utilized by company hearing conservation personnel but could end up in a court of law. Therefore the recommended guidelines for recording, maintaining and publishing noise survey data as discussed herein should be followed.

Calibration Data

Most likely the initial data to be recorded will be the results of the pre-survey calibration of the SLM. Examples of calibration logs for an SLM exhibiting A- and C-weighting capabilities and for a noise dosimeter are shown in Figures 4.2 and 4.3. In calibrating the instrumentation, adjustments should not be made unless the deviation in the indicated sound level from the expected reference level exceeds ± 0.2 dB. If it is necessary to adjust a unit, the magnitude of the adjustment should be recorded on the calibration form.

Prior to calibrating acoustic instrumentation, it is important that the equipment have an opportunity to approach the work area's ambient temperature. It should not have been left in the trunk of a car when the outside temperatures are either very low or very high. If the equipment is being transported to the plant facility, it is good practice to take the equipment to the plant the day before the survey or store the equipment in the motel room the night before in order to keep the equipment at a reasonable temperature prior to the survey. Extremes of temperature have been known to significantly affect the stability of sound measurement instrumentation.

When calibrating an SLM, a complete set of instrumentation descriptors and conditions such as those shown in Figure 4.2 should be recorded. In the case of survey instruments that have a removable microphone, the sound surveyor should check that the microphone has not been replaced by a different unit unless the surveyor is the only individual who has access to the equipment.

During the calibration of an SLM, record the date, time, serial numbers of the meter and microphone (if a separate unit), and indicate that the battery check for each unit was satisfactory. The sound surveyor should always begin with fresh batteries in each unit because the cost of the batteries is trivial compared to the cost of lost or invalid survey data. For noise dosimeters a similar set of calibration data should be recorded as shown in Figure 4.3.

At the conclusion of the noise survey, a post-survey calibration of the instrument should be completed. If the instrumentation's reference level has shifted more than ± 1.0 dBA (which is very unusual unless extreme

DATE: ____________ SOUND LEVEL METER TYPE: ____________ SN: ____________

TIME:PRESURVEY ____________ CALIBRATOR SN: ____________

BATTERY CHECK: SLM ____________ CALIBRATOR ____________

FREQUENCY; HZ	WEIGHTING A	C
	Value/Adjust.*	Value/Adjust.
125		
250		
500		
1000		
2000		

TIME:POSTSURVEY ____________

BATTERY CHECK:SLM ____________ CALIBRATOR ____________

FREQUENCY; HZ	WEIGHTING A	C
	Value/Adjust.	Value/Adjust.
125		
250		
500		
1000		
2000		

GENERAL REMARKS: ____

____

RECORDED BY: ____________

* Indicate final reading (after adjustment) and the magnitude of the adjustment.

Figure 4.2 — Sound level meter calibration log (Form A).

DATE __________ INDICATOR SERIAL NO. __________

TIME:PRESURVEY __________

	MONITOR SN	BATTERY CHECK	CALIBRATION READING	RESET-CLEAR UNIT
(1)				
(2)				
(3)				
(4)				

DATE __________ INDICATOR SERIAL NO. __________

TIME:POSTSURVEY __________

	MONITOR SN	BATTERY CHECK	CALIBRATION READING	RESET-CLEAR UNIT
(1)				
(2)				
(3)				
(4)				

GENERAL COMMENTS: ____

____

RECORDED BY: __________________

Figure 4.3 — Noise dosimeter calibration log (Form B).

environmental conditions are encountered) then the data collected during the survey should be questioned. Any unusual equipment responses or environmental conditions encountered (rain, high or low temperatures, *etc.*) should be indicated in the general response section of the calibration data form. Finally, the surveyor should sign and date all data log sheets using a non-erasable pen..

Survey Sound Level Data

During the basic sound survey, the data obtained should be sufficient to estimate minimum TWA ranges for all noise-exposed personnel. If the sound surveyor is fortunate, the measured sound levels will be "relatively constant," defined as regular variations in the SLM's indicator on slow response setting of no more than 8 dBA. For this type of noise environment, the sound surveyor will estimate the average position of the meter by eye (Delany *et al.,* 1976). This procedure should provide an estimate of the L_{OSHA} (Chapter 3) within approximately 1 dBA of the true value (Christensen, 1974). If the sound level is not regular or is varying by more than 8 dBA, then it is recommended that the sampling methodology outlined herein be utilized in order to satisfactorily predict worker TWAs.

An alternate method of sampling utilizes an integrating/averaging SLM (Chapter 3) having the appropriate dynamics and exchange rate. This type of instrument and mode of use simplify the problem of eyeball averaging and interpretation of the dial. However, the concern expressed earlier as to the appropriate instrument (SLM or dosimeter) for use in determining TWAs still applies.

Examples of data log forms that may be used to record sound level measurements are shown in Figures 4.4 through 4.7. These forms provide different options for recording data including the date, SLM type, serial number, time or duration of survey samples, microphone type and serial number if the microphone is not an integral part of the instrument, meter response selection and environmental conditions such as temperature and wind conditions and if a windscreen for the microphone is being used. It is important that if special circumstances are observed during the sound survey (such as an expected major noise source not operating or turned on or off in a manner not typical of normal production), then these observations must be indicated on the form in the space allocated.

Observe that on each log a place has been provided for the surveyor to sign the form. DO NOT at a later time transfer the recorded data to a clean form and throw away the original data. During legal hearings involving sound surveys, there have been instances of the data submitted being brought into question because the original forms could not be produced. It is difficult to explain how a sound survey could have been conducted in a dirty production area and the data end up on a clean log form or, even more unbelievably, one on which the recorded values have been typed!

DATE: ____________ SOUND LEVEL METER TYPE: ____________ SN: ____________

METER RESP.-FAST: ______ SLOW: ______ MICROPHONE TYPE: ______ SN: ______

TEMPERATURE: __________ WIND CONDITION: __________ WINDSCREEN: ______

OTHER EQUIPMENT UTILIZED: ____

____

LOCATION: ____

SURVEY NO.	LOCATION NO.	TIME	dBA	dBC	dBC-dBA	COMMENTS
(1)						
(2)						
(3)						
(4)						
(5)						
(6)						
(7)						
(8)						
(9)						
(10)						

GENERAL REMARKS: ____

____

____

RECORDED BY: ____________________

Figure 4.4 — Sound survey form (Form C).

DATE: __________ SOUND LEVEL METER TYPE: __________ SN: __________

METER RESP.-FAST: _____ SLOW: _____ MICROPHONE TYPE: _____ SN: _____

TEMPERATURE: ________ WIND CONDITION: ________ WINDSCREEN: _____

OTHER EQUIPMENT UTILIZED: ______________________________

____

LOCATION: ____

SURVEY NO./ POS.	dBA/ TIME	dBA/ T	dBA/ T	dBA/ T	dBA/ T	dBA/ T	dBA/ T	dBA/ T	dBA/ T	dBA/ T	EQ,dBA
(1)											
(2)											
(3)											
(4)											
(5)											

GENERAL REMARKS: ______________________________________

____

____

RECORDED BY: ______________

Figure 4.5 — Sound survey form (Form D).

DATE: __________ SOUND LEVEL METER TYPE: __________ SN: __________

METER RESP.-FAST: ______ SLOW: ______ MICROPHONE TYPE: ______ SN: ______

TEMPERATURE: __________ WIND CONDITION: __________ WINDSCREEN: ______

OTHER EQUIPMENT UTILIZED: ____

____

LOCATION: ____

____

TIME START SURVEY: ________
SURVEY NO. (SNO.)/POSITION (POS.)

SNO.	POS.	dBA	dBC	SNO.	POS.	dBA	dBC	SNO.	POS.	dBA	dBC
(1)				(11)				(21)			
(2)				(12)				(22)			
(3)				(13)				(23)			
(4)				(14)				(24)			
(5)				(15)				(25)			
(6)				(16)				(26)			
(7)				(17)				(27)			
(8)				(18)				(28)			
(9)				(19)				(29)			
(10)				(20)				(30)			

TIME STOP SURVEY: ________

GENERAL REMARKS: ____

____

____

RECORDED BY: ____________________

Figure 4.6 — Sound survey form (Form E).

COMPANY: ____________ FACILITY AND LOCATION: ____________

AREA SURVEYED: ____________ SURVEY CONDUCTED BY: ____________

TIME SURVEY STARTED: ____________ TIME SURVEY ENDED: ____________

SOUND LEVEL METER TYPE: ____________ MODEL: ______ SERIAL NO.: ______

CALIBRATOR TYPE: ____________ MODEL: ______ SERIAL NO.: ______

CALIBRATION LEVEL, INITIAL: __________ FINAL: __________ ΔDB (____) . . . . .

RELATIVE PRODUCTION RATE WAS (CIRCLE ONE): LIGHT MEDIUM HEAVY

DATE: __________ NUMBER OF MEASUREMENT SEQUENCES, N: __________

INDICATE EQUIPMENT NOT OPERATING DURING EACH MEASUREMENT SEQUENCE AT THE BOTTOM OF EACH COLUMN:

	MEASURED SOUND LEVELS, dBA, SLOW RESPONSE										
	STARTING TIME FOR EACH MEASUREMENT SEQUENCE, HR										
MEASUREMENT LOCATIONS											
EQUIPMENT NOT OPERATING											

Figure 4.7 — Noise evaluation work sheet.

COLLECTING THE DATA

Area Survey

Since one purpose of the basic sound survey is the identification of locations where more detailed surveys will be necessary, it is recommended that a check of the area TWA be conducted initially. Assuming that the work environment sound level is relatively constant (less than 8 dBA variation in the SLM's indicator), then during the basic sound survey a minimum of ten sound level samples should be recorded, spaced over a typical workday, using a data log format similar to the examples presented as Figures 4.4 through 4.7. Measurements should be recorded at a height of approximately 1.5 meters. In interpreting the movements of the SLM indicator, the surveyor should observe the instrument's response for approximately 5 seconds for each measurement and estimate the indicated mean sound level value.

It is also recommended that sample C-weighted sound levels be recorded. The difference between C-weighted and A-weighted levels has applications in the selection of HPDs and in certain noise control procedures.

Work Station/Job Description Survey

If the findings of the area survey predict TWAs of 85 dBA or higher, then the surveyor may elect to schedule a detailed sound survey of the area to determine the employees' TWAs. However, even if the findings of the area sound survey do not result in TWAs of 85 dBA or higher, it will still be necessary to survey specific work stations and various job descriptions in the area, because area TWAs are typically lower than actual worker TWAs, except in special instances such as when the time spent away from the job by a worker is significant or when the dominant noise source is located at a sufficient distance from the employee's work station. Typically as the workers approach their work stations the sound level will increase, unless of course the dominant noise source is coming from a different machine or other sources. As a consequence, a survey of an employee work station will normally exhibit a higher TWA than for the general work area.

In conducting sound surveys close to dominant noise sources, it is necessary that the surveyor give particular attention to the effects of the type of microphone utilized on the measurements obtained, potential effects of the surveyor's body and the changing characteristics of the sound field (Chapter 3).

EVALUATING THE DATA FOR OSHA

Area Survey

In order to demonstrate the calculation procedures for determining the area's equivalent noise dose, sound level or TWA, two examples are presented. We begin by presenting the overall general methodology of calculating the area's noise dose, TWA and OSHA equivalent sound level.

For discussion purposes we will assume that the employee's noise level-time exposure profile presented as Figure 4.8 is for an area measurement. To determine the equivalent TWA we could first determine the noise dose using the following equation (the reader is encouraged to refer to Chapters 2 and 3 for additional discussion of the equations referenced in this section):

$$D = \left[\frac{(C_1)}{(T_1)} + \frac{(C_2)}{(T_2)} + \ldots\ldots + \frac{(C_N)}{(T_N)} \right] \times 100, \text{ percent} \qquad \textbf{(4.1)}$$

where D is the noise dose in percent, C_N is the exposure duration for the Nth sound level and T_N is the corresponding allowed noise exposure. For the OSHA criterion, Table G-16a in Appendix I presents values of T_N for different sound levels. However, if the exposure duration needed is not listed, then it can be calculated using the following equation:

$$T_N = (8)/(2^{(L_N - 90)/5}) \text{ hours} \qquad \textbf{(4.2)}$$

where L_N is the sound level corresponding to the C_N exposure duration.

Once the noise dose has been determined, the equivalent TWA can be found by evaluating the following equation:

$$TWA = 16.61 \log_{10} [D(\%)/100(\%)] + 90, \text{ dBA} \qquad \textbf{(4.3)}$$

where D is the noise dose for the whole shift. If the sampled period is for a time period **less than** the actual work shift, then the associated dose must be corrected to a whole-shift equivalent value, that is,

$$D(H^*) = D(\text{measured}) \times H^*/T^*,$$

where T* is the actual period sampled and H* is the whole shift duration.

However, different approaches to the problem are possible. One would be to directly determine the equivalent (OSHA) sound level for the period of noise exposure using the following equation:

$$L_{OSHA} = 16.61 \log_{10} \left[\frac{1}{T} \sum_{i=1}^{N} (t_i \times 10^{(L_i/16.61)}) \right] \text{dBA} \qquad \textbf{(4.4)}$$

where t_i is the duration of exposure to level L_i and T is the exposure duration. Then, knowing the OSHA allowed exposure time for the L_{OSHA} level calculated, the employee's noise dose and equivalent TWA can be determined using Equations 4.1 and 4.3.

Unfortunately, the hearing conservationist is often presented sound survey data calculated using different approaches and different formats. Therefore it is important that the surveyor understand how the data were developed and be able to verify their accuracy.

In presenting sound survey data, values should be rounded off to the nearest dB, since measurement accuracy does not justify reporting the values to tenths of a decibel. However, in order that the reader can verify the sample calculations presented herein we have elected to maintain at least one figure beyond the decimal. An exception to this procedure of rounding off occurs during the actual calculation process. It is always necessary to maintain an additional number of digits to prevent accumulating large errors in the final predicted values, and this is especially true when evaluating numbers raised to a power such as in equation 4.4.

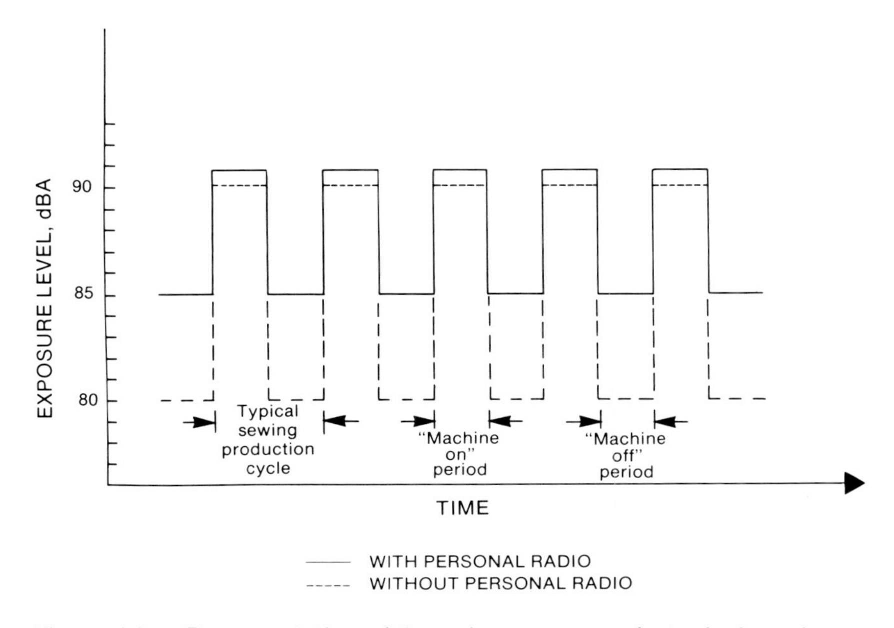

Figure 4.8 — Representation of the noise exposure of a typical employee with and without the use of a personal radio.

Example No. 1 (OSHA Criterion):

A basic sound survey was carried out in a synthetic fiber production department by sampling the area noise environment ten times during a typical workday. The measurements recorded were as follows: 106, 107, 105, 106, 107, 107, 105, 104, 106 and 105 dBA.

To determine the work area's equivalent TWA it is necessary to take into account the employee's duration of noise exposure. For the production area studied, the employees worked 9-hour (540-minute) shifts each day excluding time for lunch and breaks. The lunch and break areas exhibited sound levels less than 80 dBA.

The noise dose for this work area can be calculated using equation 4.1. Either hours or minutes may be used, provided that the same units are used for each term in each ratio. That is, each (C_N/T_N) ratio may be calculated using either hours or minutes. The problem can be approached by treating each sound level measurement as a separate noise exposure of 540/10 = 54 minutes, yielding ten terms (C_N/T_N) to evaluate in equation 4.1. Alternatively the equivalent sound level over the work period can be estimated as follows. If the range of the measured sound levels (maximum reading minus the minimum reading, or for this problem 107–104=3 dBA) is 5 dBA or less, then the OSHA equivalent 9-hour sound level can be approximated by calculating the arithmetic average of all the samples, or 105.8 dBA (Irwin and Graf, 1979; Peterson, 1980).

Therefore the employees working in the area are exposed to an area equivalent level of 105.8 dBA for 9 hours (540 minutes) each day. According to the information presented in Table G-16a, Appendix I, the employees are allowed a daily exposure to a noise level of 105.8 dBA for 53.7 minutes. The user may interpolate between the values given or use equation 4.2. The evaluation of equation 4.3 provides the daily noise dose in percent, or

$$D = (540/53.7) \times 100 = 1006\%.$$

The corresponding equivalent daily TWA can now be determined either by using the information presented in Table A-1, Appendix I, if the noise dose is equal to or less than 999%, or by evaluating equation 4.3:

$$TWA = 16.61 \log_{10} (1006/100) + 90 = 106.7 \text{ dBA}.$$

(Note that the 106.7 dBA TWA, based on eight hours of exposure, should not be confused with the 105.8 dBA value that was the average level over nine hours).

Comments On Example No. 1:

The OSHA criterion for allowed noise exposures is based on an 8-hour work day. For many industrial job functions, the actual work day will be either more or less than eight hours. Employees typically work 7 hours with two 15-minute breaks and a 30-minute lunch period. If these break periods are in areas where the equivalent sound level is less than 80 dBA, then the noise exposures during this period will not be included in the calculated TWA. However, if the break and lunch areas are in the workplace, as is sometimes the case, the noise exposure accumulated during those periods must be included.

It is also fairly common for employees to work shifts longer than 8 hours. In these instances the calculated TWA must include the total number of hours of noise exposure during each 24-hour period. The calculated TWA will represent the OSHA equivalent sound level that would produce the same noise dose in 8 hours as what the employee received over the total work shift.

Example No. 2 (OSHA Criterion):

The following set of measured sound levels were obtained in a textile spinning department by sampling the middle of the aisles at various times during the first shift: 90, 92, 91, 94, 90, 93, 92, 91, 90 and 92 dBA. Assume that for this work area the employees are given a 30-minute lunch period and two 15-minute breaks in areas where the sound levels are less than 80 dBA. Therefore the daily duration of exposure is taken to be 7 hours or 420 minutes.

Approach A: One approach to determining the work area's equivalent TWA is to first determine the noise dose by treating each measurement as representative of its equivalent portion of the total noise exposure time, or (420 minutes/ 10 samples) = 42 minutes per sample, or C_N. To determine T_N we can utilize the data presented in Table G-16a, Appendix I, or use equation 4.2. Then the noise dose is calculated using equation 4.1:

$$D = (42/480 + 42/364 + 42/418 + \ldots\ldots + 42/364) \times 100,$$

$$D = (1.095) \times 100 = 109.5\%.$$

Now to convert the noise dose in percent to an equivalent TWA we use equation 4.3,

$$TWA = 16.61 \log_{10} (109.5/100) + 90 = 90.7 \text{ dBA}.$$

Approach B: An alternate approach to determining the area's TWA would be to first determine an OSHA equivalent sound level for the seven-hour work shift, then use equation 4.1 to obtain the area's noise dose, and finally equation 4.3 to determine the TWA. Of course, we could determine the area's TWA directly by evaluating equation 4.4 for an eight-hour exposure. However, for this example we will solve the problem by the former method.

As pointed out in Example No. 1, if the range of the sound level measurements (maximum value minus the minimum value) is 5 dBA or less (for this problem 94–90=4 dBA), then the OSHA equivalent seven-hour sound level can be approximated by calculating the average of all 10 samples, or 91.5 dBA.

However, if one desires to follow a formal procedure, or if the range in measured sound levels is greater than 5 dBA, then equation 4.4 may be used with T = 7 hours, or

$$L_{OSHA} = 16.61 \log_{10} \{(1/7)\times[(7/10)\times 10^{(90/16.61)} + (7/10)\times 10^{(92/16.61)} + \ldots$$

$$\ldots + (7/10)\times 10^{(92/16.61)}]\} = 91.6 \text{ dBA},$$

Now that we know the OSHA equivalent seven-hour sound level is 91.6 dBA, we can use equation 4.1 to determine the daily area noise dose. First from Table G-16a, Appendix I, or equation 4.2 we determine that for an exposure of 91.6 dBA, the worker is allowed an exposure of 385 minutes. The actual exposure duration is seven hours or 420 minutes. Therefore using equation 4.1 we obtain:

$$D = (420/385) \times 100 = 109.2\%.$$

Now using Table A-1, Appendix I or equation 4.3 the equivalent TWA is found to be 90.6 dBA, or

$$TWA = 16.61 \log_{10} (109.2/100) + 90 = 90.6 \text{ dBA}.$$

Comments On Example No. 2:

Two approaches were used to determine the area's TWA. First, each survey sample was treated as an exposure level, and the duration of exposure was set equal to the number of samples divided by the typical employee's daily work duration. This procedure established a daily noise dose which was then converted to an equivalent TWA. The second approach involved first determining the seven-hour equivalent sound level, then calculating the noise dose and corresponding TWA. Either approach is acceptable.

Work Station/Job Description Survey

When measurements are being made in the immediate vicinity of sources, a large range of sound levels will often be observed. For areas in which the sound levels vary by more than 8 dBA, the surveyor will have to estimate the varying levels of exposure and corresponding time durations or else await the completion of the detailed sound survey using a different instrument or a more elaborate sampling plan. Consider the following example:

Example No. 3:

Approximately 50% of the employees at a production facility involved in a sewing operation had elected to use personal radios as a way of alleviating job boredom. However, management at this facility had expressed concern over the potentially harmful nature of the additional noise exposure resulting from radio usage (Skrainar, 1985).

Presented in Figure 4.8 is the idealized typical employee exposure level (based on sound survey samples) with and without the effect of using a personal radio. Initially the employees' TWA will be calculated without the effect of the personal radio, *i.e.,* for the noise exposure indicated by the dashed line. The employees typically worked a seven-hour day, not including the lunch period and two break periods allowed each day. During the break periods their exposure level was less than 80 dBA.

The idealized "machine on" and "machine off" periods as shown in Figure 4.8 are of equal time durations and the corresponding levels of exposure are 80 and 90 dBA respectively. Therefore the daily dose is determined using equation 4.2 and then equation 4.1:

$$D = \left(\frac{C_1}{T_1} + \frac{C_2}{T_2}\right) \times 100 = \left(\frac{3.5 \text{ hrs}}{32 \text{ hrs}} + \frac{3.5 \text{ hrs}}{8 \text{ hrs}}\right) \times 100,$$

or $$D = (0.11 + 0.44) \times 100 = 54.7\%.$$

The corresponding daily TWA may then be estimated by evaluating equation 4.3:

$$\text{TWA} = 16.61 \log_{10} (D/100) + 90, \text{ or}$$

$$\text{TWA} = 16.61 \log_{10} (55/100) + 90 = 85.6 \text{ dBA}$$ (without the effect of a personal radio)

If the effect of the personal radio is considered, then for "machine-on" and "machine-off" exposures of 91 and 85 dBA respectively, the following predictions are obtained based on the assumption that the work schedules of the two groups are identical.

$$D = \left(\frac{3.5 \text{ hrs}}{16 \text{ hrs}} + \frac{3.5 \text{ hrs}}{7 \text{ hrs}}\right) \times 100 = (0.22 + 0.5) \times 100$$

$$D = (0.72) \times 100 = 72\%.$$

$$TWA = 16.61 \log_{10} (72/100) + 90 = 87.6 \text{ dBA}.$$

Therefore the increase in the employees' predicted daily TWA due to the use of personal radios is 87.6 - 85.7 = 1.9 dBA.

Comments On Example No. 3:

We have assumed that the level-time exposure profile presented in Figure 4.8 is typical for both noise-exposed populations. In making this determination it is necessary to sample the noise levels at a sufficient number of employee work stations. Normally the recording of ten samples during the basic sound survey is sufficient for our purposes.

The Detailed Sound Survey

PURPOSE OF THE DETAILED SOUND SURVEY

The data collected during the basic sound survey should permit the identification of the work areas, work stations and/or job classifications at the plant site where more detailed sound survey information is needed. A sketch or drawing of the plant facility should now exist that can be referred to during the detailed sound survey. In addition, some work areas can be classified based on the findings of the basic sound survey. That is, all work areas where the basic sound survey findings clearly predict TWAs less than 85 dBA should now be indicated on the sketch or drawing of the plant facility. It is also recommended for those areas with TWAs less than 85 dBA that the typical and maximum sound levels measured also be indicated on the sketch. In addition, for those areas with TWAs above 85 dBA for which the data obtained during the basic sound survey were sufficient, the appropriate TWA can also be indicated on the drawing or sketch.

The purpose of the detailed sound survey is to complete the classification of the remaining areas. At the conclusion of the detailed sound survey a plant drawing or sketch and/or tabulation list should exist that defines the typical noise exposure level for all company employees.

COLLECTING THE DATA

Sampling Methodology

Noise Dosimeter

The microphone of the noise dosimeter should be placed roughly in the center of the shoulder on the side that would result in the higher estimated daily noise dose. The microphone holder, if provided by the manufacturer, should be used to support the microphone. One difficulty that arises in practice is that employees often wear very thin and loose T-shirts, or no shirt at all, or other garments that do not readily support some of the dosimeter microphone holders. However, the more modern noise dosimeters have smaller microphones that exhibit less body-reflection and sound-field-directivity effects and are easier to support. For these smaller microphones the orientation is not the critical issue, but the position of the microphone on the body must be maintained.

In a pinch, we have used safety pins to support the microphone slightly above the surface of the shoulder and ignored the fact that as the worker moved about, the microphone did not always remain in the desired orientation. The resulting data have been in reasonable agreement with simultaneous noise measurements taken using an SLM. An attempt should be made to attach the microphone lead-in cable to the worker's clothes in order to minimize the possibility that the cable may get caught in the production equipment or on other protruding objects. When practical, an even safer solution is to route the cord underneath the shirt or top clothing.

The noise dosimeter case should be firmly attached to the employee's belt in a position that will minimize interference with work and susceptibility to impact. If the unit is placed up front it may be uncomfortable to the employee when he leans toward the equipment. The position that seems to work the best is slightly behind either the left or right side. If the employee normally does not wear a belt and the clothing will not adequately support the unit, then a supporting belt may have to be provided.

An alternate method for supporting the dosimeter and microphone is to use a vest made of a light porous material such as lycra. This procedure mitigates problems caused by extra belts which may, for some job functions, hinder the workers in carrying out their job assignments.

A critical factor to be considered when the subject is in a very cold environment is protection of dosimeter batteries from temperatures which will derate them or even render the instrument inoperative. An effective solution is to place the dosimeters inside protective clothing at locations where body heat will keep them at a satisfactory temperature.

Ideally, the data collection should start and stop at the beginning and end of the work shift. This implies that both the employee(s) who will be sampled and the sound surveyor are willing to arrive soon enough before the work shift begins so that the worker can be fitted with the dosimeter and properly instructed in its use, and to stay long enough beyond the end of the work shift so that the dosimeters can be collected and the employees questioned about their use. However, some employees will refuse to report to work early or stay later unless they are paid for the extra time. If this compensation cannot be provided, the dosimeter must be fitted after the work shift begins and collected before it ends, in which case it will be necessary to account for the reduced sampling time relative to the typical employee workday.

The surveyor should keep in mind that for many job assignments the employees' noise exposures during the first and last fifteen minutes to an hour of the work shift, as well as during the morning and afternoon work breaks and the lunch period, can be significantly lower than for the remaining work periods. Therefore TWAs predicted from dosimetry data for sampling times that do not include these time periods may be higher than the employee's actual TWA. Although it is possible to account for these contributions to the employee's noise exposure by analysis and judgment, for OSHA enforcement purposes the noise dosimeter should be worn during the entire work period.

Before issuing the unit, the surveyor should carefully instruct the employee in its use and care. Hopefully, as part of the educational phase of the company's HCP the employees will already have been partially informed about the importance of the noise survey and of their responsibilities in ensuring the validity of the results. The instructions to the employee might include how to handle the unit when using the rest room (more than one microphone has fallen into the toilet) and what to do if the microphone becomes entangled in the equipment, if the case is dropped, or if the microphone or case is unexpectedly watered down.

Employees are hesitant to inform the surveyor of instances that might have affected the dosimeter or the readings obtained. This reluctance is usually not because of hostility, but stems from a fear that if the unit or data were adversely affected while the employee was wearing the unit, then the employee's job status could be jeopardized. The employee is more likely to provide this type of information if, at the end of the sampling period, when the noise dosimeters are being collected, the surveyor simply asks a non-intimidating question such as whether any difficulties were experienced while using the unit.

The sound surveyor should generally stay in the vicinity of the employees who are using the noise dosimeters until a reasonable knowledge of the noise level characteristics for the environment has been established. Since it is recommended that dosimeter data be checked by SLM measurements, the surveyor will most likely collect these data at the same time that the noise dosimeter samples are being collected.

If the sound surveyor is conducting a resurvey of a production area and obtains data that are significantly different from previous results, the surveyor should be suspicious of the data unless the differences can be accounted for. Current state-of-the-art profiling dosimeters are available which collect and read out detailed time history records of worker noise exposures (Chapter 3). Such records provide valuable information with which to assess the validity of the measured noise exposures. For example, in an area where the maximum sound level is known to be 100 dBA, measurements which show instantaneous or short-term average values of 110 dBA certainly demand further investigation.

Ordinarily, after the employees in the work area being sampled get over any initial effects of wearing the noise dosimeters and/or seeing other employees using them, they begin to ignore the units (Shackleton & Piney, 1984). Sound survey findings obtained during the adjustment period should be carefully analyzed for atypical values before they are included in the permanent sound survey data base.

The general guidelines for recording data discussed in the previous section should be followed when obtaining dosimeter data. Figure 4.9 presents one type of dosimeter data log. It is important to know the times at which sampling began and ended, and the name, location and/or job classification associated with the survey sample. Any pertinent information that may bear on the predicted TWA should be indicated in the general remarks section of the data log. Examples would include unusual down times for the operator's equipment or other significant noise sources in the work area, the need for the employee to spend an unusual amount of time at a location not normally visited, and any other information that the surveyor feels could affect the data. Of course, all noise dosimeter units should be calibrated before and after each daily use.

Sound Level Meter

Whereas the use of a noise dosimeter to record sound survey data is relatively straightforward, employing an SLM to obtain estimates of the actual TWA for those types of environments that could not be sampled during the basic sound survey is more difficult, especially when the employee's exposure exhibits large variations in the measured sound levels.

In sampling the sound field in the vicinity of a worker, the SLM's microphone should ideally be placed at the position in the sound field where the center of the noise-exposed employee's head would normally be, but with the employee removed from the immediate area. However, in

DATE: __________ INDICATOR SN: __________ TIME: __________

EXCHANGE RATE: ________ RESPONSE MODE: FAST ________ SLOW ________

CRITERION LEVEL: ______________ THRESHOLD LEVEL: ______________

WEIGHTING SELECTED: A: ________ C: ________ TEMPERATURE: __________

DEPARTMENT/LOCATION SAMPLED: ______________________________

____

____

SAMP. NO.	UNIT SN.	RESET /ON	START TIME	ENDING TIME	EXPOSURE TIME(MIN)	PERCENT INDICA.	DAILY DOSE	EQ. LEVEL
(1)								
(2)								
(3)								
(4)								
(5)								
(6)								
(7)								
(8)								
(9)								
(10)								

GENERAL REMARKS: ____

____

____

RECORDED BY: ______________

Figure 4.9 — Noise dosimeter data log (Form F).

many instances this is not a practical approach, since the operator may need to be present for the machine to operate properly. In such cases, the microphone is positioned approximately in the vicinity of the worker's ear that is receiving the highest exposure level. Ear placement differences are only critical if the sound field changes significantly with the position and orientation of the microphone.

A simple way to detect significant variations in the instrument's indicated sound level associated with position is to move the microphone around. When significant variations (greater than ± 3 dBA in the hearing zone of the employee, see Chapter 3) are detected, the surveyor should attempt to place the microphone within six inches of the worker's ear. In selecting the orientation of the microphone, the surveyor should follow the manufacturer's recommendations.

When using an SLM, the surveyor must not allow the operating condition of pertinent noise sources (either up or down) to affect a preselected sequence for taking samples. Often sound surveyors neglect to record the level at a work station if the equipment is not running at the scheduled sample time, even though the observed occurrence was normal for the production area being surveyed. Such a procedure will bias the data toward higher TWAs than should have been recorded or would have been indicated if a noise dosimeter had been employed. However, if the equipment is experiencing an unusually high number of failures, then this fact should be noted on the survey form.

If the sound surveyor is using an SLM and is surveying several workers in an area, then the periods between measurements usually will be sufficiently randomized due to environmental constraints placed on the surveyor's movements.

Sampling Procedure Examples

Presented in Table 4.2 are the results of a sound survey of a supervisor at a sawmill using both a noise dosimeter and an SLM. The SLM sampling format shown and data analysis outline presented in this section are similar to the format presently used by some OSHA inspectors (OSHA, 1981).

Initially the employee was fitted with the noise dosimeter and properly instructed in its use and care. Then the SLM sampling was initiated. Since other employees were sampled during the same time period, the time intervals between the SLM measurements naturally varied over the sampling period.

In order to determine the OSHA TWA, each measured sound level is initially given a relative weighting. The relative weighting is equivalent to the noise dose (divided by 100%) that would have been predicted if the

TABLE 4.2
Exposures for One Employee as Measured by SLM and by Dosimeter

JOB FUNCTION: Supervisor/Sawmill Operation
WORKSHIFT: 08:00 to 16:30 with one half-hour break for lunch and no scheduled break periods

TIME	SOUND LEVEL, dBA	RELATIVE WEIGHTING	COMMENTS
08:37	84	0.44	Planer coasting to a stop
09:08	78	0.00	Change over on planer
09:46	110	16.00	Standing next to top head
09:50	106	9.19	Standing next to top head
09:55	88	0.76	Planer idling-sharpening head
09:56	108	12.13	Next to planer, making adj.
09:59	87	0.66	Planer coasting to stop
10:08	64	0.00	Changing heads
10:14	64	0.00	Changing heads
10:20	106	9.19	Standing next to planer, 2′
10:27	87	0.66	Planer coasting to stop
10:32	67	0.00	In maintenance room
10:42	106	9:19	Planer running, making adj.
10:55	70	0.00	In maintenance room
11:22	107	10:56	~2′ from planer, running
12:20	105	8.00	~2′ from planer, running
12:24	-	-	Lunch
13:01	101	4.59	Feeding planer
13:10	100	4.00	Feeding planer
13:30	88	0.76	Planer down

Summed Rel. Wts. = 86.13
No. of samples = 19 (not including the lunch sample)

DOSIMETER DATA:
Time On: 07:54 Time Off: 13:37 Readout (%): 320.7

employee had been exposed to the measured sound level for eight hours. As an example, referring to Table 4.2 for the first sound level measured (08:37 -84 dBA), interpolation of Table A-1, Appendix I, provides a relative weighting for this measurement of 44%/100% or 0.44. The relative weighting can also be calculated using the inverted form of equation 4.3:

$$\text{Relative Weighting} = 10^{[(L_A-90)/(16.61)]} \quad \textbf{(4.5)}$$

where L_A is the measured A-Weighted sound level. For this example,

$$\text{Relative Weighting} = 10^{[(84-90)/16.61]} = 10^{(-0.361)} = 0.44$$

When equation 4.5 is being used, the contribution of sound levels below 80 dBA is zero since the OSHA threshold is 80 dBA.

Once the relative weightings for all samples are determined, the average of all the relative weightings is determined. For the example problem presented, 86.13/19(samples) = 4.53. This number multiplied by 100% is now the estimated employee's daily noise dose, or 453%.

If desired, the TWA can be determined either from Table A-1, Appendix I, or by using equation 4.3:

$$TWA = 16.61 \log_{10} (453\%/100\%) + 90 = 100.9 \text{ dBA}.$$

The duration of the sampling period for the noise dosimeter sample was from 07:54 to 13:37, including a thirty-minute lunch period, or an actual sampling time of 5.2 hours. Since this supervisor typically worked a seven-hour day, the daily TWA for the measured dose of 320.7% is:

$$TWA = 16.61 \log_{10} [(320.7/100)\times(7/5.2)] + 90, \text{ or}$$

$$TWA = 100.5 \text{ dBA}.$$

Even though for this problem the SLM-measured sound levels indicate very strong variability during the sampled work period (approximately 40 dBA), the predictions of the employee's TWA by the two sampling instruments are very close: 100.9 dBA for the SLM measurements versus 100.5 dBA for the dosimeter measurements. Also it is important to note that we have assumed that the data collected during the sampled periods are in fact representative of the employee's noise exposure during the period that was not sampled.

We could have employed a different approach in determining the TWA for the SLM measurements presented in Table 4.2. First assume that the time period between each measurement is constant (which is not the case for this actual example). Since the duration of the sampling period was from 08:37 to 13:37 less thirty minutes for lunch, or 263 minutes, and 19 samples were obtained, the sampling time represented by each sample is 263/19 or 13.8 minutes. Therefore in using equation 4.4, t_i = 13.8 minutes and T = 263 minutes. However, since the duration between each level is assumed constant, it would also be correct to let t_i = 1 and T = 19 units of time.

Now evaluating equation 4.4 yields:

$$L_{OSHA} = 16.61 \log_{10} \{(1/19)\times[1\times10^{(84/16.61)} + 0 \text{ (less than 80)} +$$

$$1\times10^{(110/16.61)} + 1\times10^{(106/16.61)} + \ldots\ldots + 1\times10^{(88/16.61)}]\}, \text{ or}$$

$$L_{OSHA} = 100.9 \text{ dBA}.$$

Sound Level Contours

Presented as Figure 4.10 is a sound level contour for one floor of a turbine building (DiBlasi, Suuronen, Horst and Bradley, 1983). Sound level contours may be used to illustrate to workers the degree of the noise hazard in their work areas, as an easy-to-understand educational tool during presentations to management, to identify dominant noise sources in a work area or even to identify hot spots (high noise levels) close to dominant noise sources, and to approximate the employees' noise dose.

To construct a sound level contour, the surveyor starts with a set of plant layout drawings or sketches. Then measurement positions are selected on approximately 3-meter grid patterns. If the spacing of the facility's main structural beams is only slightly different from 3 meters, then the existing spacing should be utilized. When the observed sound levels vary significantly, then a smaller grid pattern will be necessary. In the neighborhood of dominant noise sources, it usually will also be necessary to shorten the grid spacing due to the more rapidly changing sound level.

The contour lines drawn in Figure 4.10 are based on 2-dBA changes in the measured sound level. However, if the purpose of creating sound level contours is for employee or management educational purposes, we would recommend that the contours be drawn based on 5-dBA increments in order to be consistent with the classification scheme presented in Table 4.1.

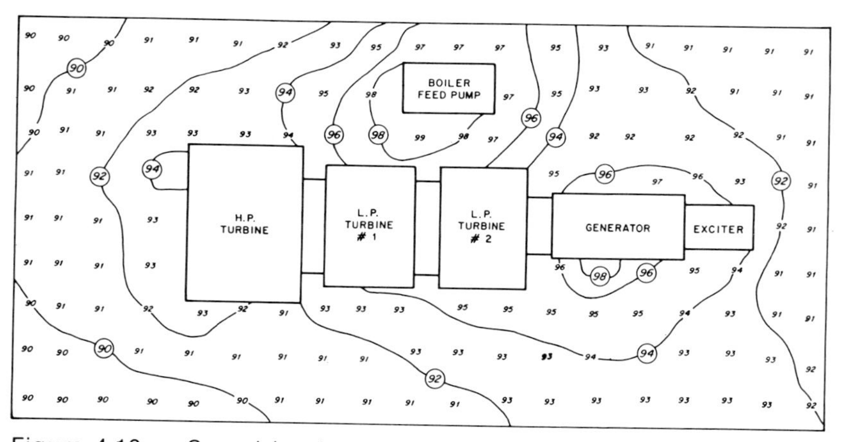

Figure 4.10 — Sound level contours - operating level turbine building. From Di Blasi *et al.* (1983).

The Engineering Noise Control Survey

PURPOSE OF THE ENGINEERING NOISE CONTROL SURVEY

Since the focus of this handbook is on practical guidelines for implementing effective HCPs, this section is not intended to provide a complete education on how to conduct general engineering noise control surveys. Rather, its purpose is to provide fundamental guidelines for the surveyor and other hearing conservation personnel in identifying major noise sources for which engineering noise control measures may be implemented. The reader is also encouraged to study the contents of Chapter 12 to gain greater insight into the need for and benefits expected from conducting this type of survey.

Experience has demonstrated that effective noise control sound surveys can be conducted and resulting noise control projects implemented by individuals with only limited, if any, formal training in noise control techniques. Examples include company nurses and audiometric technicians who, armed only with a basic SLM and calibrator and the benefits of a two-hour noise control lecture presented during an audiometric technician training course, have carried out effective noise control programs. They simply followed recommended procedures for identifying the dominant noise sources, then brought together company personnel with the best available background experience and technical capability to solve the noise problem, and finally, they stayed with the project until it was completed.

It is important to realize that for most of U.S. production facilities with noise problems, personnel with formal training in noise control technology are not available and most likely never will be, simply because of financial constraints. Therefore it is critical that hearing conservation personnel be educated in the basic sound survey techniques for noise control purposes by all means possible.

DETERMINING THE DOMINANT NOISE SOURCES

During the basic and detailed sound surveys, it is likely that several possible dominant noise sources will have been identified. In some instances the dominant noise source will be obvious. An example would be a work area where the typical sound level is less than 90 dBA except when a grinding operation is in progress, at which time the level rises to 95 dBA. A second example would be a work station where the measured sound level is 90 dBA with the equipment in operation but where the level at the work station and surrounding area is typically less than 85 dBA with the equipment inoperative.

Area Sources

We will discuss two stages in the identification of noise sources. The first is to determine the dominant noise source, or sources. Examples of dominant noise sources include individual pieces of production equipment, HVAC (heating, ventilating and air conditioning) systems, and conveyor systems. The second stage of the survey concerns the identification of the component noise sources for each dominant source. As an example, the initial survey in a production area might identify two or more packaging machines and a Syntron vibrator bowl as the dominant noise sources. Then a followup study of each of these noise sources would attempt to identify the component sources for each piece of equipment. After the dominant and the component noise sources have been identified, a priority list for noise control efforts can be established.

Once the noise survey has established typical exposure levels, the surveyor needs to set up a team who will work together to identify the dominant noise sources. Ideally this team should include an experienced machine operator, a mechanic and other personnel who the surveyor feels would contribute to the successful completion of the planned study. Personnel from the "professional" pool, such as management, engineers, or designers, are not desirable members unless they can contribute to the solution or are needed for political reasons.

The ideal procedure consists of running the individual machines in the production area while making measurements at specified locations. The surveyor may also elect to run each individual piece of equipment separately and to add together the effects of each successive piece of equipment.

It is desirable to conduct the noise control surveys at a time when all production equipment and other noise sources can be turned off, so that the surveyor has complete control of the background sound level. However, this situation often exists only during the second or third shift, the weekend, or once a year when the plant is closed down for annual maintenance. As a result, the surveyor will find it necessary to be flexible in selecting the survey periods.

To demonstrate the procedures for implementing the two different types of noise control surveys, an actual noise source identification effort at a dairy packaging facility is discussed. For the dairy production area surveyed, the background sound level was measured as less than 70 dBA as indicated in Table 4.3. If the background sound level is not at least 10 dB below the level being evaluated, the measured level should be corrected for the effects of background noise level (see Figure 3.6). Next, each piece of production equipment was operated independently, resulting in the equipment sound levels indicated by the center column in Table 4.3. As a

TABLE 4.3
Sound Level Measurements Made to Define the Dominant Machine Noise Sources at a Dairy Packaging Plant Workstation

NOISE SOURCE	SOUND LEVEL, dBA	
	Individual Piece of Equipment	Addition of New Pieces of Equipment
Background	<70	<70
Machine No. 1	80.4	80.8/(No. 1)
Machine No. 2	83.0	85.0/(1+2)
Machine No. 3	83.4	87.3/(1+2+3)
Machine No. 4	83.0	88.6/(1+2+3+4)
All Units Operating	88.6	

check on the sound measurements made for individual machine operations, all four pieces of equipment were put into operation successively and the sound level measurements repeated as each additional machine was added.

The sound level data measured for each of the four pieces of equipment can now be combined to check the sound level measured for all units operating simultaneously by using the following equation:

$$L_{sum} = 10 \ \mathrm{Log}_{10} \sum_{i=1}^{N} (10^{(L_i/10)}), \qquad \textbf{(4.6)}$$

where L_{sum} is the level resulting from the contributions of all levels, $L_{i=1 \text{ to } N}$. Also see equation 2.11 and associated discussion in Chapter 2.

For the data presented in Table 4.3, column two, the combined calculated level for all four machines running simultaneously, neglecting the contribution from the background sound level, is

$$L_{sum} = 10 \ \mathrm{Log}_{10} \left(10^{(80.4/10)} + 10^{(83/10)} + 10^{(83.4/10)} + 10^{(83/10)}\right)$$

or $$L_{sum} = 88.6 \text{ dBA}.$$

The calculated overall level for the sources measured individually agrees with the data obtained by running all units at the same time.

For some production lines, it is not practical to operate individual pieces of equipment because the product of one machine is needed by a following machine in order for the latter to operate properly. Under these circumstances it is necessary to repeat the sound survey measurements as new pieces of production equipment are turned on. This situation produces the type of measurements shown in the third column of Table 4.3.

In order to determine the contribution by each unit, one sound level is subtracted from another. This is because the sound level measured for Machine No. 1 and No. 2 operating simultaneously (column three) includes the contribution from both machines. As a consequence, the effect of Machine No. 1 plus background must be subtracted from the effect of both machines operating simultaneously in order to obtain the sound level attributed to Machine No. 2 operating independently.

Subtraction of levels can be achieved using the following equation:

$$L_{sub.} = 10 \text{ Log}_{10} \left(10^{(L_1/10)} - 10^{(L_2/10)}\right), \quad \textbf{(4.7)}$$

where L_1 is the larger of the two levels being subtracted.

For the data presented in column three of Table 4.3, if 80.8 dBA is subtracted from 85.0 dBA, using equation 4.7, we obtain:

$$L_{sub.} = 10 \text{ Log}_{10} \left(10^{(85.0/10)} - 10^{(80.8/10)}\right),$$

or $$L(\text{Machine No. 2}) = 83.0 \text{ dBA}.$$

A word of caution: theoretically, there are certain types of acoustic signals (non-random with respect to phase) for which equations 4.6 and 4.7 are not valid (Irwin & Graf, 1979). However, during fifteen years of conducting noise surveys we have encountered only two instances where the above equations did not produce reasonable predictions. In both of these cases, the dominant noise sources were HVAC fans that exhibited strong approximately equal pure tone, or single frequency, components.

As an alternative to using equations 4.6 and 4.7, the addition and subtraction of levels can be achieved by using the data presented in Table 2.1 and Figure 3.6. However, the use of an electronic calculator to evaluate these types of equations can be easily mastered, and should be.

Contributors to Each Dominant Noise Source

Assuming that the dominant noise sources have been identified, the surveyor may attempt to identify the contributing noise sources for each dominant source. For this type of sound survey effort, it is essential that a qualified machine operator and mechanic be available. It is relatively simple to operate production units for sound measurement options, but it is usually an order of magnitude more difficult to operate the various components of the equipment independently.

The procedure for studying the individual machine components is very similar to the procedures described above. However, it is the norm when studying contributing noise sources within a piece of production equipment that the measurements be made as components of the unit are disengaged or as the machine is reengaged. In general it is recommended that both approaches be utilized and the findings compared.

A word of caution is warranted at this point. When the sound levels created by the contributing noise sources of a production unit are studied by either disengaging or reengaging the machine's components, greater differences between the predicted (from individual machine components) and measured overall machine sound levels are to be expected due to the fact that the loads within the piece of equipment are changing. However, the effect of changing loads on the actual contribution made by individual machine components is normally minor in comparison to the overall effect of these individual sources.

As an example, consider the sound survey findings for a dairy packaging machine as presented in Table 4.4. For this packaging machine, it was relatively easy to turn on the major machine components, since the test procedure followed was similar to the normal start-up procedure. The resulting measured sound levels are presented in the second column. Each level presented in the second column includes the effect of all previous machine components.

To predict the contribution to the measured sound levels from each of the components of the machine, the procedure described in the previous section may be used: that is, by using equation 4.7 to subtract the contribution from the background + air sources from the measured level for the background + air + vacuum pump. The result is 65.9 dBA, as shown in the third column of Table 4.4. Continuing in this manner would result in an estimation of the contribution of each component of the machine to the overall sound level (83 dBA). A final check on the calculations can be

TABLE 4.4
Defining the Contributing Noise Sources for a Dairy Packaging Machine

EQUIPMENT/CONDITION	SOUND LEVEL, dBA	
	Measured Level	Calculated Individual Component Levels
Background + Air	62.7	62.7
+ Vacuum Pump	67.6	65.9
+ Burners	73.6	72.3
+ Idle Speed	74.0	63.4
+ Defoamer (pump)	76.0	71.7
+ Jaws (clamping mechanism)	83.0	82.0
Combined Levels	----	83.0

obtained by using equation 4.6 to combine the contributions from each component, the third column of Table 4.4.

The data presented in the third column of Table 4.4 clearly identify the jaw mechanism of this packaging machine as the principal noise source. Now we are able to ask the basic question: If the principal source can be adequately controlled, what would be the net effect on the measured level at the measurement position in question? This can be estimated by adding the contributions of the remaining noise sources (background, air, vacuum pump, burners, idle speed and defoamer) to the predicted controlled level for the jaw mechanism. Using equation 4.6 we find a predicted level of 76 dBA (excluding the jaws). If the level contributed by the jaw mechanism were reduced by 10 dBA, the predicted sound level then would be 72 + 76 = 77.4 dBA, or a reduction of 5.6 dBA.

Noise source identification procedures have been used extensively in industrial environments by various company personnel with a reasonable degree of success. Obviously there are pieces of production equipment for which the best noise control consultants have not been able to achieve a cost-effective solution. Therefore one would not normally expect less trained personnel using the procedures outlined above to succeed in such situations. However, experience has clearly demonstrated that solutions to problems do not necessarily come only from the formally educated population.

INFORMATION TO RECORD

The basic guidelines for recording sound survey data described in previous sections of this chapter also apply to sound surveys aimed toward noise control applications. However, additional information such as room volume, surface characteristics, relative location of source(s) and receivers and other pertinent information as discussed in Chapter 12 should be recorded on the noise control sound survey data sheet.

It should be noted that information recorded during noise control surveys may also be useful supporting data when the time comes for management to demonstrate to OSHA that a sincere noise control effort is part of the company's HCP.

SPECIAL NOISE CONTROL SURVEYS

There are several situations in which the surveyor will find it necessary or desirable to obtain additional information concerning the frequency range over which the acoustic energy radiated by a noise source or sources is located. One use for such information is to more effectively select materials to be used in the construction of an enclosure to control the noise radiated by a source. A second purpose would be to investigate the sound at a work station in order to obtain a better estimate of the attenuation needed by HPDs (Chapter 10).

The primary instrument of choice to obtain this type of additional information is the octave-band analyzer (Chapter 3). Instruments of greater frequency resolution, such as the one-third and one-tenth octave band filters, are also available. However, greater frequency resolution seldom provides significant additional useful and cost-effective information for general hearing conservation purposes, although for in-depth noise control applications greater resolution is often beneficial, if not essential.

In order to demonstrate the use of an octave-band analyzer, a sample problem is presented. Shown in Table 4.5 are the findings of an octave-band analysis at one employee's work station. While collecting these data, the surveyor noticed that the noise created by a major source seemed to contribute most of its acoustic energy over a narrow range of frequencies. In addition, it was observed that the machine would cease operation for short time periods. Therefore the surveyor decided to conduct a second octave-band analysis but with the SLM's FAST response option selected in order to be able to quickly measure the sound levels when the noise source was not operating. The estimated octave-band SPLs with and without the noise source in operation are presented as steps (1) and (2) in Table 4.5. A comparison of the two sets of measurements verified the surveyor's subjective evaluation in that the only change observed was in the 250-Hz octave band.

TABLE 4.5
Weighted and Octave-Band Sound Pressure Levels, and the Predicted Effects of Spectrum Modification

STEP NO.	OCTAVE-BAND SOUND PRESSURE LEVELS, dB							
	Octave-Band Center Frequency, kHz							
	.063	.125	.250	.5	1.0	2.0	4.0	8.0
(1) (spectrum/ 100.4 dBC, 93.3 dBA)	75	70	100	85	80	85	80	70
(2) (modified spectrum)	75	70	90	85	80	85	80	70
(3) (A-weighting corr.)	−26	−16	−9	−3	0	+1	+1	−1
(4) (2) + (3)	49	54	81	82	80	86	81	69
(5) $L(A) = 10 \text{ Log}_{10} (10^{(49/10)}+10^{(74/10)}+ \ldots 10^{(69/10)}$ (equation 4.6) $L(A) = 89.6$ dBA.								
(6) Reduction in L(A) = 93.3 − 89.6 = 3.7 dBA								

Once the contribution from the suspected noise source has been estimated, then the potential effect of a noise control application can be calculated. Begin by assuming that the noise control application would reduce the employee's exposure level at the octave-band center frequency of 250 Hz from 100 dB to 90 dB. Next apply the A-weighting corrections (Chapter 3) as indicated by step (3) to calculate the A-weighted octave-band SPLs shown as step (4). Then the A-weighted octave-band SPLs are summed using equation 4.6, as indicated by step (5), to predict a controlled exposure level of 89.6 dBA.

Finally the predicted reduction in the A-weighted sound levels as a result of the assumed noise control modification is obtained by subtracting the initially measured weighted sound levels from the predicted sound levels after adequate noise controls are implemented, step (6), or 3.7 dBA.

The procedures described for identifying dominant and component noise sources in the previous sections only provided an A-weighted sound level for each source investigated. However, it is sometimes advantageous to know the range of frequencies over which each dominant or component noise source contributes acoustic energy. Presented in Figure 4.11 are the octave-band SPLs for different component noise sources of a packager. The frequency resolution provided by the octave-band filter yields additional insight into the component contributors of this machine.

At the two lower octave-band frequencies investigated (63 and 125 Hz), the major component is the background + air + burners + vacuum pump combination. At the octave-band center frequencies of 250, 500 and 1000 Hz the defoamer dominates, and at the remaining three frequencies the idle mechanisms (motors, drive train, conveyors) dominate. Since the A-weighted filter places the most emphasis on the frequency range covered by the octave-band center frequencies of 500 to 4000 Hz, the defoamer and idle noise sources should be given top priority for noise control. In order to provide a visual picture reflecting the effect of A-weighting on the measured octave-band SPLs and thus implying which component should be reduced, the A-weighted octave-band SPLs are often plotted.

The results of the octave-band analysis reveal that overall the packager radiates acoustic energy predominantly at and below 1000 Hz. This fact implies that constraints exist on the options that would be most effective for noise control. As an example, if the measured reverberant field (normally several of these units are located in an acoustically hard room) is to be controlled by the application of acoustical absorbing materials, then the fact that the octave-band analysis reveals that this machine radiates significant acoustic energy at the lower frequencies dictates that the materials

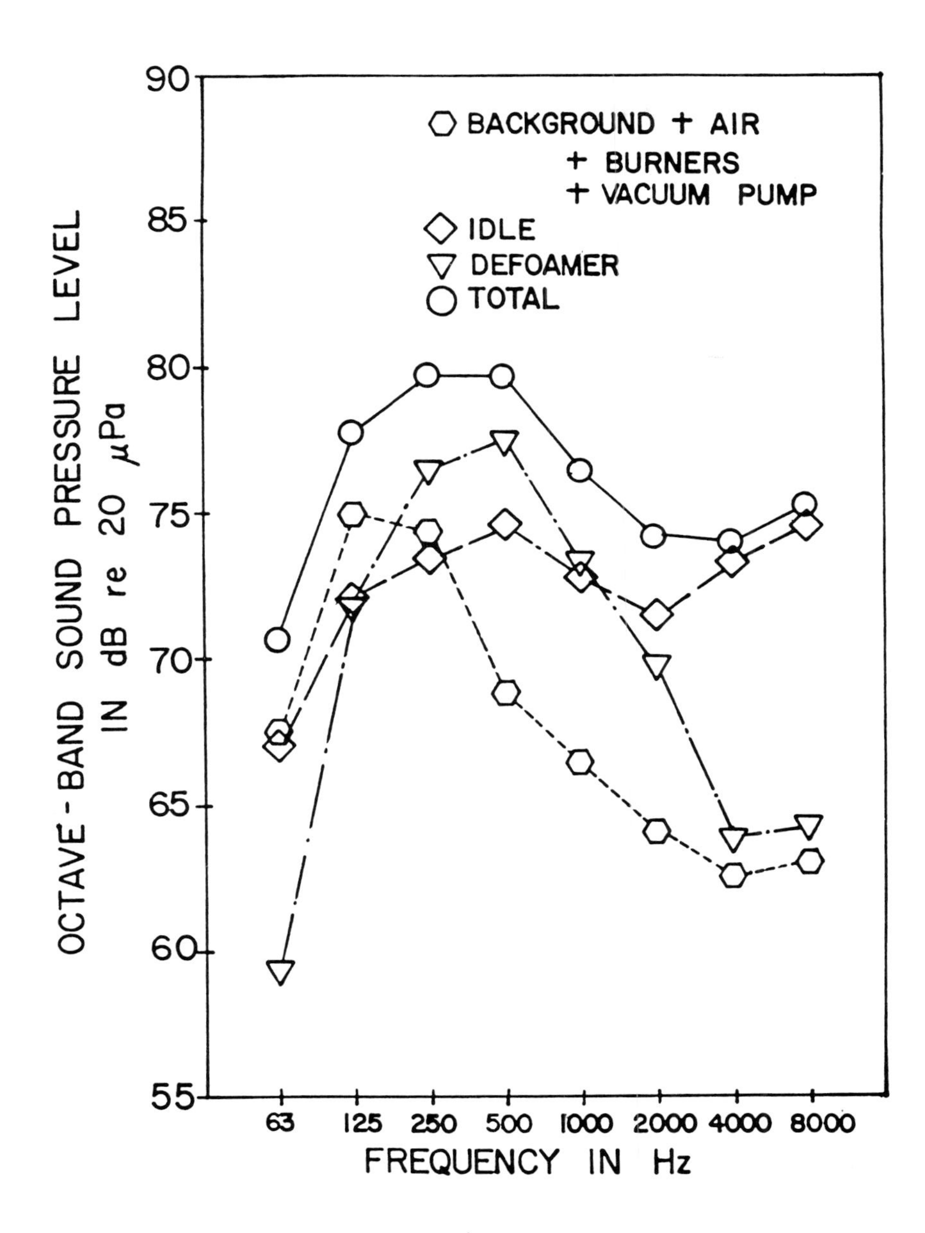

Figure 4.11 — Reverberant field sound level contributions for components of a packaging machine.

used must be capable of significant absorption at low frequencies. Of course, the C-weighted minus A-weighted sound level from the basic or detailed sound survey data would also indicate, though with less accuracy, that most of the acoustic energy is located in the lower frequencies. Presented as Figure 4.12 is a data log format for use in recording octave-band SPLs in addition to weighted SPLs.

Statistical Factors in Sound Survey Methods

DRAWING CONCLUSIONS FROM SAMPLE DATA

Sound surveyors hope to make inferences about (draw conclusions from) the noise exposures of employee populations by measuring exposures for small samples of workers. Well-planned sampling procedures and statistical analysis of sound survey data can help the surveyor describe the noise environment and employees' noise exposures with more confidence than by looking at raw sound level measurements. This section will briefly summarize the application of selected statistical techniques to answer questions such as whether any employees' TWAs ever exceed the 85 dBA OSHA action level or the 90 dBA OSHA PEL, or what the range of the 95% confidence interval around the mean TWA for workers in a particular job classification might be. Readers should consult additional references to determine whether the techniques presented are appropriate for their specific situations and to obtain more detailed guidelines, especially for sampling procedures (Natrella, 1963; Snedecor and Cochran, 1967; Leidel, Busch and Lynch, 1977; DiBlasi, Suuronen, Horst, and Bradley, 1983).

Noise exposure measurements vary due to random factors which are accounted for statistically as well as systematic factors which introduce undesired additional variability or bias into measurement values (Leidel *et al.,* 1977). Random factors can include hour-to-hour fluctuations in the noise levels produced by processes or machinery, worker mobility and/or job task changes, and unavoidable measurement error. Non-random sources of variation may include calibration errors, technical errors in measurement procedures, and systematic changes in exposure level due to source factors (such as regular production changes) or employee factors (such as an operator's removal of noise control mufflers from machinery, or an individual's tendency to lean closer than usual to the machine, or another person's habit of taking extra-long breaks), as well as measurement system factors (such as differences between individual SLMs or between SLMs and dosimeters). Statistics cannot distinguish between random and non-random sources of variation, so measurement mistakes must be avoided and systematic source or employee factors must be identified and handled through sampling techniques (Mellott, 1978).

DATE: __________ SOUND LEVEL METER TYPE: __________ SN: __________

TIME: ______ MICROPHONE TYPE: _______ SN: _______ WINDSCREEN: ______

METER RESP. - FAST: _______ SLOW: _______

OCTAVE BAND FILTER TYPE: ________ SN: ________ TEMPERATURE: ________

OTHER EQUIPMENT UTILIZED: ______________________________

____

LOCATION: ______________________________

____

LOCAT. NO.	SURV. NO.	dBA	dBC	dBL	OCTAVE-BAND SOUND PRESSURE LEVELS, dB									
					Octave-Band Center Frequencies, Hz									
					32	63	125	250	500	1k	2k	4k	8k	16k

GENERAL REMARKS: ______________________________

____

____

RECORDED BY: ________________

Figure 4.12 — Octave-band sound pressure level data log (Form G).

Because of the variability of the noise environment, each worker has a distribution of exposures over time rather than a fixed exposure level. Similarly, a group of workers performing the same job has a different distribution of exposures than that for any one group member. The entire population of workers across job classifications also has a distribution of noise exposures that is probably even wider. We want to select the distribution which is appropriate for answering the question we have in mind, then describe that distribution based on measurements for a sample of its members. For example, we infer the range of the population mean from the measured sample mean. Because the sample is only a fraction of its parent population and due to measurement uncertainty, our inferences are not exact, but rather are probability statements made at a specified level of confidence.

SAMPLING CONSIDERATIONS

One requirement for valid statistical inferences is random sampling, which means that at each stage in sample selection each available member of the sampling group has an equal chance of being selected for measurement. However, the sound surveyor must be familiar enough with the production processes and employees' job tasks to choose appropriately homogeneous sampling groups. For example, in one production department including workers with job classifications Y and Z, if job-Y workers stay on one end of the room operating the loudest machinery while job-Z employees are on the opposite end doing a quiet task, then the surveyor should take each job category as a separate sampling group rather than combining the entire department into a single sampling group.

By choosing homogeneous sampling groups and then selecting subjects randomly within groups (stratified random sampling), the surveyor can reduce the variability of noise exposure measurements within the sample, and thereby make more accurate estimates of the population characteristics. DiBlasi *et al.* (1983) discuss selection of stratified groups based on a noise contour map of the plant developed from SLM noise measurements (see their section 4 and our section on sound level contours).

Besides group member selection, other critical sampling factors include the timing and duration of measurements and the number of observations sampled. If noise levels are relatively steady, then simple random selection of sampling times is adequate. See section 4.4.3. of DiBlasi *et al.* for detailed instructions on how to randomly select sampling times. However, if periodic variations are suspected (such as morning versus afternoon, first shift versus second shift, or summer versus winter), then stratified random sampling should be used to draw proportionate numbers of sampling times from each period. DiBlasi *et al.* suggest spreading the measurements out over a duration of 3-6 months in order to tap random noise level fluctuations over time.

The duration of each measurement sample and the number of observations comprise a cost-benefit trade-off. Greater numbers of observations allow the confidence interval around the mean exposure to be defined more narrowly, but at a greater cost of time and effort. However, if on each day of sampling an instrument is used to measure two or more observations, then the sample size may be increased at little extra cost *if* it can be assumed that the noise environment does not vary systematically through the work-shift. For example, one dosimeter can be used to obtain 4-hour doses for two separate employees (which can then be converted to expected 8-hour doses), rather than an 8-hour dose for a single worker. Sampling a larger number of employees has the added advantage of reducing the influence of any "extreme outlier" observations which might be included in the sample, such as a worker who always leans closer to the machinery than average, or an employee who leaves the workstation more than average. Extreme cases cannot be neglected, but in very small samples it may not be clear which values are more typical.

In contrast, if typical noise doses over time for small groups of individuals are of interest, then the same workers can be sampled repetitively. Behar and Plener (1984) sampled the same employees each day for an entire work week, then integrated the daily TWAs using the OSHA 5-dB exchange rate to find an equivalent weekly exposure. Although OSHA currently does not address the issue of time-weighted average levels for periods longer than a single day, this idea has some merit. However, it would be more protective to use the equal-energy 3-dB exchange rate.

Unrelated to OSHA purposes, the sound surveyor may need to establish long-term equivalent exposures in order to evaluate the adequacy of worker protection. An equivalent exposure level is needed if one wishes to predict potential hearing damage by using a model of noise-induced hearing loss such as the ISO DIS 1999. When integrating noise levels from different days to compute an equivalent exposure level for use in the ISO DIS 1999 model, the equal energy method (3-dB exchange rate) must be used.

The most common question sound surveys can answer is whether selected workers' exposures exceed criterion values such as the TWA=85 dBA action level, the TWA=90 dBA PEL, or a company policy criterion such as a restriction in choice of HPDs at or above TWA=95 dBA. To determine whether certain job classifications of workers should be placed in an HCP, the surveyor should sample the persons most likely to have the highest noise exposures. If only a few employees are at high risk, then the TWAs can be measured for each one (a non-random survey of the complete subpopulation rather than a sample). However, if the individuals most at risk cannot be singled out intuitively, then the surveyor should randomly sample the entire job classification group. Leidel *et al.* have provided a guide for the sample size required to attain the desired probability (90% or 95%) of including within the sample at least one of the employees with the highest exposures (either the top 10% of exposures or the top 20% of exposures) based on groups of differing size. The required sample sizes for the 95% confidence level are presented as Tables 4.6-A and 4.6-B. As group size increases the required sampling proportion decreases. These tables can also be used to determine the minimum sample size for other stratified groups which clearly will be included in the HCP, though of course larger samples always provide better inferences.

TABLE 4.6-A
Sample Size Needed to Ensure at the 95% Confidence Level that the Sample Will Include One or More Observations for Employees with Exposures in the Top 10% of the Distribution (from Leidel, Busch, & Lynch, 1977)

Size of Group (N)	12	13-14	15-16	17-18	19-21	22-24	25-27	28-31	32-35	36-41	42-50	∞
Required No. of measured employees (n)	11	12	13	14	15	16	17	18	19	20	21	29

TABLE 4.6-B
Sample Size Needed to Ensure at the 95% Confidence Level that the Sample Will Include One or More Observations for Employees with Exposures in the Top 20% of the Distribution (from Leidel, Busch, & Lynch, 1977)

Size of Group (N)	7-8	9-11	12-14	15-18	19-26	27-43	44-50	51-∞
Required No. of measured employees (n)	6	7	8	9	10	11	12	14

The OSHA Hearing Conservation Amendment (HCA) is based on exposures for individuals, requiring that every person whose TWA equals or exceeds the action level of 85 dBA on any single day be placed in an HCP. **Therefore, use of a statistical sampling procedure in place of individual exposure monitoring does not alleviate the employer's responsibility to identify all those workers with doses of 50% or more.** However, if the employer's classification of employees into the HCP based on statistical sampling and analysis is **at least as protective** as the HCA requires, then individual monitoring can be replaced by appropriate sampling of the population.

USE OF CONFIDENCE INTERVALS

Because the sample mean is based on only a portion of the entire population, and because noise exposures vary, the mean TWA for a sample cannot be applied indiscriminately. If other individuals had been sampled, or if the same persons had been sampled at other times, then a different sample mean would have been obtained. Setting a confidence interval around the sample mean gives the surveyor a range of values within which the true population mean is expected to lie. Usually the 95% confidence level is used, meaning that of all possible samples, only 5% would give intervals that would fail to include the true population mean. However, the choice of a confidence level is not dictated by scientific principles, and the surveyor may apply a more or less strict confidence level in certain cases.

Depending on the purpose, the surveyor may need to use a two-sided confidence interval or a one-sided confidence interval. Two-sided confidence intervals extend both below and above the sample mean to indicate the range of mean values expected for different samples. One-sided confidence intervals, which extend only in one direction either below or above the sample mean, are used when only the larger or smaller expected mean values are of interest. One-sided confidence intervals are therefore useful in evaluating criterion compliance questions.

The use of confidence intervals to make criterion compliance decisions is illustrated in Figure 4.13 (adapted from Leidel *et al.*). If the upper confidence limit (UCL), or limit of the upper one-sided confidence interval, on the mean TWA for a single worker is below 85 dBA, then the sampled exposures are considered in compliance and the employee is not required to be included in an HCP (case A). If the UCL on the mean TWA is above the 85-dBA criterion, then the employee should be included in the HCP, whether the particular sample's mean TWA is above the cutoff (case B) or below the cutoff (case C). However, Leidel *et al.* suggest that an OSHA

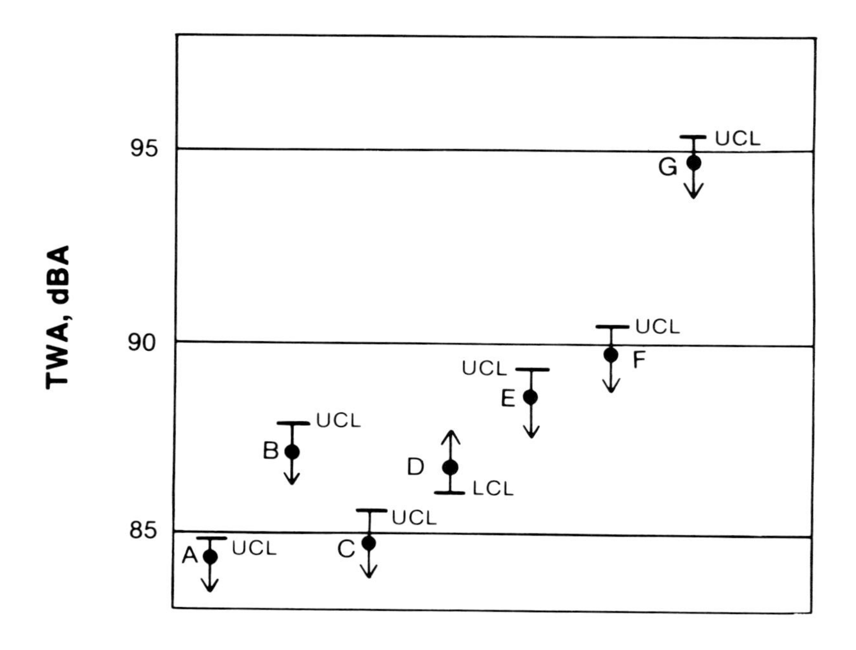

● = mean
UCL = upper confidence limit (upper limit of upper one-sided confidence interval).
LCL = lower confidence limit (lower limit of lower one-sided confidence interval).

Figure 4.13 — Comparison of boundaries for one-sided confidence intervals to criterion levels relevant for making compliance decisions (adapted from Leidel, Busch, & Lynch, 1977).

compliance officer probably should not issue a citation for non-compliance unless the lower confidence limit (LCL), or limit of the one-sided lower confidence interval, is above the cutoff (Case D). Cases E and F illustrate similar uses of the UCL in relation to the 90-dBA cutoff for mandatory HPD utilization: HPDs would not be mandatory for case E, but would be for case F. Case G illustrates a situation in which the UCL is above a TWA of 95 dBA, so the company might adopt a policy of requiring HPDs with high real-world attenuation in this case.

USE OF TOLERANCE LIMITS

In addition to determining confidence intervals for the mean TWA, it is helpful to calculate an upper tolerance limit of the TWA distribution below which a selected percentage of observations would fall (see Natrella, section 2.5). Typically the tolerance limit is calculated to include 75%, 90%, 95%, or 99% of the observations. If the TWA distribution being analyzed is exposures over time for an individual, then the 90% tolerance limit indicates the level below which the person's TWA would be expected to fall 90% of the time. If the distribution under consideration is for a homogeneous group of employees in a situation where every worker performs every potential task on an equal basis, then the 90% tolerance limit indicates the level below which there is a 90% probability that any individual's daily exposure would fall. However, if different individuals in a group were more likely to perform one task than another, then the tolerance limit would not be interpretable with respect to individuals, but only for the group of worker exposures.

Tolerance limits may be useful in making criterion compliance decisions for groups of workers. Although OSHA does not address the issue of infrequent exposure, the company might want to adopt a policy that employees would be placed in the HCP only if their probability of receiving a dose of 50% or more exceeded some selected level, such as 5% or 10%.

EXAMPLE STATISTICAL APPLICATIONS

The following paragraphs outline steps in analyzing sound survey data for homogeneous sampling groups of employee exposures by completing sample calculations for an example situation.

A basic sound survey in one department of a paper manufacturing plant yielded levels of approximately 85-95 dBA, so a detailed sound survey was scheduled. Thirty employees worked in the area on each shift, and there were no known production differences between shifts. Using Table 4.6-A to determine the minimum sample size needed to insure at the 95% confidence level that at least one employee with exposure in the top 10% would be included, the surveyor found that a minimum of 18 of the 30 employees should be sampled. In order to check for unsuspected exposure differences between shifts, 18 workers on first shift and 18 on second shift were sampled, for a total N of 36.

Basic Descriptive Techniques

Step 1: Tabulating the Data

The TWAs obtained are tabulated in ascending order for each shift with associated frequencies. As shown below, the ranges of TWA values for each shift are similar, and in each case most observations fall between 87 dBA and 90 dBA. A *t*-test of the difference between the two means showed no significant difference (see Snedecor and Cochran, Chapter 4), so the data for the two shifts were combined.

First Shift TWA	Frequency	Second Shift TWA	Frequency
82	1	83	1
85	2	85	1
86	1	87	2
87	3	88	2
88	2	89	5
89	1	90	3
90	3	91	1
91	1	93	2
92	2	94	1
93	1		
95	1		
mean = 88.7 dBA		mean = 89.1 dBA	
standard deviation = 3.3		standard deviation = 2.7	

Step 2: Plotting the Data

Drawing a simple histogram or frequency polygon of the exposure values is a basic step in data evaluation. A graph allows the surveyor to see if the distribution is bimodal (having two peaks), which might mean that dissimilar job classifications or time periods had mistakenly been combined in a single sampling group.

A histogram for our sample data, shown as Figure 4.14, displays an approximation of the normal distribution's bell-shaped curve.

Step 3: Calculating the Sample Mean

The mean of the sample's TWA values (the simple arithmetic average value) is an indicator of the center of the data. For a normal distribution, as shown in Figure 4.15, the mean is the exact center of the distribution.

$$\text{mean} = \sum_{i=1}^{N} x_i / N \qquad \textbf{(4.8)}$$

$$= 82+83+\ldots\ldots+94+95/36$$

$$= 88.92 \text{ dBA}$$

Step 4: Calculating the Sample Standard Deviation

The standard deviation is an indicator of the spread of the data around the mean. For normal distributions with a wide range and great variability, the standard deviation is larger than for more tightly clumped normal

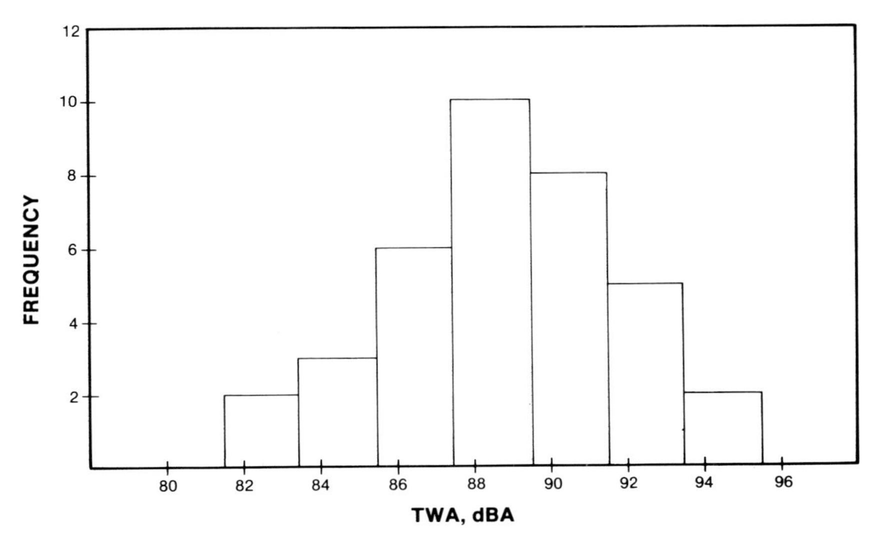

Figure 4.14 — Frequency histogram for the sample TWA data.

distributions. However, for all normal distributions the standard deviation defines the proportion of observations under various parts of the bell-shaped curve. As shown in Figure 4.15, 68.2% of observations fall within the range from one standard deviation below the mean to one standard deviation above the mean, and 95.4% fall within plus or minus two standard deviations of the mean.

To calculate the standard deviation of the sample TWA values:

$$\text{s.d.} = \sqrt{[\sum_{i=1}^{N} x_i^2 - (\sum_{i=1}^{N} x_i)^2 / N)]/(N-1)} \quad \textbf{(4.9)}$$

$$= 2.96 \text{ dBA}$$

Step 5: Checking for Normality of the Distribution

The surveyor can check the normality of the sample's distribution by applying the Chi-square goodness-of-fit statistic. If the normality assumption is met, then parametric statistical techniques can be used to analyze the sample; if not, then a log transformation may be applied to the data in an attempt to achieve normality before proceeding (Snedecor and Cochran, section 3.12; DiBlasi *et al.,* Appendix B). The Chi-square statistic is based on the difference in expected class frequencies between a bell-shaped

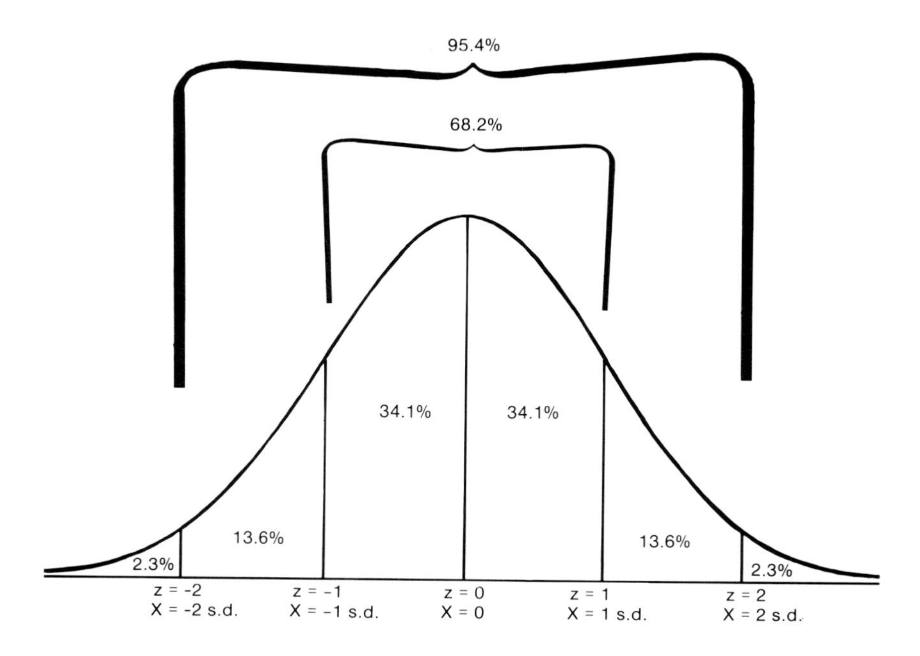

Figure 4.15 — Unit normal distribution showing the percentages of observations within one and two standard deviations of the mean, and corresponding standard scores (Z scores).

normal distribution and a flat distribution. The range of values is divided into classes, each of which must have an expected frequency of at least 5 observations (*i.e.* the total number of observations divided by the number of classes must be 5 or greater when class boundaries are chosen to have equal probabilities). The number of observations within each class is counted, and the statistic is calculated using the observed and expected frequency counts per class. If the calculated Chi-square does not exceed a critical value then the distribution may be considered normal at the 95% confidence level.

Class boundaries for the Chi-square statistic are computed from standard scores, or Z scores, found in Table 4.7. Z scores represent the distance from the score to the mean in terms of the standard deviation (score - mean/s.d.). The unit normal distribution has a mean of 0 and s.d. of 1; this special distribution is used to define the proportion of observations with values of Z or less, as shown in Table 4.7.

TABLE 4.7
Cumulative Unit Normal Distribution Showing the Proportion P of the Population with Standard Scores (Z Scores) Equal to or Less Than the Indicated Z Values. The Value of P for a Negative Z Score (- Z) equals 1.0 Minus the Value of P for the Corresponding Positive Z (+ Z). For Example, the P for Z= -1.62 Equals 1 - .9474, or 0.0526.

z_P	.00	.01	.02	.03	.04	.05	.06	.07	.08	.09
.0	.5000	.5040	.5080	.5120	.5160	.5199	.5239	.5279	.5319	.5359
.1	.5398	.5438	.5478	.5517	.5557	.5596	.5636	.5675	.5714	.5753
.2	.5793	.5832	.5871	.5910	.5948	.5987	.6026	.6064	.6103	.6141
.3	.6179	.6217	.6255	.6293	.6331	.6368	.6406	.6443	.6480	.6517
.4	.6554	.6591	.6628	.6664	.6700	.6736	.6772	.6808	.6844	.6879
.5	.6915	.6950	.6985	.7019	.7054	.7088	.7123	.7157	.7190	.7224
.6	.7257	.7291	.7324	.7357	.7389	.7422	.7454	.7486	.7517	.7549
.7	.7580	.7611	.7642	.7673	.7704	.7734	.7764	.7794	.7823	.7852
.8	.7881	.7910	.7939	.7967	.7995	.8023	.8051	.8078	.8106	.8133
.9	.8159	.8186	.8212	.8238	.8264	.8289	.8315	.8340	.8365	.8389
1.0	.8413	.8438	.8461	.8485	.8508	.8531	.8554	.8577	.8599	.8621
1.1	.8643	.8665	.8686	.8708	.8729	.8749	.8770	.8790	.8810	.8830
1.2	.8849	.8869	.8888	.8907	.8925	.8944	.8962	.8980	.8997	.9015
1.3	.9032	.9049	.9066	.9082	.9099	.9115	.9131	.9147	.9162	.9177
1.4	.9192	.9207	.9222	.9236	.9251	.9265	.9279	.9292	.9306	.9319
1.5	.9332	.9345	.9357	.9370	.9382	.9394	.9406	.9418	.9429	.9441
1.6	.9452	.9463	.9474	.9484	.9495	.9505	.9515	.9525	.9535	.9545
1.7	.9554	.9564	.9573	.9582	.9591	.9599	.9608	.9616	.9625	.9633
1.8	.9641	.9649	.9656	.9664	.9671	.9678	.9686	.9693	.9699	.9706
1.9	.9713	.9719	.9726	.9732	.9738	.9744	.9750	.9756	.9761	.9767
2.0	.9772	.9778	.9783	.9788	.9793	.9798	.9803	.9808	.9812	.9817
2.1	.9821	.9826	.9830	.9834	.9838	.9842	.9846	.9850	.9854	.9857
2.2	.9861	.9864	.9868	.9871	.9875	.9878	.9881	.9884	.9887	.9890
2.3	.9893	.9896	.9898	.9901	.9904	.9906	.9909	.9911	.9913	.9916
2.4	.9918	.9920	.9922	.9925	.9927	.9929	.9931	.9932	.9934	.9936
2.5	.9938	.9940	.9941	.9943	.9945	.9946	.9948	.9949	.9951	.9952
2.6	.9953	.9955	.9956	.9957	.9959	.9960	.9961	.9962	.9963	.9964
2.7	.9965	.9966	.9967	.9968	.9969	.9970	.9971	.9972	.9973	.9974
2.8	.9974	.9975	.9976	.9977	.9977	.9978	.9979	.9979	.9980	.9981
2.9	.9981	.9982	.9982	.9983	.9984	.9984	.9985	.9985	.9986	.9986
3.0	.9987	.9987	.9987	.9988	.9988	.9989	.9989	.9989	.9990	.9990
3.1	.9990	.9991	.9991	.9991	.9992	.9992	.9992	.9992	.9993	.9993
3.2	.9993	.9993	.9994	.9994	.9994	.9994	.9994	.9995	.9995	.9995
3.3	.9995	.9995	.9995	.9996	.9996	.9996	.9996	.9996	.9996	.9997
3.4	.9997	.9997	.9997	.9997	.9997	.9997	.9997	.9997	.9997	.9998

Calculating the Chi-square goodness-of-fit statistic:

If we divide 36 observations into 7 classes we would expect 5.14 observations per class.

Class boundaries (6 boundaries yield 7 classes):

Boundary 1: Probability expected = 1/7 = 0.143
From Table 4.7 the Z score for p=0.143 = -1.07
Class boundary = mean + (- 1.07 × s.d.)
= 88.92 + (-1.07 × 2.96) = 85.74

Boundary 2: Probability expected = 2/7 = 0.286
From Table 4.7 the corresponding Z score = -0.57
Class boundary = 88.92 + (-0.57 × 2.96) = 87.24

Boundary 3: Probability expected = 3/7 = 0.429
From Table 4.7 the corresponding Z score = -0.18
Class boundary = 88.92 + (-0.18 × 2.96) = 88.38

Boundary 4: Probability expected = 4/7 = 0.571
From Table 4.7 the corresponding Z score = 0.18
Class boundary = 88.92 + (0.18 × 2.96) = 89.45

Boundary 5: Probability expected = 5/7 = 0.714
From Table 4.7 the corresponding Z score = 0.56
Class boundary = 88.92 + (0.56 × 2.96) = 90.58

Boundary 6: Probability expected = 6/7 = 0.857
From Table 4.7 the corresponding Z score = 1.07
Class boundary = 88.92 + (1.07 × 2.96) = 92.09

Observed frequency counts for each class are:
Class 1 (less than 85.74): 5 observations
Class 2 (85.75 to 87.24): 6 observations
Class 3 (87.25 to 88.38): 4 observations
Class 4 (88.39 to 89.45): 6 observations
Class 5 (89.46 to 90.58): 6 observations
Class 6 (90.59 to 92.09): 4 observations
Class 7 (92.1 and above): 5 observations

CAUTION: If data are rounded to whole units (grouped data), it is best to modify the class boundary to conform to the real limits of the measurement units, which are the midway mark between integers (Steel and Torrie, 1980). Otherwise empty classes may occur if some class widths are too narrow to include any integer values. After modifying the boundaries and combining too-narrow classes with adjacent classes, determine the Z scores of the new boundaries using Table 4.7, calculate the expected number of observations for the new classes from the Z scores (same procedure as Step 10), and proceed with the Chi-square computation.

$$\text{Chi-square} = \sum_{i=1}^{N} (\text{observed} - \text{expected})_i^2 / \text{expected}$$
$$= [(5-5.14)^2 + \ldots\ldots + (5-5.14)^2]/5.14$$
$$= 4.857/5.14$$
$$= 0.95 \text{ with 4 degrees of freedom (d.f.)}$$
(d.f. = number of classes minus three)

Determine the critical value of Chi-square from Table 4.8.
For 4 d.f. the value is 9.49.

Compare the calculated value to the critical value.
The critical value is greater than the calculated value, so the distribution may be considered normal at the 95% confidence level.

Making Inferences

Step 6: Setting a Confidence Interval Around the Mean

The 95% confidence interval around the mean is the range of values constructed from the sample in such a way that it has a 95% chance of including the population mean. Since the value of sample means depends on the composition of each sample, the confidence interval around the mean gives the surveyor a way of estimating the true population mean. The alpha probability of error is split into an alpha/2 probability that the true

TABLE 4.8
Critical Values for the Chi-Square Goodness-of-Fit Test for Normality of a Distribution, Using the 95% Confidence Level. If the Calculated Chi-Square Value for the Sample Does Not Exceed the Critical Value, Then the Sample May be Considered a Normal Distribution. (Adapted from Snedecor & Cochran, Table A 5.)

Number of Classes	Degrees of Freedom	Critical Value of Chi-Square
4	1	3.84
5	2	5.99
6	3	7.81
7	4	9.49
8	5	11.07
9	6	12.59
10	7	14.07
11	8	15.51
12	9	16.92
13	10	18.31
14	11	19.68
15	12	21.03
20	17	31.41

mean will be lower than the confidence interval's lower bound (2.5% error in the lower tail) and an equal probability that the true mean will be higher than the confidence interval's upper bound (2.5% error in the upper tail).

Calculating a Two-Sided Confidence Interval for the Mean:

Choose desired confidence level and determine the value of the term 1 – (alpha/2).

Desired level of confidence is 95%
Alpha = 1 – .95 = .05
1 – (alpha/2) = 1 – .025 = .975

Determine t for the 1 – (alpha/2) column and correct degrees of freedom (d.f.) from Table 4.9.

Find the t value for the .975 column in Table 4.9.
The degrees of freedom is (n – 1), or 35 for our example.
The $t_{.975}$ with 35 d.f. = 2.032 (interpolated)

Determine upper bound of confidence interval:

$$\text{UCL} = \text{mean} + t(\text{s.d.}/\sqrt{n}) \tag{4.10}$$
$$= 88.92 + 2.032(2.96/6)$$
$$= 89.92 \text{ dBA}$$

Determine lower bound of confidence interval:

$$\text{LCL} = \text{mean} - t(\text{s.d.}/\sqrt{n}) \tag{4.11}$$
$$= 88.92 - 2.032(2.96/6)$$
$$= 87.92 \text{ dBA}$$

Therefore we are 95% confident that the true mean of the TWA values for the population sampled falls within the range from 87.9 dBA to 89.9 dBA (rounded).

Step 7: One-Sided Upper Confidence Interval for the Mean

If the surveyor is concerned only about how high a value the mean might take, then a one-sided upper confidence interval is appropriate. Rather than splitting the 1-alpha probability of the mean's falling outside the confidence interval between an upper tail and a lower tail, the entire 5% probability of error is placed in a single upper tail. This provides a more conservative test which may be appropriate for compliance-related decisions (as discussed in reference to Figure 4.13).

Calculating One-Sided Confidence Interval Above the Mean:

Choose desired confidence level and determine the value of the term (1 – alpha).

We will use the 95% confidence level, that is an alpha of .05.
Therefore,
1 – alpha = 1 – .05 = .95

TABLE 4.9
Percentiles of the t Distribution
(from Natrella, Table A-4)

df	$t_{.90}$	$t_{.95}$	$t_{.975}$	$t_{.99}$	$t_{.995}$
1	3.078	6.314	12.706	31.821	63.657
2	1.886	2.920	4.303	6.965	9.925
3	1.638	2.353	3.182	4.541	5.841
4	1.533	2.132	2.776	3.747	4.604
5	1.476	2.015	2.571	3.365	4.032
6	1.440	1.943	2.447	3.143	3.707
7	1.415	1.895	2.365	2.998	3.499
8	1.397	1.860	2.306	2.896	3.355
9	1.383	1.833	2.262	2.821	3.250
10	1.372	1.812	2.228	2.764	3.169
11	3.363	1.796	2.201	2.718	3.106
12	1.356	1.782	2.179	2.681	3.055
13	1.350	1.771	2.160	2.650	3.012
14	1.345	1.761	2.145	2.624	2.977
15	1.341	1.753	2.131	2.602	2.947
16	1.337	1.746	2.120	2.583	2.921
17	1.333	1.740	2.110	2.567	2.898
18	1.330	1.734	2.101	2.552	2.878
19	1.328	1.729	2.093	2.539	2.861
20	1.325	1.725	2.086	2.528	2.845
21	1.323	1.721	2.080	2.518	2.831
22	1.321	1.717	2.074	2.508	2.819
23	1.319	1.714	2.069	2.500	2.807
24	1.318	1.711	2.064	2.492	2.797
25	1.316	1.708	2.060	2.485	2.787
26	1.315	1.706	2.056	2.479	2.779
27	1.314	1.703	2.052	2.473	2.771
28	1.313	1.701	2.048	2.467	2.763
29	1.311	1.699	2.045	2.462	2.756
30	1.310	1.697	2.042	2.457	2.750
40	1.303	1.684	2.021	2.423	2.704
60	1.296	1.671	2.000	2.390	2.660
120	1.289	1.658	1.980	2.358	2.617
∞	1.282	1.645	1.960	2.326	2.576

Entries originally from Table III of *Statistical Tables* by R.A. Fisher and F. Yates, 1938, Oliver and Boyd, Ltd., London.

Determine t for the 1 – alpha column and correct degrees of freedom from Table 4.9.

Find the t for the .95 column and 35 d.f. in Table 4.9.

Interpolated between 30 d.f. and 40 d.f., t = 1.691

Determine the upper bound of the one-sided confidence interval above the mean.

$$\begin{aligned} UCL &= \text{mean} + t(\text{s.d.}/\sqrt{N}) \\ &= 88.92 + 1.691(2.96/6) \\ &= 89.75 \text{ dBA} \end{aligned}$$

Therefore we are 95% confident that the true population mean is less than a TWA of 89.8 dBA (rounded).

Step 8: One-Sided Tolerance Limit of TWA Distribution

Table 4.10 (from Natrella) presents the **K** factors for the 95% confidence level; these **K** factors correspond to Z scores, but with adjustments for the uncertainty associated with sample size, as shown by the left-hand **N** column. By using a tolerance limit the surveyor may determine at the selected level of confidence the TWA level below which a selected proportion P of the distribution falls, meaning that only 1 – P % of employee exposures would be greater than the tolerance limit.

Calculating an Upper Tolerance Limit for a Distribution:

Choose the desired confidence level.

We will use the 95% confidence level.

Choose the percentage of the population distribution which you wish to be included within the tolerance limit.

We will choose 90%, or the .90 proportion.

Find the appropriate **K** value in Table 4.10 for the selected proportion and the appropriate d.f.

The **K** factor for the .90 proportion and **N**=35 (close enough to our true **N** of 36) is 1.732.

$$\begin{aligned} \text{Upper Tolerance Limit} &= \text{mean} + K(\text{s.d.}) \\ &= 88.92 + 1.732(2.96) \\ &= 94.05 \text{ dBA} \end{aligned} \qquad \textbf{(4.12)}$$

Therefore we can say that there is a 95% probability that 90% of worker TWAs will fall below 94 dBA (rounded), while the remaining 10% would be at this level or higher.

Step 9: Proportion of Observations Above a Certain Value

Often the surveyor is more interested in knowing what percentage of TWAs fall at or above a predetermined criterion (such as 85 dBA or 90 dBA) than in setting a tolerance limit for a certain proportion of the

TABLE 4.10
K Factors for One-Sided Tolerance Limits for Normal Distributions, Using the 95% Confidence Level, Where N is the Number of Observations in the Sample and P is the Selected Proportion of the Population (from Natrella, Table A-7)

N \ P	0.75	0.90	0.95	0.99	0.999
3	3.804	6.158	7.655	10.552	13.857
4	2.619	4.163	5.145	7.042	9.215
5	2.149	3.407	4.202	5.741	7.501
6	1.895	3.006	3.707	5.062	6.612
7	1.732	2.755	3.399	4.641	6.061
8	1.617	2.582	3.188	4.353	5.686
9	1.532	2.454	3.031	4.143	5.414
10	1.465	2.355	2.911	3.981	5.203
11	1.411	2.275	2.815	3.852	5.036
12	1.366	2.210	2.736	3.747	4.900
13	1.329	2.155	2.670	3.659	4.787
14	1.296	2.108	2.614	3.585	4.690
15	1.268	2.068	2.566	3.520	4.607
16	1.242	2.032	2.523	3.463	4.534
17	1.220	2.001	2.486	3.415	4.471
18	1.200	1.974	2.453	3.370	4.415
19	1.183	1.949	2.423	3.331	4.364
20	1.167	1.926	2.396	3.295	4.319
21	1.152	1.905	2.371	3.262	4.276
22	1.138	1.887	2.350	3.233	4.238
23	1.126	1.869	2.329	3.206	4.204
24	1.114	1.853	2.309	3.181	4.171
25	1.103	1.838	2.292	3.158	4.143
30	1.059	1.778	2.220	3.064	4.022
35	1.025	1.732	2.166	2.994	3.934
40	0.999	1.697	2.126	2.941	3.866
45	0.978	1.669	2.092	2.897	3.811
50	0.961	1.646	2.065	2.863	3.766

distribution. By converting criterion values such as 85 dBA or 90 dBA into standard scores or Z scores (expressed in units of the sample standard deviation), the surveyor can predict what percentage of workers' TWAs would fall above the selected criterion. This procedure is discussed by DiBlasi *et al.* (see their section 6). Unlike the tolerance limit procedure, the use of Z scores includes no allowance for sample variation, so there is no level of confidence associated with these predictions. The mean and standard deviation of the sample are simply taken as the true values for the population, when actually they are only estimates.

Table 4.7 (from Natrella's Table A-1) gives the Z scores at or below which differing proportions of a distribution are expected to fall based on the normal distribution concept with no adjustments made to correct for the uncertainty of variance estimation due to the sampling process and sample size.

Predicting the Proportion of TWAs at 90 dBA or Higher:

Since Table 4.7 gives Z values at or below which a proportion falls, the desired criterion of 90 dBA must be changed to a value just barely smaller: 89.9 dBA. The table will tell us the proportion of the distribution up through this value, and the remainder falls above it (that is, at 90 dBA or above).

Convert the criterion score to a Z score.

$$Z = (\text{criterion} - \text{mean})/\text{s.d.} \tag{4.13}$$
$$= (89.9 - 88.92)/2.96$$
$$= .33$$

Locate the obtained Z score in Table 4.7.

The closest value below .33 is .3. Go across the .3 row to the .03 column to find the P value for Z=.33

Read the associated proportion P from the table.

The table value for Z=.33 is P=.6293.

Therefore, 62.9% of employee TWAs are estimated to fall below 90 dBA, while 37.1% are estimated to fall at 90 dBA or above.

Step 10: Predicting the Proportion of TWAs in a Range

Often the surveyor may want to know what percentage of employees' TWAs are in the range 85-89.9 dBA. This answer is obtained using the same method as Step 9. The proportion in the range 85-89.9 dBA is simply the proportion which falls at 89.9 dBA or below (from step 8) minus the proportion which falls at 84.9 dBA or below.

Predicting the Proportion of TWAs from 84.9 dBA to 89.9 dBA:

Convert the criterion value 84.9 dBA to a Z score.

$$Z = (\text{criterion} - \text{mean})/\text{s.d.}$$
$$= (84.9 - 88.92)/2.96$$
$$= -1.36$$

Locate the obtained Z in Table 4.7. For negative Z scores the proportion P equals one minus the P value for a positive Z of the same size. Therefore we look up the P for Z=+1.36.

Read the proportion P for the desired Z.

The P for Z=+1.36 is .9131.

The P for Z=−1.36 is (1.00 − .9131) or .0869.

The proportion P in the desired range 84.9 dBA through 89.9 dBA is P for 89.9 (from Step 8) minus P for 84.9.

$$P = .6393 - .0869$$
$$= .5524$$

Therefore 55.2% of the employee TWAs are estimated to fall in the range from 85 dBA through 89.9 dBA.

USING STATISTICS

The preceding section has provided basic statistical information for consideration by the sound surveyor in analyzing the data base and making predictions based on its content. However, it is essential that the surveyor seek out and study the suggested references in order to apply these techniques properly. The availability of personal computer software for statistical applications (Statistical Graphics Corporation, 1985) greatly increases the sound surveyor's ability to use statistics easily and quickly in decision-making, provided of course that the software user understands the procedures being applied and can check their validity.

Report Preparation

GOALS

There are at least three important facets of a successful sound survey report. First, the objective(s) established prior to the survey must be clearly stated in the report and the text should demonstrate the attainment of those objectives. Second, the report should be complete. A sufficient condition for completeness is the requirement that if at a later date an individual different from the author desired to reproduce the report's contents, this could be accomplished provided that the characteristics of the noise environment had not changed. Third, the sound surveyor's report should exhibit an acceptable level of technical writing skill (Brusaw, Alred and Oliu, 1976).

POTENTIAL AUDIENCE

When preparing the report it is important to keep in mind not only the survey objectives but also the potential audience. It is an unfortunate waste of resources to spend days conducting a sound survey only to fail to communicate the findings effectively or to antagonize the reader of the report. As a consequence, management, the key individual or company nurse may fail to properly implement the recommendations of the report.

For example, if the sound surveyor forgets that the primary reason for writing the report is for OSHA compliance purposes, essential information to demonstrate compliance may not be included or adequately covered. However, if the report is to be used as evidence in workers' compensation proceedings, the sound surveyor must keep in mind that different data may be necessary depending upon the state in which the claim has been filed.

CONSIDER THE POLITICAL CONSTRAINTS

When preparing a report careful consideration must be given to local management attitudes and potential political constraints. It is unfortunate, but true, that at some plant sites the report will have to be written so as to account for political realities. Examples include purposely omitting from the report breakdowns in some production areas of the enforcement of HPD utilization, or praising a production manager who has not strongly supported the company's HCP. Failure to do so could irritate the reader, who in turn might limit any potential for further sound survey efforts, noise control modifications, or other projects at the plant site.

In other words, playing the political game effectively is extremely important in achieving significant progress in the hearing conservation area. The sound survey report through its indicated findings and resulting recommendations is in fact a significant political document. The effective sound surveyor will make maximum utilization of this fact.

REPORT FORMAT

A general outline for preparing a sound survey report is as follows:

(a) Title Page
(b) Summary/Acknowledgments
(c) Background and/or Introduction
(d) Data Collection/Measurement Methods/Instrumentation
(e) Analysis of the Data
(f) Results/Conclusions/Recommendations
(g) References
(h) Appendix

This list provides guidance for topics that could be included in a sound survey report. However, the two most important sections of the report will be the summary/acknowledgment and results/conclusions/recommendations sections. The types of individuals who normally implement the findings often will take the time necessary to read only these two sections. Therefore if these sections are not suitably short, limited to two to three pages (doubled spaced), the report may be effectively suppressed by top management. Likewise, if the sound surveyor makes many recommendations, management may be overwhelmed and disregard the report. Therefore in writing these two sections remember to be brief, striving for the most important actions desired and being sure to keep the political implications in mind.

Do not forget to acknowledge, at the end of the summary page, all individuals who provided special assistance during the sound survey. This takes very little effort and can reap substantial benefits.

The background and/or introduction sections should include information such as why the survey was conducted, the time period over which the survey took place and any other pertinent information that would help the serious report reader to better understand the contents therein.

A detailed discussion of the data collection and measurement procedures should be included. The instrumentation used and calibration procedures followed should be adequately explained. It is important that the reader of the report be able to verify the analysis of the data that is provided by the author of the report. If the analysis involves detailed calculations and many tables, then only sample calculations and tables should be included in the main body of the report. The remaining information should be put in an appendix where it can be found and studied if desired, but where it does not detract from the readability of the report.

If the surveyor selects materials and equations extensively from other reports, then proper references should be included. Likewise, if the author elects to use a sound survey procedure different from what is commonly understood as acceptable practice, then the source of the justification for a modified procedure should be stated. Reports that include statements such as "The existing noise levels will cost the company $2,000,000 over the next 10 years" without adequate supporting evidence will fail to achieve credibility during the review process.

If the sound survey was conducted in order to investigate a special problem, such as a request from a manager concerning the noise levels in a computer room, then it is acceptable to limit the reporting effort to a letter-report format or some other type of short summary document. This type of report would typically include a one- or two-paragraph description of the type of sound survey conducted and its purpose, a summary of the findings and any recommendations that would be appropriate.

If the sound survey report is a result of the semi-annual or annual sound survey, then it is not necessary to include information such as sound level measurement, calculation, data recording procedures, etc. that have not changed since the previous survey. The surveyor should simply refer to an earlier report that is on file which described the procedures followed. The annual update report should include the findings from this most recent survey and point out clearly any significant differences that have been found in accordance with the general classification scheme presented earlier in Table 4.1.

Record Keeping

It is important for the surveyor to establish reasonable estimates of employee TWAs, at least within the framework of the classification system defined herein. However, it is equally important that adequate records of the actual measurements upon which the TWA estimations are based, including the results of calibration checks, be maintained.

Recall that the sound survey findings are not only for the purpose of satisfying governmental regulations, such as those of OSHA, but that they also may end up being involved in other legal proceedings such as workers' compensation hearings and negligence suits against the company. With respect to workers' compensation claims for noise-induced hearing loss, the records should be kept for the duration of employment for an employee plus the potential length of time after employment during which the employee could file a claim.

As a consequence, it is recommended that all sound survey records be kept indefinitely. Unfortunately, some companies that had conducted annual sound surveys for several years elected to discard all survey findings except for those obtained every five or ten years. The inability to establish the employee's TWA over exposure periods during which significant noise-induced hearing loss is possible, can be used against the company during compensation hearings. All data should be kept on file until such time as it is determined that no potential future use of the data exists.

Guidelines for Updating Survey Findings

The decision as to the need for updating sound survey findings will normally depend upon several factors including: (1) the severity of the potential noise hazard, (2) possible significant shifts in the predicted TWAs for specific job classifications, workers and work locations due to changes in the rate of production, equipment modifications, equipment additions or deletions, etc., (3) observed abnormal number of significant threshold shifts for the noise-exposed population or the finding of a potentially ineffective HCP based on ADBA procedures (Chapter 9), (4) legal requirements specified by federal or state OSHA programs, workers' compensation boards or governing agencies, and union contracts, or (5) the sound survey guidelines as specified by the company's HCP policy manual.

Our general recommendations for conducting and updating sound survey findings are as follows: For all employees, work stations and job classifications for which the initial sound survey yielded a TWA classification of B or C (85 to 94 dBA, Table 4.1), annual sound surveys are recommended.

Where the predicted TWA classification is D or higher (≥95 dBA), it is recommended during the first year of sound surveying that semi-annual sound surveys be conducted. After the completion of the initial survey and the subsequent two semi-annual surveys, then only annual sound surveys would be required. If, at an existing plant location, two or more annual sound surveys have been conducted, a semi-annual sound survey is not necessary if the existing sound survey data represent a consistent set of predicted classifications: that is, if less than 25% of the classification grades have changed one letter and if no classification grade has changed more than one letter.

There are several important reasons for conducting the initial semi-annual sound survey for predicted TWA exposures of 95 dBA or higher. First, the potential of the noise environment to produce significant noise-induced hearing loss over a relatively short time span, six months to a year, increases significantly for noise-sensitive employees whose predicted TWAs approach or exceed 95 dBA (ISO, 1983). Second, the guidelines for conducting sound surveys presented herein do not recommend extensive daily sampling procedures in order to correct possible low estimated TWAs at the time of the initial sound survey.

For plant locations, job descriptions and worker locations that were initially classified as A, it is recommended that these areas be resurveyed every five years unless of course there is reason to suspect that the characteristics of the noise environment have changed significantly, for example because of the addition of new equipment.

The End

The recommendations and guidelines presented herein are a framework for the development and implementation of a sound survey program. Real-world constraints necessarily play an important part in determining the final form of the company's survey efforts. As long as the participants in the HCP are being adequately protected from on-the-job noise exposures, then the sound survey efforts, along with all other hearing conservation efforts, must be judged adequate. Finally, it is often true that the impact of the sound surveyor's efforts on the level of protection provided by the company's HCP will depend not only upon the technical skills exhibited, but also upon the level of concern expressed by this individual toward the noise-exposed population.

Acknowledgments

The authors wish to acknowledge general industry for the sound survey experience gained over the years which provided the background knowledge needed for this chapter. The authors are also grateful for the help of all those persons who contributed to this chapter by criticizing draft versions.

Special thanks is due to Dr. Peter Bloomfield, Professor in the Department of Statistics, North Carolina State University, for his review of the statistical section of the chapter.

References

Alpaugh, E.L. (1975). "Sound Survey Techniques," in *Industrial Noise and Hearing Conservation,* edited by J.B. Olishifski and E.R. Harford, National Safety Council, Chicago, IL.

ANSI (1976). "Methods for the Measurement of Sound Pressure Levels," ANSI S1.13-1971(R1976), American National Standards Institute, New York, NY.

Behar, A. and Plener, R. (1984). "Noise Exposure — Sampling Strategy and Risk Assessment," *Am. Ind. Hyg. Assoc. J. 45(2)*,105-109.

Berger, E.H. (1985). "EARLog #15-Workers' Compensation for Occupational Hearing Loss," *Sound and Vibration 19(2)*,16-18.

Brief, R.S. and Confer, R.G. (1975). "A Commentary on Noise Dosimetry and Standards," Inter-Noise '75, Noise Control Foundation, Poughkeepsie, NY.

Bruce, R.D., Jensen, P., Jokel, C.R., Bolt, R.H., and Kane, J.A. (1976). "Workplace Noise Exposure Control — What Are the Costs and Benefits?," *Sound and Vibration 10(9)*,12-18.

Bruel, P.V. (1983). "Sound Level Meters — The Atlantic Divide," *Noise Control Engineering J. 20(2)*, 64-75.

Brusaw, C.T., Alred, G.J., and Oliu, W.E. (1976). *Handbook of Technical Writing*, St. Martin's Press, NY.

Christensen, L.S. (1974). "A Comparison of ISO and OSHA Noise Dose Measurements," Technical Review 4,14-22, Bruel & Kjaer Instruments, Denmark.

Delany, M.E., Whittle, L.S., Collins, K.M., and Fancey, K.S. (1976). "Calibration Procedures for Sound Level Meters to be Used for Measurements of Industrial Noise," NPL Acoustics Report Ac 75.

DiBlasi, F.T., Suuronen, D.E., Horst, T.J., and Bradley, W.E. (1983). *Statistical Audio Dosimeter Guide for Use in Electric Power Plants*, Empire State Electric Energy Research Corporation, New York, NY.

Erlandsson, B., Hakansson, H., Ivarsson, A., and Nilsson, P. (1979). "Comparison Between Stationary and Personal Noise Dose Measuring Systems," *Acta Otolaryngol, Suppl. 360*,105-108.

Gasaway, D.C. (1985). *Hearing Conservation: A Practical Manual and Guide*, Prentice-Hall, Englewood Cliffs, NJ.

GenRad (1976). *Instruction Manual: Type 1954 Personal Noise Dosimeter*, GenRad, Concord, MA.

Irwin, J.D. and Graf, E.R. (1979). *Industrial Noise and Vibration Control*, Prentice-Hall, Inc., Englewood Cliffs, NJ.

ISO (1982). "Draft International Standard ISO/DIS 1999: Acoustics-Determination of Occupational Noise Exposure and Estimation of Noise-Induced Hearing Impairment," International Standards Organization, Geneva, Switzerland.

Jones, C.O. and Howe, R.M. (1982). "Investigations of Personal Noise Dosimeters for Use in Coalmines," *Ann. Occup. Hyg. 25(3)*, 261-277.

Leidel, N.A., Busch, K.A., and Lynch, J.R. (1977). Occupational Exposure Sampling Strategy Manual, HEW (NIOSH) Publication no. 77-173, U.S. Department of Health, Education, and Welfare, Cincinnati, Ohio.

Mellott, F.D. (1978). "Noise Exposure Sampling: Use with Caution," Inter-Noise '78, Noise Control Foundation, Poughkeepsie, NY.

Natrella, M.G. (1963). *Experimental Statistics*, National Bureau of Standards Handbook 91, U.S. Department of Commerce, Washington, D.C.

OSHA (1974). "Occupational Noise Exposure; Proposed Requirements and Procedures," Fed. Reg. 39, 37773-37778, U.S. DOL, Occupational Safety and Health Administration, Washington, DC.

OSHA (1981). "Occupational Noise Exposures; Hearing Conservation Amendment," Fed. Reg. 46, 4078-4179, U.S. DOL, Occupational Safety and Health Administration, Washington, DC.

OSHA (1983). "Occupational Noise Exposure: Hearing Conservation Amendment; Final Rule," Fed. Reg. 48, 9738-9785. U.S.,DOL, Occupational Safety and Health Administration, Washington, DC.

OSHA (1984). "Chapter VI — Noise Survey Data," OSHA Industrial Hygiene Technical Manual, OSHA Instruction CPL 2-2.20A, Office of Health Compliance Assistance, U.S. DOL, Occupational Safety and Health Administration, Washington, DC.

Peterson, A.P.G. (1980). *Handbook of Noise Measurement*, Ninth Edition, GenRad, Inc., Concord, Massachusetts.

Royster, L.H. (1980). "Calibration of OSHA-NC Noise Survey Instrumentation," in *Noise Survey Procedures-Phase II,* edited by L.H. Royster. Proceedings of a special session at the fall 1980 meeting of the N.C. Regional Chapter of the Acoustical Society of America. Available from D.H. Hill Library, NCSU, Raleigh, N.C.

Royster, L.H., Royster, J.D., and Berger, E.H. (1982). "Guidelines for Developing an Effective Hearing Conservation Program," *Sound and Vibration 16(5)*,22-25.

Royster, L.H. and Royster, J.D. (1984). "Hearing Protection Utilization: Survey Results Across the USA," *J. Acous. Soc. Am. Suppl. 1, 74*,S5.

Royster, L.H. and Royster, J.D. (1985a). "An Overview of Effective Hearing Conservation Programs," *Sound and Vibration 19(2)*,20-23.

Royster, L.H. and Royster, J.D. (1985b). "Hearing Protection Devices," Chapter 6 in *Hearing Conservation in Industry,* edited by A.S. Feldman and C.T. Grimes, Williams and Wilkins, New York.

Shackleton, S. and Piney, M.D. (1984). "A Comparison of Two Methods of Measuring Personal Noise Exposure," *Ann. Occup. Hyg. 28(4)*,373-390.

Shadley, J., Gately, W., Kamperman, G.W., and Michael, P.L. (1974). "Guidelines for a Training Course in Noise Survey Techniques," National Academy of Sciences — National Research Council Committee on Hearing, Bioacoustics, and Biomechanics, Report of Working Group 70, Office of Naval Research Contract No. N00014-67-A-0244-0021.

Skrainar, S.F. (1985). The Effects on Hearing of Using a Personal Radio in an Environment Where the Daily Time-Weighted Average Sound Level is 87 dBA. M.S. Thesis, N.C. State University, Raleigh, N.C.

Skrainar, S.F., Royster, L.H., Berger, E.H., and Pearson, R.G. (1985). "Do Personal Radio Headsets Provide Hearing Protection?," *Sound and Vibration 19(5)*,16-19.

Snedecor, G.W. and Cochran, W.G. (1967). *Statistical Methods,* sixth edition, The Iowa State University Press, Ames, Iowa.

Statistical Graphics Corporation (1985). *STATGRAPHICS*™ Statistical Graphics System, STSC, Inc., Rockville, MD.

Steel, R.G.D. and Torrie, J.H. (1980). *Principles and Procedures of Statistics — A Biometrical Approach* (2nd Edition), McGraw-Hill, New York, NY.

Svensson, J. (1978). "Dosimeter Response to Impulsive Noise — Measurement Errors and Their Consequences," Inter-Noise '78, Noise Control Foundation, Poughkeepsie, NY.

Walker, D.G., Jr. (1979). Noise Control Efforts in the Dairy Packaging Industry, M.S. Thesis, N.C. State University, Raleigh, N.C.

Wells, R. (1979). "Noise Measurements: Methods," in *Handbook of Noise Control,* C.M. Harris, ed., McGraw Hill Book Company, New York, NY.

Yerges, L.F. (1979). "Do We Correctly Measure Worker Noise Exposure?," *Sound and Vibration 13(5)*,8-12.

Noise and Hearing Conservation Manual, edited by
E.H. Berger, W.D. Ward, J.C. Morrill and L.H. Royster

5 Anatomy & Physiology of the Ear: Normal and Damaged Hearing

W. Dixon Ward

Contents

Page

The Normal Ear 177
Outer Ear 177
Middle Ear 178
Inner Ear 179
Pathology of Hearing Damage 181
Auditory Sensitivity 184
Loudness 186
Hearing Threshold Level 188
0 dB HL: Typical Hearing or Normal Hearing? 189
Presbyacusis, Sociacusis and Nosoacusis 190
Age Correction Curves 192
References 195

The Normal Ear

The human ear is the special organ that enables man to hear, transmitting the information contained in sound energy to the brain, where it is perceived and interpreted. Because the deleterious effects of excessive noise occur within the ear, some knowledge of its structure and function is desirable. Figure 5.1 shows the ear in cross section.

The action of the ear involves three stages: (1) modification of the acoustic wave by the outer ear, (2) conversion of the modified acoustic wave to vibration of the eardrum that is transmitted through the middle ear to the inner ear, and (3) transformation of the mechanical movement to nerve impulses.

OUTER EAR

The outer ear consists of the pinna (commonly called "the ear") and the auditory canal, an open channel leading directly to the eardrum (tympanic membrane). Both the pinna and the auditory canal modify the acoustic wave, so that the spectrum of the sound impinging on the eardrum is not quite the same as the sound that originally reaches the pinna. The pinna is to some extent a "collector" of sound; because of the exact dimensions of the convolutions of the pinna, certain sound frequencies are amplified, others attenuated, so that each individual's pinna puts its distinctive imprint on the acoustic wave progressing into the auditory canal. This information is used in the recognition and localization of sounds.

A very important role in modifying the acoustic wave is played by the auditory canal. Because of its shape and dimensions, it greatly amplifies sound in the 3-kHz region. The net effect of the head, pinna and ear canal is that environmental sounds in the 2-4-kHz region are amplified by 10 to 15 decibels. It is not surprising, therefore, that sensitivity to sounds is greatest in this region and that noises in this frequency range are the most hazardous to hearing.

MIDDLE EAR

The eardrum, or tympanic membrane, in addition to providing protection against invasion of the middle and inner ears by foreign bodies, vibrates in response to the pressure fluctuations in the sound wave, and this vibration is transmitted by the small bones of the middle ear—the malleus, incus and stapes (hammer, anvil and stirrup, respectively)— to the oval window at the entrance to the inner ear. The function of the middle ear is to efficiently transform the motion of the eardrum in air to motion of the

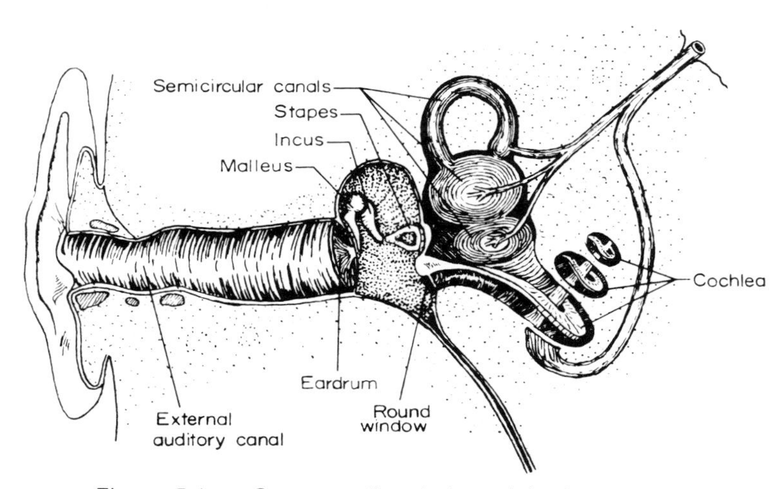

Figure 5.1. — Cross-sectional view of the human ear.

stapes, which must in turn drive the fluid-filled inner ear. When sound travels from one medium to another, only part of the energy will be transmitted, and this fraction is lower, the greater the difference between the media in density. Without the transformer action of the middle ear, only about 1/1000th of the acoustic energy in air would be transmitted to the inner-ear fluids, a loss of 30 decibels.

The middle ear enhances the energy transfer in two main ways. First, the area of the eardrum is about 17 times as large as that of the oval window (stapes footplate), so that the effective pressure (force per unit area) is amplified by this amount. Second, the ossicles are so constructed that they produce a lever action that further amplifies the pressure. As a result, most of the energy entering the normal ear via the eardrum is transmitted to motion of the stapes and hence to stimulation of the inner-ear system.

The middle ear also has two muscles that are attached to the malleus and the stapes: the tensor tympani and the stapedius muscles, respectively. These muscles, which act in opposite directions, are activated either by the process of vocalization (speaking or singing) or by loud sounds (above about 80 dB SPL), which results in stiffening the whole middle-ear system, thereby reducing the transmission of low-frequency energy (500 Hz and below). Although this muscular activity shows partial adaptation with time, it provides some protection against sustained high-intensity noise. However, because about a tenth of a second is needed for the muscles to reach full contraction, they are of no importance in protection against sudden impulsive noises such as gunfire.

A third important feature of the middle ear is the eustachian tube, a channel that is connected to the nasal air passages by means of a valve that is supposed to open periodically in synchrony with swallowing. When this does not occur, a difference between the static middle-ear pressure and the outside pressure may develop, so that the eardrum may be displaced inward or outward; in either case, the efficiency of the middle ear is reduced, and less energy will be transmitted to the inner ear.

INNER EAR

The sensory receptors actually responsible for the initiation of neural impulses in the auditory nerve consist of approximately 25,000 hair cells in the inner ear, or cochlea. The cochlea, deeply embedded in the temporal bone, is a coiled tube in the form of a snail with three spiral turns. Two membranes, Reissner's membrane and the basilar membrane, run the length of this spiralling tube, dividing it into three compartments: scala vestibuli, scala media, and scala tympani. Figure 5.2 shows a cross-sectional view of one turn of the cochlea. The hair cells are the most important component of the organ of Corti, which rests on the basilar membrane. A third membrane, the tectorial membrane, lies on top of the fine hairs that extend from each hair cell.

In the normal cochlea, initiation of neural impulses occurs when the basilar membrane moves up or down; because a shear force is developed between the tectorial membrane and the organ of Corti, the hairs of the hair cells will be bent, and neural impulses will be developed in one or more of the some 31,000 sensory neurons that are connected to the hair cells.

Let us return to the progression of acoustic energy to the inner ear. When the stapes footplate, set in the oval window that opens onto the scala vestibuli, begins to move in response to an incoming wave, the resulting pressure is transmitted throughout all the fluids of the inner ear almost instantaneously. However, since these fluids, unlike air, are essentially incompressible, the footplate can actually move inward only if something else moves outward. In the cochlea this is accomplished by an elastic membrane, the round window, which separates the scala tympani from the middle ear. When the oval window is pushed inwards into scala vestibuli, the round window will bulge outward, and vice versa. This involves the actual movement of fluid from scala vestibuli to scala tympani, so there is a

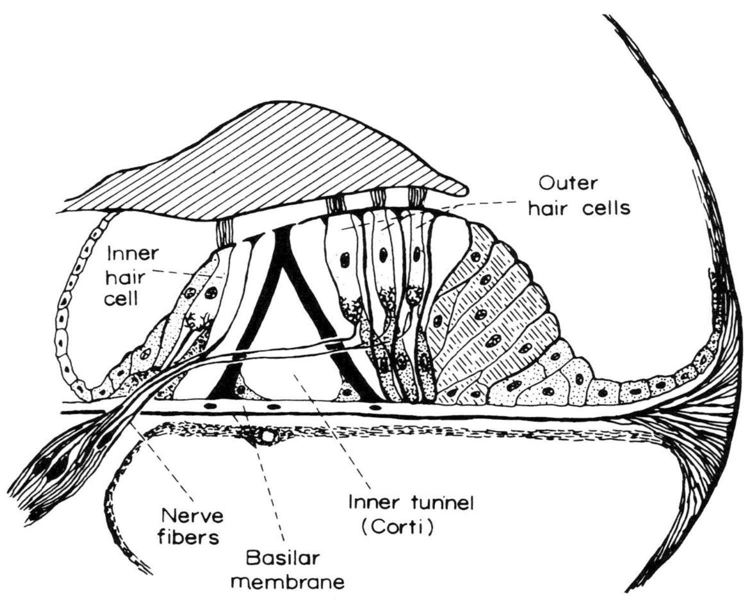

Figure 5.2. — Cross-sectional view of the cochlea.

corresponding force tending to deflect the basilar membrane toward scala tympani ("downward" in Figure 5.2). This force initiates a wave of movement that begins near the oval window (at the basal end of the cochlea) and travels away from it (toward the apex of the cochlea, or apically) because the basilar membrane, loaded by the organ of Corti, has different physical characteristics along its length, being narrower and stiffer at the base than at the apex. The amplitude of this wave of movement gradually builds up to a maximum and then rapidly dies out. The distance at which the movement reaches its maximum value depends on the speed with which the oval window moved: if it were pushed in very rapidly, as would be the case with high-frequency acoustic stimulation, the maximum displacement will occur near the base of the cochlea; with slower movement, as in low-frequency stimulation, the maximum will occur farther toward the apex.

Deflection of the basilar membrane and the resultant bending of the hair cells leads to initiation of nerve impulses, as described earlier. There are two main characteristics of the pattern of neural firing aroused by a pure tone. First, the firing of any particular fiber will be synchronized with the frequency of the tone: if the pure tone has a frequency of 100 Hz, then the neural unit will be fired at most 100 times each second, or once every 10 milliseconds. If it fails to be fired by one maximum of the input wave, it may be fired by the next one; in any event, successive firings will be separated by 10 msec, or 20, or 30, or 40, etc. Second, the firing will be localized along the basilar membrane, high frequencies giving rise to maximum activity near the base, and low frequencies giving rise to maximum activity near the apex, although this pattern becomes increasingly broad as the intensity of the sound is raised.

Both the place of maximum neural excitation and the frequency (timing) of neural discharge represent information about the acoustic stimulus that somehow is used by the central nervous system in interpreting the overall pattern of neural activity. Place is more important for high frequencies, timing for low frequencies, with both principles apparently being active in the range from 200 to 4000 Hz. However, the exact process whereby a particular pattern of activity in 31,000 different nerve fibers, transmitted from the cochlea to the cerebral cortex via several intermediate neural centers, gives rise to a specific sensation is still unknown.

Pathology of Hearing Damage

Normal hearing, then, requires a normal outer, middle and inner ear. Abnormality or malfunction of the outer and middle-ear systems will generally decrease the amount of sound energy conducted to the inner ear; the term *conductive* hearing loss is therefore applied to such conditions. Conductive hearing loss is associated with excessive ear wax, a ruptured or heavily-scarred eardrum, fluid in the middle ear (otitis media), dislocated or missing elements of the ossicular chain, or otosclerosis, which is an abnormal growth of bone in the middle ear. Employment-related conductive hearing losses are not common, although they may occur occasionally as the result of an accident (eardrum rupture or ossicular chain disarticulation by a head blow, an explosion, or a rapid pressure change in a decompression chamber; penetration of the eardrum by a sharp object or fragment of metal).

Because conductive hearing losses are generally reversible, either by medical or surgical means, they are not as serious as *sensory* hearing losses associated with irreversible damage to the inner ear, or *neural* deficits due to damage to higher centers of the auditory system. Noise produces primarily sensory hearing loss. Study of the inner ears of experimental animals exposed to noise has shown that after even moderate exposures, subtle effects may be observed such as twisting and swelling of the hair cells, disarray of the hairs on top of the hair cells, detachment of the tectorial membrane from the hairs, and reduction of enzymes and energy sources in the cochlear fluids; these are all conditions that would reduce the sensitivity of the hair cells to mechanical motion. The system at this point is in a state of auditory fatigue: in order to initiate neural activity, more acoustic energy must enter the cochlea than before the noise exposure, and the mechanical motion of the cochlear partition is greater.

As the severity of the noise exposure increases, these changes increase in degree and eventually become irreversible: hairs become fused into giant cilia or disappear, hair cells and supporting cells disintegrate, and ultimately even the nerve fibers that innervated the hair cells disappear (Lim and Dunn, 1979). Figures 5.3A and 5.3B illustrate the appearance of a normal cochlea and one severely damaged by noise exposure, respectively.

These effects are accentuated in acoustic trauma, the aftermath of a single noise exposure of relatively short duration but very high intensity, such as an explosion. In this case the picture that emerges is that of a system that has been vibrated so violently that its "elastic limit" has been exceeded. Attachments of the various elements of the organ of Corti are disrupted, hair cells are torn completely from the basilar membrane on which they normally rest, and a temporary rupture of the reticular lamina may occur, allowing intermixture of fluids within the cochlea and thus poisoning those hair cells that may have survived the mechanical stress of the explosion.

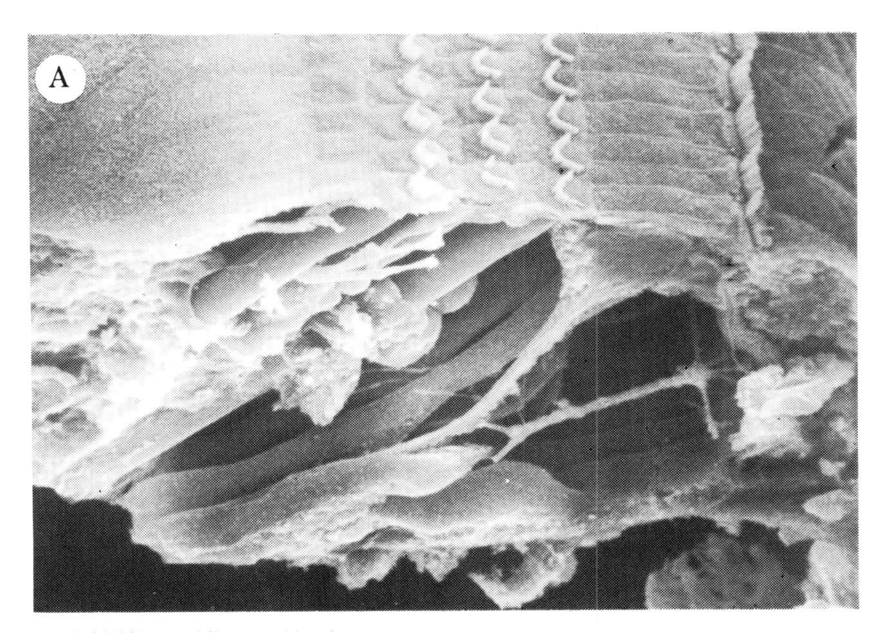

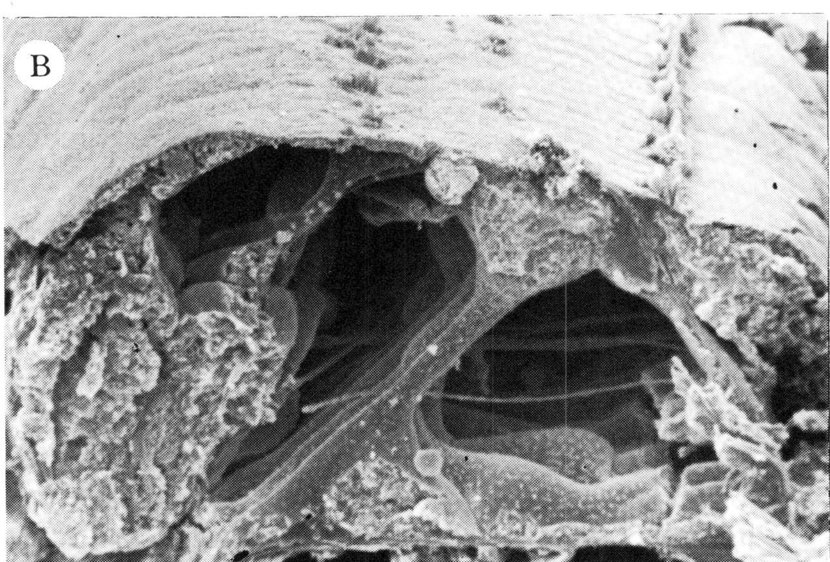

Figure 5.3 — Scanning electromicrographs of (A) the normal cochlea, and (B) a cochlea severely damaged by noise exposure. Note the disarray of the hairs of the inner hair cells (IHC) and the near-complete destruction of outer hair cells (OHC). Photographs courtesy I. Hunter-Duvar.

Auditory Sensitivity

Irreversible damage to the cochlea as described above cannot, in man, be measured directly. Loss of hearing has therefore traditionally been assessed by measuring auditory sensitivity, the ability to detect weak pure tones. The auditory absolute threshold is defined as the level of a sound that can just be heard, in quiet surroundings, some specific percentage of the time, generally 50%. In the normal ear, the localization of activity on the basilar membrane discussed earlier is accentuated in the case of weak tones. It appears that a weak 1000-Hz tone, for example, will actually trigger the release of energy stored in the organ of Corti, but only in a very restricted area, so that only the receptors tuned to 1000 Hz, and not those tuned to 950 to 1050 Hz, for example, will respond vigorously. Because of this specificity of response, the auditory capability of an individual ear is usually assessed by measuring its sensitivity to these weak pure tones, even though there is only a moderate correlation between hair-cell destruction and loss of auditory sensitivity in individual ears.

The auditory threshold depends on all the conditions under which it is tested, including the test procedure used, the instructions to the listener, the listener's tendency to guess when in doubt, the duration of the test tone, the amount of internal and external noise that may mask the test tone, and even on the procedure that is used to determine the intensity of the tone that was just barely heard. Two classical examples of sensitivity to long sustained pure tones, in which two different methods were used to specify the average sound level that was just heard by a group of people with presumably "normal" hearing, are shown in Figure 5.4. For the solid curve, thresholds were determined by having the test persons listen with open ears in a completely quiet room (Sivian and White, 1933; Berger, 1981). The average values shown represent the just-audible sound levels measured at the position of the center of the listener's head, but with the listener removed from the situation; such a sensitivity curve is called a minimum audible field (MAF). The dashed curve, by contrast, represents an experiment in which the tones were presented to the listeners by means of an earphone (Sivian and White, 1933). The sound level in this case is the output of a microphone at one end of an "artificial ear," a device designed to stimulate the human ear, when the earphone, positioned at the other end of this artificial ear, is driven with the same electrical signal that produced a just-audible signal. This sensitivity curve is called a minimum audible pressure (MAP) curve. For these persons, sounds down to 20 Hz and up to 20 kHz in frequency can be heard, although only at high intensities at these extremes. The range of hearing is therefore often said to be 20 to 20,000 Hz.

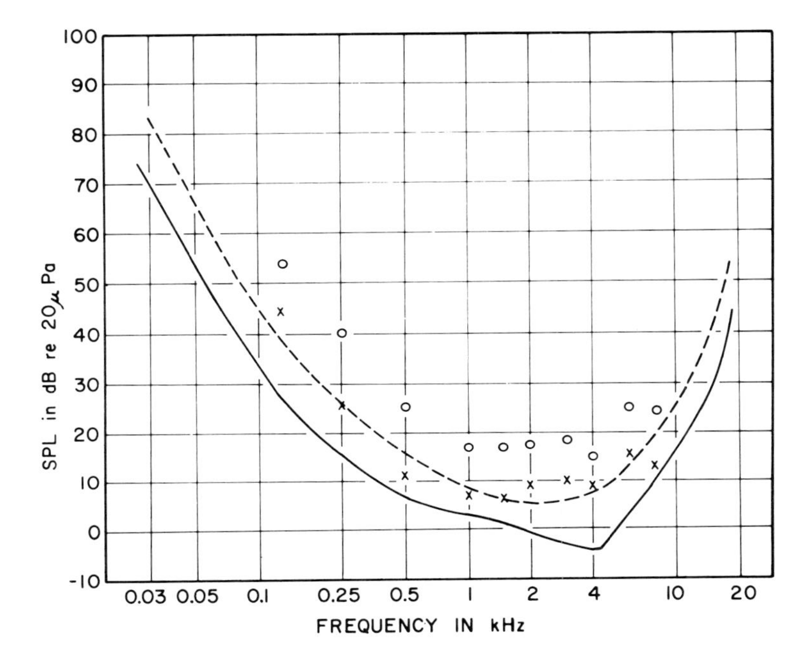

Figure 5.4. — Auditory thresholds associated with "normal" hearing. The solid curve indicates sensitivity to pure tones for young adults listening with open ears (minimum audible field), the dashed curve listening with earphones (minimum audible pressure). Circles represent the SPLs that defined 0 dB HL at standard audiometric frequencies under the old audiometric standard (ASA 1951); crosses denote 0 dB HL under the present standard (ISO 1964; ANSI 1969).

In neither of the curves of Figure 5.4 do the values indicate the actual sound level at the eardrum, *i.e.*, the sound being transmitted to the cochlea. The MAF curve is influenced by the amplification of the signal provided by the resonance characteristics of the head and outer ear, so that the occurrence of the lowest threshold at 3000 Hz does not imply that the 3000-Hz hair cells are intrinsically more sensitive than say the 1000-Hz receptors, but merely that sound of that frequency more easily reaches the inner ear. The 3000-Hz minimum is absent in the MAP measurements, because putting on an earphone not only eliminates the effect of head and pinna on the sound, but also radically alters the effect of the ear canal. More recent measurements of the sound level actually present at the eardrum indicate that field and earphone listening result in the same values for threshold, except at low frequencies, where MAP measures are elevated because wearing an earphone increases the level of low-frequency body noises, which may produce some masking (Killion, 1978).

Loudness

Because the auditory mechanism is less sensitive to extremely high and low frequencies, the loudness of audible sounds in those frequency regions will in general be less than that of middle-frequency sounds of the same sound presure level. For example, Figure 5.4 implies that a tone of 40 dB SPL will be heard clearly by a normal individual when its frequency is 1000 Hz, but will be barely audible if its frequency is 100 Hz, and will arouse no sense of loudness at all if it is a 50-Hz tone, as it will be inaudible. Therefore if it is desired to develop an instrument that will indicate the loudness of sounds rather than their physical intensities, it will be necessary to somehow deemphasize the sound components having these extreme frequencies. In order to accomplish this, equal-loudness contours must be determined: curves that indicate the intensity and frequency of tones judged to be equal in loudness. If one can assume that, at threshold, all tones sound equally loud, both of the curves of Fig. 5.4 are equal-loudness contours—one for earphone listening, the other for listening with open ears. Other contours can be determined by asking listeners to adjust a tone of some particular frequency to have the same loudness as a standard tone of a fixed frequency and intensity.

Equal-loudness contours for suprathreshold tones are shown in Figure 5.5. These are open-ear field data, because most judgments of loudness of noises in real life are not made while wearing earphones. Each contour defines tones that have been judged equal in loudness to a 1000-Hz tone of a specific SPL; the *loudness level* of a particular contour, in *phons,* is numerically equal to the SPL of the 1000-Hz standard. Thus, for instance, the following tones each have a loudness level of 40 phons: 1000 Hz at 40 dB SPL (by definition), 4000 Hz at 37 dB SPL, 8000 Hz at 51 dB SPL, 100 Hz at 60 dB SPL, and 50 Hz at 70 dB SPL. So if the sound-measuring device is supposed to indicate the relative loudness of rather weak sounds, it is necessary to reduce the 50-Hz components of the sounds by 30 decibels (70–40), the 100-Hz components by 20 dB, and the 8000-Hz components by 10 dB, while the 4000-Hz components should be amplified by 3 dB, relative to 1000 Hz. The A-weighting network on the standard sound-level meter does apply these adjustments, so that when the sound-level meter is set at "A-weighting," the reading on the meter, in dBA, will correctly indicate the overall loudness level of the weak sound in question.

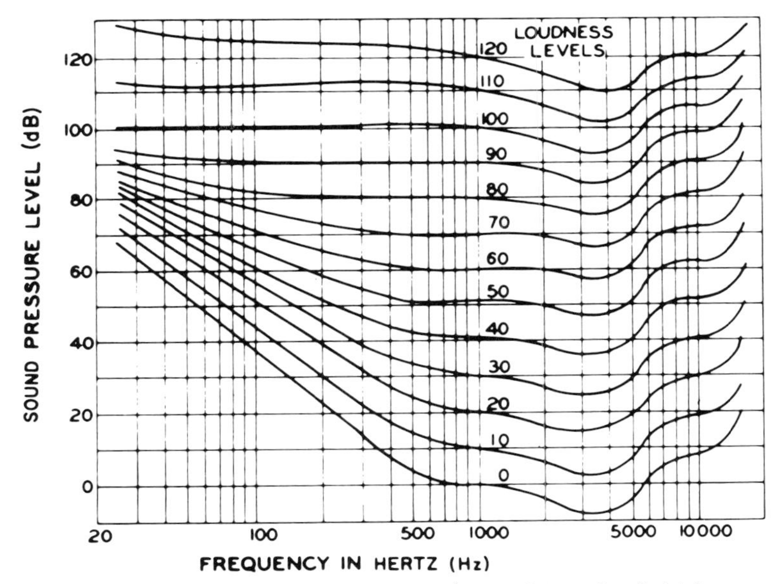

Figure 5.5. — Equal-loudness contours of pure tones for field (open-ear) conditions. The numbers indicate the loudness level, in phons, of the tones that fall on each contour.

If the contours for equal loudness were parallel, A-weighting could serve to indicate the relative loudness of sounds at all levels. Unfortunately, Figure 5.5 shows clearly that the auditory system of humans does not behave in this way. By the time the level of components reach 100 dB SPL, equal SPLs give rise to nearly equal loudnesses, except for the greater efficacy of the 3000-4000-Hz components associated with the resonance of the outer ear at these frequencies. So to evaluate the overall loudness of very loud sounds, a different weighting network, the C-weighting, one that provides little attenuation or amplification of different frequencies, is used. Finally, B-weighting is designed to indicate the loudness level of "moderately loud" sounds, and so has filtering characteristics that tend to agree with the 70-phon contour of Figure 5.5.

As indicated in Chapter 3, the use of A-weighting to characterize the relative hazard of high-intensity complex noises would appear paradoxical at first glance, since A-weighting was designed to assess weak sounds. However, the hazardousness of a sound is not indicated by its loudness. The empirical fact is that low frequencies are less hazardous, for whatever reason, even at 100 dB SPL, than the middle frequencies, and of the various weightings available on the sound-level meter, A-weighting happens to deemphasize the low frequencies by approximately the right amount.

It is worth noting that loudness levels indicate only an ordinal relation between sounds. If one sound, X, has a loudness level of 50 phons, while another sound, Y, has a loudness level of 40 phons, then we know only that X will be judged louder than Y, but not how much louder; it is not true, for instance, that the louder is 50/40=1.25 times as loud as the softer. Similarly, the fact that the 50-phon sound has about 10 times as much intensity as the 40-phon sound does not imply that it would be judged to be 10 times as loud. The fact of the matter is that most people would say that two sounds that differ in loudness level by 9-10 phons have a loudness ratio of about 2:1. That is, a sound that is 10 phons higher in loudness level than another will appear to be twice as loud. The *sone* has been defined as the loudness perceived by a normal listener when presented with a 1000-Hz tone at 40 dB SPL (40 phons), so a 50-phon sound would have a loudness of 2 sones, a 60-phon sound a loudness of 4 sones, and so on.

Hearing Threshold Level

Returning to the measurement of auditory threshold, it must be noted that sound levels at the eardrum are difficult to measure, and MAF determinations require a large sound-free room, so threshold measurements have been, and probably will continue to be, expressed in terms of MAP, although in a somewhat indirect manner. Instead of sound levels relative to a reference pressure of 20 μPa and a reference intensity of 10^{-12} W/m^2 (both of which are implied by the term "dB SPL") as in Figure 5.4, an individual's sensitivity is expressed in terms of "Hearing Threshold Level" (HTL): the number of decibels by which, to be heard, a pure tone at a particular frequency must be raised above a reference SPL, specific to the test frequency concerned, designated 0 dB HL (Hearing Level). Audiometers, the instruments with which thresholds are measured using earphones, are calibrated so that when the HL dial is set to zero, this reference level is generated by the earphone. To illustrate, if the reference SPL at a particular frequency were 10 dB SPL, and a worker requires an SPL of 25 dB to hear this tone, then he could do so only when the HL dial was set at "15" or higher; the worker would be said to have a Hearing Threshold Level (HTL) of 15 dB because he can just hear a tone whose Hearing Level is 15 dB. For reasons that have been lost in antiquity, the graph relating HTL to frequency, the audiogram, is so constructed that 0 dB (hearing ability corresponding to the reference) is at the top of the graph, with HTL increasing in the downward direction. Therefore the audiogram is, in essence, an inverted representation of the relative auditory sensitivity of the individual ear as a function of frequency.

0 dB HL: TYPICAL HEARING OR NORMAL HEARING?

Only one major step would appear to remain in order to standardize measurement of auditory sensitivity: selection of suitable values of SPL to represent 0 dB HL at the audiometric frequencies. Unfortunately, this has not proved as easy to accomplish as it might appear, because of a lack of agreement, among those who set standards, as to whether this hearing standard should represent "typical" hearing or "normal" hearing. The first attempt to establish 0 dB HL adopted the former objective, using values of the average threshold SPLs of 20-29-year-old persons in a random sample of the U.S. population tested in 1934 (Beasley, 1938). These values for frequencies from 125 to 8000 Hz are shown by the open circles in Figure 5.4, and are referred to as "0 dB HL (ASA 1951)" because they were approved by the American Standards Association in 1951. For several reasons, including the fact that nobody was excluded from this sample, so that people with damaged hearing were included, these sound levels were somewhat higher than they otherwise would have been. Consequently, most young adults showed negative values of HTL—that is, they had hearing that was better (more sensitive) than 0 dB on the audiometer dial. For example, the group whose MAP thresholds are shown by the dashed curve in Figure 5.4 would have negative HTLs at all frequencies.

The fact that young people with normal ears showed negative values of HTL, coupled with the fact that in those days "HL" was often referred to as "Hearing Loss," was a source of distress to those who believed that 0 dB HL should mean "normal; in its unsullied state; absolutely unaffected by anything." How, they complained, can anyone have better than normal hearing? So in 1954 British standardizers proposed a new set of reference SPLs that were about 10 dB lower than ASA 1951, and these were adopted, with minor changes, by the International Standards Organization in 1964 and by the American National Standards Institute in 1969. These reference SPLs (ISO 1964 or ANSI 1969) are shown in Figure 5.4 by the crosses. It must be noted that both the ASA and ISO—ANSI values apply only to one particular earphone-cushion combination, namely, TDH-39 earphones with MX41/AR cushions; for other earphones and/or cushions, the SPLs that produce an equivalent effect may differ by a few decibels.

Unfortunately, the new ISO—ANSI norms are modal values of thresholds, gathered under optimum laboratory conditions, of a very highly selected group of young intelligent listeners. Excluded from consideration were not only persons who, as far as could be determined by questionnaire, might be suffering from hereditary hearing loss, or had been exposed to high-intensity noises or various diseases known to sometimes produce deterioration of hearing, but also those whose sensitivity was, for unknown reasons, worse than the majority of other listeners—that is, they were rejected "because they obviously are not normal." The results of this questionable procedure is that the ISO audiometric standard is as unrealistically stringent as the ASA is unrealistically lax, so that now the typical young adult with no prior exposure to industrial noise will show *positive* values of HL.

This fact is particularly crucial when audiometric survey data are used to infer whether or not a particular work environment is causing damage to hearing. For example, if a group of 18-25-year-old workers in a particular noise environment are found to have average HTLs (ISO) of 3 or 4 dB, this is completely "normal" (here, in the sense of "to be expected"), and hazard has not been demonstrated.

The same uncertainty over the hazard associated with a particular work environment is even more accentuated in the case of older workers. Loss of auditory sensitivity can result from many causes other than industrial noise; therefore, the older the group of workers in question, the greater will be the net effect of these other causes, and the greater must the average HTL be in order that the industrial noise exposure be judged hazardous. Just how great the HTLs must be to justify the inference that a hazard exists is a problem that forces a closer examination of the causes of hearing loss.

PRESBYACUSIS, SOCIACUSIS AND NOSOACUSIS

Hearing loss may be categorized not only as being conductive, sensory, or neural, but also in terms of possible cause. One suitable system of classification (Ward, 1977) divides sensory hearing loss into the following:

(1) ***presbyacusis,*** loss caused by the aging process per se;
(2) ***noise-induced hearing loss,*** which must itself be divided into
(2a) ***industrial hearing loss*** caused by work-related noise exposure, and
(2b) ***sociacusis,*** losses that can be ascribed to noises of everyday life; and
(3) ***nosoacusis,*** losses attributable to all other causes, such as hereditary progressive deafness; diseases such as mumps, rubella, Meniere's disease; ototoxic drugs and chemicals; barotrauma; and blows to the head.

Although these causes of sensory hearing loss can be studied separately in laboratory animals, it is obvious that in man they are inextricably mixed.

Therefore, in order to determine how much damage has been produced by a particular industrial noise environment, the hearing of the workers concerned must be compared with that of a control group of individuals who have never worked in a noisy industry, but who are matched to the workers not only in age but also in histories of exposure to sociacusic and nosoacusic influences.

Ideally, for each noise-exposed worker there should be a control subject who is the same age, has the same noisy hobbies, is exposed to the same diseases and industrial chemicals, does as much hunting as the worker in question, and so on. However, such a matched-pair procedure would be so expensive—if, indeed, it is even possible—that it has never been attempted. In practice, the thresholds of the workers must be compared either with those of a non-industrial-noise-exposed group of employees of the same age in the same industry or with a set of HL norms that purport to indicate "typical" thresholds for persons of that age.

Use of the non-noise-exposed employees of the same industry has the advantage that they will have been tested with the same procedure and under the same conditions as the noise-exposed workers. In addition, since they live in the same area, some of the sociacusic and nosoacusic influences may, on the average, be nearly the same for both groups. However, this equivalence should not be taken for granted. All persons must be administered the same questionnaire, so that it can be shown that it is not mere conjecture that the groups are comparable in exposure to chain saws, gunfire, loud music, mumps, ototoxic drugs, head blows, and so on. Experience has shown that they will not be comparable if there is a disparity in sex between the groups: men are, on the average, exposed to more sociacusic influences than women, so they will inevitably have higher average HTLs than women at frequencies above 1000 Hz. Care must therefore be taken to ensure that the control group has the same proportion of males as the noise-exposed workers; to compare the hearing of male steel workers with that of secretaries would clearly be inappropriate.

The other alternative for making a comparison of HTLs, *i.e.,* using a set of normative curves, is even more likely to lead to invalid results. To avoid coming to erroneous conclusions, it must be the case that all details of the audiometric procedures used to test the noise workers are the same as those used to test the persons on whom the norms were established: type of audiometer, instructions, procedure, order of frequencies tested, criterion for defining threshold, etc. In addition, sociacusic and nosoacusic influences in the two groups are unlikely to be equivalent. Of the many proposals for "Age-corrected HL norms" that have been suggested, most differ from each other in the rules of exclusion of individuals on the basis of their history. Indeed, often these rules were so vague (*e.g.*, "otological abnormality" or "excessive exposure to gunfire") that an equivalent exclusion rule cannot be applied to the data from the industrial workers.

AGE CORRECTION CURVES

In an attempt to circumvent these problems, at least in part, it has been common practice to derive average "age correction" curves based on studies on non-industrial-noise-exposed populations by comparing, in each study, the median hearing of a particular age group to that of the 20-year-old individuals. The main assumption here is that in that age group, all of the populations should have the same median thresholds, so that any differences in measured HTLs between the 20-year-olds in different surveys that are found may be attributed to differences in audiometric calibration or technique.

Such a set of age correction curves, for males and females separately, was derived by Spoor in 1967 from all the data then available. Although sometimes called "presbyacusis corrections," they did involve also some sociacusis and nosoacusis. Subsequent studies of non-industrial-noise-exposed populations have generally substantiated their accuracy. Unfortunately, however, these curves have often been erroneously indicated to be "Hearing Levels"(rather than "Age Corrections Relative to Age 20.") Such a label will be correct only if the median 20-year-old in the non-industrial-noise-exposed population concerned, tested with the audiometric procedures actually involved, has HTLs of precisely 0 dB. As indicated earlier, this will be the case only for audiometric tests, under optimum conditions, of a highly screened group of listeners. All actual field studies of non-industrial-noise-exposed 20-year-olds have found median HTLs of at least 3-5 dB at most test frequencies. It is clear, therefore, that tables or graphs that purport to show realistic HTLs to be expected in a group of workers at a particular age if their industrial noise exposures have had no effect must consist of Age Corrections *added to the actual HTLS of the 20-year-olds.*

Such a set of Expected Hearing Level norms are shown in Figures 5.6 and 5.7. Here the Spoor age corrections have been combined with the median actual thresholds of several field studies of the hearing of 20-year-old persons as they enter the work force (Glorig and Roberts, 1965; NIOSH, 1972; Burns *et al.*, 1977; Royster and Thomas, 1979; Gatehouse *et al.*, 1982). These contours indicate the most probable values of HTL, at a particular age, to be expected in males and females who have not been exposed to workplace noise and who are otologically normal, in the sense that the ears appear normal to visual inspection (*e.g.*, no excessive ear wax or scarring of the eardrum) and that there is no history of severe ear problems linked with some particular obvious cause.

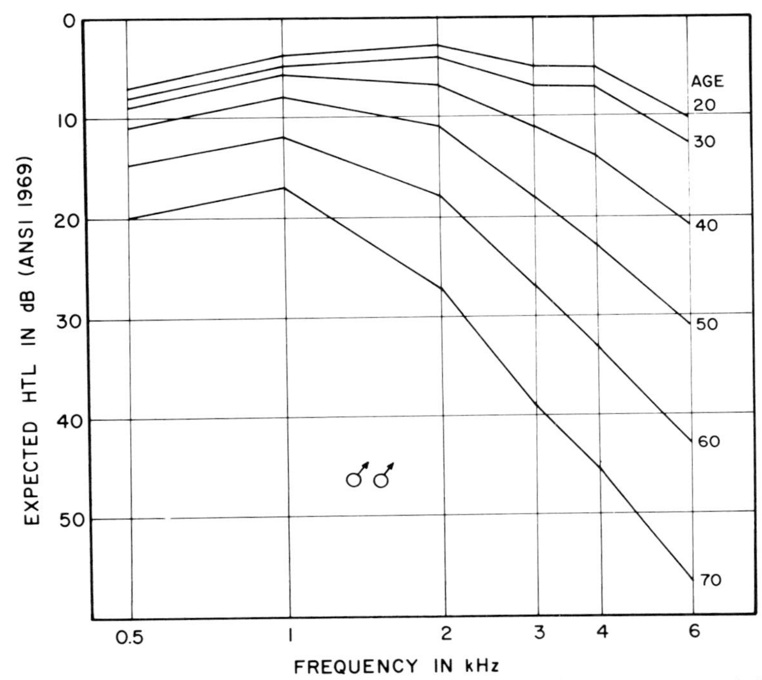

Figure 5.6. — Expected Hearing Threshold Levels in an industrialized society, for males not exposed to workplace noise, as a function of age. Average of right and left ears. Age corrections after Spoor (1967) have been added to actual median HTLs of the 20-year-old non-noise-exposed workers.

As indicated earlier, sets of norms similar to those of Figures 5.6 and 5.7 will differ if they are based on different rules of exclusion. For example, two random samples of the hearing of the population of the USA (Glorig and Roberts, 1965; Rowland, 1980) show HTLs for age groups above 20 years that are slightly higher in value than those in Figures 5.6 and 5.7, because these surveys included individuals with industrial noise exposure. On the other hand, a study of non-noise-exposed workers was made by NIOSH in which an attempt was made to exclude individuals with "no significant noise exposure on the job, off the job, or during military service;" thus, strenuous efforts were made to reduce sociacusis from the sample, and the values of expected HTL are somewhat smaller than in Figures 5.6 and 5.7. These NIOSH curves, expressed in tabular form, have been incorporated into federal rules regulating noise exposure as non-mandatory Appendix F of OSHA Rule 29 CFR 1910 (see Appendix I).

However, if these tables are to be used correctly, the criteria for determining what constitutes "significant noise exposure off the job" must be made public so that similar exclusions can be applied to the workers in question.

Eventually, a survey of a random sample of the total population of the USA will be performed in which a detailed history of each individual's exposure to all auditory hazards is taken. If those exposed to industrial noise, and only those, are then eliminated, the remainder will provide a firmer basis on which to estimate the effect of noise in a group of workers by eliminating sociacusis and nosoacusis. The resultant curves, however, will probably differ only slightly from those in Figures 5.6 and 5.7.

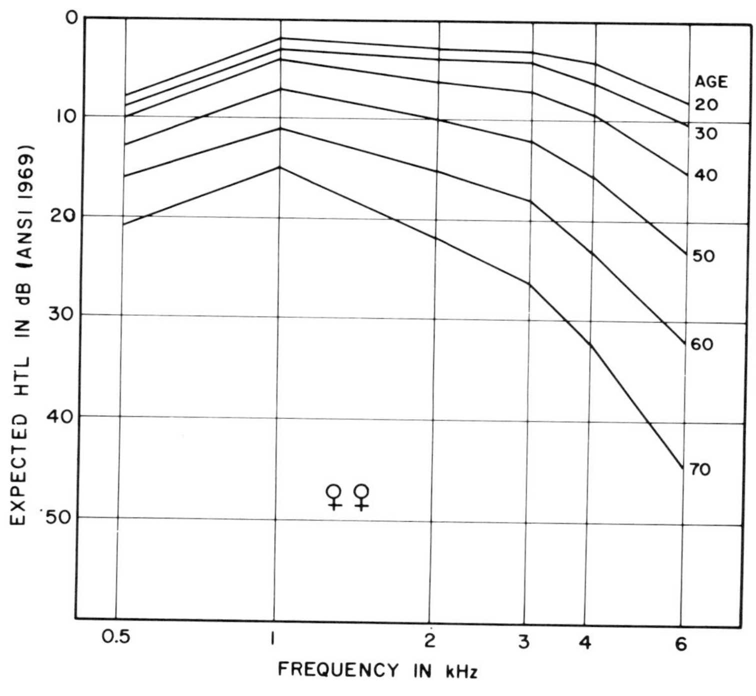

Figure 5.7. — Expected Hearing Threshold Levels in an industrialized society, for females not exposed to workplace noise, as a function of age. Average of right and left ears. Age corrections after Spoor (1967) have been added to actual median HTLs of the 20-year-old non-noise-exposed workers.

References

ANSI (1969). "Specifications for Audiometers. S3.6-1969," American National Standards Inst., New York, NY.

ASA (1951). "American Standard Specification for Audiometers for General Diagnostic Purposes, Z24.5-1951," American National Standards Institute, New York, NY.

Beasley, W.D. (1938). "National Health Survey (1935-36)," Preliminary Reports, Hearing Study Series, Bulletins 1-7. U.S. Public Health Service, Washington, DC.

Berger, E.H. (1981). "Re-examination of the Low-frequency (50-1000 Hz) Normal Threshold of Hearing in Free and Diffuse Sound Fields," *J. Acoust. Soc. Am. 70(6),* 1635-1645.

Burns, W., Robinson, D.W., Shipton, M.S. and Sinclair, A. (1977). "Hearing Hazard from Occupational Noise: Observations on a Population from Heavy Industry," NPL Acoustics Report Ac 80, National Physical Laboratory, Teddington, England.

Gatehouse, S., Haggard, M.P., and Davis, A.C. (1982). "Implications of a Population Study of Hearing Thresholds and Noise Exposure," *J. Acoust. Soc. Am. 72,* Suppl.1, S107.

Glorig, A. and Roberts, J. (1965). "Hearing Levels of Adults by Age and Sex," Vital and Health Statistics. Public Health Service Publication 1000, Series 11, No. 11. U.S. Govt. Printing Office, Washington, DC.

ISO (1964). "Standard Reference Zero for the Calibration of Pure Tone Audiometers," R389, International Organization for Standards, Geneva, Switzerland.

Killion, M.C. (1978). "Revised Estimate of Minimum Audible Pressure: Where is the 'Missing 6 dB'?" *J. Acoust. Soc. Am. 63,* 1501-1508.

NIOSH (1972). "Occupational Exposure to Noise. Criteria for a Recommended Standard," Nat. Inst. for Occup. Safety and Health, U.S. Dept. of Health, Education, and Welfare, Washington, DC.

Rowland, M. (1980). "Basic Data on Hearing Levels of Adults 25-74 Years," United States, 1971-75. DHEW Publication No. (PHS) 80-1663, Vital and Health Statistics: Series 11, No. 215. US PHS, National Center for Health Statistics, Hyattsville, MD.

Royster, L.H. and Thomas, W.G. (1979). "Age Effect Hearing Levels for a White Nonindustrial Noise Exposed Population (NINEP) and Their Use in Evaluating Industrial Hearing Conservation Programs," *AIHA J. 40,* 504-511.

Sivian, L.J., and White S.D. (1933). "On Minimum Audible Sound Fields," *J. Acoust. Soc. Am. 4,* 288-321.

Spoor, A. (1967). "Presbycusis Values in Relation to Noise Induced Hearing Loss," *Internat. Audiol. 6,* 48-57.

Ward, W.D. (1977). "Effects of Noise Exposure on Auditory Sensitivity," in *Handbook of Physiology, Vol. 9: Reactions to Environmental Agents,* edited by D.H.K. Lee, American Physiological Society, Bethesda, MD, pp. 1-15.

Noise and Hearing Conservation Manual, edited by
E.H. Berger, W.D. Ward, J.C. Morrill and L.H. Royster

6 Auditory Effects of Noise

W. Dixon Ward

Contents

Page
Introduction 197
Social Handicap 198
Tinnitus 199
Paracusis 199
Speech Misperception 200
Physiological Measures of Damage 200
NIPTS 201
Susceptibility 201
The Sound Conduction Mechanism 201
Characteristics of the Inner Ear 201
Sex 202
Skin Color 202
Right or Left Ear? 202
Age and Experience 203
Initial HTL 203
Health and Vices 203
INIPTS from Steady Noise 204
INIPTS from Non-steady Noise 208
Equal Energy 209
Equal TTS 210
OSHA Regulations 211
Other Equivalent-Exposure Regulations 212
INIPTS from Pure Tones 213
INIPTS from Very High Levels 213
References 214

Introduction

The most undesirable effect of exposure to noise is generally agreed to be permanent hearing loss. The primary goal of any industrial noise control program is usually the prevention of this damage, even though there are other effects of noise, both concomitant with and subsequent to exposure, that are also relevant to industrial health and productivity, so that an aggressive noise control program may produce desirable side effects such as an increase in productivity, a decrease in accident rate, and/or an improvement in worker morale. "Prevention of hearing loss," however, is a rather vague objective, and so it is necessary to examine the concept in somewhat greater depth.

Social Handicap

Basically, what society would like to avoid is production of hearing loss sufficient to produce social handicap. Unfortunately, there is no widespread agreement on how "social handicap" is to be defined and measured. Until the middle of this century a hearing loss was regarded as handicapping, and therefore compensable as the result of an occupational hazard, only if it was disabling, in the sense that it led to a loss in earning power of the individual, a situation whose existence could be relatively easily and unambiguously determined. Since then, however, there has been a gradual acceptance of the principle that a worker is entitled to compensation for any material impairment suffered as a result of employment; thus, handicap has come to be any condition that interferes with everyday living.

A typical example of this viewpoint is the definition of handicap adopted by the American Academy of Otolaryngology: "an impairment sufficient to affect a person's efficiency in the activities of daily living" (Anon. 1979). Since "activities of daily living" include so much, yet vary so widely from one person to another, recent practice has been to attempt to measure auditory handicap in terms of a reduction in the individual's ability to "understand ordinary speech," a simplification that ignores, among other things, the social significance of the perception of warning signals, sounds of nature, and music.

However, even this simplification fails to solve the problem, because there is no accepted definition of "ordinary speech." Speech consists of messages of various degrees of complexity and redundancy, spoken by talkers differing in age, sex, ethnic background, education, and dialect, at a large range of sound levels, in quiet and in the presence of a near-infinite variety of interfering noises. Any test that claims to measure an individual's ability to understand ordinary speech would have to include enough test items to provide a representative range of all these parameters — providing that agreement could be reached on what "representative" means. No such direct test has yet been developed.

In view of the lack of a direct measure of social handicap, auditory dysfunction has traditionally been assessed indirectly, in terms of threshold sensitivity to pure tones as described in Chapter 5, because a relation does exist between pure-tone thresholds and the ability to detect specific speech sounds in quiet. Most of the information-carrying energy in speech occurs in the range from 500 to 4000 Hz, and so the ability to detect speech sounds will depend primarily on auditory sensitivity in that region also, with perception of vowels depending mainly on lower frequencies and consonants on higher frequencies. Hence the avoidance of social handicap is at present equated with preservation of normal sensitivity to pure tones.

Before proceeding to a discussion of the relation between noise exposure and a change in pure-tone sensitivity, it is worthwhile to summarize other manifestations of damage to the auditory system, if only to make it clear why threshold sensitivity continues to be the main indicator of damage.

Tinnitus

A common accompaniment of a loss of ability to hear weak pure tones is, in a sense, its opposite: the hearing of sounds that do not exist. Tinnitus, a ringing in the ears, often occurs in conjunction with hearing loss. Although intractable tinnitus is often distressing to the individual concerned, measurement of its loudness is difficult. Furthermore, only seldom does noise cause a permanent tinnitus without also causing hearing loss. So although the person with tinnitus should probably be considered more impaired than someone with the same amount of hearing loss but without tinnitus, the question of how much more impairment a particular tinnitus represents has not yet been answered.

Paracusis

In addition, some sounds may be heard, but heard incorrectly. Musical paracusis exists when the pitch of tones near a region of impaired sensitivity due to noise is shifted; that is, a tone is heard, but one having an inappropriate pitch. Unfortunately, direct measurement of paracusis is possible only in highly musical persons, so the phenomenon has received little attention. If the paracusis is greater in one ear of a given individual than in the other, then binaural diplacusis will be found: a particular tone will give rise to different pitches in the two ears, and the magnitude of this difference can be inferred by having the listener adjust the frequency of a tone in one ear to match the pitch of a fixed tone in the other. Again, however, noticeable degrees of paracusis or diplacusis that are attributable to noise exposure occur only in conjunction with a considerable loss of sensitivity, so the importance of paracusis *per se* in determining social handicap is still unknown.

Speech Misperception

A complete failure to hear certain speech sounds, or phonemes, is not the only effect on speech perception caused by a loss of pure-tone sensitivity: some phonemes may be heard, but heard incorrectly. As indicated earlier, phonemes differ in frequency composition, vowels having energy predominantly in the low-frequency range, consonants in the high-frequency range (above 1500 Hz), with more subtle differences among specific closely-related consonants such as s, f, sh and th, or p, k and t. A particular configuration of hearing loss may result in only part of the spectral energy being perceived so that the individual hears—often very plainly—the wrong phoneme. For example, in a study designed to find differences between auditory characteristics of ears with high-frequency losses presumably caused by steady noise and those of ears with similar losses caused by gunfire (none was found), individuals with a high-frequency loss that began at 2000 Hz consistently heard an initial 't' as a 'p'; for example, when given the word 'tick', with no opportunity to see the lips of the speaker, they almost always responded 'pick' when forced to choose between 'tick' and 'pick' (Ward *et al.,* 1961). It is fairly obvious that the individual who mistakenly believes that something has been correctly heard is usually much worse off than someone who knows that a particular message has not been understood. So in recent years, this misperception of speech sounds has received increasingly greater attention. Eventually some standardized consonant-confusion test may be adopted as a part of a battery of speech tests designed to assess the evanescent "ability to understand ordinary speech". As yet, however, no such test has met wide acceptance.

Physiological Measures of Damage

Physiological indicators of damage, including destruction of the hair cells of the inner ear, have already been described in Chapter 5. As indicated there, these characteristics cannot be observed directly in the intact human, so it is necessary to infer the presence of damage from industrial noise by less direct methods. The foregoing discussion should have made it clear why the indicator of choice is a permanent change for the worse in sensitivity to pure tones—*i.e.,* an increase in Hearing Threshold Level (HTL) that constitutes a noise-induced permanent threshold shift (NIPTS), over and above changes that can be attributed to the action of the noises of everyday life (sociacusis), to results of infections and otological diseases and of blows to the head (nosoacusis), and to the aging process *per se* (presbyacusis), as discussed in Chapter 5 and illustrated in Figures 5.6 and 5.7.

NIPTS

At first glance, the problem of protection of workers against noise-induced hearing loss might appear to be simple: it is necessary only to determine the relation between noise exposure and the resultant NIPTS and then limit exposures to those that produce less than some tolerable amount of NIPTS. Unfortunately, the simplicity is quite illusory, as both "noise exposure" and "NIPTS" are multidimensional, and the question of how much NIPTS is "tolerable" is still undecided.

Noises exist in infinite variety. They may be sinusoids of constant or varying frequency (whistles, sirens), narrow bands of frequencies (the hiss of escaping steam), broad bands of frequencies ("static" on the radio, most industrial noises), impulses (explosive release of gas, as in gunfire), or impacts (a hammer striking a steel plate, as in drop forges), of short or long duration. They may be steady or fluctuating regularly or irregularly in level. In short, noise exposures vary widely in frequency, intensity, and temporal pattern.

Similarly, the PTS produced by a given noise exposure will depend not only on the audiometric frequency, but also on a host of characteristics of the individual whose NIPTS is being measured. Some of the most important of these individual characteristics that contribute to differences in susceptibility to NIPTS are described below.

Susceptibility

THE SOUND CONDUCTION MECHANISM

Other things being equal, the more acoustic energy that reaches the inner ear, the greater will be the effect (this may be the only principle that can be stated without qualification about noise damage). Therefore the structural characteristics of the external and middle ears, which determine how much power finally is transmitted into the cochlea, must play a role: characteristics such as the size and shape of the pinna and the length of the auditory canal, the area of the eardrum and of the footplate of the stapes, the mass of the ossicles, and the strength of the middle-ear muscles that contract in the presence of high-intensity noise and thereby reduce the transmission of sound.

CHARACTERISTICS OF THE INNER EAR

Once the sound has reached the cochlea, specific properties of the inner ear, both structural and dynamic, must also play some role in determining susceptibility. For example, the stiffness of the cochlear partition, the thickness of the basilar and tectorial membranes, the blood supply to the cochlea, the rate of oxygen metabolism, and the density of afferent and efferent innervation doubtless have some effect.

SEX

In industries in which men and women work side by side, audiometric surveys invariably show that, on average, women have significanly better hearing than men (*e.g.* Berger *et al.,* 1978; Welleschik and Körpert, 1980). This could be due to average differences in some of the dimensions of the auditory system listed above, in which case women would have to be considered to have 'tougher' ears. On the other hand, it could simply indicate that women generally are exposed to fewer and less severe sociacusic influences, especially gunfire. Or it may bear some relation to absentee rate, which is higher in women than men, so that the former actually get less exposure. It is also possible that women are ordinarily freer to leave a noisy job if the noise bothers them, so that only the least susceptible women are the ones who continue to work and hence get included in the survey. Regardless of the underlying cause, however, the difference between men and women cannot be ignored.

SKIN COLOR

A recent survey comparing the hearing of white and black workers in the southeast section of the USA has indicated that blacks have slightly better average hearing than whites (Royster *et al.,* 1980). Again, it is not yet clear whether this difference is due to a disparity in sociacusic exposure (*e.g.,* whites shoot rifles more often than blacks) or instead implies that dark skin pigmentation, presumably correlated with similar pigmentation in the cochlea, somehow makes the individual less susceptible to damage.

RIGHT OR LEFT EAR?

Large-scale surveys (Glorig and Roberts, 1965; Rowland, 1980) consistently show that average thresholds of the left ears are worse than those of the right ears, by a decibel or two. Whether this can be ascribed to inherent structural differences or differences in sociacusic exposure (*e.g.,* the left ear of a right-handed rifleman receives noise levels several dB higher than the right ear), care must be taken to ensure that comparisons between populations use the same ear(s): either the right ear only, the left ear only, both ears, or the "better" ear.

AGE AND EXPERIENCE

Although common sense might suggest that the "young, tender" ear of a teenager would be more easily damaged than the ear of a middle-aged worker, it is just as sensible to propose that the "young, resilient" ear of the teenager should be *less* easily damaged. There is little evidence for either view; Hetu *et al.* (1977) have shown that susceptibility to temporary hearing loss is not different in 12-year-olds than in adults. Similarly, no convincing experimental support exists for the notion that the ears of persons who have worked in noise have gradually become "toughened" and therefore more resistant to NIPTS (Welleschik and Raber, 1978). In short, age *per se* does not seem to be a factor that directly influences the amount of hearing loss produced by a given industrial noise exposure.

INITIAL HTL

Age does have an indirect influence, because as discussed in Chapter 5, the effect of presbyacusis, sociacusis, and nosoacusis does increase with age, so that the older the individuals are at the beginning of the exposure being studied, the higher will be their median HTLs, as shown in Figures 5.6 and 5.7. To the extent that a change in HTL depends on its initial value, then, age will be an important factor. The evidence in this case indicates that the higher the initial HTL, the smaller will be the TTS from a particular exposure (Ward, 1973); presumably the same will be true of NIPTS. If it were not—that is, if PTS were independent of the initial HTL—then the growth of PTS with time of exposure would be linear; for example, if 1 year of exposure to a particular noise produced 15 dB of PTS, then 2 years would produce 30 dB, 3 years 45 dB, and so on. Instead, both animal and human studies agree that growth is exponential—rapid at first, and then slowing down (Glorig *et al.,* 1961; Burns and Robinson, 1970; Herhold, 1977). So the initial HTL is an important population parameter.

HEALTH AND VICES

As is the case for nearly all human ills, attempts have been made to link susceptibility to hearing loss with every type of bad habit such as smoking, dietary deficiencies or excesses, artificial food additives, use of social drugs or stimulants, poor posture, lack of exercise, over-exercise, promiscuity, sexual inactivity, and so on. Such studies, however, are seldom adequately controlled, so no causal relationship between any of these characteristics and hearing loss has been established.

It is clear that no simple answer can be expected to questions as vague as "How much PTS is caused by noise exposure?" Questions must be posed that are much more specific, *viz.*: "How much PTS at frequency F is caused, in the average individual of sex G, skin color C, and initial HTL of H dB, by an exposure for time T to a noise of spectrum S at an intensity level L and temporal pattern P?" None of the infinite number of forms of this question have yet been convincingly answered by use of actual industrial audiometric measurements. Indeed, even if attention is confined to the simplest industrial temporal patterns possible—*i.e.,* a daily 8-hour exposure to a steady noise of fixed spectrum, 5 days/week, 50 weeks/year—the data are inadequate. In many studies, the audiometric measurements have been characterized by faulty technique (such as the use of screening audiometry), by the presence of masking noise in the testing room, or by temporary threshold shifts that inflated the inferred PTSs. In nearly all cases, the initial HTL of the workers was unknown, so that PTS had to be estimated either (1) by comparison of HTLs with those of a control group of non-noise-exposed workers whose age, sex, history of exposure to nosoacusic influences and habitual exposure to sociacusic agents might or might not bear some resemblance to those of the noise-exposed workers, (2) by assuming that the workers would have had hearing that was typical for their age if they had not been exposed to the industrial noise (that is, by applying correction factors such as those shown in Figures 5.6 and 5.7) or even (3) by assuming that the measured HTLs represent industrial noise-induced PTSs (INIPTSs) directly, thus neglecting completely the effects of presbyacusis, nosoacusis and sociacusis.

INIPTS from Steady Noise

Despite the foregoing inadequacies of estimates of INIPTS from steady exposure, a compilation in 1968 by Passchier-Vermeer of these estimates displayed a surprising degree of consistency. Figure 6.1 shows the inferred INIPTS at 4 kHz, the frequency most severely affected by noise, after 10 years of exposure to a steady industrial noise environment whose A-weighted level is given on the abscissa.

The use of A-weighting to reduce different spectra to a single index reflects evidence that low- and very-high-frequency components of a noise are less hazardous than the middle frequencies, in the sense that temporary effects are less severe; of the weightings available on a standard sound level meter, A-weighting comes closest to reflecting the actual relative hazard. The reason for the greater hazard of the middle frequencies may be complex, but at least a large part of the story is the fact that the outer ear canal is resonant at about 3 kHz, so that in broadband noises, the most energy will be transmitted to the cochlea in the 3-kHz region; this fact, coupled with the observation that the frequency affected by a narrow-band noise may be

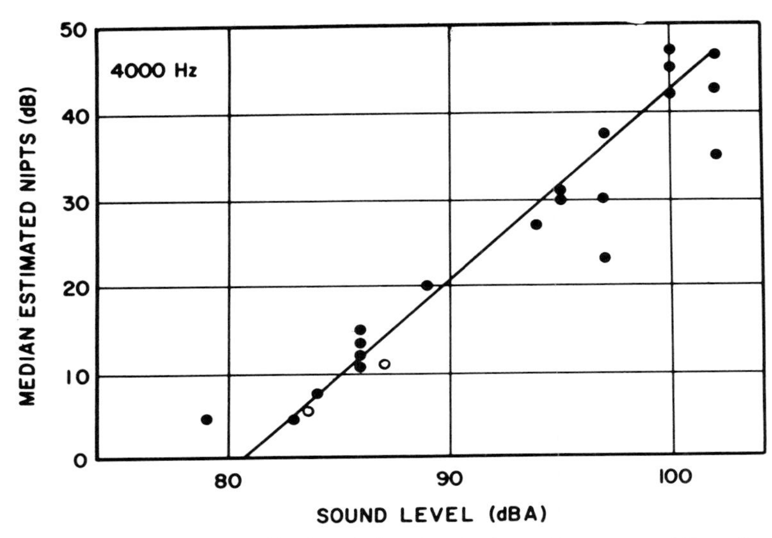

Figure 6.1—Estimated industrial-noise-induced permanent threshold shift at 4 kHz produced by 10 years or more of exposure to noise at the indicated A-weighted level, 8 hours/day, 250 days/year. Solid points indicate values cálculated from the literature by Passchier-Vermeer (1968), and open points represent subsequent studies by Robinson *et al.* (1973) and Yerg *et al.* (1978).

half an octave or so above the frequency of the noise itself, is also responsible for the fact that the 4-kHz region is the first and most affected by the broadband noises that are typical of industry.

The accuracy of this summarization is indicated by the two open circles, which represent the results of two subsequent large-scale studies of workers employed in levels of 80 to 90 dBA. Robinson *et al.* (1973) found that textile workers from an 83-dBA environment showed a loss about 5 dB greater than a control group who worked in 70 dBA or below, and a later study involving several industries demonstrated a loss of 11 dB in workers whose daily A-weighted exposure levels were about 87 dB (Yerg *et al.*, 1978).

Figure 6.2 shows the entire set of functions similar to that of Figure 6.1 for all of the audiometric frequencies normally tested. These curves indicate that, for steady 8-hour exposures, (1) 80 dBA is innocuous; (2) 85 dBA will result in an average INIPTS of around 10 dB at the most noise-sensitive audiometric frequencies of 3,4 and 6 kHz, which is the smallest change in HTL that can be regarded as significant in the individual ear; (3) only at 90 dBA and above do the average INIPTSs reach values that, when

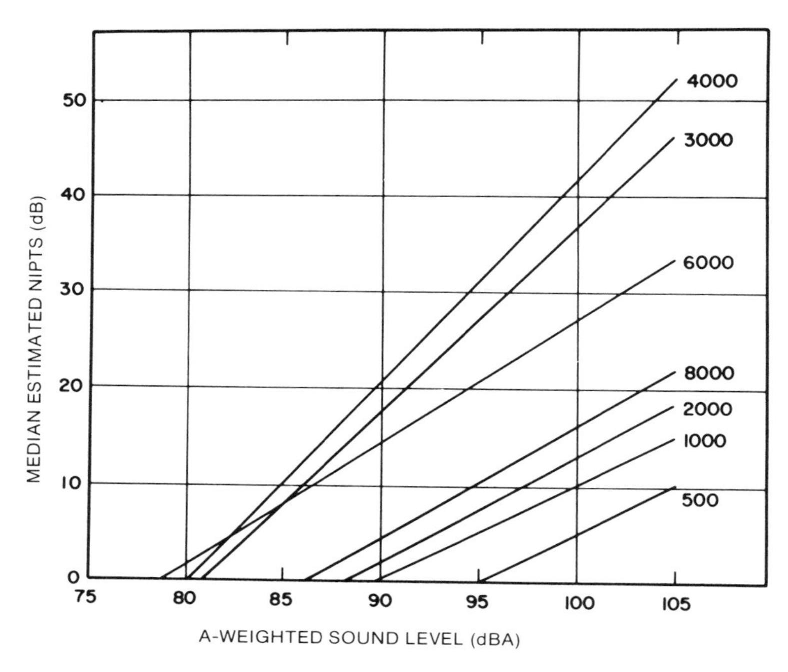

Figure 6.2—Estimated industrial-noise-induced permanent threshold shifts at various frequencies produced by 10 years or more of exposure to noise at the indicated A-weighted level, 8 hours/day, 250 days/year. After Passchier-Vermeer (1968).

added to the inevitable changes in HTL caused by presbyacusis, sociacusis and nosoacusis, will produce a loss of hearing that can be noticed by the worker in question.

Burns and Robinson (1970) published results of a careful study of 759 workers in various industries, again those with a uniform environment. Figure 6.3 shows what they consider to be the median HTLs at 4 kHz to be expected in workers exposed to various levels for up to 45 years, assuming that they began work with 0 dB HTL. Inspection of these contours confirms the implication of Figure 6.1 that 80 dBA produces a negligible effect; the contours also indicate that INIPTS, at least at this frequency, grows rapidly in the first few years of exposure, reaching a near-asymptote by the end of 10-15 years, after which HTLs continue to deteriorate no more rapidly in these workers than in the "no noise" group. The INIPTSs are somewhat smaller than is indicated in Figure 6.1, but this might be expected in view of the fact that the Burns and Robinson data did exclude some sources of error that were operating in the studies summarized by

Passchier-Vermeer, particularly temporary threshold shifts. Therefore the somewhat smaller values of Figure 6.3 are probably more valid.

These data are consistent with the adoption of 90 dBA as the 8-hour exposure limit in many countries, although 85 dBA is advocated by those who feel that a 15-dB loss at high frequencies is too great to be tolerable, arguing that the figure of 15 dB is only an average that does not take individual differences into account. When the average INIPTS is 15 dB, some workers will have lost 20 dB, and a few even 30 dB. In order to protect the most sensitive individuals, it is contended, the exposure limit should be set at a value lower than merely what is necessary to protect the average worker from the average amount of INIPTS.

For the case in question, that is, where the median HTL shows a shift, this line of argument is reasonable. It is, however, sometimes extended to situations where the median NIPTS is zero, for example at 85 dBA for test frequencies of 500, 1000, and 2000 Hz in Figure 6.2: the allowable limit should be lower than 85 dBA, it is argued, because even though the average person is not affected by 85 dBA, the most susceptible persons will be. But the only way in which a shift in the HTLs of the most susceptible individu-

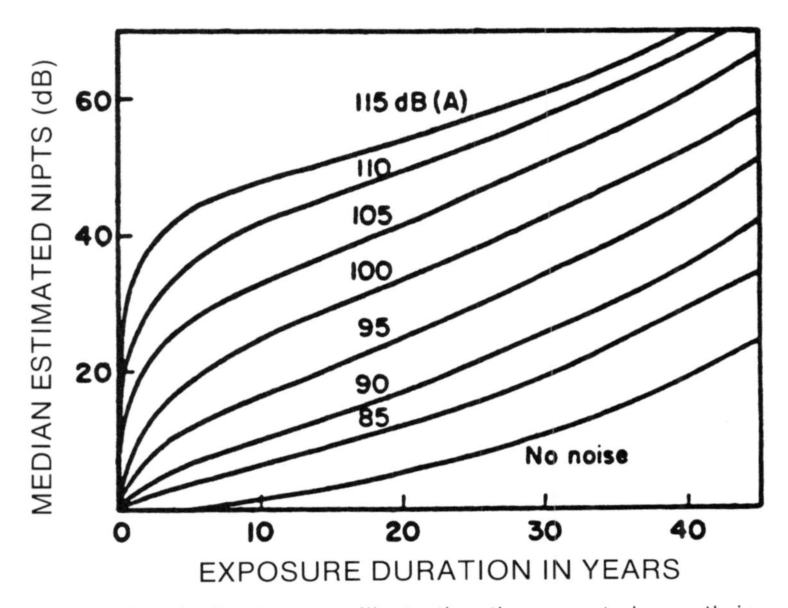

Figure 6.3—Idealized curves illustrating the expected growth in time of industrial-noise-induced permanent threshold shift at 4 kHz as a function of workplace A-weighted noise level. After Robinson (1970).

als would not influence the median HTL would be if these "high-susceptibility" persons are exclusively those whose initial HTLs already lie above the median—that is, only if already-damaged ears are the only ones that will be affected. There is, however, no evidence to support this assumption; indeed, the opposite seems to be true, as the PHS data show that the median HTL shifts slightly more with age than the worst 10 percentile (Glorig and Roberts, 1965), which implies that ears with best hearing are the most likely to be affected (Ward, 1976). Basing exposure standards on some hypothetical '10 percentile most susceptible' individuals is therefore unjustified.

INIPTS from Non-Steady Noise

Unfortunately, Figures 6.2 and 6.3 represent nearly the total quantitative knowledge about noise exposure and INIPTS in humans: the loss in hearing that can be attributed to many years of daily 8-hour exposure to a steady noise, in an average worker with originally normal hearing (*i.e.*, HTLs of 15 dB or less) who is also exposed to an average amount of sociacusic and nosoacusic influences during those years. It is also known that acoustic trauma from a short exposure at a very high level—for example, a single firecracker—can cause the same loss, at least at high frequencies, as 10 years of daily 8-hour exposure at 110 dBA. Most of the exposures that actually occur in industry lie somewhat between these simple extremes. Industrial noises often fluctuate regularly or irregularly, and impact and impulse components may exist. Even when the noise itself is steady, the worker's job may require movement among several areas, so that the exposure involves a number of different levels. For assessing the hazard associated with these non-steady exposures, information such as that shown in Figures 6.2 and 6.3 is of little use, unless some scheme can be devised by which a particular exposure, no matter how complicated, can be expressed as being equivalent in *effectiveness* to some particular steady noise. That is, the exposure in question, if repeated every day, would eventually lead to some value of INIPTS. If this INIPTS at 4 kHz were, say, 15 dB, then from Figure 6.3, the noise exposure could be said to possess an effective level of 90 dBA in terms of its ability to cause hearing damage, because an 8-hour exposure to a steady 90 dBA would produce the same 15-dB INIPTS.

Two major approaches have been taken in attempts to develop a simple scheme that will permit the determination of this effective level. The equal-energy approach is an example of attempts to equate exposures on the basis of their physical characteristics directly, while the equal-TTS approach is based on an assumed correlation between permanent and temporary effects of noise exposure.

EQUAL ENERGY

The equal-energy approach (Robinson, 1970) makes the assumption that damage depends only on the daily amount of A-weighted sound energy that enters the ear of the worker, and that the temporal pattern during the day is irrelevant. Since this energy is the product of the A-weighted sound intensity and its duration, then in order to hold constant the energy of two exposures—and hence, presumably, their effect on hearing—it is only necessary to balance a difference in intensity by a corresponding difference in duration (in the opposite direction). Thus if an exposure of 90 dBA for 8 hours (480 minutes) is established as the standard daily noise dose, the equal-energy theory postulates that an exposure of 110 dBA (20 dBA higher than, and hence possessing 100 times the intensity of, a 90-dBA noise) will be just as hazardous if its duration is 4.8 minutes (480/100). Furthermore, the hazard would be the same whether this exposure came in one 4.8-minute burst or instead consisted of 10 bursts, each of 0.48 minute in duration, spread over the 8-hour workday.

The equal-energy principle leads to the so-called "3-dB Rule" in evaluation of the hazard associated with exposures involving levels other than whatever level is adopted as the permitted 8-hour limit. In such a system, the duration must be cut in half when the A-weighted level is increased by 3 dB, because a 3-dB increase means that the intensity has doubled. If 90 dBA for 8 hours constitutes the permitted daily exposure, then the 3-dB rule postulates that 93 dBA can be tolerated for 4 hours, 96 dBA for 2 hours, 99 dBA for 1 hour, and so on, regardless of the distribution of the energy within the workday. All of these exposures are said to possess an *equivalent level* of 90 dBA. In formal terms, the *8-hour equivalent level* $L_{eq(8h)}$ of an exposure is that level that, if maintained for 8 hours, would transfer the same amount of energy to the ear as the actual exposure.

The equal-energy principle makes calculation of the presumed effective level of the exposure relatively easy, because in this case the effective level is simply the equivalent level. Even if the exposure is hopelessly complicated, involving multiple or continuously-varying levels, it is easy to construct devices—dosimeters—that will integrate the instantaneous intensity over time in order to provide a measure of the total energy, expressed either in terms of $L_{eq(8h)}$ or as a fraction or multiple of the standard daily dose.

Because of this simplicity, the 3-dB rule has been adopted by the International Standards Organization. In ISO/DIS 1999 (1984), it is stipulated that exposures be measured only in terms of $L_{eq(8h)}$, a step that automatically commits prediction of hazard to use of the equal-energy principle.

Unfortunately, the equal-energy theory has one serious defect: it is true only for single steady uninterrupted exposures, because the daily temporal pattern does play a role in INIPTS. In the chinchilla it is not the case, for instance, that a 48-minute exposure to 110 dBA and a series of 40 1.2-minute bursts, also at a level of 110 dBA but separated by 10.8 minutes of quiet, will produce the same damage (Ward *et al.,* 1983). The short bursts produce much less damage. The short-term recovery that occurs during the quiet periods of the intermittent exposure serves to reduce the permanent effect. By contrast, the second principle that has been used to evaluate non-steady exposures, the equal-TTS approach, does not make the mistake of ignoring temporal pattern.

EQUAL TTS

The equal-TTS approach to the problem of determining the equivalence of steady and time-varying noise exposures is based on the hypothesis that daily exposures that produce the same temporary effects will eventually produce the same permanent effects. For example, if the standard daily dose produces a temporary threshold shift of 15 dB, then any daily exposure that also produces a 15-dB TTS should ultimately cause the same INIPTS as does the standard daily dose. Because TTS, which can be studied under controlled conditions in the laboratory, behaves fairly lawfully, it is a relatively simple matter to determine combinations of level, duration and temporal pattern that produce the same TTS as the standard daily noise dose.

There is, of course, a wide variety of TTS measures that could be used in these determinations. The TTS index about which the most empirical information existed in 1964, when Working Group 46 of the NAS/NRC Committee on Hearing, Bioacoustics and Biomechanics (CHABA) attempted to develop standard equal-effect contours, was TTS_2, the TTS at a particular audiometric frequency measured 2 minutes after the termination of the noise, so this index was the one they used. However, indices that would appear to have equal face validity include, in addition to TTS at any arbitrarily chosen time following exposure other than 2 minutes, the maximum value of TTS that occurred at any time during the exposure, or the time required for complete recovery, or the integral over time of the instantaneous value of TTS from the beginning of the exposure to the end of the exposure or to the time of complete recovery (Kraak, 1982). Even more complicated possible measures would include those involving integration of each of the foregoing indices over frequency.

The most well-known set of formal damage-risk contours based on TTS, however, are those finally proposed by CHABA (Kryter *et al.,* 1966; Ward, 1966). These are contours that, according to the most accurate data then available, showed the levels and durations of octave-band noises that would just produce a TTS_2 of 10 dB at 1000 Hz or lower frequencies, 15 dB at 2000 Hz, or 20 dB at 3000 Hz or higher frequencies, in the average normal-hearing person. These are values of TTS_2 that are produced by 8-hour continuous exposures to noises of 85 to 95 dBA, depending on the actual spectrum of the noise in question.

Equal-effect contours were developed not only for exposures shorter than 8 hours but also for those involving various patterns of intermittence. The end result was a set of ten graphs that, if used properly, would indicate whether or not a particular exposure was more hazardous or less hazardous than the standard daily dose of 8 hours at 90 dBA, provided that TTS_2 were in fact a valid indicator of relative hazard. The procedure for using these contours was so complicated that they were never widely used. Nevertheless, the implication of these contours was clear: intermittence did reduce the TTS, quite drastically under certain conditions, so that considerably more energy could be tolerated in an intermittent exposure without exceeding the TTS_2 criterion described above.

OSHA REGULATIONS

Consequently, when it became necessary to establish exposure standards for industry in the U.S.A., the committee that attempted to find a simplification such that equivalent exposure could be calculated easily, but without assuming the validity of the equal-energy theory, decided that the best way to take into account the reduction of hazard associated with intermittence was to use a trading relation of 5 dB per halving of exposure time instead of 3 dB (Anon., 1967). In effect, this use of a fixed trading relation again ignores the effect of different patterns of intermittence; the only difference is that use of the 5-dB trading relation tacitly assumes that any exposures involving levels above 90 dBA will *be* intermittent, indeed involving only short noise bursts distributed relatively evenly throughout the workday. Thus with a standard daily dose of 90 dBA for 8 hours, a 4-hour exposure at 95 dBA was defined as equivalent in effect, as was a 2-hour exposure at 100 dBA, and so on, up to a 15-minute exposure at 115 dBA. Each of these exposures is said to possess a time-weighted average (TWA) of 90 dBA, a metric that is analogous to the $L_{eq(8h)}$ that results when the exposure is evaluated using the 3-dB trading relation. This formula for assessing the effective level of short or intermittent exposures was adopted as a federal regulation for government contractors under the Walsh-Healey

Act of 1969 (Anon., 1969) and was then extended to all firms engaged in interstate commerce under the Occupational Safety and Health Act (OSHA) of 1970 (Anon., 1971). It is still in force, and dosimeters that can evaluate exposures in terms of TWA instead of (or in addition to) $L_{eq(8h)}$ are commercially available.

OTHER EQUIVALENT-EXPOSURE REGULATIONS

The history of development of exposure standards indicates that only simple formulas for assessing equivalence of exposures will be acceptable to those who must establish and enforce these standards. Unfortunately, however, damage to hearing clearly depends on the intensity, the duration, the spectrum, and the temporal pattern of the noise in a very complex manner, and so no simple solution can possibly be correct, although it may be acceptably accurate over a limited range of exposures. The 3-dB rule of ISO overestimates the hazard of intermittent noises. On the other hand, use of the 5-dB rule of OSHA will underestimate the hazard of a steady noise, especially at the highest levels. The contrast between them is greatest at 115 dBA, the highest level permitted for more than 1 second in the OSHA regulation: the 15 minutes permitted by OSHA is nearly 10 times as long as the 1.6 minutes allowed by the equal-energy principle. A compromise between these two systems has been adopted by the armed forces of the U.S.A., who employ a 4-dB-per-halving rule (Anon., 1973), and some East German investigators have recently proposed that a 6-dB rule is appropriate (Kraak, 1982).

Another possibility that might be explored is to use one trading relation over the range of the lower levels and another for the higher levels, or even to adopt a fixed curvilinear relation that, although difficult to use directly, would pose no problem to a dosimeter. Still another solution would be to accept the 3-dB trading relation of equal energy because it is the most protective, but develop a set of correction factors, to be applied to the $L_{eq(8h)}$, that depend on the degree of intermittence of the exposure.

Which of the foregoing solutions will eventually prove to be the best is obviously still to be determined, and because very few measurements of actual auditory damage from steady *vs.* intermittent exposures in experimental animals have been made, the decision as to which is the most valid lies far in the future. Of course, it is possible that the problem may simply evaporate, if a trend now seen in noisy industries (and in the U.S. Army—cf. Anon., 1980) continues, involving establishment of noise-hazardous zones—zones in which levels above 85 or 90 dBA may occur and in which therefore hearing protectors must be worn (by analogy to "hard-hat areas"). If no exposures, in terms of sound reaching the inner ear, can involve levels above that level associated with the 8-hour standard daily dose, then protection will be complete because most workers will be overprotected.

INIPTS from Pure Tones

Some regulatory schemes prescribe a 5- or 10-dBA reduction in permitted level if the noise has a pronounced pure-tone component. The weight of evidence, however, indicates that a pure tone in the frequency range most important to hearing conservation—*i.e.,* from 1000 Hz to 4000 Hz—produces no greater damage than does an octave-wide band of noise at the same sound level (Ward, 1962). Although a pure-tone component may increase the annoyingness of a noise, it does not constitute an additional hazard to hearing.

INIPTS from Very High Levels

Most exposure regulations have some sort of ceiling on the instantaneous peak sound level, a limit that may not be exceeded regardless of duration. This concept of a "critical intensity"—an acoustic intensity that would produce some irreversible damage no matter how short the exposure might be (Rüedi, 1954)—is somewhat consistent with the notion of an "elastic limit" of the structures in the inner ear that if exceeded leads to immediate mechanical injury to tissues. The existence of such a "breaking point" would be implied by a sudden increase in the rate of growth of damage in a series of experimental exposures at increasing intensity but fixed duration. Such abrupt changes are indeed found, but the implied critical intensity nevertheless still depends on the duration: the shorter the exposure, the higher the indicated critical intensity. For example, in the case of the shortest of acoustic events, single impulses such as those produced by explosions, although the critical level inferred from measurements of TTS following exposure to impulses of about 1 millisecond duration is somewhere around 140 dB SPL (McRobert and Ward; 1973), exposures to the pulses of less than 100 microseconds generated by a cap gun imply that the critical level for them is somewhat higher. Conversely, for reverberant sounds such as the impact noises generated by drop forges, somewhere around 130 dB may represent a critical level.

Just where, if anywhere, this type of limit should be placed is still undecided. Although the present OSHA regulations state: "Exposure to impulsive or impact noise should not exceed 140 dB peak sound pressure level" (Anon., 1971), this number was little more than a guess when it was first proposed in the CHABA document (Kryter *et al.,* 1966), and no convincing supportive evidence has since appeared. While 140 dB may be a realistic ceiling for impact noises, it is inappropriate for impulses, so exposure limits in which the permitted peak level increases as the duration of the pulses becomes shorter should continue to be used (Anon., 1968).

It should be noted that the present OSHA exposure limit has a *de facto* second critical level for continuous noise: 115 dBA. Although 15 minutes of exposure at 115 dBA is permitted, more than one second at any higher level is forbidden, even though a 1-second exposure even at 130 dBA would represent a dose considerably less than 1% of the permitted daily dose if the 5-dB-per-halving trading relation were extended to include exposures at that level. Indeed, even in the ISO equal-energy system, 3 seconds of exposure would be permitted at 130 dBA, as this would be the energy equivalent of 90 dBA for 8 hours. This 115-dB ceiling is therefore highly artificial, as no human data even suggest that permanent damage can be caused by 3 seconds or less of 130 dBA.

References

Anon. (1971). "Intersociety Committee Report: Guidelines for Noise Exposure Control," *Am. Ind. Hyg. Assoc. J. 28,* 418-424.

Anon. (1968). "CHABA Proposed Damage-Risk Criterion for Noise (Gunfire). Report on Working Group 57," NAS-NRC Committee on Hearing, Bioacoustics, and Biomechanics, Washington, DC.

Anon. (1969). *Fed. Regist. 34,* 7891-7954.

Anon. (1971). "Occupational Safety and Health Standards," *Fed. Regist. 36,* no. 105, part II.

Anon. (1979). "Guide for the Evaluation of Hearing Handicap. American Academy of Otolaryngology Committee on Hearing and Equilibrium," *J. Am. Med. Assoc. 241,* 2055-2059.

Berger, E.H., Royster, L.H., and Thomas, W.G. (1978). "Presumed Noise-induced Permanent Threshold Shift Resulting from Exposure to an A-weighted L_{eq} of 89 dB," *J. Acoust. Soc. Am. 64,* 192-197.

Burns, W., and Robinson, D.W. (1970). *Hearing and Noise in Industry,* Her Majesty's Stationery Office, London, England.

Glorig, A., and Roberts, J. (1965). "National Center for Health Statistics: Hearing Levels of Adults by Age and Sex." *Vital and Health Statistics.* PHS Publication number 1000—Series 11, No. 11. Public Health Service. US Government Printing Office. Washington, D.C.

Glorig, A., Ward, W.D., and Nixon, J. (1961). "Damage Risk Criteria and Noise-Induced Hearing Loss," *Arch. Otolaryngol. 74,* 413-423.

Herhold, J. (1977). "PTS beim Meerschweinchen nach Langzeitbelastung mit stationä - rem Lärm," (PTS in the Guinea Pig after Extended Exposure to Steady Noise) in *Lärmschäden Forschung 1977* (Proceedings, Meeting 14 April 1977 of Working Group "Lärmschäden" of the KdT). Friedrich-Schiller-Universität, Jena, Deutsche Demokratische Republik, 29-36.

Hetu, R., Dumont, L., and Legare, D. (1977). "TTS at 4 kHz among school-age children following continuous exposure to a broadband noise," *J. Acoust. Soc. Am. 62,* Suppl. 1, S96.

International Standards Organization (1981). "Acoustic-Determination of Occupational Noise Exposure and Estimation of Noise-induced Hearing Impairment," ISO 1999, Geneva, Switzerland.

Kraak, W. (1982). "Investigations on Criteria for the Risk of Hearing Loss Due to Noise," in *Hearing Research and Theory,* Vol. 1, edited by J.V. Tobias and E.D. Schubert, Academic Press, New York, NY, 187-303.

Kryter, K.D., Ward, W.D., Miller, J. D., and Eldredge, D. H. (1966). "Hazardous Exposure to Intermittent and Steady-state Noise," *J. Acoust. Soc. Am. 39,* 451-464.

McRobert, H. and Ward, W. D. (1973). "Damage-risk Criteria: The Trading Relation Between Intensity and the Number of Nonreverberant Impulses," *J. Acoust. Soc. Am. 53,* 1297-1300.

Passchier-Vermeer, W. (1968). "Hearing Loss Due to Exposure to Steady-State Broadband Noise," (IG—TNO Report 35) Delft, Netherlands.

Robinson, D. W. (1970). "Relations Between Hearing Loss and Noise Exposure, Analysis of Results of a Retrospective Study," in *Hearing and Noise in Industry,* edited by W. Burns and D.W. Robinson, Her Majesty's Stationery Office, London, England.

Robinson, D. W., Shipton, M.S., and Whittle, L.S. (1973). "Audiometry in Industrial Hearing Conservation—1," National Physical Laboratory Acoustics Report Ac 71, NPL, Teddington, England.

Rowland, M. (1980). "Basic Data on Hearing Levels of Adults 25-74 Years," United States, 1971-75. DHEW Publication No. (PHS) 80-1663. Vital and Health Statistics: Series 11, No. 215. US PHS, National Center for Health Statistics, Hyattsville, MD (Jan. 1980).

Royster, L. H., Royster, J. D., and Thomas, W. G. (1980). "Representative Hearing Levels by Race and Sex in North Carolina Industry," *J. Acoust. Soc. Am. 68,* 551-566.

Rüedi, L. (1954). "Actions of Vitamin A on the Human and Animal Ear," *Acta Otolaryngol. 44,* 502-516.

US Air Force (1973). "Hazardous Noise Exposure: Air Force Regulation 161-35," USAF, Washington, DC.

US Army (1980). "Hearing Conservation. TB Med 501," Headquarters. Dept. of the Army.

Ward, W. D. (1962). "Damage-risk Criteria for Line Spectra," *J. Acoust. Soc. Am. 34,* 1610-1619.

Ward, W. D. (1966). "The Use of TTS in the Derivation of Damage Risk Criteria for Noise Exposure," *Intern. Audiol. 5,* 309-313.

Ward, W. D. (1973). "Adaptation and Fatigue," in *Modern Developments in Audiology,* edited by J. Jerger, Academic Press, New York, NY, 301-344.

Ward, W. D. (1976). "Susceptibility and the Damaged-ear Theory." in ***Hearing and*** *Davis,* edited by S. K. Hirsh, D. H. Eldredge, I. J. Hirsh and S. R. Silverman, Washington University Press, St. Louis, MO.

Ward, W.D. (1984). "Noise-induced Hearing Loss," in *Noise and Society,* edited by D. M. Jones and A. J. Chapman, John Wiley and Sons Ltd., London, 77-109.

Ward, W. D., Fleer, R. E., and Glorig, A. (1961). "Characteristics of Hearing Loss Produced by Gunfire and by Steady Noise," *J. Aud. Res. 1,* 325-356.

Ward, W. D., Turner, C. W. and Fabry, D. A. (1983). "The Total-energy and Equal-energy Principles in the Chinchilla," Contributed paper, 4th Int. Congress on Noise as a Public Health Problem, Torino, Italy.

Welleschik, B. and Körpert, K. (1980). "Ist das Lärmschwerhörigkeitsrisiko für Männer grösser als für Frauen?" (Is the Risk of Noise-induced Hearing Loss Greater for Men than Women?), *Laryngol. Rhinol. 59,* 681-689.

Welleschik, B. and Raber, A. (1978). "Einfluss von Expositionszeit and Alter auf den lärmbedingten Hörverlust," (The Influence of Exposure Time and Age on Noise-induced Hearing Loss), *Laryng. Rhinol. 57,* 681-689.

Yerg, R. A., Sataloff, J., Glorig, A., and Menduke, H. (1978). "Inter-industry Noise Study; The Effects Upon Hearing of Steady State Noise Between 82 and 92 dBA," *J. Occup. Med. 20, 351-358.*

Noise and Hearing Conservation Manual, edited by
E.H. Berger, W.D. Ward, J.C. Morrill and L.H. Royster

7 Visual Evaluation of the External Ear and Eardrum

Kathy A. Foltner

Contents

	Page
Introduction	217
Anatomy	218
Abnormalities	220
Examination	222
Training of Examiners	223
Instrumentation	223
Performing Evaluations	224
Acting on Screening Results	225
Policy	225
Case History	226
Documentation and Referral	226
Follow-up Benefits	227
References	227

Introduction

Visual evaluation of the external ear, including the eardrum, should be a component of all industrial hearing conservation programs. The principal reasons for evaluating the condition of the ear are: to detect abnormalities that may contraindicate or interfere with the use of hearing protection devices (HPDs), to detect abnormalities that may affect audiometric test results, to help identify and document existing ear disease, and to provide a potential health benefit for the employee.

Several studies have investigated the incidence of various ear abnormalities in different populations (Hoadley and Knight, 1975; Hopkinson, 1981;

Hawke *et al.,* 1984), but information relating to such incidence is limited (Berger, 1985). It is important to recognize that some external ear abnormalities require medical referral and medical clearance before an employee should be permitted to wear HPDs. Some external ear abnormalities (*i.e.,* external ear infection) may have an increased incidence in the summer or in warm environments and may be aggravated by the presence of a foreign material such as an ear mold (Hoadley and Knight, 1975; Hawke *et al.,* 1984).

In one investigation, otoscopic findings were reviewed for a total of 68,647 industrial workers undifferentiated by geographic location, age, sex, type of industry, or hearing protection utilization. Six percent (6%) were found to have active ear disease, existing perforations, or occluding wax. Seven and one-half percent (7.5%) were found to have partial obstruction of the ear canal with wax (Foltner, 1984).

In a second investigation, otoscopic findings were reviewed for 101 industrial workers located in a mid-Michigan rural community. This industry had not established a hearing conservation program and the employees were not using HPDs. Eight percent (8%) were found to have abnormal findings including old perforation, mastoid cavity, mould in the external ear canal, external ear infection, or occluding wax in the external ear canal. Specifically, the incidence of active ear infection was one percent (1%), the incidence of existing middle ear pathology was one percent (1%), and the incidence of occluding wax in the ear canal was six percent (6%) (Foltner, 1984).

Available data indicate that approximately ten percent (10%) of the industrial population may have ear abnormalities that could contraindicate the use of HPDs. Therefore, it is important to complete the visual evaluation of the ear before the distribution of HPDs so that existing conditions will not be aggravated by their use.

Anatomy

The ear consists of three major parts: the outer ear, the middle ear, and the inner ear. The outer ear includes the pinna (auricle) and the external auditory meatus (ear canal) which ends at the tympanic membrane (eardrum). The function of the outer ear is to direct acoustic energy (sound) to the eardrum. Visual evaluation of the external ear involves examination of the pinna, ear canal, and eardrum (also see Chapter 5).

The pinna, commonly called "the ear," is the part of the hearing system that is easily visible. It is primarily composed of flexible cartilage covered by skin. Figure C 7.1 (see color photographs, p. 229) shows some of the major parts of the pinna. The helix is the outer rim. The anti-helix is the inner rim. The concha is the bowl-shaped portion which leads to the external ear canal. The lobule (ear lobe) is the lowest portion, which is floppy and contains no cartilage. The tragus is the small flap just in front of the opening to the ear canal. The mastoid bone is the portion of the skull located behind the pinna.

The normal pinna may have some wrinkles, but in general the skin should be clear and free from masses, lesions, scaliness, or signs of infection such as redness or drainage. The examiner should be able to touch and move the pinna in various directions and palpate the tragus, post-auricular area, and pre-auricular area without signs of tenderness or distress from the employee.

The external ear canal is an irregularly shaped tube that begins at the concha and ends at the eardrum. It may vary in length from 25 mm to 35 mm and in diameter from 6 mm to 8 mm (Zemlin, 1968) or about an inch to an inch and a half in length, and a quarter to a half inch in diameter (Concept Media, 1977). The outer one-half to one-third of the ear canal consists of cartilage, whereas the inner one-half to two-thirds consists of bone. Both the cartilaginous and bony portions are covered with skin which should be free from any signs of infection or disease, such as redness, masses, or discharge.

The outer one-half of the ear canal is much less sensitive to touch than the inner half; this fact is important to know when one completes a visual evaluation of the ear canal to help reduce the chances of employee discomfort. The examiner should place the otoscope in the outer half of the ear canal, since touching the inner half may be uncomfortable for the employee.

The outer one-third of the ear canal also contains numerous hairs and the sebaceous and cerumenous glands, which are wax-producing glands located below the surface of the skin. It is normal to have some wax in the outer one-third of the ear canal. Wax functions to lubricate the skin of the ear canal and maintain the proper pH balance of the ear canal, but excessive wax that occludes the ear canal can cause hearing loss and may contraindicate the use of earplugs.

The eardrum is located at the innermost end of the ear canal and forms the boundary between the external ear and the middle ear. The eardrum has three layers: the epithelial (outer skin) layer, the fibrous (middle) layer, and the mucous membrane (innermost) layer. The eardrum completely seals the ear canal and protects the middle ear cavity from infection. Vibration of the intact eardrum is important to normal hearing.

Abnormalities

Abnormalities of the pinna identifiable to the qualified observer include masses, lesions, drainage, and signs of infection. Tenderness or reported pain with manipulation may also be a sign of an underlying problem. Abnormalities of the ear canal identifiable to the qualified observer include external ear infection, clear or pus-like drainage, foreign body in the ear canal, masses or growths, excessive wax, atresia (closure or absence of the ear canal), and collapsing ear canal (McLaurin, 1973; Paparella, 1973a, b, c).

The following pictures are examples of some normal and abnormal external ear conditions (Concept Media, 1977; Becker *et al.*, 1984).

Figure C 7.2 (see color photographs, p. 229) shows examples of normal pinnae. Although they vary in shape, size, and angle of attachment, each is considered to be within normal limits. There is no evidence of infection, masses, or lesions which would require a medical referral or contraindicate the use of an earmuff. If the pinna is unusually shaped the employee may not be able to properly or comfortably wear earmuffs. He may also have to use an insert-type HPD for sufficient protection.

Figure C 7.3 (see color photographs, p. 229) is an example of a mass on the anti-helix of the pinna. Such lesions may occur on, behind (post-auricular), or in front of (pre-auricular) the pinna and require medical referral. The employee should not wear earmuffs until medical clearance is received.

Figure C 7.4 (see color photographs, p. 229) is an example of atresia, a congenital condition in which the ear canal never fully develops. An employee with this condition in essence has a built-in HPD. An earmuff would not be contraindicated, but this employee would not be able to wear an insert-type HPD since there is no ear canal. A medical referral should not be required since this condition is congenital and most likely well known to the employee. A referral would be appropriate if the employee has not seen a physician for this condition or has not consulted with an audiologist regarding improving communication abilities with hearing aids.

The normal ear canal is lined with skin that should be clear of any signs of redness, infection, or drainage. Hair and minimal wax may be present in the external ear canal and are not considered abnormal unless excessive.

Figure C 7.5 (see color photographs, p. 230) is an example of an external ear infection (external otitis). The ear canal is swollen and inflamed and there is evidence of drainage. The symptoms would probably include tenderness and pain (Senturia *et al.*, 1980). A medical referral should be made. An employee with these symptoms should not be permitted to wear insert-type hearing protection until medical clearance is received. Earmuffs should be permitted unless painful, although their use may be prohibited during treatment. It is possible that use of an earmuff could aggravate this condition, therefore medical evaluation and clearance are essential.

Figure C 7.6 (see color photographs, p. 230) is an example of discharge from the external ear canal. This discharge may be coming through a perforated eardrum or may originate in the external ear canal. A medical referral should be made. Employees with this condition should not be permitted to wear insert-type HPDs. Earmuffs should be acceptable on an interim basis, but they could aggravate this condition.

Figure C 7.7 (see color photographs, p. 230) shows a bead lodged in the external auditory canal. Foreign objects are much more commonly found in children's ears than in adults' ears. A medical referral should be made for the removal of the object. Insert-type ear protection should not be permitted. Earmuffs would be advisable.

Figure C 7.8 (see color photographs, p. 230) is an example of a pressure equalization (P.E) tube located in the eardrum as well as an insect located in the external ear canal. PE tubes are usually placed in the eardrum by a physician to help prevent middle ear infection. Therefore, it is likely an employee with a PE tube has a history of ear infections. It is possible, although unlikely, that the use of HPDs may cause a recurrence of an ear infection since air circulation to the middle ear will be altered by the use of the HPDs. The employee should be aware of the PE tubes, but medical clearance to wear HPDs should be on record since they could aggravate an abnormal ear condition. The insect should be treated as a foreign object and would require medical referral for removal.

Figure C 7.9 (see color photographs. p. 231) is an example of occluding wax. The employee may or may not be aware that wax is present, since discomfort is rarely present with this condition. A medical referral should be made for removal of the wax. Insert-type HPDs should not be permitted. Earmuffs would be advisable. Employees with chronic wax accumulation problems need to take regular steps to prevent blockage; for example, they should have regular otoscopic screenings to determine if the wax is becoming excessive or occluding the ear canal. Wax removal should be completed under the supervision of a physician. The physician may choose to train a nurse to remove the wax or may recommend an over-the-counter agent to remove the wax. An over-the-counter agent should not be used unless a physician recommends it because use of such agents can cause ear problems if the individual's eardrum is not intact. Under no circumstances should bobby pins, Q-tips, keys, or pen caps be used to remove wax from the ear canal.

The external ear canal ends at the eardrum, but the more experienced examiner, if properly trained and qualified, can evaluate the status of the eardrum for certain abnormalities. These include signs of middle ear infection, eardrum perforation, and cholesteatoma.

Figure C 7.10 (see color photographs, p. 231) is an example of a normal eardrum. The eardrum is pearly gray and completely intact. Normal eardrum landmarks include the bones of the middle ear and a light reflection in the lower front section of the eardrum. In comparison, Figure C 7.11 (see color photographs, p. 231) shows a red and bulging eardrum, illustrating one type of middle ear infection. The bones cannot be seen and the color is not pearly gray. The employee may or may not feel discomfort, and there may or may not be some hearing loss associated with this condition. With appropriate supportive data from the employee interview, a medical referral should be made. The physician should decide if any insert-type HPD can be used. Earmuffs should be used until clearance is received.

Figure C 7.12 (see color photographs, p. 231) shows a hole in the eardrum. This may be the result of an ear infection or traumatic incident. If it was caused by a middle-ear infection and the infection is still active, drainage may be present in the ear canal. Even a hole without drainage should be medically evaluated by a physician since it is an open pathway for infection to enter the middle ear cavity. Earmuffs should be used until medical clearance is received.

Figure C 7.13 (see color photographs, p. 232) is an example of a cholesteatoma, which is a benign growth that may be located in the middle ear. This is a potentially serious condition and requires medical referral. Hearing loss, dizziness, or facial paralysis may be associated with this condition. Earmuffs should be used until medical clearance is received.

Visual evaluation of the ear will also help to identify collapsing ear canals as described below in the Performing Evaluations section. This condition will most likely not contraindicate the use of HPDs, but may cause a worsening of air conduction audiometric thresholds, especially in the range of 2000 to 6000 Hz. Refer to the chapter on Hearing Measurement.

Examination

It is essential that persons completing visual evaluations of the ear understand their responsibilities as well as their limitations. Any person with the desire, proper attitude, and sufficient training can complete an accurate visual evaluation of the ear. It must be understood, however, that only the physician can offer a diagnosis. The physician's responsibilities include not only review of those employees with abnormal visual evaluation results to rule out or diagnose the apparent abnormality, but also an active role in the training of persons completing the visual evaluation. The only way a person can identify what is abnormal is to fully understand what is normal. This takes time, appropriate training, and good instrumentation.

TRAINING OF EXAMINERS

Training for completion of visual evaluations of the ear should begin with a study of basic ear anatomy, particularly of the external ear and the eardrum. Evaluation of the external ear includes a visual inspection of the pinna for size, protrusion, shape, and the presence of lesions, disease, or tenderness as well as an evaluation of the ear canal for size, sensitivity, and the presence of excessive wax, foreign bodies, or possible disease. Evaluation of the eardrum includes noting the color and integrity of the eardrum.

Training to identify what is normal can begin with color pictures and slides. There are several training packages and reference manuals available which include numerous slides or pictures of many external and middle ear abnormalities as well as examples of normal conditions (Mechner, 1975; Concept Media, 1977; Becker *et al.*, 1984; Hawke *et al.*, 1984). These training packages and books can serve as the beginning stage in the training for identification of normal versus abnormal ear conditions.

Hands-on experience by visual evaluation of actual ears is the only way the examiner will become proficient in the identification of normal versus abnormal ear conditions. The entire hearing conservation team must understand this and provide sufficient training and guidance for the person completing the visual evaluation of the ear. It is the supervising professional (otolarynogologist, physician or audiologist) who must determine the examiner's level of competence with visual evaluations of the ear. Therefore, direct monitoring by the supervising professional is essential until the examiner achieves that competence.

INSTRUMENTATION

An otoscope is a small hand-held instrument which is used to illuminate and magnify the ear canal and eardrum. Disposable specula (tips) are used with the otoscope (Figure C 7.14, color photographs, p. 232). A small-diameter lightweight otoscope with fiber-optic 3.5v Halogen light source and 2.5X magnifying capabilities with disposable tips is the instrument of choice for completion of visual evaluations in the industrial setting. This will provide optimal lighting and magnification of the ear canal and eardrum and will eliminate chances of cross contamination from placing one speculum into multiple ears.

An ear light or pen light can also be used to help evaluate the entrance and the outermost portion of the ear canal, but will not provide magnification or allow good viewing of the eardrum or innermost portion of the ear canal. The examiner can visually evaluate the outer ear canal by separating the tragus and the pinna to enlarge the ear canal opening while using the ear light to illuminate the ear canal.

PERFORMING EVALUATIONS

Initially, the examiner should determine if there is a history of any ear problems such as pain, drainage, ear disease, or excessive wax build-up. The examiner may ask the employee or refer to the current employee case history if it has already been completed. If the case history is negative, the examiner should proceed with an explanation of what is about to happen and why. For example, the examiner should begin by explaining that he/she is going to touch the pinna and the circumaural regions to examine them for the presence of abnormalities and tenderness. It should be explained that the purpose of this procedure is to examine the ear for fitting the most appropriate HPD.

If the employee's history of ear disease is positive, the examiner should document and clarify the reported ear problems before proceeding. For example, if the employee reports a history of ear pain with drainage, the examiner should determine if the condition occurred in the past or is currently present. If the condition is currently present, manipulation of the ear may cause discomfort, and placing the tip of the otoscope in the ear canal may be contraindicated. If the condition occurred in the past, but is now resolved, it is doubtful that manipulation of the ear would cause discomfort.

Examination of the external ear should continue to the ear canal. The entrance to the ear canal should be carefully evaluated for the presence of drainage or a foreign body. Either of these abnormalities would not only contraindicate the insertion of a HPD, but would also contraindicate the insertion of an otoscope into the ear canal by anyone other than a qualified professional.

The examiner may proceed with the evaluation if the entrance to the ear canal appears clear. If an otoscope is to be used to help illuminate and magnify the ear canal and the eardrum, the examiner must have a good understanding of how to hold an otoscope as well as how to place it in the ear canal. First, the examiner should choose the largest tip that will fit comfortably into the ear canal. This will not only permit better viewing, but will also reduce the chance of causing discomfort to the employee. The pinna of an adult should be gently pulled upward and backward or outward to straighten the ear canal for viewing the eardrum and the inner portion of the ear canal (see Figure C 7.15, color photographs, p. 232). The otoscope should be held like a pencil between the thumb and forefinger. For the right-handed examiner, the hand should be braced securely against the employee's cheek when viewing the right ear (see Figure C 7.16, color photograph, p. 232) and against the mastoid when viewing the left ear. This will reduce the chance of causing any trauma if the employee moves suddenly. The examiner should place the tip of the otoscope approximately halfway into the ear canal and attempt to view the eardrum, as well as all portions of the ear canal. The examiner should remain cognizant of the employee's reaction and be aware of any signs of distress. Once the examiner has obtained a good view of the eardrum and all areas of the ear canal, the otoscope tip should be removed from the ear canal and disposed of. The examiner should immediately document the otoscopic findings.

To evaluate the ear canal for collapsing ear canals, the examiner should gently press the helix of the pinna toward the mastoid bone while observing the ear canal with a light source. The otoscope tip should not be placed in the ear canal for this evaluation. If the employee has a collapsing ear canal, the canal will close when the pinna is pushed toward the mastoid. Approximately four percent (4%) of the general population may have collapsing ear canals (Chaiklin and McClellan, 1982).

Visual evaluation of the ear should be completed prior to the allocation of HPDs and before the completion of audiometric testing. This is essential so that ear abnormalities that may contraindicate the use of one or more types of HPDs are identified and not aggravated by the use of HPDs, and so that ear conditions that may affect audiometric test results are identified prior to testing.

Visual evaluation of the ear should be completed at least annually in conjunction with audiometric testing, but may be required more frequently if an employee suspects that an ear problem was aggravated by the use of HPDs.

Acting on Screening Results

POLICY

A company policy should be established regarding documentation of suspected ear abnormalities, employee notification, and medical referral. If the visual evaluation of the ear is completed prior to the allocation of HPDs, it should be obvious that the suspected abnormality was not caused or aggravated by the HPD. In this case the examiner should document that an abnormality is suspected, notify the employee, and suggest a medical evaluation. Earmuffs should not be used if a pinna abnormality is suspected, and insert-type HPDs should not be used if an external ear canal abnormality is suspected until medical clearance is received.

If the visual evaluation of the ear is completed after the employee has been wearing HPDs, the examiner cannot determine if the HPD caused or aggravated the abnormality. However, the abnormality should be documented, the employee notified, and a medical referral made. Use of an inappropriate type of HPD should be discontinued until medical clearance is received.

The company should establish a policy regarding medical referrals. Some companies may choose to pay for all medical referrals as an employee benefit regardless of the cause of the abnormality. Many insurance carriers will cover the cost of these referrals, but each insurance company is different. The exact employee coverage should be investigated for each insurance company prior to establishing a company policy.

CASE HISTORY

Auditory history information can be very helpful in confirming suspected abnormalities and determining if a medical referral is necessary. The case history should include questions regarding draining ears, earaches, ear surgery, excessive ear wax, numbness in the face, ringing in the ears, and dizziness (refer to Chapter on Hearing Measurement). Earaches, numbness in the face, ringing in the ears, and dizziness may or may not be caused by an ear problem, but each of these symptoms may be an indication of an underlying ear abnormality. Auditory history information will be extremely useful to the reviewing professional to help substantiate the results of visual evaluation as well as to determine if a medical referral is necessary.

DOCUMENTATION AND REFERRAL

It is advantageous to establish an understanding with a physician prior to making referrals. It is essential that the physician (preferably an ear, nose and throat (ENT) physician, otolaryngologist or otologist) be aware that the individual's employment requires the use of HPDs and that if the individual cannot use some type of HPD, his job may be in jeopardy. The physician needs to be educated regarding the consequences of a recommendation that HPDs not be used.

If a medical referral is suggested, the examiner should provide a medical referral report for the employee to take to the physician. This report should state that a visual evaluation of the ear was completed, summarize the results, and request a diagnosis and recommendation from the physician. It is important to inform the physician that the visual evaluation was completed as part of a hearing conservation program, and that the employee is required to wear HPDs. The physician's recommendation may fail to include a statement regarding HPD use if the physician is unaware the patient is required to wear some form of HPD.

A medical referral for a suspected ear abnormality may not be necessary if the employee is already under the care of a physician. However, documentation of the suspected abnormality and medical clearance from that physician for the use of HPDs should be on record.

FOLLOW-UP BENEFITS

Visual evaluation of the ear is an excellent employee benefit that can help to increase employee awareness and understanding of aural hygiene. The results from the visual evaluation of the ear can be used to improve counselling and discipline strategies for problem or complaining employees. For example, an employee may complain that earplugs are uncomfortable or cause pain. These objections may be consistent with an external ear infection. If the visual evaluation of the ear indicates an abnormality, the appropriate course of action would be a medical referral for the treatment of the ear infection or abnormality and the use of earmuffs until the problem is cleared. It would be inappropriate to allow the use of insert-type HPDs which would likely aggravate the ear canal infection, causing more pain and longer employee absence from work. On the other hand, if the employee is complaining that earplugs are uncomfortable or cause pain and has had a normal visual evaluation of the ear, the proper course of action would be to suggest another type of HPD (either a different type or size of plug or a muff). Medical referral would not be necessary in this case.

Documentation of an individual's ear condition throughout his/her employment may be very helpful in compensation cases, especially if pathology is involved. Completion of visual evaluation of the ear will decrease the chances of causing or aggravating ear pathology with the use of HPDs and thereby should reduce potential company liability.

References

Becker, W., Buckingham, R.A., Holinger, P.H., Steiner, W., and Jaumann, M.P. (1984). *Atlas of Ear, Nose and Throat Diseases,* W.B. Saunders Company, Philadelphia, PA.

Berger, E.H. (1985). "EARLog #17 — Ear Infection and the Use of Hearing Protection," *J. Occup. Med. 27(9),* 620-623.

Chaiklin, J.B., and McClellan, M.E. (1982). "Audiometric Management of Collapsible Ear Canals," *Hearing Measurement,* edited by J.B. Chaiklin, I.M. Ventry, and R.F. Dickson, Addison-Wesley Publishing Co., Reading, MA, 122.

Concept Media (1977). "Physical Assessment: Eye and Ear, " Concept Media, Costa Mesa, CA, 77-88.

Foltner, K.A. (1984). "The Case for Otoscopic Screenings in Industrial Hearing Conservation," *Hearing J. 37(6),* 27-30.

Hawke, M., Keene, M., and Albert., P.W. (1984). *Clinical Otoscopy—A Text and Colour Atlas,* Churchill Livingstone, Edinburgh, UK.

Hawke, M., Wong, J., and Krajden, S. (1984). "Clinical and Microbiological Features of Otitis Externa," *J. Otolaryngol. 13(5),* 289-295.

Hoadley, W. and Knight, D. (1975). "External Otitis Among Swimmers and Non-Swimmers," *Arch. Environ. Health 30,* 445-448.

Hopkinson, N.T. (1981). "Prevalence of Middle Ear Disorders in Coal Miners," NIOSH, Cincinnati, OH, 1-39.

McLaurin, J.W. (1973). "Trauma and Infection of the External Ear," *Otolaryngology Volume II,* edited by M.M. Paparella and D.A. Shumrick, W.B. Saunders Company, Philadelphia, PA, 24-32.

Mechner, F. (1975). "Patient Assessment: Examination of the Ear," *Am. J. Nursing 75(3),* 1-24.

Paparella, M.M. (1973a). "Otalgia," *Otolaryngology Volume II,* edited by M.M. Paparella and D.A. Shumrick, W.B. Saunders Company, Philadelphia, PA, 33-35.

Paparella, M.M. (1973b). "Cysts and Tumors of the External Ear," *Otolaryngology Volume II,* edited by M.M. Paparella and D.A. Shumrick, W.B. Saunders Company, Philadelphia, PA, 36-43.

Paparella, M.M. (1973c). "Surgery of the External Ear," *Otolaryngology Volume II,* edited by M.M. Paparella and D.A. Shumrick, W.B. Saunders Company, Philadelphia, PA, 44-52.

Senturia, B.H., Marcus, M.D., and Lucente, F.E. (1980). *Diseases of the External Ear — An Otologic-Dermatologic Manual,* Grune and Stratton, New York, NY.

Zemlin, W. (1968). "The Ear," *Speech and Hearing Science: Anatomy and Physiology,* Prentice Hall, Inc., Englewood Cliffs, NJ, 363-367.

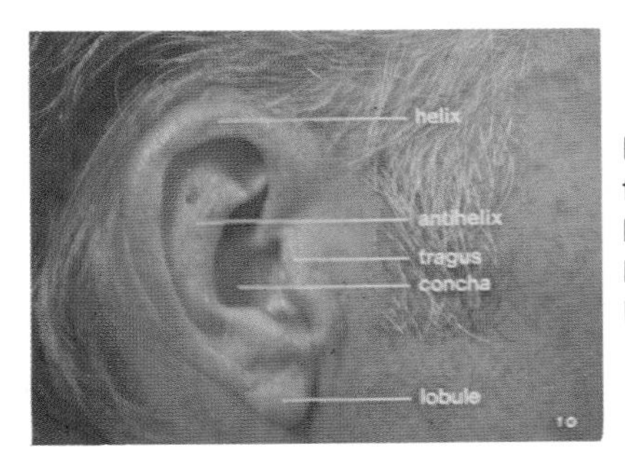

Figure C 7.1 — Principal anatomical features of the pinna including the helix, anti-helix, tragus, concha, and lobule. (Picture courtesy of Concept Media, Irvine, CA).

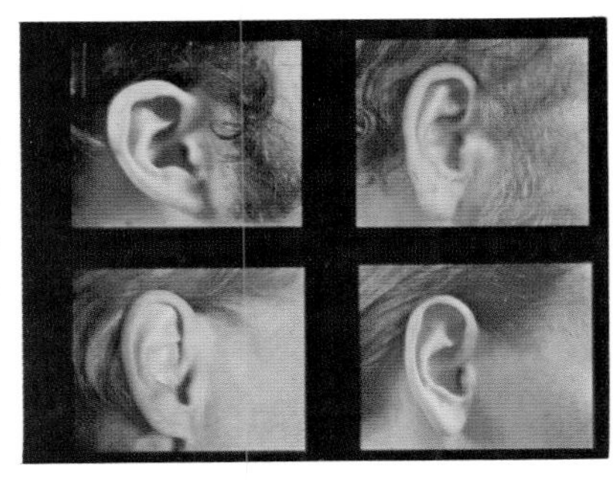

Figure C 7.2 — Examples of normal pinnae. Note the various shapes, sizes and angles of attachment. (Picture courtesy of Concept Media, Irvine, CA).

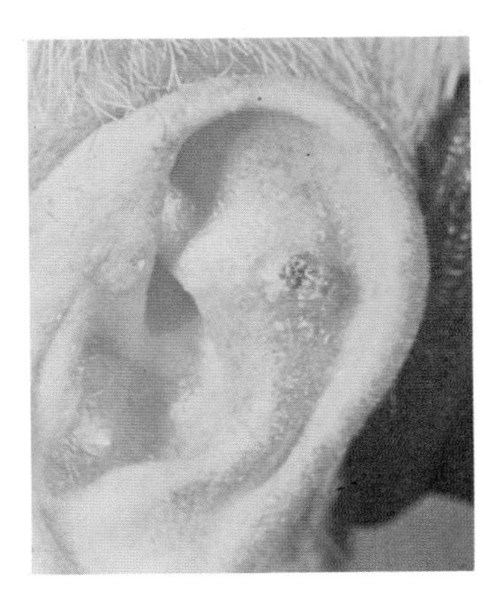

Figure C 7.3 — A mass on the anti-helix of the pinna. (Picture courtesy of Richard Buckingham, M.D.)

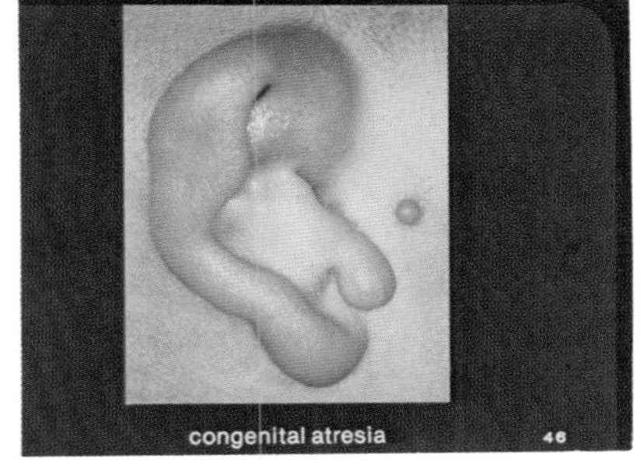

Figure C 7.4 — Congenital atresia or the complete absence of an ear canal. (Picture courtesy of Richard Buckingham, M.D.)

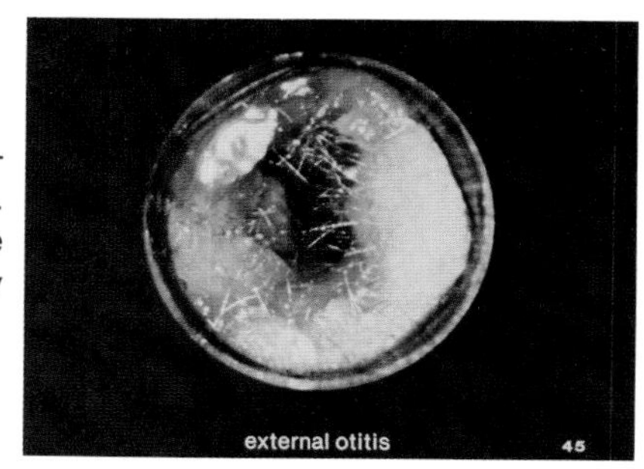

Figure C 7.5 — External otitis otherwise known as external ear infection. The canal walls are swollen and there is drainage present. (Picture courtesy of Richard Buckingham, M.D.)

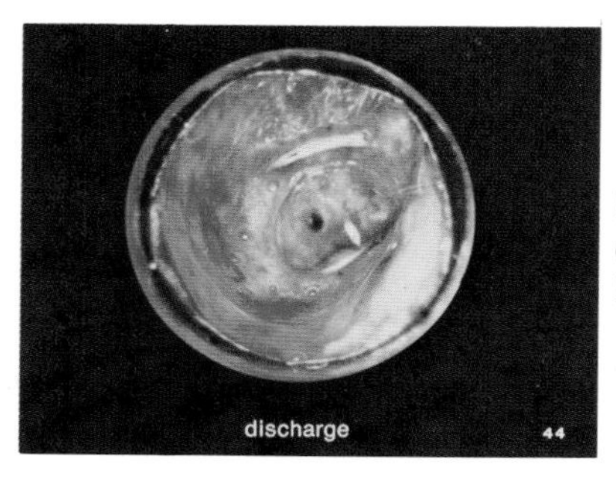

Figure C 7.6 — External ear canal discharge. (Picture courtesy of Richard Buckingham, M.D.)

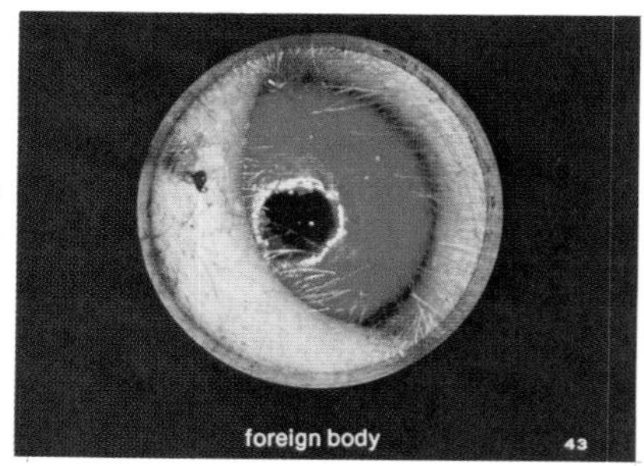

Figure C 7.7 — A bead lodged in the external auditory canal. (Picture courtesy of Richard Buckingham, M.D.)

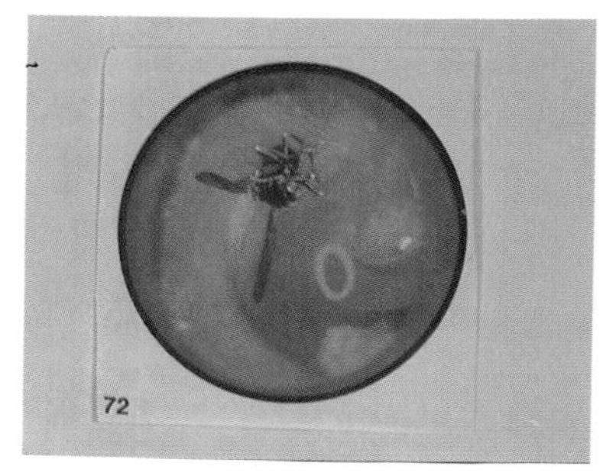

Figure C 7.8 — A clear pressure equalization (PE) tube located in the eardrum and a small insect located in the ear canal. (Picture courtesy of Becker *et al.*)

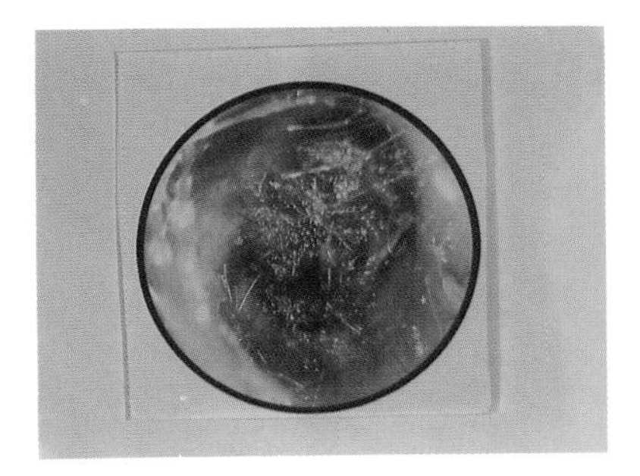

Figure C 7.9 — Occluding wax in the external auditory canal (Picture courtesy Becker *et al.*)

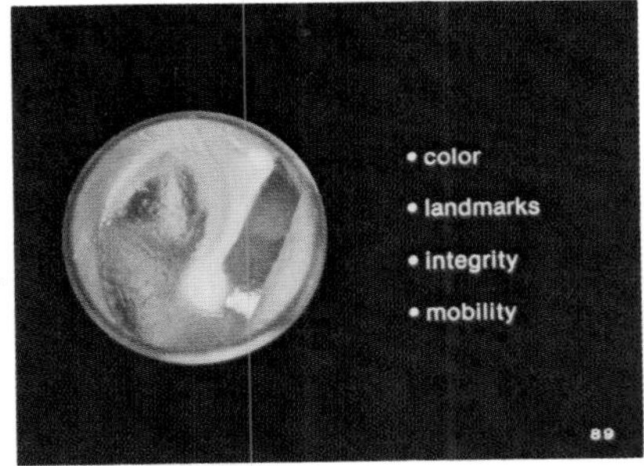

Figure C 7.10 — Principal evaluation features of the eardrum including color, landmarks (bones, light reflection), integrity, and mobility. (Picture courtesy of Richard Buckingham, M.D.)

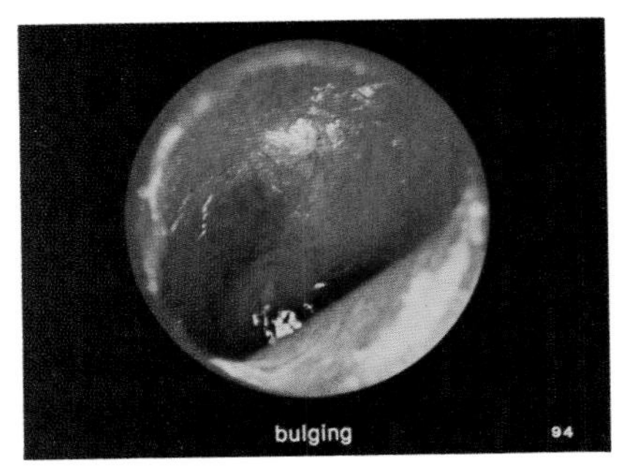

Figure C 7.11 — Red and bulging eardrum consistent with one type of middle ear infection. (Picture courtesy of Richard Buckingham, M.D.)

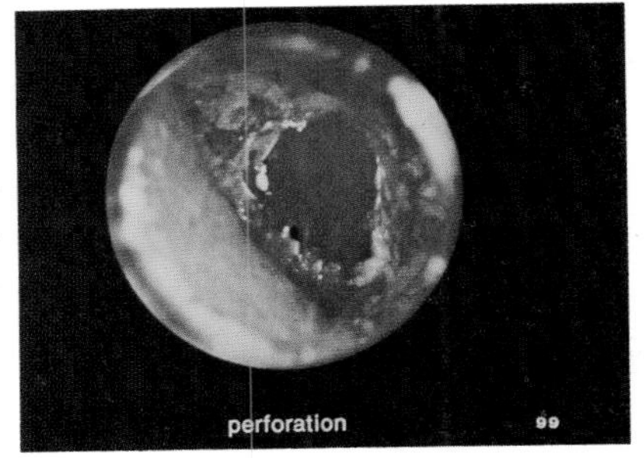

Figure C 7.12 — A hole in the eardrum otherwise known as a perforation. (Picture courtesy of Richard Buckingham, M.D.)

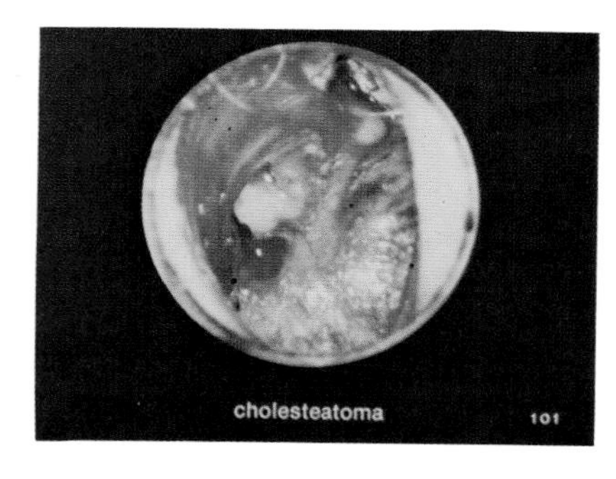

Figure C 7.13 — A cholesteatoma which is a benign growth which may be located in the middle ear cavity. (Picture courtesy of Richard Buckingham, M.D.)

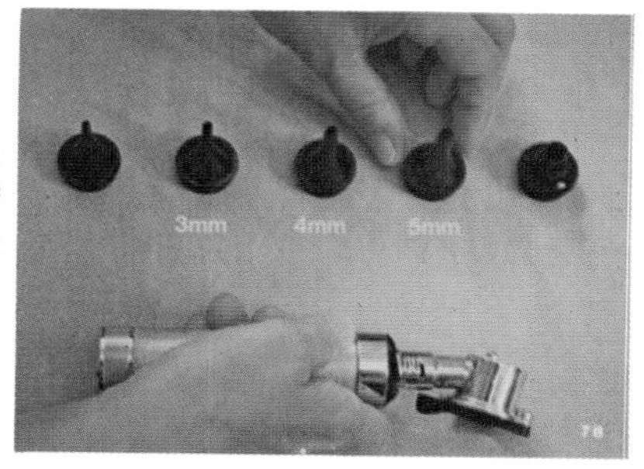

Figure C 7.14 — A portable hand-held otoscope shown with various sizes of specula. (Picture courtesy of Concept Media, Irvine, CA.)

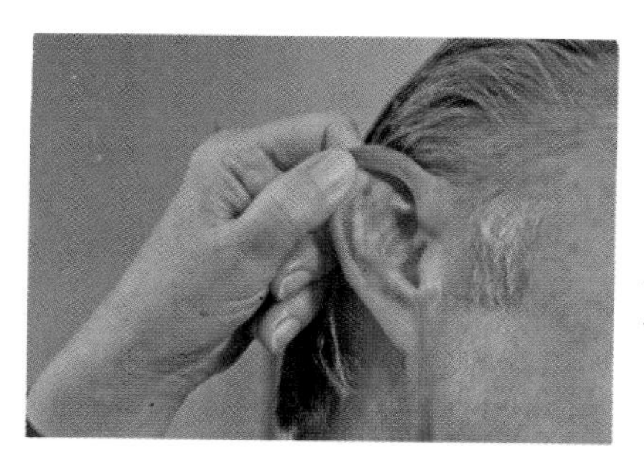

Figure C 7.15 — The pinna must be pulled upward and backward or outward to straighten the ear canal for viewing the eardrum. (Picture courtesy of Concept Media, Irvine, CA.)

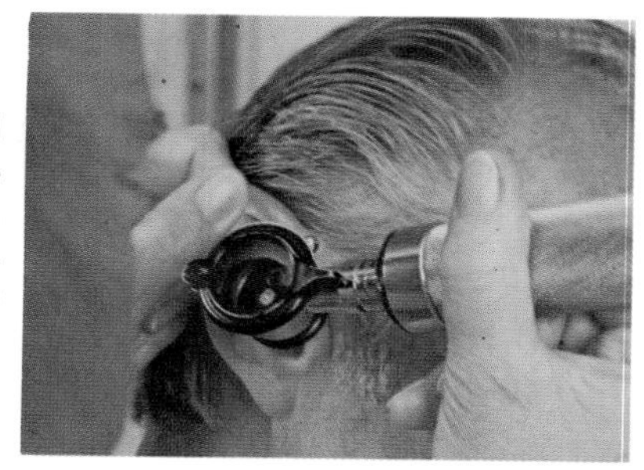

Figure C 7.16 — The examiner's hand should be braced securely against the employee's cheek or mastoid when using an otoscope for a visual evaluation of the ear. (Picture courtesy of Concept Media, Irvine, CA).

Noise and Hearing Conservation Manual, edited by
E.H. Berger, W.D. Ward, J.C. Morrill and L.H. Royster

8 Hearing Measurement

Jeffrey C. Morrill

Contents

Page

Introduction 234

Audiometric Test Environment 235
Noise Level Considerations 235
Controlling the Audiometric Test Environment 237
Single Station Test Rooms 237
Multiple Station Test Rooms 238
Selecting the Test Area and the Audiometric Testing Room 240
Noise-Attenuating Earphone Enclosures 241
Documentation of Noise Levels in the Testing Environment 244

Pure-Tone Audiometry: Instrumentation and Technique 245
Manual Audiometry 246
Self-Recording Audiometry 250
Scoring (Grading) Self-Recording Audiograms 256
Computer-Controlled Audiometry 257
Differences in Thresholds Across Procedures 261

Audiometer Calibration 264
Functional Calibration Checks 265
Acoustic Calibration Checks 266
Exhaustive Calibration Checks 268
Calibration Records 268

Training for Audiometric Testing Personnel 269
Audiometric Technicians 269
Occupational Hearing Conservationists 269
Hearing Professionals and Physicians 271

Preparing Employees for Audiometric Testing 272
Pre-Test Quiet Period for Baseline Testing 272
Options to Achieve 14 Hours of Quiet for Baseline Testing 272
Utilization of Protection for Periodic (Annual) Hearing Tests 273
Pre-Test Information for Employees to Ensure Valid Tests 274
Test Instructions to Enhance Quality 275
Earphone Placement 276
Employees with Testing Problems 277

Supportive Test Procedures 279
Employee Aural History 279
Employee Noise History 280
Otoscopic Examination 280
Review and Treatment of Audiogram(s) 280
Technician's Review and Determination of "Problem Audiograms" 283
Determination of Standard Threshold Shift 283
Hearing Norms and Age Adjustment of Audiograms for Presbycusis 284
Professional Review of Problem Audiograms 284
Employee Follow-up Action Recommended when STS Occurs 285
Non-OSHA-Related Referrals 286
Summary 289
References 290

Introduction

Hearing measurement is central to the industrial hearing conservation program (HCP), since all follow-up activities and determination of program effectiveness are based upon the test results. It is the yardstick against which all programs are measured.

At best, the accurate measurement of human hearing is difficult and the consistency of measured responses rarely complements the level of precision of the audiometric instruments employed. Both the tester and the employee must be properly trained and prepared if valid and accurate results are to be achieved.

Hearing measurement for industry is accomplished by conducting pure-tone audiometry. The primary purpose of audiometry in industry is to detect changes in hearing, or threshold shifts. Initially, a baseline audiogram is established to which future periodic audiograms may be compared. Periodic audiograms should be conducted annually. Reliability in audiometric testing requires a skilled tester, a controlled test environment, instrument accuracy and employee cooperation. Understandably, the possibility of obtaining erroneous test results is always present.

Certain aspects of the audiometric evaluation procedure are easily controlled in the process of achieving valid test results: the physical parameters of instrumentation and test environment. Somewhat more difficult to control are the techniques applied by the tester in administering the tests. The least controllable element is the employee. This chapter will explore both the problems in and possible solutions to these three critical areas, plus required and recommended follow-up activities which strengthen the overall hearing conservation effort.

Compliance with OSHA's requirements for the audiometric testing aspects of the noise regulation may be the foremost consideration in the implementation and maintenance of an HCP. Apart from the OSHA issue, there are significant benefits for both the employer and the employee *if* proper quality controls, documentation and follow-up measures are maintained. For the employer's benefit, the efficacy of the audiometric testing program can be substantiated for medicolegal purposes and thus noise-related hearing losses can be contained. For the employees' benefit, further hearing loss can be prevented, for example, by encouraging the employees to adopt the strict use of hearing protection. Without proper quality controls, the value of the test data are diminished for either prevention or loss control purposes.

In addition to conforming with the legal necessity to prove the integrity of both instruments and test environment, adherence to a set of quality controls will emphasize the importance of precision and accuracy to the individual performing the test and will help promote employee confidence in the overall program. These quality controls are discussed in the following sections.

Audiometric Test Environment

NOISE LEVEL CONSIDERATIONS

There is both a physical and a psychological need for quiet in the testing area. The ambient noise level in the testing environment should be low enough not to interfere with accurate hearing threshold measurement. The maximum permissible noise level during air conduction audiometry (ANSI, 1977) is shown in Table 8.1. Strict adherence to the levels shown in the table is vital in that ambient noise above these levels can adversely affect hearing thresholds through their masking effect on the test signals — thus resulting in higher (poorer) apparent hearing threshold levels (HTLs).

TABLE 8.1
American National Standard for Permissible Ambient Noise During Audiometric Testing, ANSI S3.1-1977

Octave Band Center Frequency (Hz)	500	1000	2000	4000	8000
Octave-band SPL* (dB re 20 μPa)	21.5	29.5	34.5	42.0	45.0

*Permissible noise for ears covered with MX-41/AR cushions.

Valid hearing thresholds are critical to the accurate detection of shifts in hearing due to noise; therefore, the quieter the testing room the better.

It is a common argument that since most industrial workers possess some degree of hearing loss, higher ambient noise levels in the test environment should be allowed. Indeed, this argument apparently influenced the decision by OSHA to permit background noise levels in audiometric test rooms that are 10 to 15 dB higher than those specified in Table 8.1. Table 8.2 shows these OSHA-permitted noise levels. However, if noise levels of this magnitude actually exist, they will not permit accurate measurement of hearing of employees whose hearing at some frequencies is still normal. While the ambient noise may not be intense enough to mask the test signals where hearing loss is present, it may be sufficient to require test signals which are above the individual's actual HTLs at frequencies with normal hearing. Thus, invalid test results will be obtained. Therefore, at least at some test frequencies, most employees will require low ambient noise levels in order to eliminate the problem of masking.

In addition to the effect on hearing threshold, ambient noise in the testing room can have a distracting effect on the employee as well. That is, the process of hearing evaluation (*i.e.,* test environment, preparation, instruction, *etc.*) leads the employee to expect the test room to be extremely quiet. Thus, when ambient noise can be heard, the employee can easily become distracted or annoyed.

Unfortunately, hearing measurement is often referred to as a hearing "test," so employees may fear the consequences of not doing well. (The proper preparation of employees is covered in later sections.) Any outside interference may be perceived as a challenge to perform well on the "test."

TABLE 8.2
Maximum OSHA-Allowed Octave-Band SPLs for Audiometric Test Rooms*

Octave-band Center Frequency (Hz)	500	1000	2000	4000	8000
Octave-band SPL (dB re 20 μPa)	40	40	47	57	62

*The original hearing conservation amendment issued January 1981 (OSHA, 1981) specified the implementation of the noise levels shown in Table 8.1, with 500 Hz modified to allow 27 dB, but permitted the values shown in Table 8.2 on an interim basis until April 1983. However, that aspect of the regulation was changed with the publication of March 1983 amendment (OSHA, 1983) so that Table 8.2 became law on a continuing basis.

This may cause anxiety which can seriously affect the validity of the test results. Even under satisfactory background noise conditions, employees often complain about these transient noises. Body sounds, such as heartbeat or earphone cords rubbing against clothing, are also frequent employee complaints even in an adequately quiet test room.

CONTROLLING THE AUDIOMETRIC TEST ENVIRONMENT

The most desirable approach to controlling test noise is to start with a quiet area away from manufacturing noise, street noise, railroad tracks and busy office activity where conversation is prevalent. In addition to a quiet area, a commercially manufactured audiometric test booth will greatly enhance the reduction of ambient noise, reduce many structural and airborne transients and therefore contribute to more accurate audiometric test results.

Single Station Test Rooms

Both "mini" booths and "standard" prefabricated testing booths are available as commercially manufactured products. The "mini" booth is typically of two-inch wall construction, has magnetic door seals and a circulation fan to bring air from outside to inside the booth. These "mini" rooms can be purchased with casters for greater mobility and are designed to pass through doorways of standard size (Figure 8.1). Their main disadvantage, apart from their limited capability to reduce ambient noise, is that their small size may present problems of entrance and egress for employees with generous dimensions, and may also enhance feeling of claustrophobia in some people.

The "standard" booth adds substantial noise attenuation through walls of 4″ thickness, double-paned windows and isolation rails or pads. This is a more permanent installation which requires disassembly to move to another location if standard doorways are involved (Figure 8.2). In general, the standard room offers greater noise reduction.

Of course, neither of these booths is "soundproof". That is, improper selection and placement of these booths may not satisfy the ANSI (1977) requirements for the ambient noise level in audiometric test rooms, and thus they may be inadequate for testing. In fact, a recent survey (Siegenthaler, 1981) of audiometric test rooms indicated the inadequacy of a number of booths for hearing testing. Out of 46 audiometric test rooms in current use, only 18 met the ANSI (1977) specified levels for ambient noise for testing with ears covered (*i.e.*, for pure tone air-conduction testing) and five met the specified levels for testing with ears uncovered (*i.e.*, bone-conduction testing). This study, furthermore, indicated that of six double-walled rooms, five met the specified levels for tests such as the pure-tone

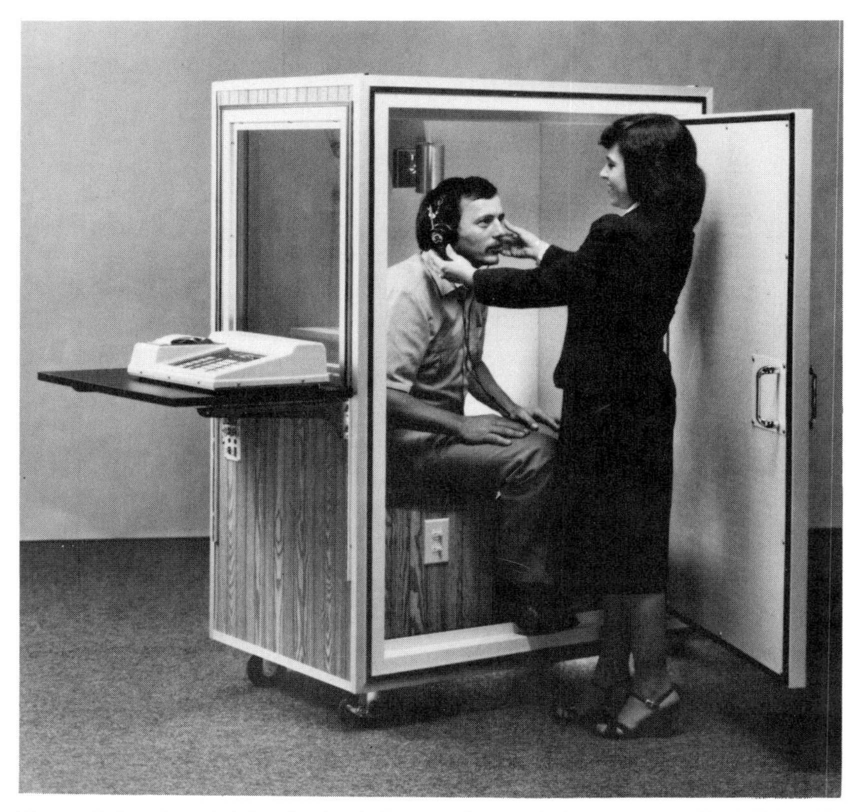

Figure 8.1 — A portable, single station audiometric test room, or "mini" booth. Construction is typically of 2" wall thickness. Photo courtesy of Tracor Instruments, Austin, Texas.

air-conduction testing. It is obvious that the selection of the testing area must be made carefully prior to purchase and installation of the audiometric test booth.

Multiple Station Test Rooms

Often a testing booth with multiple test stations is employed due to the need to test large numbers of people in a short time. Such booths will typically accommodate two to six employees to be tested at a time (Figure 8.3). They can be constructed for as many test stations as one desires; however, it becomes increasingly difficult to obtain reliable test results as the size of the group increases. The presence of several employees may create added ambient noise from coughing, the shuffling of feet, and other noises from body movement. Perhaps the most significant potential distraction is the

Figure 8.2 — A stationary single station audiometric test room. "Standard" booths are typically constructed with 4" wall thickness. Photo courtesy of Industrial Acoustics Company, New York, New York.

presence of loud test signals from other employees' earphones; employees with significant hearing loss may require such loud signals that other employees may hear and even respond to them. It is also difficult, in administering multiple tests, to detect testing problems as they occur. Thus the probability of testing error increases with the number of employees being tested, as the tester simply cannot devote enough time to each individual employee to provide adequate instruction and to monitor the test.

Although it would seem advantageous to isolate the employees within the test booth by means of curtains, this may serve only to add problems to the group test setting. Curtains may visually separate the employees from

Figure 8.3 — A multiple station audiometric test room. These booths are typically constructed of 4″ wall thickness with internal partitions or curtains to visually separate the employees. Photo courtesy of Industrial Acoustics Company, New York, New York.

one another, but they also prevent the tester from observing the employees. Open seating, with good two-way visual contact between test monitor and employees, is preferred over separation by curtains. This will also minimize the claustrophobic feeling experienced by some employees in the test booth.

Selecting the Test Area and the Audiometric Testing Room

The OSHA permissible noise levels (Table 8.2) may actually be achieved in a "quiet" room; however, one must consider the potential problems associated with this approach, since "quiet" rooms generally represent poorly controlled acoustical environments. Furthermore, the political compromises involved in developing the OSHA-permissible levels were such that those levels are not stringent enough to assure that the HTLs of listeners with normal hearing can be accurately measured (cf. Tables 8.1 and 8.2). Thus, merely meeting the OSHA-permitted levels may result in employee hearing loss going undetected at critical speech frequencies, in spurious test results, or in the challenging of program credibility if invalid test results are detected.

The best approach to selecting the area of testing and the proper test booth is to take octave-band sound pressure level (SPL) measurements to establish maximum and typical noise levels. The test booth manufacturers can supply noise attenuation data specifications for the various prefabricated rooms, as shown in Table 8.3. However, these data have been obtained under optimum laboratory conditions, and thus they may differ from the performance of the test room as installed (Singer, 1984). Therefore, a 5- to 10-dB safety factor should be subtracted from the laboratory data before they can be utilized to predict actual room performance (ANSI S3.1-1977).

NOISE-ATTENUATING EARPHONE ENCLOSURES

Situations often exist where the addition of circumaural (around the ear) earphone enclosures may further reduce the ambient noise in the testing environment that reaches the listener's ear. These enclosures (Figure 8.4) are designed to employ the standard earphone and ear cushion (Figure 8.5), which are suspended within the noise-reducing cup. A foam-filled outer cushion seals the device tightly to the head.

There are additional advantages to these earphone enclosures when more than one employee is being tested in the same room. They assist in reducing distracting noises from within the test booth. In addition, when test tones must be presented at high levels due to worker hearing loss, these

TABLE 8.3
Representative Noise Reduction Data for Audiometric Testing Rooms, dB

	Frequency						
	125	**250**	**500**	**1000**	**2000**	**4000**	**8000**
Model AR-200HD[1]	15	35	40	44	52	>52	>52
Model 401-A-SE[2]	31	39	50	57	61	68	62
Model AB-200[3]	15	31	34	42	49	50	51
heavy duty option	24	31	41	46	50	49	50

NOTE: In this context, noise reduction is defined as the difference between the 1/3-octave-band SPLs outside and inside the booth under specified conditions. See ASTM E596-78.

[1]Data supplied by Tracor Instruments, Inc.

[2]Data supplied by Industrial Acoustics Company

[3]Data supplied by Eckel Industries, Inc.

enclosures will provide additional attenuation of the test signals to prevent other employees in the test room from being exposed to them.

The added quiet achieved through use of these devices can help the employees concentrate on their own tests. If testing is conducted in a room where intermittent airborne sounds (such as street noise) create employee distractions, the application should also strengthen the employee's confidence in the testing procedure.

These devices also exhibit significant potential problems. The supra-aural earphones that are incorporated into the circumaural cups can be calibrated using the standard 6-cc coupler (Melnick, 1979). However, finding an earphone to be in proper calibration prior to mounting in the circumaural cup does not assure that it will be properly calibrated after mounting (Martin, 1986). The point is that the standard 6-cc couplers are

Figure 8.4 — Circumaural earphone enclosures are designed to provide additional noise attenuation in the audiometric test area. Photo courtesy of American Overseas Trading Corporation, New Orleans, Louisiana.

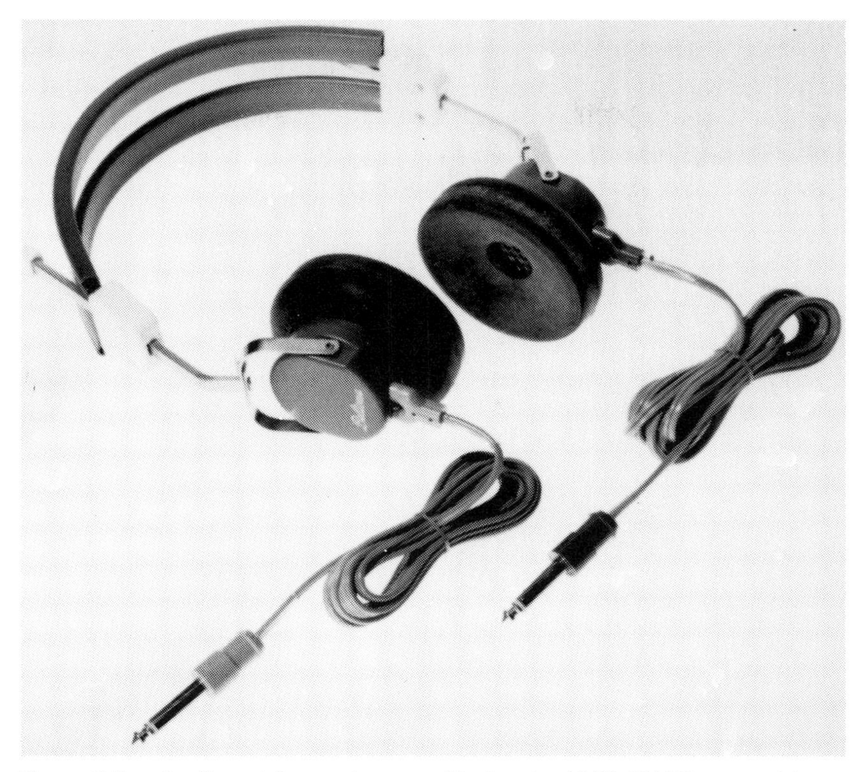

Figure 8.5 — Audiometric earphones with standard MX-41/AR ear cushions. The cushion size and shape is designed to provide a uniform cavity for audiometer calibration. Photo courtesy of Beltone Electronics, Chicago, Illinois.

not designed for calibration of circumaural earphones; thus, use of circumaural earphones should await standardization of these calibration procedures. There are, however, techniques other than the coupler technique for assessing these earphones (Melnick, 1979).

Earphone placement is of particular concern, as the circumaural headset is heavy and it is difficult to see the actual ear cushion when placing the headset over the ear. If the ear cushion is not directly over the ear canal, invalid measures of HTL may be obtained.

Earphone enclosures should only be used by testers who can repeatedly demonstrate proper placement of the earphones and who utilize readjustment when improper placement is suspected. Testers using these earphones should experiment with different placement positions while conducting actual tests on an employee with known and stable hearing levels. The range of HTLs will very convincingly show the importance of consistent earphone placement.

If earphone enclosures are to be utilized to reduce noise to meet the requirements of OSHA's Table D-1 for background noise levels, the ambient noise level must be measured inside the earphone enclosure. However, it is difficult to measure the octave-band SPL inside the earphone enclosure on the employee's head. Therefore, at best, they should only be used as a supplemental noise reduction device.

DOCUMENTATION OF NOISE LEVELS IN THE TESTING ENVIRONMENT

Measurement and documentation of the test room noise level should be performed at least annually. This measurement should be made with a sound level meter with octave filter measuring capacity and should be taken at the place where the employee's head will normally be located during testing (Wilber, 1983). A form suitable for recording this information is shown in Table 8.4.

There are also scanning devices now available which will constantly check the octave-band levels according to preset maximum sound pressure levels. These instruments simply indicate when any of those maximum SPLs are exceeded rather than showing the actual noise level present. They provide an excellent quality control for test settings where marginal noise levels exist, and should help satisfy the acceptable noise level criterion and reduce employee complaints about noises in the "soundproof" room (Figure 8.6). However, the instrument's microphone must be placed in the immediate vicinity of the employee's head to provide a valid check of this kind. If this is not possible, use of a correction factor may be necessary.

The use of either A- or C-weighted measurement scales (dBA and dBC, respectively) or the difference between these two measurement scales is not appropriate for specifying audiometric rooms' noise levels (Siegenthaler, 1981). Sound levels should be measured by equipment conforming to at least the Type 2 requirements for the American National Standard Specifications for Sound Level Meters, S1.4-1983 and to the Class II requirements of American National Standard Specifications for Octave, Half-Octave, and Third-Octave Band Filter Sets, S1.11-1971 [R 1976]. However, once the ambient noise characteristics have been adequately measured, then the overall level can be monitored by means of the A- or C-weighted level, provided that no change occurs in the source(s) of the ambient noise.

TABLE 8.4
Audiometric Testing Facility
Octave-Band Noise Measurements, dB SPL

Location ______________________________ Date ______________

Time(s) (1) __________ am/pm (2) __________ am/pm (3) __________ am/pm

SLM Type ______________________________ Serial # ______________

OB Filter Type ______________________________ Serial # ______________

Microphone Type ______________________________ Serial # ______________

Recorded by ______________________________

	Octave-Band Center Frequency (Hz)					
	250	**500**	**1000**	**2000**	**4000**	**8000**
ANSI S3.1-1977	23.0	21.5	29.5	34.5	42.0	45.0
OSHA (Table D-1)	--	40	40	47	57	62
ACTUAL (1)*						
ACTUAL (2)*						
ACTUAL (3)*						

*Measurements should be taken in region normally occupied by subject's head, during at least 3 intervals representative of typical maximum noise conditions. ANSI values are recommended to assure unmasked threshold measurements; OSHA values are those permitted by law.

Pure-Tone Audiometry: Instrumentation and Technique

The purpose of pure-tone air conduction audiometry is to determine hearing sensitivity at different frequencies. Various methods and techniques are available for this purpose. However, perhaps one of the most difficult decisions to be made in setting up an audiometric testing program

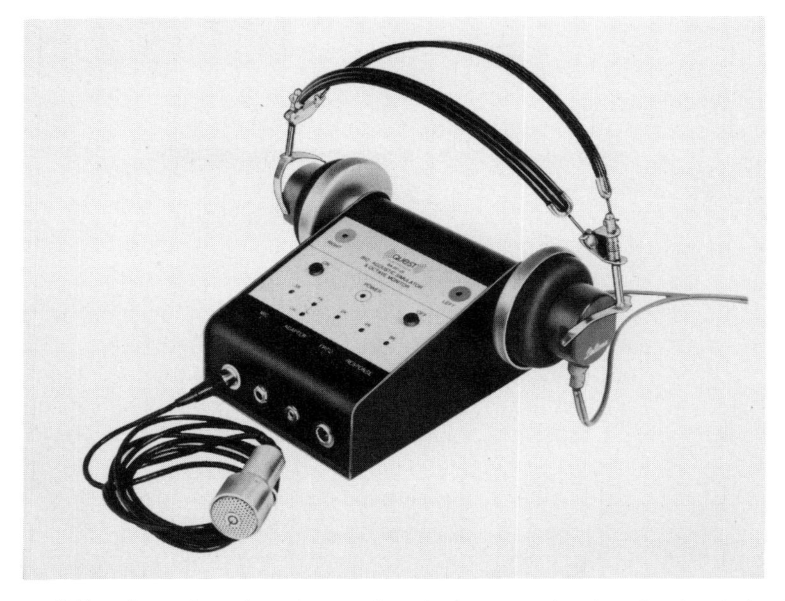

Figure 8.6 — An octave-band scanning device can check noise levels in the test room during audiometric testing. This device can also be used to check earphone output for the daily biological checks. Photo courtesy of Quest Electronics, Oconomowoc, Wisconsin.

is the selection of an audiometer that will best suit both company and employee needs. This is an especially important decision if one method of testing has been employed previously and another type of instrument is currently being considered.

Unfortunately, there can be a difference in employee response, and thus, in test results, associated with the three basic types of audiometric testing instruments. The three types are (1) manual, (2) self-recording (also known as "Bekesy"; formerly referred to as "automatic"), and (3) computer-controlled ("microprocessor") audiometers.

MANUAL AUDIOMETRY

The standard audiometric procedure, for clinical purposes, is manual audiometry (ASHA, 1978). A typical manual audiometer is shown in Figure 8.7. The basic controls of this instrument permit the manual selection of (1) the frequency to be tested, (2) the intensity, (3) right or left earphone, (4) continuous or pulsed tone, followed by (5) presentation of the test signal to the listener. Other features may be added to the manual audiometer, such as provisions for measuring bone conduction, masking, and speech functions. However, these are features that are used in diagnostic testing and are not intended for use by the industrial technician.

The accepted standard procedure for recording of test results is in the form of an X for the left ear and an O for the right ear on the audiogram, as shown in Figure 8.8 (ASHA, 1978). Individual audiograms can be stored in the employee's permanent personnel or medical file. The audiogram form is an excellent educational tool to use in communicating the test results to the employee. The graphic record is easily explained and one can superimpose previous test results to illustrate changes in hearing.

Another recording method is the "serial" audiogram (Figure 8.9). This is a recording form on which many consecutive audiograms can be displayed. Although convenient, the form is more difficult to use as an educational tool than is the traditional audiogram. One disadvantage to this recording procedure is that previous test data can be seen by the individual administering the audiogram. This may influence the tester to accept a response by the worker simply because it occurred at an HTL corresponding to the earlier audiogram. It is recommended that audiometric data be transferred to the serial audiogram only after the test is completed.

The manual audiometer is the least expensive apparatus for testing hearing. Another advantage of manual audiometry is that the one-on-one testing situation, being less impersonal than in automatic audiometry, is conducive to establishment of a good rapport between the employee and

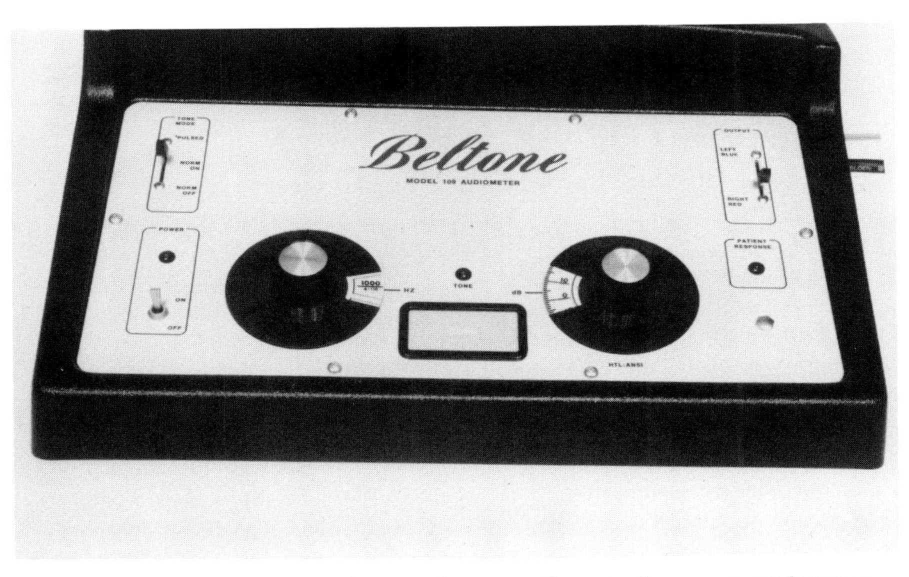

Figure 8.7 — A manual audiometer has specific selection controls for tone level, frequency, and presentation modes. Photo courtesy of Beltone Electronics, Chicago, Illinois.

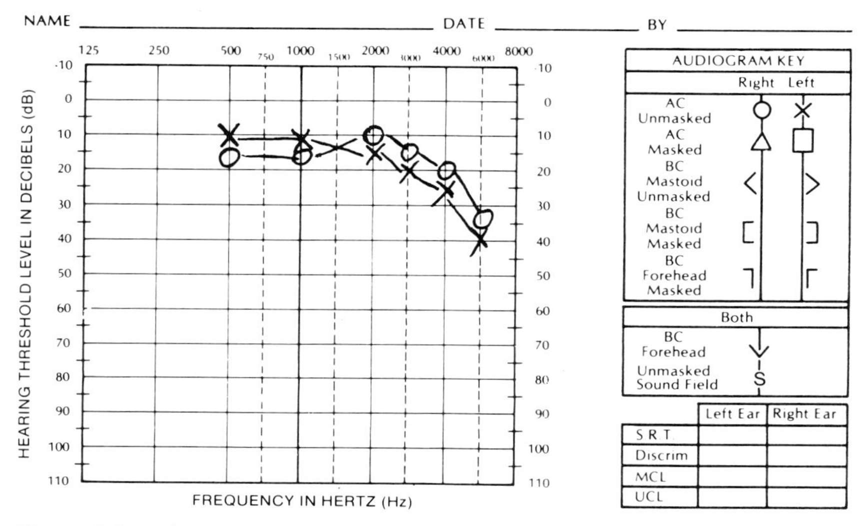

Figure 8.8 — A manual audiogram form includes the test frequencies, and decibels of HTL. Although many audiogram forms include 125 Hz and 250 Hz, those test frequencies are not tested in industrial programs. The HTL for the right ear is recorded with a O, and the HTL for the left ear is recorded with an X.

the tester, which will result in better cooperation on the part of the employee and thus provide more valid test results. However, manual audiometry does have some drawbacks. With several control switches to manipulate, the possibility of error on the part of the tester exists. Anyone who has performed manual audiometry has discovered himself testing the same ear twice by forgetting to manually select the second ear, or recording an employee response at the wrong place on the audiogram. In addition, the tester has the problem of deciding when an acceptable response to the tone has occurred: how soon after the tone is turned on the employee must raise his finger or press a response button if the response is to be accepted as valid, and how long the finger must be kept raised or the button pressed. Testers will often interpret an eye squint or facial grimace as a positive response, thus injecting additional subjectivity into the test.

Technicians using the manual audiometer may develop their own self-tailored technique which they believe works best for them. A lack of standardization from tester to tester and plant to plant may create a challenge to the validity of audiograms conducted by the manual test technique.

Three suggested methods for determination of pure-tone thresholds are: ascending, descending and ascending-descending (bracketing) procedures (Hodgson, 1980).

AUDIOLOGICAL EXAMINATIONS

Division

Plant | Social Sec. Number (1-9)

Last Name | First Name | Middle Int. | Date of Birth (10-6) | Sex (17) Male ☐ Female ☐

Type of Audiometer | Model No. | Serial No. | Audiometer Calibration Standard

Test No.	Type of Test	Test Year	RIGHT EAR						Retest	LEFT EAR						Retest	Time of Test to nearest hour 24 hour system	No. of hours since Recent Noise Exposure	No. of hours exposed to noise	Ear Protectors worn during Recent Noise Exposure	Ear Protectors Worn past year?	Employees hearing evaluation	Hobbies during past year?	Second job during past year yes/no		Audiogram Classification	Employees Signature
			500	1000	2000	3000	4000	6000	1000	500	1000	2000	3000	4000	6000	1000				Yes No		L R					
18-19		20-21	22-23	24-25	26-27	28-29	30-31	32-33	34-35	36-37	38-39	40-41	42-43	44-45	46-47	48-49	A	B	C	D	E	F	G	H	I	K	
1																											
2																											
3																											
4																											
5																											
6																											
7																											
8																											
9																											
10																											
11																											
12																											
13																											
14																											
15																											
16																											
17																											
18																											
19																											
20																											

CODES:

Type of Test: Annual (Air) 1; Pre-employment (Air) 2; Recheck (Air) 3; Recheck (Bone) 4; Termination (Air) 5; 6 ......; 7 ......; 8 ......; 9 ......

Column No. E: Always A; Most times M; Seldom S

Column No. F: Good G; Fair F; Poor P

Hobbies: Motorcycle 1; Boating 2; Hunting 3; Woodwork 4; Loud Music 5; 6 ......; 7 ......; 8 ......; 9 ......

Figure 8.9 — The serial audiogram form is utilized to record multiple test results and history information.

In the ascending method, hearing thresholds are determined by initially presenting test tones at intensity levels that are inaudible to the listener and gradually increasing the intensity of the tones to audible levels; in the descending method, thresholds are determined by presenting the test tones at intensity levels that are audible and gradually decreasing the test tone to inaudible levels; in the bracketing procedure, both ascending and descending trials are utilized for threshold determination (Hodgson, 1980). These methods as well as the other psychophysical procedures (Wilber, 1979) can be utilized for threshold determination. The standard audiometric procedure, however, is the modified Hughson-Westlake (Carhart and Jerger, 1959) which is based on the ascending technique. Carhart and Jerger (1959) compared the Hughson-Westlake ascending procedure with descending and ascending-descending procedures on a group of normal hearing listeners. This study found that the average threshold differences among the three procedures were less than 2 dB (with the descending procedure resulting in more sensitive thresholds than the other two procedures). The variability in thresholds was also the same across procedures. The Hughson-Westlake procedure, with some modifications, was recommended for clinical utilization. This procedure has been widely accepted and used for conducting most hearing tests for children and adults (Martin, 1986; Wilber, 1979). The procedural steps are outlined in Table 8.5.

SELF-RECORDING AUDIOMETRY

The Bekesy-type or self-recording audiometer was originally introduced by Bekesy (1947). This type of audiometer offers a different way of measuring the employee's hearing threshold levels than manual audiometry.

The self-recording method is a tracking procedure in which the employee is allowed to track his own thresholds by pressing and releasing a response switch. The audiometer automatically increases the intensity of the test signal until the listener presses the button, at which point the intensity of the signal begins to decrease. The listener is required to continually press the button as long as he hears the test signal. When he no longer hears the signal, he releases the response button which again automatically increases the signal intensity. A chart pen, which is controlled by the subject's response button, tracks this increase-decrease response to intensity and thus provides tracing of the subject's adjustment over time. Hearing thresholds are determined from these tracing excursions.

McMurray and Rudmose (1956) described an automated discrete frequency self-recording audiometer (ARJ-3), for industrial audiometric purposes. This audiometer incorporated six test frequencies (0.5, 1, 2, 3, 4, and 6 kHz) with each test frequency presented for 30 seconds. Self-recording audiometers similar to this have been extensively used in industrial settings (Figure 8.10).

An important aspect of the self-recording procedure is that every tone presentation and employee response is recorded, which provides the tester and the supervising professional documentation for interpretive purposes (Figure 8.11). This documentation offers a "signature" to the audiogram which can add credence to the validity of the test, or document a questionable or invalid test. The step-by-step procedure for operation of a self-recording audiometer is given in Table 8.6.

There are also automatic functions to this instrument which eliminate some aspects of human error prevalent in manual audiometry. More specifically, the audiometer switches from the left to the right ear automatically. The employee's response is also automatically recorded at the correct test frequency and level.

With the self-recorder, testing problems can be identified by observing the employees' graphical responses in the form of "tracings". There are many visual indications of test problems including: (1) excursions of the

TABLE 8.5
Manual Audiometric Testing

I. Introduction

A. Description: A manual audiometer is an instrument capable of producing pure tone stimuli of varying intensity and frequency, and directing these stimuli to either ear via earphones.

B. Employee Response: The employee responds by raising his hand or pressing a response switch when he detects a pure-tone signal.

C. Operation: The operator varies the intensity of the pure-tone signal in a systematic manner until hearing threshold is determined at each target frequency, for each ear.

D. Data: Hearing thresholds are plotted on a graph (an audiogram) following testing.

E. Maintenance, Calibration, and Troubleshooting: Refer to the audiometer instruction manual for this information.

II. Operation

A. Preparation
1. Make sure that noise in the test area is not excessive (refer to ANSI S3.1-1977 criteria for permissible ambient noise during audiometric testing).
2. Turn on the audiometer and check its operation carefully.
3. Allow the audiometer a warm-up period before use, if required.
4. Perform a functional check (listening check) and record the result.
5. Obtain an adequate supply of audiogram forms to record results.

B. Pretest
1. For baseline hearing test, fourteen hours must have elapsed since the subject was last exposed to high noise levels.
2. Perform an otoscopic examination to detect possible obstructions or medically referable conditions.
3. Glasses, hair ribbons, headbands, clips, gum, etc., should be removed prior to testing.
4. Obtain a complete case history from the employee (see Figure 8.19).

C. Instructing the Employee:
1. Seat the employee facing away from the dials of the audiometer, but where his face may be observed. Avoid eye contact while testing.
2. Keep instructions very simple and consistent.
3. Tell the employee that he will hear a series of tones.
4. He should respond the instant he hears a tone or thinks he hears one.
5. He should respond every time he hears a tone.
6. Answer any questions before testing is begun.

Continued on next page

TABLE 8.5 — continued

D. Administration of the Test

1. Place the headphones on the subject very carefully, making sure the diaphragm of the phone is directly over the ear canal (red: right ear; blue: left ear).

   DO NOT allow the employee to handle the headphones.

2. Begin testing at 1000 Hz in the better ear, or the left ear, if the better ear is not known.
3. Present the tone at 40 dB HL, with a duration of 1-2 seconds. (The duration of test tone presentation must be 1-2 seconds in all the following procedures.)
4. If no response at this initial presentation, present the tone at 50 dB HL and successively increase the intensity in 10-dB steps until a "clear" response is obtained or the maximum limit of the audiometer is reached.
5. Following the initial response, decrease the intensity in 10-dB steps until the listener is unable to hear the tone and thus fail to respond.
6. After each failure to respond to the test tone, increase the intensity in 5-dB steps until a response to test signal is observed. Then decrease the tone in 10-dB steps. Remember and follow this rule (up 5, down 10 dB) strictly.
7. Repeat the "up 5, down 10 dB" procedure until threshold has been determined. Threshold is defined as the lowest level at which the listener has been able to correctly identify the test tone in at least half of a series of ascending trials with a minimum of three responses required at a single level (ASHA, 1978).
8. Test frequencies in this order:
   1000 Hz,
   2000 Hz,
   3000 Hz,
   4000 Hz,
   6000 Hz,
   1000 Hz (retest) and 500 Hz.
9. Avoid rhythmic presentations.
10. If retest at 1000 Hz is ± 10 dB or more, then reinstruct and retest.
11. Test the other ear using the same procedure.
12. Remove the headphones yourself.
13. Record hearing thresholds on the audiogram or audiometric record.
14. Write the employee's name, the date of the test and your signature on the audiogram. The employee should also sign the original record.
15. Always indicate whether the instrument is ANSI or ISO calibrated.

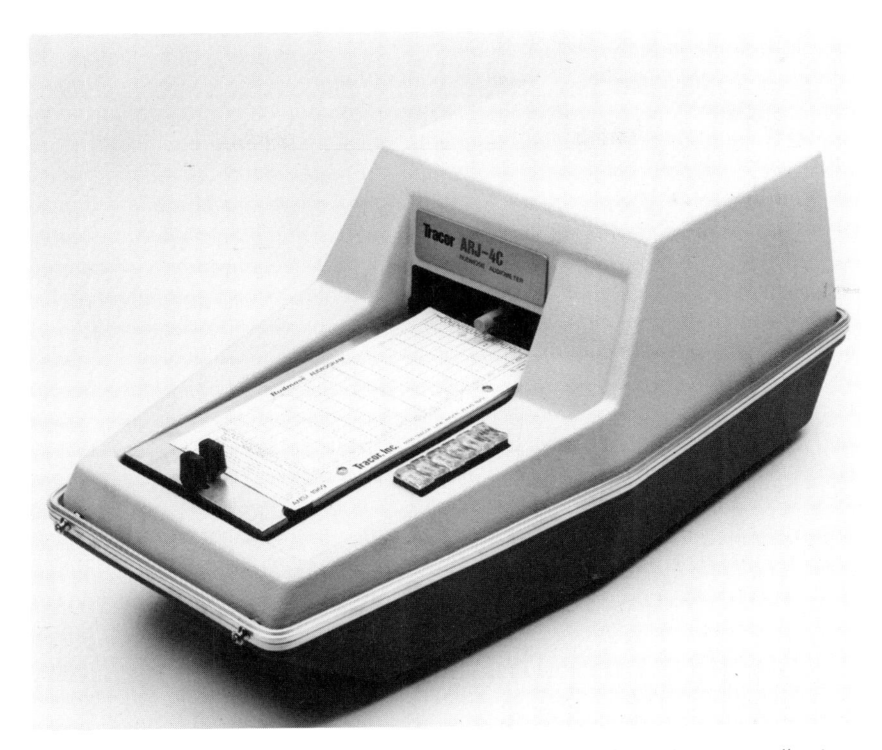

Figure 8.10 — A self-recording audiometer, often referred to as an "automatic" audiometer, records the employee response patterns in the form of "tracings" as the audiogram passes along each 30-second test frequency for both ears. Photo courtesy of Tracor Instruments, Austin, Texas.

tracings from top to bottom that are too wide, (2) a lack of consistency of tracings throughout each 30-second test frequency, (3) lack of repeatability of prior tracings by retest, and (4) failure to respond to automatic and manual validity tests.

Figure 8.11 illustrates acceptable and unacceptable tracings in terms of excursion characteristics. In addition, tests can be performed which help substantiate the validity of the test. The "minus 10" function is a validity test that can be manually administered by the tester. When the minus 10 switch is activated, it stops the horizontal motion of the pen and reduces the level to the earphone by 10 dB. Thus, if the employee is responding at threshold level, there should be a corresponding drop in the tracing by approximately 10 dB as is shown by the spike at 2000 Hz in Figure 8.12. If this does not occur, the employee should be reinstructed and the test started again.

Figure 8.11 — This audiogram displays audiometer tracings that represent typical response patterns. The consistent tracings on the left are acceptable. The tracings on the right are either not uniform in excursion length or in response level; thus, they are unacceptable.

TABLE 8.6
Self-Recording (Automatic) Audiometric Testing

I. Introduction

A. Description: The self-recording audiometer is automatic. Tones are presented to the listener in regulated order. The test time is approximately 7 1/2 minutes.

B. Employee Response: During automatic testing, the employee has control of the sound intensity by use of a handswitch.

C. Operation: The operator may vary usage of the audiometer by selection of various automatic control functions, such as the following:

1. Hold: Inhibits the frequency drive motor with the employee still in control of intensity, so that the operator may manually change the frequency.
2. Stop: Halts the test.
3. Talkthru: Stops the test and allows the operator to reinstruct the employee through a built-in intercom.
4. -10 dB: Allows the operator to reduce the output level 10 dB in order to validate the threshold. Horizontal pen motion is stopped.

D. Maintenance, Calibration and Troubleshooting: Refer to instruction manual.

II. Operation

A. Preparation:

1. Noise in the test area should not be excessive (see Table 8.1).
2. Turn on audiometer and check operation carefully.

Continued on next page

TABLE 8.6 — continued

3. Set the audiometer for a pulsed tone and consult the owners manual to set an appropriate slew rate (no greater than 6 dB/sec).
4. Place audiogram card in position.
5. Perform a functional check (listening check) and document this.

B. Pretest:

1. For baseline hearing test, fourteen hours must have elapsed since the employee was last exposed to high noise levels.
2. Perform an otoscopic examination to detect obstruction or medically referable conditions.
3. Glasses, hair ribbons, headbands, clips, gum, *etc.*, should be removed prior to testing.
4. Obtain a complete case history from the employee (see Figure 8.19).

C. Instructions to the Employee:

1. Seat the employee where he is facing directly away from the audiometer.
2. Keep instructions very simple and consistent.
3. Tell the employee that he will hear a series of tones, and that he can control their loudness with the handswitch.
4. Instruct the employee to press the switch when he hears the tones and to release the switch when he doesn't hear them. The employee's response should be immediate: as soon as he hears them, press; when he no longer hears them, release.
5. Answer any questions before you begin the test.

D. Administration of the Test:

1. Place the headphones on the employee very carefully, making sure the diaphragm of the phone is directly over the ear canal (red: right ear; blue: left ear).

   DO NOT allow the employee to handle the earphones.
2. Press the TEST button to begin the test (the motor may not start until the second response, depending on the type of audiometer).
3. The audiometer will automatically present seven pure-tone signals (each sustained for 30 seconds), first to the left ear and then to the right ear.
4. Conduct a -10 dB validity check for each ear during the course of the test, preferably at 500 or 1000 Hz. The tracing should drop down approximately 10 dB. If not, stop and reinstruct.
5. If you are not satisfied with the subject's response (wide excursions, erratic tracings, poor validity checks or unreliable retests), then stop the test and reinstruct the employee. Erratic tracings may indicate confusion, malingering, drowsiness, or tinnitus problems. After reinstruction, retest the frequencies showing poor tracings.
6. If the employee's responses continue to be inappropriate, a manual test is recommended. You may do this yourself if you have the appropriate equipment. Otherwise, it may be necessary to refer the employee to an outside source.

Continued on next page

TABLE 8.6 — continued

7. The audiometer will shut off automatically when the test is finished. The audiogram should be complete with reliable tracings from 500 to 6000 or 8000 Hz.

III. Evaluation of Results

A. Be sure to monitor the employee's behavior and the audiometric tracings throughout the test. OSHA requires that the horizontal midline indicating audiometric threshold cross the tracing at least six times. If you do not have six tracing excursions at a given frequency, you must retest that frequency.

B. Retest any frequencies which demonstrate inconsistent tracings, or show thresholds over 25 dB at 500, 1000 or 2000 Hz.

C. Scoring of each frequency should be done by the technician administering the test, before the employee leaves the test setting. Refer to section on Scoring (Grading) Self-Recording Audiograms in this chapter. Each frequency should receive a numerical score at the midpoint of the excursions.

Perhaps the most valid reliability test is to retest the employee at any frequencies that are in question, as shown in Figure 8.12. The employee response should be within 10 dB of the previous test just as in the manual audiometry retest procedure.

Many common test problems can also be identified on the self-recording audiometer: tinnitus, possible malingering, coordination problems, fatigue and instruction problems (Figure 8.13). However, the tester should recognize these abnormal test patterns as only an indication to retest the employee, not to diagnose the problem. (Note: The tone selector for "pulsed tone" versus "continuous tone" should always be used in the pulsed mode. This will help to ameliorate many of the test problems mentioned here.)

The employee who cannot, after sufficient reinstruction and assistance, complete the self-recording audiogram successfully should be retested on a manual audiometer. However, unless the employee simply lacks the coordination to respond quickly enough to the test signal on the self-recorder, the manual approach will probably offer no better validity. Referral to an audiologist is recommended for these employees.

SCORING (GRADING) SELF-RECORDING AUDIOGRAMS

The grading of self-recording audiograms introduces the same potential for human error as recording of manual audiometric tests. The tester must visually estimate the midpoints of the tracings and place a horizontal line through them to determine threshold. The OSHA rule (OSHA, 1983) states in its Appendix C (see Appendix I of this volume):

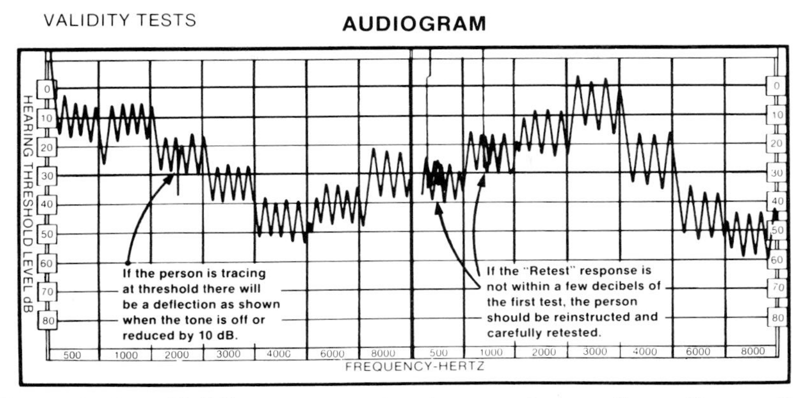

Figure 8.12 — Validity tests can be documented on the self-recording audiometer. The employee response to manually-induced validity checks or to retesting at a frequency should be recorded as is indicated on this audiogram.

> It must be possible at each test frequency to place a horizontal line segment parallel to the time axis on the audiogram, such that the audiometric tracing crosses the line segment at least six times at that test frequency. At each test frequency, the threshold shall be at the midpoints of the tracing excursions.

Figures 8.11 (left ear) and 8.12 (both ears) illustrate tracings that are acceptable for scoring. The first tracing in Figure 8.11, 500 Hz left ear, has 10 excursions. A horizontal line through that time segment would be considered a 5-dB threshold. The remaining test frequencies for that ear would be graded at 15, 10, 25, 35, 50 and 40 dB, respectively, if rounded to the nearest 5 dB.

The technician must adhere to the 6-crossing rule and the practice of placing a horizontal line through the tracing midpoint. If less than six tracings pass through this horizontal line, a retest must be conducted at that test frequency.

COMPUTER-CONTROLLED AUDIOMETRY

The availability of the microprocessor chip has resulted in manufacturers developing a manual audiometer which is controlled by a small computer chip. Such "microprocessor audiometers" are similar in operation to the computer-assisted audiometer originally described by Brogan (1956). The chip is programmed to present a routine manual test, record employee responses and, based upon a programmed software algorithm, determine

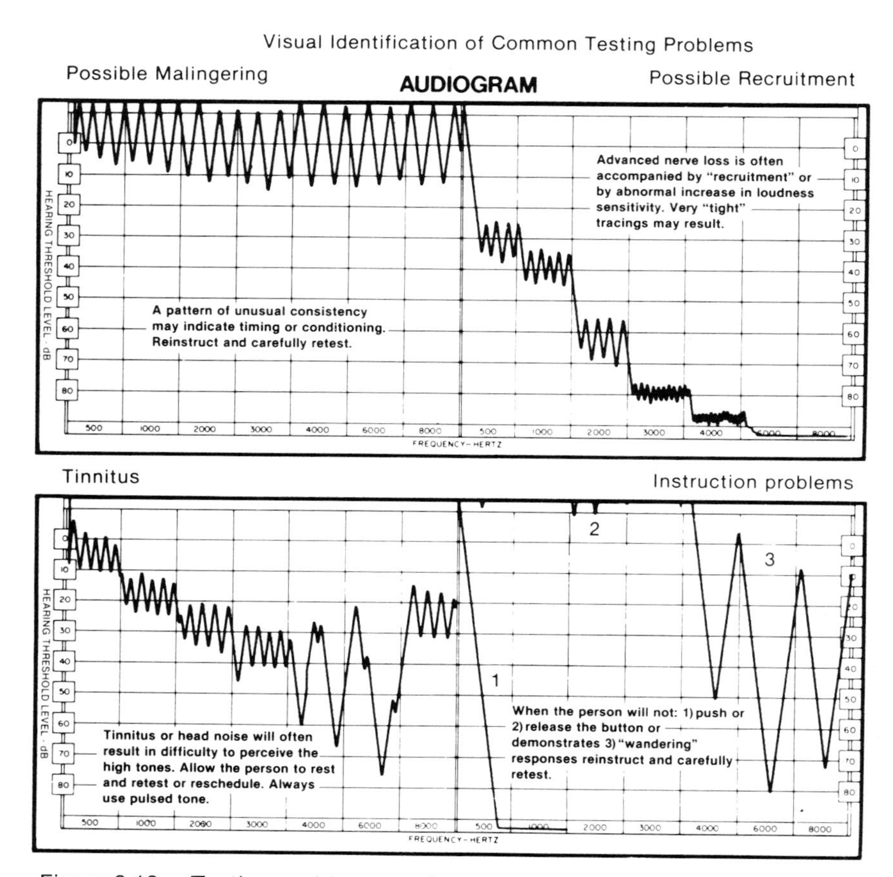

Figure 8.13 — Testing problems can be seen on the self-recording audiogram as illustrated by these tracings. These are typical of common testing problems such as tinnitus and misunderstood test procedures.

the employee's hearing threshold. The microprocessor audiometer, an example of which is shown in Figure 8.14, offers significant advances over traditional manual audiometry as it, like the self-recording audiometer, automatically switches ears, records at the appropriate test frequency and hearing level, and administers preprogrammed validity checks and retests.

In addition, the microprocessor does not adopt a self-stylized procedure and it will perform many functions with consistency. For example, if the employee fails to respond, or threshold is not achieved within the procedures of the program, the audiometer may proceed with the testing of other test frequencies and return to the invalid frequency and attempt to perform a retest. Other microprocessors may be programmed to require a manual test at those failed test frequencies.

One significant advantage of the microprocessor is that its electronic construction makes it more stable than the more mechanical self-recording and manual instruments. The test results can be obtained from a visual display window or a printer (Figure 8.15).

Variation among microprocessors raises some question as to potential problems in test validity. This is particularly important if different microprocessor instruments are to be used on the same employee population, because differences between test procedures may produce divergent employee thresholds. Employees who have been tested by the manual or self-recording means are accustomed to that procedure and may experience difficulty in adapting to the administration of test signals by the microprocessor. Some microprocessors present the tones quickly in sequence and this may condition the employee to respond in anticipation of the test signal. Employees may experience anxiety and feel pressured to respond.

As long as everything goes well with the test and the employee is responding properly, the microprocessor is an outstanding alternative to manual testing. If, on the other hand, the employee does experience problems or cannot keep up with the instrument, the test must be conducted manually.

It is important that procedures for test administration be standardized for the microprocessor instrument. Currently, one of the points upon

Figure 8.14 — Microprocessor audiometers generally have a printer to record the HTLs. Except for the printer, mechanical parts and manual switches are replaced by more dependable electronic components. Photo courtesy of Tracor Instruments, Austin, Texas.

```
CALIBRATED TO
ANSI S3.6-1969

L 1K40-60+50+40+30+
 20+10-15-20+10-15+
 05-10-15+
L.5K25-45+35+25+15-
 20-25+15-20-25+
L 1K05-25+15+05-10-
 15+05-10-15-20+10-
 15-20+
L 2K30+20+10+00-05-
 10+00-05-10+
L 3K20+10+00-05+00-
 05+
L 4K15+05-10+00-05-
 10+
L 6K20-40+30+20+10+
 00-05-10+00-05-10+
R.5K40+30+20+10-15-
 20-25+15-20+10-15+
 05-10-15-20-25-30-
 35-
R 1K40-60-80-95+85+
 75-80-85-90-95-95-
R 2K40-60+50+40+30-
 35-40+30-35-40+
R 3K50-70+60-65-70+
 60-65+55-60-65-70+
R 4K80-95+85-90+80-
 85+75-80-85+
R 6K95+85+75-80+70-
 75-80+
R.5K40-60+50-55-60-
 65-70-75-80+70+60-
 65-70-75+65-70-75+
R 1K40+30-35-40-45-
 50-55-60-65+55-60+
 50-55-60+

FREQ.  L DB R DB
  500HZ  25   75
 1000HZ  15   60
 2000HZ  10   40
 3000HZ  05   70
 4000HZ  10   85
 6000HZ  10   80
THRESHOLD AVERAGE
.5,1,2K  17   58
 2,3,4K  08   65
 3,4,6K  08   78
```

Figure 8.15 — The microprocessor may print HTLs, test averages, total test events, and employee data for permanent records.

which purchase decisions are made is the speed of the test. Since accuracy can be influenced by speed, it would seem appropriate to sacrifice speed for a procedure which would ensure a higher degree of accuracy and validity.

With the continued trend toward more compact and less expensive computers, use of a complete computer rather than just a microprocessor to run and record audiometric testing is becoming more practicable. The advantage of complete computer control (Figure 8.16) is that it can be programmed to do anything; its disadvantage, aside from its cost, is that it must be programmed to do everything that a real tester does. A computer can present test signals in any sequence, taking into account any or all of the prior responses, so that it can simulate any technique whatever, including any form of manual or self-recording audiometry. However, it must also be programmed to make rigorous definition of what constitutes a valid response or series of responses, so that unsatisfactory performance by the employee being tested can be identified in order to initiate corrective action.

DIFFERENCES IN THRESHOLDS ACROSS PROCEDURES

Do different procedures for measuring hearing produce equivalent results? This question is obviously important if HTLs determined with one technique are to be compared with those obtained using a different procedure, such as might be the case for a worker who changed jobs or for an entire industry if a change in procedure occurred at some point.

Clearly, the value of HTL that is the final product of an audiometric procedure depends on a host of conditions beside the state of the sensitivity at a particular frequency of the ear being tested. From the foregoing description of specific procedures to follow in the two main forms of audiometry (Tables 8.5 and 8.6), together with the earlier discussion of problems in even defining "normal" thresholds (see the sections on "Auditory Sensitivity" and "Hearing Threshold Level" in Chapter 5), one can infer that these conditions include:

(1) the specific type of earphone and cushion employed,
(2) the precise position of the earphone on the pinna,
(3) the presence or absence of hair between the earphone and the pinna,
(4) the headband tension (which determines the force with which the earphone is pressed against the pinna),
(5) the duration of the test tone, and
(6) whether it is pulsed or steady,

Figure 8.16 — The computer-controlled or assisted audiometry system also incorporates a personal computer to manage test records and employee data. The audiometer and computer can be designed to communicate with other systems for more detailed analysis, and reporting of data.

(7) the exact instructions to the listener (especially when these either encourage or discourage responding when in doubt),

(8) the order in which test stimuli are presented,

(9) the difference in level between successive test items,

(10) the criterion of what constitutes threshold using the various procedures (*e.g.*, the level responded to 50% of the time, or 2/3 of the time, or 3/4 of the time in the case of manual audiometry; for

self-recording audiometry, whether threshold should be the mean or the median of the peaks of the tracings, the valleys, or, as specified in the OSHA Amendment [see Appendix I], halfway between them), and

(11) for manual and computer-controlled manual procedures, what constitutes a response (how soon after initiation of the test tone a response must occur to be classed as valid).

In view of the number of these conditions which, if not held constant, will affect the measured thresholds, it is not surprising that there is no unequivocal answer to the general question of the extent to which manual, self-recording, and computer-controlled audiometry give equivalent results. There exist too many different variants of each. Indeed, since no particular version of any procedure has yet been declared to be the method that determines the "true" threshold, one cannot even ask which conditions produce results that are "valid" (equivalent to those produced by that selected method).

If, of course, a computer-controlled procedure faithfully duplicates a manual technique in all possible respects, including use of the same earphones in the same position, one would expect completely equivalent results, and this has indeed been demonstrated (Jerlvall, Dryselius and Arlinger, 1983; Cook and Creech, 1983). The same would be true of a "real" self-recording procedure and a computer-simulated one.

The question that remains, then, is the degree of equivalence of the "typical" manual and "typical" self-recording audiogram. On the whole, comparisons of manual and self-recording techniques have shown a slight advantage to the latter: under conditions not remarkably different from those specified in Tables 8.5 and 8.6, the midpoint of the peaks and valleys on the self-recording tracing will occur at an HTL a few dB less than the manually-determined threshold (Burns and Hinchcliffe, 1957; Knight, 1966; D.A. Harris, 1979a, b). This slight difference can readily be wiped out, for example, by instructing the listeners in the self-recording tests to "keep the tone always just audible" (Corso, 1955).

The most recent well-designed research dealing with the effect of method *per se* was reported by J.D. Harris (1980), who compared thresholds determined by three different fully computer-controlled procedures: two of the "manual" type and one "self-recording". In this case, the self-recording HTLs were 5 dB more sensitive than those from one of the manual procedures, but only 1.5 dB lower than the other, which means that the two manual methods differed by 3.5 dB. In other words, changes in procedure within the realm of manual audiometry can produce larger average changes in the indicated thresholds than the changes that can be ascribed to the difference between manual and self-recording techniques. In this particular instance, there were several differences between the two manual techniques; however, the 3.5-dB difference in HTLs was probably mainly due to the fact that in the procedure showing the highest HTLs, the employee not only had to press the response button within 1 second after initiation of the tone, but also had to release it within a quarter of a second after the tone ended, if the tone were to be regarded as "heard".

The conclusion that emerges from this evidence is that there is no simple answer to the question of equivalence of results using different audiometric techniques. Differences in indicated HTLs can arise if the two methods differ in *any* way — in stimulus characteristics, in the definition of the response, or in experimental procedure. Furthermore, if they differ in several ways, it may often be difficult to predict even the direction of the net effect.

In short, different audiometric techniques will usually lead to different average HTLs. Therefore, comparison of average audiograms from year to year in order to estimate the overall success of the HCP (see Chapter 9) will be justified only if the technique has not been altered in any respect, because in this case, differences of 2-3 dB become highly important.

The implications of a change in procedure for the annual audiogram of the individual worker are not quite as extreme, because a difference of only 2-3 dB is smaller than the test-retest variability of ±5 dB to be expected even when the procedures have been held completely invariant. However, the probability that the worker will erroneously be identified as showing a standard threshold shift (STS) (a change from the baseline audiogram of 10 dB, averaged over 2000, 3000 and 4000 Hz) will be increased by any procedural change that results in higher HTLs at all frequencies.

Audiometer Calibration

Conducting audiometric tests for industrial hearing conservation purposes demands not only a competent examiner as well as a controlled acoustic test environment, but also exact instrumentation integrity. That is, a calibrated audiometer is a definite requirement for conducting an accurate hearing test. A calibrated audiometer refers to an audiometer which (a) emits the test signal at the frequency and level which the respective dials show the audiometer to be producing; (b) directs the test signal only to the earphone to which it is set to deliver the signal; and (3) produces a test signal free from contamination by extraneous noises or unwanted byproducts of the test signal (Harford, 1967).

Although there is no certification of personnel conducting audiometer calibration procedures, both OHSA (1983) and ANSI (1969) specifications exist for audiometer calibration procedures and instruments.

FUNCTIONAL CALIBRATION CHECKS

A "functional check" is synonymous with the "biological check" which is recommended by OSHA to be performed at daily testing intervals. Functional checks consist of obtaining hearing threshold levels of a person with known stable thresholds and comparing this test to the baseline for that individual. If changes of 10 dB or greater (in either direction) occur, then a more extensive electroacoustic calibration is required. Of course, the possibility exists that the person's hearing threshold has actually changed; therefore, it is advisable to have more than one person with stable thresholds on whom to conduct the audiometer check.

The functional check is also intended to identify any unwanted or distracting sounds and to ensure that the various automatic or manual functions are working properly. These vary with the make and model of the audiometer. Therefore, the manufacturer's guide should be reviewed and a checklist prepared to document that each of the operational features is working properly.

Often mechanical or electronic switches in the audiometer itself become worn and will produce crackling, hum, distortion or other unwanted sounds which the employee may respond to instead of the desired test signal. By manipulating the dials and various test function switches on the audiometer while listening through the earphones, these unwanted noises can often be detected. It is very important to make sure that the test signal is not emitted through the non-test earphone. The problem of the test signal being present in the non-test earphone is referred to as crosstalk (ANSI, 1969). This check, which must be done at each audiometric frequency, consists of setting the HL dial to at least 60 dB and unplugging the phone to which the signal is delivered and listening to the non-test earphone (Hodgson, 1980). The presence of any audible sound in the non-test earphone will indicate the problem of crosstalk, and, thus, probably invalid test results.

In addition, the earphone cords should be manipulated during the functional check to detect broken or frayed cords. It is also important to remember that earphones should not be brought from the test rooms and directly connected into the audiometer. This is true for any check or calibration, since the connecting cables and plugs themselves must be included in any calibration procedure.

Functional checks should be performed frequently to detect any significant calibration or instrument problems that would affect the audiogram. The record (audiograms) of these checks should be dated, signed and maintained to document the instrument's performance. Although this procedure may seem excessive to some testers, it is certainly less time-consuming and costly than repeating hearing tests after it is discovered that the audiometer has been out of calibration or not functioning properly for some undetermined period of time.

ACOUSTIC CALIBRATION CHECKS

Recommended practice, as well as OSHA regulations (OSHA, 1983; see Appendix I of this manual) requires annual acoustic calibration on audiometers used for industrial hearing conservation purposes. Also, as discussed in the previous section, acoustic calibrations are required whenever deviations in the functional calibration of 10 dB or greater are detected.

It should be noted that calibration checks must be made with the specific set of earphones for that audiometer. The calibration procedure is for the earphones and the audiometer as an integral set. It is not possible to check calibration or calibrate one without the other or to change either without a calibration check.

With the proper equipment, it is possible for the program supervisor, technician or other trained personnel within the industry to check the audiometer calibration acoustically. Since shipping can affect calibration, it is preferable to check calibration at the test location rather than to send the audiometer out for this procedure.

Electroacoustic calibration includes measurement of the SPL output from both earphones and of attenuator linearity. In order to check the SPL output of the earphones, the earphone which is housed in a standard earphone cushion (*i.e.*, MX-41/AR cushion) is coupled to a condenser microphone via the NBS 9A (6 cm^3) coupler (Figure 8.17). A 500-gram force is applied to the top of the earphone to assure an adequate seal.

Hodgson (1980) recommends the following procedure for correct positioning of the earphone on the coupler: (a) present a 250-Hz tone at 70 dB HL (or 500 Hz for industrial audiometers with limited test frequencies); (b) adjust the earphone on the coupler so that the maximum output SPL is obtained; (c) repeat the steps (a) and (b). You should be able to achieve the same reading as the original. Following this adjustment, the earphone should remain on the coupler until calibration for that earphone is com-

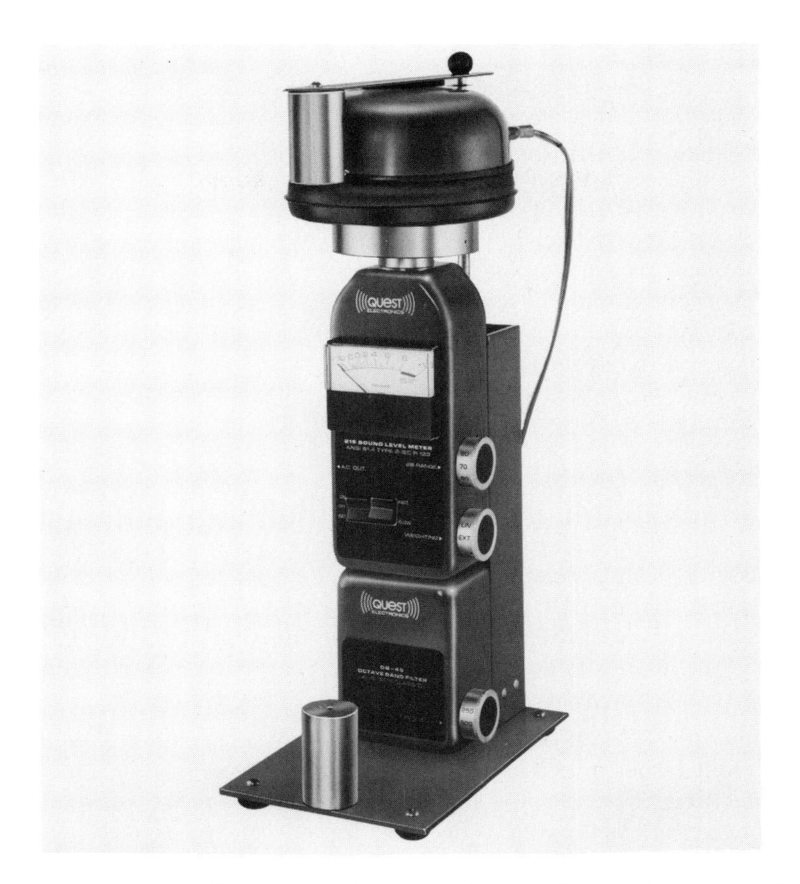

Figure 8.17 — Calibration systems are designed to perform acoustical checks, and can also be utilized for noise surveys. Photo courtesy of Quest Electronics, Oconomowoc, Wisconsin.

pleted. As with the first test frequency, the audiometer hearing level dial should be set at 70 dB HL for all other frequencies. The SPL output of the earphones read on the sound level meter is recorded on a calibration chart. These values should be compared to those proposed by ANSI (1969). If deviations of 10 dB or more exist, an exhaustive calibration should be performed by an authorized factory service representative. It is also important to remember that the SPL values given by ANSI (1969) holds true only for earphones housed in supra-aural (MX-41/AR) cushions, not for circum-aural earphones.

Following the completion of the SPL output check, do not remove the earphone from the coupler; instead, proceed with an attenuator linearity check. This can be accomplished either acoustically or electronically. It is preferable to carry out the linearity check acoustically. To do this, set the frequency dial at 1000 Hz and adjust the octave-band filter to the appropriate value. Adjust the hearing level dial to its maximum output level. Obtain and record the SPL output from the sound level meter. Decrease the hearing level dial in 10-dB steps and record each value until the ambient noise and the internal noise of the measuring instrument interferes with the measurement, at which point the check has to be done electronically. The same procedure should be repeated for the other earphone. The measured difference between the hearing level dial settings (in two successive settings) should not differ from the indicated difference by more than 1 dB or 30 percent of the indicated difference, whichever is smaller.

EXHAUSTIVE CALIBRATION CHECKS

The exhaustive calibration includes the sound pressure level and linearity checks of the standard acoustic calibration (see previous section) and also requires evaluation of the frequency accuracy and distortion of the test tones, measurement of channel crosstalk and any other residual or unwanted sounds in the earphones, and verification of the rise and fall times of the tone switch used for signal presentation on manual audiometers. Test frequencies below 500 Hz and above 6000 Hz may be omitted.

Because of the need for specialized equipment, exhaustive calibrations will usually be performed in a calibration laboratory or by the audiometer manufacturer. Whenever the audiometer has been removed from the test site for an exhaustive calibration check, it should be checked acoustically upon its return, if at all possible. At a minimum, a careful functional calibration should be completed to determine that the audiometer was not damaged in shipment.

OSHA regulations dictate that exhaustive calibrations shall be performed at least every two years in accordance with the relevant sections of the ANSI S3.6-1969, as cited in OSHA (1983; see Appendix I this text). They are also recommended, as mentioned in the previous section, if an acoustic calibration indicates a deviation of 10 dB or more at any frequency, although OSHA does not require them until the deviations equal or exceed 15 dB.

CALIBRATION RECORDS

The tester must maintain calibration records to support situations where audiometric records are to be used as evidence (Figure 8.18). In a worker's compensation situation, the audiometric data are virtually worthless without the calibration records. Likewise, OSHA inspectors demand calibration records.

It is recommended that the calibration records and audiograms be maintained indefinitely (Gasaway, 1985). Many state compensation laws allow claims to be filed after termination of employment or retirement. Again, without the calibration record, background ambient noise levels in the test area, the audiogram itself and employee noise exposure history, the employer will have little protection in the event of a workers' compensation claim. The calibration records could be the weakest link in that claim if not documented properly.

It is recommended that both pre-calibration and post-calibration records be maintained to document all deviations that may have existed. It will be important to the supervising professional to be aware of any instrument changes when interpreting data or trying to account for trends observed when using audiometric data base analysis procedures (see Chapter 9).

Training for Audiometric Testing Personnel

AUDIOMETRIC TECHNICIANS

Although audiometric testing presents a difficult challenge in terms of establishing consistent, valid test results, almost anyone can legally perform industrial testing without formal training or certification.

Technicians must be responsible to either an audiologist or a physician. The ultimate responsibility lies with the professional in charge. The quality of tests and the program results will reflect the technique, procedures and quality controls specified by the professional.

OCCUPATIONAL HEARING CONSERVATIONISTS

The certification of audiometric technicians is a concept which dates back to 1965. The first group of professionals to express a concern was the American Association of Industrial Nurses (AAIN) (now the American Occupational Health Nurses Association). In 1965, a committee of professionals representing the American Industrial Hygiene Association (AIHA), American Speech-Hearing-Language Association (ASHA), Industrial Medical Association (IMA) and AAIN developed a syllabus listing objectives, material and criteria essential to training audiometric technicians. This was the beginning of the Intersociety Committee. The administration of the program, dissemination of materials and primary leadership was

Acoustical Calibration Form

Company ______________ Date ______________ Time ______________

Location ______________________________ By ______________

Audiometer

Make ______________

Model ______________

Serial No ______________

1 Physical Condition of Equipment:

Audiometer ______________

Cushion ______________

Earphone Cords ______________

Electrical Cords ______________

2 Listening Check: ______________

3 Acoustical Check:

Earphones ______________

Cushions ______________

Response Switch ______________

Audiocups ______________

Other ______________

		Frequency-Hz						
		500	1000	2000	3000	4000	6000	8000
Expected SPL at 70 dB HTL (ANSI S3.6-1969)	TDH-39	*81.5*	*77.0*	*79.0*	*80.0*	*79.5*	*85.5*	*83.0*
	TDH-49 & 50	*83.5*	*77.5*	*81.0*	*79.5*	*80.5*	*83.5*	*83.0*
	Tolerances: (ANSI S3.6-1969)	± 3	± 3	± 3	± 3	± 4	± 5	± 5

Left Earphone	Actual SPL							
	Deviation							

Right Earphone	Actual SPL							
	Deviation							

4 Attenuator Linearity Check

Attenuator Setting dB HTL at 1000 Hz	*110*	*100*	*90*	*80*	*– 70 –*	*60*	*50*	*40*	*30*
SPL Deviation from Reference (70dB HTL)									

Figure 8.18 — The acoustical calibration record form shown here displays the expected SPL values for the ANSI prescribed calibration procedures. The output tolerance levels are also displayed and other parameters for functional checks are also listed.

placed with the AAIN throughout the years of the Intersociety group. This group also published a Guide for the Training of Industrial Audiometric Technicians in the May/June 1966 issue of the *American Industrial Hygiene Association Journal* (Intersociety Committee on Industrial Audiometric Technician Training, 1966).

In addition to the original organizations represented on the Intersociety Group, the American Academy of Occupational Medicine, the American Council for Otolaryngology and the National Safety Council are now also represented. In February of 1973, the Intersociety changed its name to the Council for Accreditation in Occupational Hearing Conservation (CAOHC). During the first CAOHC course director's training session at Colby College in 1973, it was decided that the term "audiometric technician" should be changed to "Occupational Hearing Conservationist". Therefore, a distinction is made between CAOHC certified personnel and audiometric technicians not having the benefit of the standardized CAOHC course.

CAOHC is currently the only professional group involved in the development and review of an approved course (presently 20 hours) which leads to a recognized certification. In addition, there is a program for screening and approval of CAOHC course directors, plus a periodic recertification program for both conservationists and course directors. CAOHC certification is recommended as a means to maintain uniformity in the training of audiometric technicians.

HEARING PROFESSIONALS AND PHYSICIANS

The OSHA regulation allows a professional audiologist, otolaryngologist, or physician to administer audiograms and provide general program supervision. This includes training, supervision and direction of the audiometric technician.

Although licensure for professional audiologists exists in most states, in general the requirements for licensure are that the audiologist possess all of the proper academic training requirements set by ASHA. Therefore, either ASHA certification or state licensure as a professional audiologist is acceptable. However, where state licensure exists, audiologists must be licensed in the state if they are going to test hearing.

The otolaryngologist is a physician who specializes in the practice of otology (ear disease), rhinology (nose disease) and laryngology (throat and larynx diseases). Ordinarily, the otolaryngologist's practice is diagnosis, evaluation and medical treatment while relying on the audiologist to perform the actual hearing testing and assessment. However, some otolaryngologists prefer to also do the actual testing, training and technician supervision.

The physician who does not specialize in diseases of the ear, nose and throat is nevertheless stipulated by the OSHA regulation (OSHA, 1983) as being capable of conducting audiometric tests and program supervision. In order to ensure program consistency, many physicians have elected to become CAOHC course directors and/or certified occupational hearing conservationists themselves.

In order to ensure this program consistency throughout the industry, it is recommended that any professional responsible for technician training and program supervision consider certification as a course director by CAOHC.

Preparing Employees for Audiometric Testing

PRE-TEST QUIET PERIOD FOR BASELINE TESTING

The immediate effect of noise exposure is a temporary loss of hearing sensitivity, or temporary threshold shift (TTS), which depends upon (1) the intensity level of the noise, (2) the spectrum of the noise, (3) the duration and temporal pattern of the noise exposure, and (4) the hearing sensitivity of the person exposed (Ward, 1973; Melnick, 1984). Rather than attempt to quantify the degree of TTS that might be expected for various levels of increased exposure, it will suffice to say that a measurable degree of TTS will probably occur when employees are exposed to noise levels greater than 80 dBA without hearing protection. Thus it is important to minimize employee noise exposure prior to performing baseline audiometry, in order to establish true baseline HTLs uncontaminated by TTS. A period of at least 14 hours away from noise is required by OSHA before testing, because TTS ordinarily disappears by this time.

OPTIONS TO ACHIEVE 14 HOURS OF QUIET FOR BASELINE TESTING

There can be no guarantee that the employee will not experience noise exposure away from the workplace during the fourteen-hour period between leaving the workplace and arrival on the day of the test. Later sections outline suggested procedures to guard against this occurring. Therefore, the only real control that can be applied and documented is the use of measures to prevent noise exposure that would produce on-the-job TTS on the day of the baseline hearing test.

There are two ways to control the employee's noise exposure prior to the test session. One is to prohibit employees from entering noisy production areas until after they have been tested. This will ensure that no known workplace noise exposure has occurred, although it will not guarantee that noise exposure did not occur on the way to work, or during the preceding 14-hour period. The other approach is to use hearing protectors to control noise exposure during work on the day of the baseline hearing test. Since it is possible that some degree of TTS may still be present from the previous day's exposure at the workplace, it is recommended that hearing protection be worn the previous day also.

This application of hearing protection to control exposure has many advantages and one potential disadvantage. First, the employee can be at work and productive during the waiting time for the hearing test; thus, the cost of the testing program is greatly reduced. In addition, the emphasis for controlling noise exposure is placed upon the utilization of the hearing protection device. This approach strengthens the concept of the effectiveness and importance of hearing protection as a viable means to control employee noise exposure.

The disadvantage to utilization of hearing protection to achieve quiet is the possibility that it will not be properly fitted and worn, so that some TTS may be incurred. If TTS has contaminated the baseline, subsequent annual tests will require greater amounts of real shift to demonstrate a standard threshold shift. Therefore, if hearing protection is to be utilized for baseline testing, care must be taken to ensure that all employees' protective devices are properly fitted, inspected and supervised until the baseline hearing test is taken. While this is an administrative problem, hearing protection can be effective and will still provide economic gains over the option of prohibiting employees from entering production areas before the testing.

If effective hearing protection measures cannot be implemented on the day of the test, it is not likely that the day-to-day use of hearing protection devices will be successful either.

UTILIZATION OF PROTECTION FOR PERIODIC (ANNUAL) HEARING TESTS

The option of employing special hearing protection or noise avoidance measures outlined above for baseline tests should *not* be implemented during the annual tests. This attention to hearing protection will prevent TTS for employees who do not properly utilize their protectors on a day-to-day basis. Thus, the opportunity for identifying workplace TTS, which is an indicator of impending PTS, will be eliminated. In this case the only shifts that can be detected will be those of a permanent nature and thus the opportunity for intervention to prevent the PTS will be lost.

When testing is conducted after exposure to workplace noise, if the hearing protection is not properly utilized, then a TTS will occur (Royster,

1980; Royster, Royster and Cecich, 1984). If the TTS occurs, then it is evident that the employee needs to be evaluated and fitted (or refitted) for proper hearing protection. Therefore, the intentional exposure to noise under the use of "normal" hearing protection will produce extremely valuable information on whether or not proper utilization is being achieved by the employee. This procedure is acceptable for periodic tests only.

Identification of employees experiencing TTS will indicate those who are likely to experience PTS whether or not the TTS is work-related. This allows the plant staff to concentrate their preventive measures and follow-up activities on those employees. Certainly if employees do not experience TTS when exposed to noise under normal hearing protection, there will not likely be any PTS due to noise exposure, and that is the purpose of hearing conservation.

PRE-TEST INFORMATION FOR EMPLOYEES TO ENSURE VALID TESTS

It is evident from the previous discussion that the objectives for the baseline test and the follow-up annual test are quite different. The baseline test must be conducted under careful conditions to ensure that no noise exposure has occurred prior to the test. The objective is to guarantee no TTS. However, the annual tests following the baseline should be conducted where every opportunity for noise exposure has been afforded; thus, no extraordinary protective measures should be taken. Just the normal day-to-day use should be implemented.

This presents an administrative problem for advance notification of employees relative to the pre-test noise exposure. Those employees who will be tested for the first time (baseline tests) must be informed of the requirement for no exposure for 14 hours prior to coming to work and the necessity of careful use of hearing protection on the day of the test until the actual test is taken. On the other hand, employees who already have their baseline test should not be told of the need to avoid noise during the previous 14 hours and on the test day.

This type of notification is not too difficult to handle for in-house testing programs, but it can present a problem where tests are performed through mobile testing. In situations where all employees are to be tested in as short a time as possible, baseline and periodic tests alike, the challenge of informing people appropriately is presented. One solution might be simply to test the employees for baselines first, before they enter the work place, thus eliminating the special hearing protection emphasis for that group. The 14-hour notification will still prevail for them, however.

In addition, all employees should be told the reasons for the hearing evaluation and that there are *no* negative consequences for taking a hearing check. Many employees believe that poor performance on their "test" may affect their job, their career opportunities, or even their right to file a claim for occupational hearing loss. Without pre-test indoctrination, these concerns may affect the test results, which may therefore be difficult to evaluate.

The employee should be informed through an employee education program well in advance of the test (1) that they cannot fail the test, that in fact, it is not a "test", but rather only an assessment of how well they hear, (2) that the results will be evaluated and if any problems exist, they *will* be informed, (3) precisely what will occur during the testing procedure, and (4) that the program is actually an employee benefit provided at no cost to the employee. At a minimum, employees should be informed of the purpose of audiometric testing, and be given an explanation of the test procedure as part of the annual employee training program. If this is done properly, it will improve employee cooperation which is critical to achieving valid and consistent audiometric data.

TEST INSTRUCTIONS TO ENHANCE QUALITY

The pre-test information to employees should be designed to alleviate fears and provide information on test procedures which will help set the stage for the final test instructions. Hopefully, if employee training has been properly conducted, they will come to the test setting with the understanding that the audiometric assessment is a valuable benefit and that no negative consequences can result.

Being thrust into the extreme quiet of a testing booth directly from a very noisy environment and having to listen for tones which are barely audible as opposed to "thundering" manufacturing noises can cause some unusual psychological sensations. Employee distraction and failure to concentrate during the test are often a result of the dramatic change in environment and the specific tasks associated with the hearing test. Employee preparation just prior to the hearing test and the annual training program should alleviate some of the common apprehensions which can contribute to frustration during the test.

The instructions on the test should be preceded by a statement about the test environment. A brief explanation of the expected transient noises which are specific to the particular test location will make them less distracting when they actually occur. Assure the employee that these transients are known to be present, that they have been measured, and that occasional transient noises will *not* affect the test. Of course, the complete absence of these annoying noises would be best; however, few industrial settings are completely free of all transient or vibration-induced noises. In addition, inform the employee that the actual test will be temporarily stopped if the noise occurrences become too frequent or too loud. (This will require a method of constant monitoring of the background noise in the audiometric test room.)

The test instructions should be kept as simple and precise as possible. For example, in manual audiometry, instruction should be as follows:

> "You are going to hear a series of tones. [pause] I want you to push this button (or raise your hand) each time there are tones. [pause]
>
> "Release the button (or lower your hand) when the tone goes away. [pause] Remember to push the button (or raise your hand) each time you hear a tone no matter how faint it is. [pause]
>
> "Are there any questions?".

Notice that this instruction does not encourage the employee to guess. Rather, he is just asked to respond no matter how faint the tones are. Encouraging the subject to guess has been shown to increase the false alarm response, which may in turn affect the validity of the measured thresholds (Dancer, Ventry and Hill, 1976).

Avoid any discussion of left ear, right ear, high pitch, low pitch, description of what the tones sound like or any other instructions which will only confuse the employee being tested. Only inform the employee what to listen for and what to do when the test signal is heard.

When the employee is seated in the test booth, just prior to earphone placement, briefly repeat the basic instructions as a final reinforcement. In addition, demonstrate the use of the handswitch at that time: "Remember, I want you to push the button like this (or raise your hand) every time you hear a tone."

EARPHONE PLACEMENT

The earphone should never be placed over the ears by the employee taking the test. It is very difficult to accurately place the earphones over the ear canal without visually observing the earphone opening and the ear canal. What is comfortable for the employee may not necessarily represent proper earphone placement. If earphones are not properly placed, the test results may reflect greatly elevated HTLs.

Improper earphone placement is perhaps the most common technician error. In addition to carefully placing the earphones, the technician should reinspect them before beginning the test. Another problem in placement is that the earphones can easily be switched around, right for left. Although earphones are color-coded, red for right and blue for left, the error is common and will produce invalid test results.

Even if the earphones are placed properly on the pinnae, the pressure from the earphones can cause the ear canals to collapse (close). If this happens, audiometric air conduction thresholds may be raised (worsened) especially in the range of 2000 to 6000 Hz. It is important to rule out collapsing ear canals since STS is based on an average change in hearing at 2000, 3000, and 4000 Hz, the frequencies at which audiometric air conduction thresholds may be affected most by collapsing ear canals (see Chapter 7).

An earlier discussion of circumaural earphone enclosures (Noise Attenuating Earphone Enclosures) gave the advantages of these devices in reducing ambient and transient test room noises. It should be reiterated that these earphones are very difficult to place properly on the employee due to the size and weight of the headsets, the visibility of the earphone opening and the headband tension adjustment. It is essential to reinspect the apparent placement of each earphone opening *after* the final headband adjustment is made.

Poor earphone placement may cause elevated thresholds, spurious test results, false shifts in hearing and unexplained wandering audiograms, from one test to the next.

Once the earphones are in place, the employee should be instructed not to adjust or touch the earphones even though they may not feel comfortable. If the employee does in fact move the earphones, a replacement and re-instruction by the tester should be accomplished.

EMPLOYEES WITH TESTING PROBLEMS

Various problems are encountered during the audiometric testing program. These problems are as follows:

(1) Physiological Problems, including:

a. tinnitus (head noises such as ringing, hissing, or humming in the ear)
b. ear-hand coordination (latency in the manual response to the test signal)
c. fatigue
d. ear disease and (or) wax impaction
e. existing hearing loss (moderate and severe loss)

f. TTS due to noise exposure
g. unilateral hearing loss

(2) Response Problems, including:
a. intelligence and comprehension
b. under the influence of prescribed medication
c. under the influence of illegal drugs
d. under the influence of alcohol
e. anxiety, fear or confusion
f. improper pre-test preparation

(3) Malicious Intent, including:
a. labor relations problems
b. malingering
c. clowning behavior
d. compensation problems

The notion that establishing valid industrial audiometric test results is simple, or that untrained technicians utilizing computerized equipment can achieve valid hearing testing without proper supervision, is true for only a relatively small segment of the industrial population. In fact, obtaining valid industrial audiometric test results is not as simple as it may appear.

An important consideration with these test problems is that the employee exhibiting them may pass through the test with recorded responses that are not valid. The technician is not trained to identify these test problems in the limited formal training program for industrial technicians; therefore, the problems will only be evident if the audiometric tests are carefully evaluated by a professional and analyzed in comparison to all previous hearing test results on the employee concerned. Employees who exhibit these testing problems must be referred for audiological testing and medical evaluation in order to establish valid threshold levels.

It is the responsibility of the professional supervisor to establish testing procedures, employee interview forms, data review and analysis systems to ensure that employee test problems are identified at the earliest possible opportunity. The section in this chapter on Pure-Tone Audiometry: Instrumentation and Technique outlines some ways to identify and/or avoid some of these testing problems.

Supportive Test Procedures

Additional information can be gathered on the employee to: (1) identify critical testing problems, (2) assist in determining the need for audiological and (or) medical referral and (3) plan the hearing conservation program for the employee. This information can be as valuable as the hearing evaluation itself in achieving effective hearing conservation for OSHA and loss-prevention purposes. An axiom should be that if employees are asked what is wrong, they will tell you and, if they are not willing to, you will not find out except through medical, audiological and/or legal investigation.

EMPLOYEE AURAL HISTORY

The purpose of taking the history is only to establish the existence of general circumstances, conditions, or events which may be a contributing factor when hearing loss or testing problems are present. The technician should administer a "high-risk history" (Figure 8.19) as opposed to a detailed history. This high-risk history establishes general categories which will lead to a detailed investigation by the physician or audiologist upon referral. A history of hobby shooting, for example, does *not* substantiate or establish cause for a unilateral hearing loss.

However, an indication of tinnitus, ringing in the ear or head noises, will certainly alert the tester that "wandering responses" to the test signals may

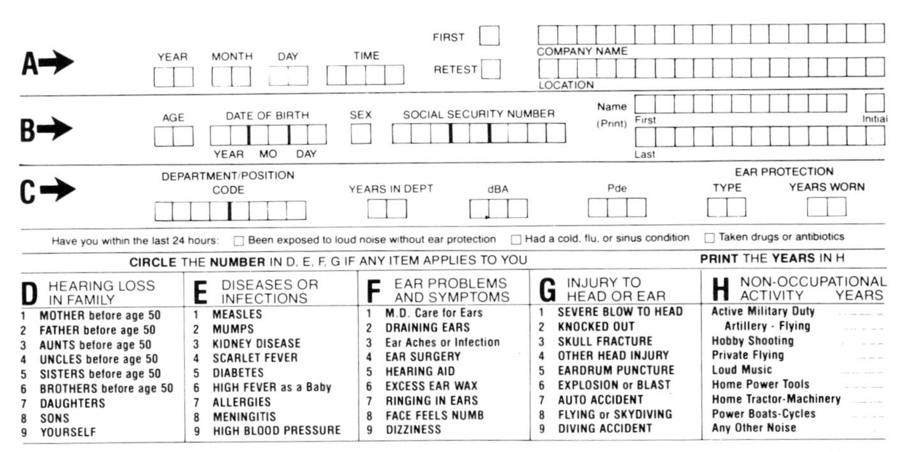

A→ YEAR MONTH DAY TIME FIRST RETEST COMPANY NAME LOCATION

B→ AGE DATE OF BIRTH (YEAR MO DAY) SEX SOCIAL SECURITY NUMBER Name (Print) First Initial Last

C→ DEPARTMENT/POSITION CODE YEARS IN DEPT dBA Pde EAR PROTECTION TYPE YEARS WORN

Have you within the last 24 hours: ☐ Been exposed to loud noise without ear protection ☐ Had a cold, flu, or sinus condition ☐ Taken drugs or antibiotics

CIRCLE THE **NUMBER** IN D, E, F, G IF ANY ITEM APPLIES TO YOU — **PRINT** THE **YEARS** IN H

D HEARING LOSS IN FAMILY	E DISEASES OR INFECTIONS	F EAR PROBLEMS AND SYMPTOMS	G INJURY TO HEAD OR EAR	H NON-OCCUPATIONAL ACTIVITY YEARS
1 MOTHER before age 50	1 MEASLES	1 M.D. Care for Ears	1 SEVERE BLOW TO HEAD	Active Military Duty
2 FATHER before age 50	2 MUMPS	2 DRAINING EARS	2 KNOCKED OUT	Artillery - Flying
3 AUNTS before age 50	3 KIDNEY DISEASE	3 Ear Aches or Infection	3 SKULL FRACTURE	Hobby Shooting
4 UNCLES before age 50	4 SCARLET FEVER	4 EAR SURGERY	4 OTHER HEAD INJURY	Private Flying
5 SISTERS before age 50	5 DIABETES	5 HEARING AID	5 EARDRUM PUNCTURE	Loud Music
6 BROTHERS before age 50	6 HIGH FEVER as a Baby	6 EXCESS EAR WAX	6 EXPLOSION or BLAST	Home Power Tools
7 DAUGHTERS	7 ALLERGIES	7 RINGING IN EARS	7 AUTO ACCIDENT	Home Tractor-Machinery
8 SONS	8 MENINGITIS	8 FACE FEELS NUMB	8 FLYING or SKYDIVING	Power Boats-Cycles
9 YOURSELF	9 HIGH BLOOD PRESSURE	9 DIZZINESS	9 DIVING ACCIDENT	Any Other Noise

Figure 8.19 — The screening history should contain information required by the OSHA regulation, and non-occupational experience which may also contribute to hearing loss. Courtesy of Impact Hearing Conservation, Inc., Kansas City, Missouri.

be expected and care must be taken in administering the test. The test session should be stopped and rescheduled if difficulties are encountered in achieving consistent test results. Also, if employees indicate they have experienced noise exposure prior to the test, it should not be surprising that thresholds are depressed, head noises are present and testing difficulty exists. The evaluation should be rescheduled if it is the baseline audiogram which must be conducted after no noise exposure for 14 hours.

EMPLOYEE NOISE HISTORY

A tremendous resource of information exists on the "damage-risk criteria" of noise exposure and resulting hearing loss (NIOSH, 1972). In fact, the research conducted in the past has created an undeniable case for the need for regulation and hearing conservation requirements.

The existence of accurate noise measurement records have only begun to be established by most industries during the past decade. An effort should be made to maintain detailed records of employee noise exposure from the point of first monitoring. This information is necessary in determining probable cause of existing hearing loss in compensation cases. In addition, by carefully monitoring exposure, audiometric testing results and hearing protection utilization, industry will be able to demonstrate the effectiveness of the hearing conservation program, or determine the necessity for additional protective measures.

OTOSCOPIC EXAMINATION

The aural case history, employment history and otoscopic exam is essential to the professional reviewing the audiograms in making a determination for further evaluation, retest or referral. Without this information, only the noise exposure is suspected as a contributing factor to testing problems and changes in hearing.

Audiometric testing only provides information on how well the employee can hear. It is not possible to identify external ear problems, eardrum or middle ear disorders by audiometric testing alone. In fact, only the more serious varieties of these disorders will cause any hearing loss. Therefore, otoscopic examination, or at least a visual inspection of the ear canal, should be conducted before every hearing test (see Chapter 7).

Review and Treatment of Audiogram(s)

The purpose of audiometric testing in industry should go beyond the minimum requirements of the OSHA rule, which is to detect changes in hearing threshold levels. In fact, where OSHA is concerned, no actions are recommended or required for employee follow-up unless a standard threshold shift (STS) occurs.

The OSHA rule defines an STS as: "A change in hearing threshold relative to the baseline audiogram of an average of 10 dB or more at 2000, 3000 and 4000 Hz in either ear." That is, an average of these three test frequencies is made from the baseline hearing test results and the same average made of the current test results. The current average is compared to the baseline average and if the difference is 10 dB or more (poorer hearing) on the current test, then an STS exists.

The audiogram can offer an abundance of information that is unrelated to shift in hearing and which can be utilized to help improve employee safety, off-the-job enjoyment, career opportunities and protection against employer liability.

Employees who have hearing losses which result in poor speech communication ability (Figure 8.20) present a unique problem that the OSHA rule does not address. These employees create a potential safety hazard when two-way human communication is required to perform the job safely. Unfortunately, this type of hearing loss may display no further change at STS frequencies, as advanced high-frequency hearing loss stabilizes or shifts very little; therefore, STS may never occur and these

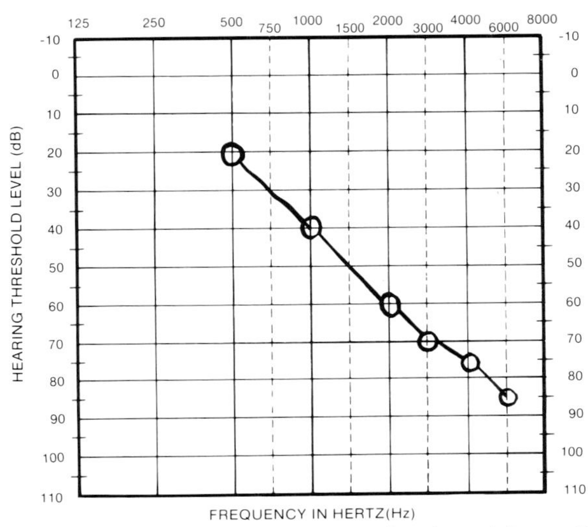

Figure 8.20 — Advanced noise-induced hearing loss (above 1000 Hz) will generally affect an employee's ability to understand speech, particularly when background noise is present.

employees will go unnoticed if the company's policy on hearing conservation is dictated solely by OSHA regulations.

It is commonly believed that many industrial accidents have occurred because the employee did not hear a warning signal that preceded the accident, or did not understand instructions which would have prevented the accident from occurring. These are hearing-related problems which may be alleviated through hearing protection, communication devices, hearing aids (in the absence of noise), safety procedures and employee counseling.

In addition to simply identifying the presence of an STS, it is equally important to recognize existing hearing loss and design a program to accommodate the special problems which both STS and hearing loss present. This is similar to dealing with other special problems which employees with partial disabilities present. Problems such as back injuries, visual disorders, loss of limb and other partially disabling situations are dealt with very successfully by employing safety procedures, prosthetic devices or equipment to ensure that the employee can be productive and safe in the work place. Hearing impairment is no different and, in all probability, easier and less expensive to ameliorate.

Virtually all audiograms which demonstrate hearing loss should be reviewed by the audiologist or otolaryngologist and evaluated to determine if a potential risk exists in relation to the employee's specific job tasks. This can only be accomplished effectively if the professional supervisor has detailed knowledge of the job and potential hazards. The factors which should be carefully reviewed are: (1) the need for understanding speech communication to perform the job tasks safely and productively, (2) the amount of noise and the type of noise in the specific work area, (3) the type of hearing protection being utilized to control the noise exposure, (4) the supervisor's evaluation of the employee's current job performance, and (5) the employee evaluation of his need for better hearing on the job.

Unfortunately, employees with existing hearing losses often express a resigned attitude that "It's too late for me, I've already lost my hearing," or the notion "I can't hear my machine or speech with the hearing protectors in." Therefore, unless attitudes are changed, the hearing loss will progress at test frequencies where an STS will not be detected. (Typically, loss progresses from high to low frequencies.) In addition, the employer's potential liability increases from a workers' compensation point of view as the speech frequencies become involved. The worker also becomes more impaired for communication and less capable to perform his job tasks safely and productively.

The involvement of a hearing professional in the hearing conservation program will provide an opportunity to control many of these problems which the OSHA rule does not address.

TECHNICIAN'S REVIEW AND DETERMINATION OF "PROBLEM AUDIOGRAMS"

The technician must carry out procedures which the professional supervisor has established in all phases of the testing program, including the initial identification of "problem" audiograms. It is the responsibility of the professional, not the technician, to review problem audiograms. The technician should only perform a comparison of the current audiogram to the baseline to identify the existence of an STS and determine if the audiogram may be invalid or otherwise require referral.

The technician cannot determine that a test may be invalid without a specific set of criteria established by the professional to whom they are responsible. The most protective case would be for the audiologist to review all audiograms; however, properly trained technicians or computer programs can screen audiograms effectively in order to identify those that may be invalid.

The technician's role is to conduct testing, implement validity check procedures established by the professional, record test values and compare test results to the baseline to detect problem audiograms and STS (if another means such as computer applications is not being utilized), then provide the professional with the data on these "problem audiograms" for review. Technicians, certified or not, are not legally qualified to interpret test results.

DETERMINATION OF STANDARD THRESHOLD SHIFT

When the determination of STS is made, the real question is whether or not the STS is valid and if so, whether it is of a temporary or permanent nature. If the shift is either invalid or temporary, then the cause for the fluctuating audiogram should be carefully evaluated. The only effective means to determine the nature of the shifts (whether spurious, TTS or PTS) is to perform a retest to identify a persistent shift or a change in either direction from the previous test.

The OSHA rule also states that the employer can administer another audiometric test within 30 days to substitute for the annual audiogram which produced the STS results.

HEARING NORMS AND AGE ADJUSTMENT OF AUDIOGRAMS FOR PRESBYCUSIS

An allowance can be made for change of hearing due to aging (presbycusis) when averages of the current and baseline tests are compared. These presbycusis adjustments are quite small from one year to the next, but can make a significant difference when hearing test data across a period of years are compared. For example, the average difference between normative HTLs for a 40- and 60-year-old male at the three frequencies where STS is calculated is 13 dB.

Therefore, if the employee demonstrated audiometric thresholds of 10 dB each at 2000, 3000 and 4000 Hz at 40 years of age (the average being 10 dB) and audiometric thresholds of 30 dB each at 2000, 3000 and 4000 Hz at 60 years of age (the average being 30 dB), there would be a 20-dB real difference before age adjustment. If the OSHA formula for age adjusting was applied, one would subtract the 13-dB average difference (which exists between the 40 and 60 year age norms as outlined above) from the current audiometric average. The current test average of 30 dB now becomes an age adjusted average of 17; therefore, only a 7-dB shift exists compared to the baseline and thus, no STS.

The OSHA rule does not require that the age adjustment formula be applied, but it does allow it. It is up to the supervising professional to determine if the formula will be used in determining STS. If one believes that aging contributes to the total change in hearing, then the application of the age adjustment seems appropriate. The OSHA data for age norms and the method of computation are found in its Appendix F: Calculations and Application of Age Corrections to Audiograms (see Appendix I of this manual).

The procedure for age-adjusting audiograms can also serve as an excellent educational tool for employees. It is important to point out that there can be many causes of change in hearing, including the aging process.

PROFESSIONAL REVIEW OF PROBLEM AUDIOGRAMS

The audiologist or otolaryngologist should review problem audiograms and determine whether there is a need for further evaluation. Problem audiograms are those that demonstrate STS, questionable validity or exhibit hearing loss. In addition, audiograms with significant improvement or other test problems (as defined by the professional responsible for the program) should also be reviewed as problem audiograms.

The various records that are important to the professional as part of the audiogram review are: (1) a copy of the current OSHA regulation, (2) the baseline and most recent audiograms, (3) the noise levels of the testing room, and (4) audiometric calibration records. Additional records which will be helpful in making an appropriate review of the need for further evaluation are: (5) all previous hearing test records, (6) otoscopic screening exam results, (7) employee non-occupational noise and aural history information and (8) record of hearing protection utilization including type, brand and documentation on fitting dates.

The physician or audiologist can estimate the likelihood that the problem audiogram is work-related or aggravated by occupational noise exposure if they have the information listed above. This determination is important in the event of workers' compensation claims; therefore, it should be carefully approached. The determination is very difficult to establish accurately and the professional must rely on the employee's reported history in addition to the employment noise exposure records.

EMPLOYEE FOLLOW-UP ACTION RECOMMENDED WHEN STS OCCURS

When an employee demonstrates an STS, he should be notified personally and the follow-up process begun as soon as possible. Since follow-up procedures are both time-consuming and expensive, it is recommended that the employee be retested first to see if the STS persists. The retest should be conducted using the same procedures as those recommended for baseline testing to control for noise exposure prior to the test.

The retest will either confirm the shift in hearing or establish that it was TTS or a spurious STS. In any event, if the STS is not confirmed, the follow-up activities will be modified considerably. If no STS is present on the retest, the employee should still be counseled on the use of hearing protection and proper fitting demonstrated. It will be helpful to try to establish a possible cause for the previously indicated STS to avoid its recurrence.

If the STS persists, it is important to investigate the probable cause. There are five possible causes of STS which must be examined: (1) instrument or technician error, (2) employee response problems, (3) occupational noise exposure, (4) non-occupational noise exposure and (5) medical ear disorders. With proper direction from the professional supervisor, the technician can effectively evaluate each situation. Probable solutions will require selection, fitting and implementation of hearing protection and/or referral for audiological or medical evaluation. Medical referral is appropriate if a personal ear disorder is suspected or if no other plausible explanation for a persistent shift is determined. Audiological referral is appropriate if it is impossible to establish consistent HTLs.

If the employer suspects that an ear problem is present and that it may have been caused or aggravated by the wearing of hearing protection (possibly contributing to the STS), the employee must be referred to an otolaryngologist for examination.

NON-OSHA-RELATED REFERRALS

Referral for follow-up medical and/or audiological evaluation is also important for employees who have hearing problems other than shifts of HTL. The purpose of referral for these employees is to treat hearing loss or improve the employee's ability to cope with the disability which the hearing loss creates. With the exception of workers' compensation cases and difficult-to-test employees (as described earlier in this chapter), referral is ordinarily not necessary to simply confirm hearing loss.

If all employees with hearing loss and symptoms of ear problems were referred, the cost of evaluation and lost work would be prohibitive. Therefore, a referral policy should be established consistent with the objective of referral for those employees who may receive some real benefit from treatment. This will also provide employer benefits in terms of reduced risk in the work place.

In general, referral will be of benefit to employees with serious losses of hearing who may improve their job performance through the use of a hearing aid. The hearing aid must not be worn in the presence of high-level noise; therefore, if the employee is in a noisy job, the referral should be considered to be one that will benefit the employee away from the job, or perhaps in a new job role in a quiet environment.

Noise-induced hearing losses are generally characterized by loss of hearing in both ears with relatively good hearing at 500 Hz and 1000 Hz followed by a "ski slope" sensorineural loss of varying degree in the mid- and high frequencies (Figure 8.21). The degree of difficulty in the activities of daily life experienced by such persons will vary; however, the employee complaints are typically that they experience difficulty in hearing and understanding conversation in crowds.

Sensorineural losses may be caused or aggravated by noise exposure; however, the determination of cause is very difficult to make and the referral is intended to solve rather than create problems. It is important to establish a good relationship with the physicians and audiologists to whom these employees will be referred. Unless the professional is certain of the cause, a diagnosis of noise-induced loss can cause serious problems for the company where workers' compensation is involved.

Referrals are also important for employees whose symptoms suggest ear disease of a conductive rather than sensorineural nature. If ear disease is

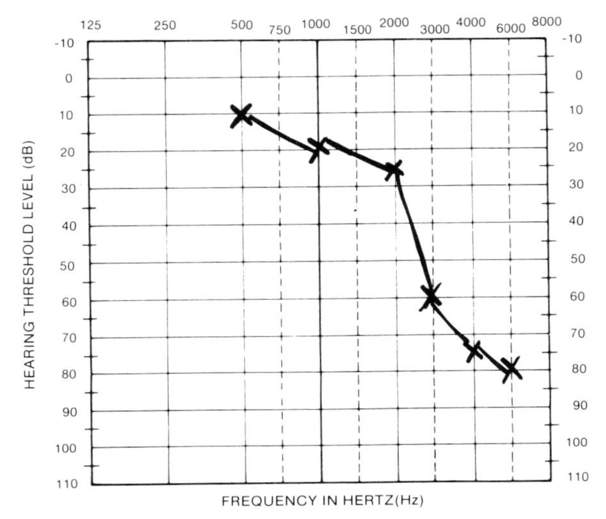

Figure 8.21 — Noise-induced hearing loss is generally characterized by poorer hearing in the high tones and relatively good hearing in the lower test frequencies. Note the difference in HTL at 1000 Hz and 2000 Hz between this figure and for the more advanced loss shown in Figure 8.20. These two figures illustrate how noise-induced loss generally progresses with continued exposure.

present or has previously been experienced, and damage to the ear drum or middle ear bones exists, the probability of restoration of hearing through medical treatment may be good. Conductive losses are typically unrelated to noise in the work place but referral is nevertheless essential to good health. The extent of conductive hearing loss may not be sufficient to detect by audiometry alone. A combination of audiometry, otoscopy and case history will ordinarily reveal the need for referral. The conductive hearing loss configuration is typically characterized by uniformly elevated HTLs at most frequencies (Figure 8.22).

Employee complaints of ear pain, drainage of pus or fluid from the ear, reduced hearing, a sense of fullness or blockage in one or both ears are important medical referral criteria.

Some employees report unusual sensations when using insert hearing protectors. Although these situations can most often be resolved through experimenting with different types of protectors, occasionally medical referral is necessary. While ear muffs may be a general solution for these employees, some job applications may preclude the use of ear muffs.

Many employees complain of tinnitus, and this ringing can affect the hearing test. Since tinnitus is a classic symptom of noise exposure, one

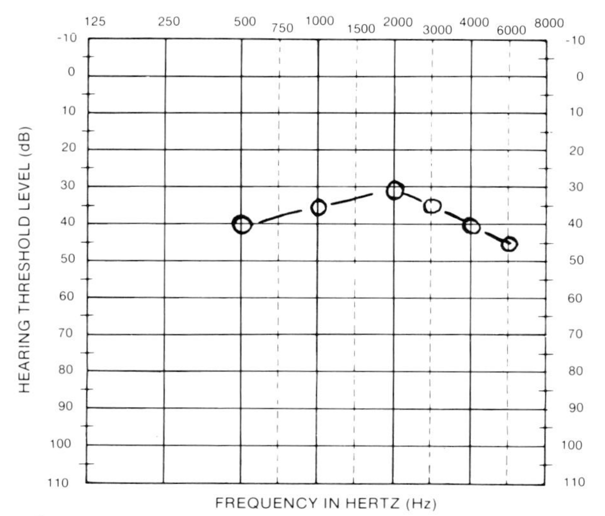

Figure 8.22 — Conductive hearing loss is generally characterized by elevated HTLs at most frequencies.

might suspect initially that the employee is still experiencing a lack of protection at some time. If careful evaluation of the potential for noise exposure fails to confirm this suspicion, the continued sensation of constant tinnitus is an indication for medical referral. Since tinnitus is a fairly common complaint and experienced by many employees, discretion must be used in which cases to refer. If employees complain of loss of sleep, job interference, psychological problems or other severe symptoms related to the tinnitus, they should certainly be given consideration for referral.

The symptom of dizziness may or may not be related to the ear; however, employees who experience a sudden onset of dizziness or loss of balance present a significant safety hazard and so should be medically evaluated. The frequency of these attacks will dictate the importance and timing of referral. It should be noted that many employees complain of dizziness when standing up quickly, changing body positions from lying to standing, or from stooping to standing. Those are not typically symptoms which demand referral. The employee who experiences severe dizziness and loss of balance will be able to lucidly describe the experience and will leave little doubt about its reality. In these cases, referral to an otolaryngologist or a neuro-otologist is recommended.

Unilateral hearing loss or a difference in hearing loss between ears is also an indication for possible referral. If an employee has a significant hearing loss in one ear only or a significant difference in hearing between the ears, the loss creates an inability to "localize" a source of sound. Unilateral hearing loss creates a unique problem if the employee must be able to recognize where the sound is coming from. One can simulate this loss of "stereo" hearing by plugging one ear completely and attempting to locate sounds by hearing out of one ear only. It is nearly impossible.

Additionally, unilateral hearing loss or a difference in hearing between ears may be an indication of middle ear problems, wax obstruction or even some neuro-otological disorders. Of course, the classic cause of unilateral hearing loss is "shooter's ear" and referral of all employees with unilateral symptoms will include an abundance of these cases. If unilateral loss is combined with complaints of dizziness, ringing or the symptoms outlined for the possible conductive problems, referral for unilateral loss will be important in identifying conditions that may be treatable.

All criteria for referral must be established by the professional responsible for the overall program. It is advisable to establish a close working relationship with the otolaryngologist and audiologist to whom referrals will be made.

Summary

In summary, the difficulty inherent in accurate measurement of hearing in the industrial setting has often been greatly underestimated. Lack of reliability of audiometric data is a problem which will go unnoticed unless a detailed examination of data is performed where each hearing test conducted is compared to all previous test data on the employee concerned. A visual review, or a simple comparison of current tests to baseline audiometric tests, will only illuminate the gross differences and obvious problems. Therefore, evaluation of all existing data should be undertaken after each test to detect patterns of change which will identify testing problems.

If the analysis is designed properly, certain testing problems will be identified, such as (1) excessive noise in the test room, (2) poor audiometric test technique, (3) instrument calibration errors, (4) employee response maladies, (5) improper hearing protection, and (6) medically referrable conditions.

With the exception of physiological disorders, which must be controlled medically, virtually all of the testing complications can be eliminated through proper planning and program maintenance. The quality controls implemented by the supervising professional will determine the accuracy that can be achieved in the testing. The preparation of the employees and the manner in which the testing procedures are administered by the tester will greatly influence the test reliability.

While OSHA requirements offer a set of minimum guidelines to follow, there exists an immediate need to also deal with employees who have loss of hearing, but will never be recognized as a result of OSHA's criterion for STS. This relates to risk-management concepts and offers a more protective approach than merely complying with OSHA guidelines.

Undoubtedly, the effect of workers' compensation for occupational deafness and safety risks, which are a direct result of noise interference, far outweigh the economic consequences of an OSHA citation. However, if the premise of the OSHA noise regulation is followed, *i.e.,* protection against hearing loss for the worker, the employer will benefit also through the ultimate control of both of these risks. Hearing conservation is truly a win-win endeavor if performed properly.

References

ANSI (1966). "Specification for Octave, Half-Octave, and Third-Octave Band Filter Sets," S1.11-1966(R1971), New York, NY.

ANSI (1969). "Specification for Audiometers," ANSI S3.6-1969, New York, NY.

ANSI (1977). "Criteria for Permissible Maximum Ambient Noise During Audiometric Testing," ANSI S3.1-1977, New York, NY.

ANSI (1983). "Specification for Sound Level Meters," S1.4-1983, New York, NY.

ASHA (1978). "Guidelines for Manual Pure-Tone Threshold Audiometry," *ASHA 20,* 297-301.

Bekesy, G.V. (1947). "A New Audiometer," *Acta Oto-Laryngol. 35,* 411-422.

Brogan, F.A. (1956). "An Automatic Audiometer for Air Force Classification Centers," *Noise Control* 2, pp. 58, 59, and 67.

Burns, W. and Hinchcliffe, R. (1957). "Comparison of Auditory Threshold as Measured by Individual Pure-Tone and Bekesy Audiometry," *J. Acoust. Soc. Am. 29(12),* 1274-1277.

Campbell, R.A. (1974). "Computer Audiometry," *J. Speech Hear. Res. 17,* 134-140.

Carhart, R. and Jerger, J.F. (1959). "Preferred Methods for Clinical Determination of Pure-Tone Thresholds," *J. Speech Hear. Disord. 24,* 330-345.

Cook, G. and Creech, H.B. (1983). "Reliability and Validity of Computer Hearing Tests," *Hearing Instruments 34(7),* pp. 10, 12, 13, and 39.

Corso, J.F. (1955). "Effects of Testing Methods on Hearing Thresholds," *Arch. Otolaryngol. 63,* 78-91.

Dancer, J.E., Ventry, I.M., and Hill, W. (1976). "Effects of Stimulus Presentation and Instructions of Pure-Tone Thresholds and False-Alarm Responses," *J. Speech Hear. Disord. 41,* 315-324.

Gasaway, D.C. (1985). "Documentation: The Weak Link in Audiometric Monitoring Programs," *Occup. Health Saf. 54(1),* 28-33.

Green, D.S. (1978). "Pure-Tone Air Conduction Testing," in *Handbook of Clinical Audiology,* 2nd edition, edited by J. Katz, Williams and Wilkins Company, Baltimore, MD, 98-108.

Harford, E.R. (1967). "Audiometer Calibration," Maico Audiological Library Series, Vol. III, Reports 5 and 6, Maico Hearing Instruments, 14-23.

Harris, D.A. (1979a). "Microprocessor Versus Self-Recording Audiometry," *J. Aud. Res. 19,* 137-149.

Harris, D.A. (1979b). "Microprocessor, Self-Recording, and Manual Audiometry," *J. Aud. Res. 19,* 159-166.

Harris, J.D. (1980). "A Comparison of Computerized Audiometry by ANSI, Bekesy Fixed-Frequency, and Modified ISO Procedures in an Industrial Hearing Conservation Program," *J. Aud. Res. 20,* 143-167.

Hodgson, W.R. (1980). *Basic Audiologic Evaluation,* Williams and Wilkins Company, Baltimore, MD.

Intersociety Committee on Industrial Audiometric Technician Training (May/June 1966). "Guide for Training of Industrial Audiometric Technicians," *Am. Ind. Hyg. Assoc. J. 27,* 303-304.

Jerlvall, L., Dryselius, H., and Arlinger, S. (1983). "Comparison of Manual and Computer-controlled Audiometry using Identical Procedures," *Scand. Audiol. 12,* 209-213.

Knight, J.J. (1966). "Normal Hearing Threshold Determined by Manual and Self-Recording Techniques," *J. Acoust. Soc. Am. 39(6),* 1184-1185.

Martin, F.N. (1986). *Introduction to Audiology,* 3rd edition, Prentice-Hall, Inc., Englewood Cliffs, NJ.

McMurray, R.F. and Rudmose, W. (1956). "An Automatic Audiometer for Industrial Medicine," *Noise Control 2,* 33-36.

Melnick, W. (1979). "Instrument Calibration," in *Hearing Assessment,* edited by W.F. Rintelmann, University Park Press, Baltimore, MD, 551-586.

Melnick, W. (1984). "Auditory Effects of Noise Exposure," *Occupational Hearing Conservation,* edited by M.H. Miller and C.A. Silverman, Prentice-Hall, Inc., Englewood Cliffs, NJ, 100-132.

National Institute for Occupational Safety and Health (1972). "Criteria for a Recommended Standard . . . Occupational Exposure to Noise," U.S. Dept. of HEW, Report No. HSM 73-11001, Cincinnati, OH.

Occupational Safety and Health Administration (1981). "Occupational Noise Exposure; Hearing Conservation Amendment," *Fed. Regist. 46(11),* 4078-4181.

Occupational Safety and Health Administration (1983). "Occupational Noise Exposure; Hearing Conservation Amendment," *Fed. Regist. 48(46),* 9738-9783.

Royster, L.H. (1980). "An Evaluation of the Effectiveness of Two Different Insert-Types of Ear Protection in Preventing TTS in an Industrial Environment," *Am. Ind. Hyg. Assoc. J. 41(3),* 161-169.

Royster, L.H. (1984). "An Evaluation of the Effectiveness of Three Hearing Protection Devices at an Industrial Facility with a TWA of 107 dB," *J. Acoust. Soc. Am. 76(2),* 485-497.

Siegenthaler, B.M. (1981). "A Survey of Hearing Test Rooms," *Ear and Hearing 2(3),* 122-126.

Singer, E.E. (1984). "Physical Environment for the Performance of Basic and Follow-up Audiometric Studies," in *Occupational Hearing Conservation,* edited by M.H. Miller and C.A. Silverman, Prentice-Hall, Inc., Englewood Cliffs, NJ, 62-74.

Ward, W.D. (1973). "Adaptation and Fatigue," in *Modern Developments in Audiology,* 2nd edition, edited by J. Jerger, Academic Press, New York, NY, 301-344.

Wilber, L.A. (1979). "Threshold Measurement Methods and Special Consideration," in *Hearing Assessment,* edited by W.F. Rintelmann, University Park Press, Baltimore, MD, 1-28.

Wilber, L.A. (1984). "Calibration of Instruments Used in Occupational Hearing Conservation Programs," in *Occupational Hearing Conservation,* edited by M.H. Miller and C.A. Silverman, Prentice-Hall, Inc., Englewood Cliffs, NJ, 75-99.

Noise and Hearing Conservation Manual, edited by
E.H. Berger, W.D. Ward, J.C. Morrill and L.H. Royster

9 Audiometric Data Base Analysis

Julia Doswell Royster
Larry H. Royster

Contents

Page

Introduction 293
ADBA Options 296
Population Comparisons 296
Evaluation Criteria 300
The Learning Effect 301
Test-Retest Statistics 304
Significant Threshold Shift Criteria 307
Application of ADBA Concepts 309
Example 1: Determination of HCP Effectiveness 309
Population Characteristics 309
The Baseline Test-Retest and Learning-Related Statistics 310
The Variability-Related Statistic 312
Example 2: Evaluation of Hearing Protector Effectiveness 314
Concluding Remarks 315
References 316

Introduction

The primary goal of industrial hearing conservation programs (HCPs) should be to prevent significant on-the-job noise-induced hearing loss (L.H. Royster, Royster, & Berger, 1982). Because the employee's hearing ability is affected by numerous factors including presbycusis, pathology, and nonoccupational noise exposures, especially gunfire (Ward, 1980), it is difficult to separate the occupational component of an employee's hearing loss. Individuals vary in sensitivity to noise-induced hearing loss, and small early shifts of 5-10 dB for an employee cannot easily be distinguished from normal measurement variability. Therefore, the ability to react to early warning signs of potential noise-induced hearing loss in an individual is limited.

In contrast to audiogram review, audiometric data base analysis (ADBA) evaluates hearing trends for employees as a group (L.H. Royster & Royster, 1984a). As a supplement to audiogram review, ADBA can detect early signs of occupational hearing loss in the population of exposed employees so that corrective actions can be taken before large threshold shifts occur for many individuals. By allowing the professional to identify ineffective HCPs, ADBA provides an opportunity to halt the progression of on-the-job loss by improving HCP effectiveness. In addition, the average hearing changes for groups of employees can serve as a reference comparison for interpretation of shifts for an individual during audiogram review.

The top portion of the flow chart shown as Figure 9.1 illustrates the reciprocal relationship between ADBA and individual audiogram review. A review of each individual's audiogram is necessary to identify significant hearing changes and trigger appropriate followup actions. However, the results from audiogram reviews cannot easily be used to assess how well the HCP is protecting the exposed population as a whole. In contrast, ADBA procedures identify population hearing trends so that the professional can evaluate the overall success of the HCP in preventing on-the-job hearing loss.

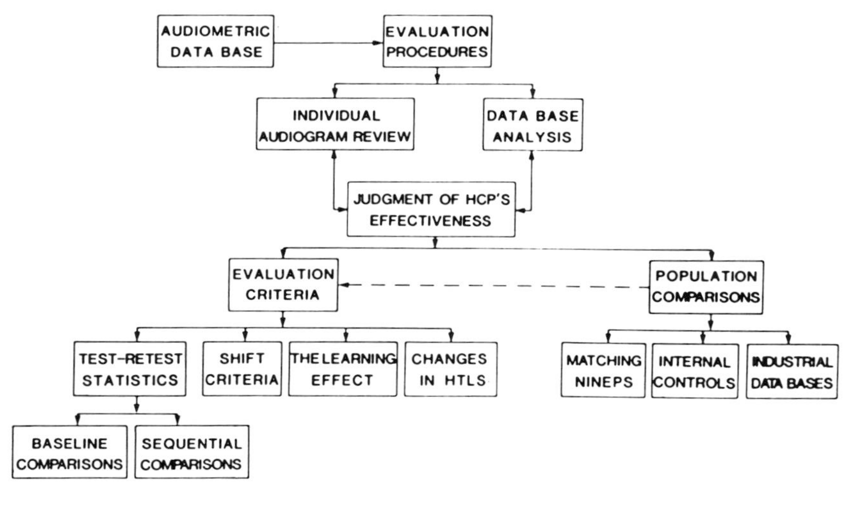

Figure 9.1. Flow chart illustrating the various component processes involved in using audiometric data base analysis to evaluate the effectiveness of the hearing conservation program.

The purpose of applying ADBA procedures is to evaluate the *overall* effectiveness of industrial HCPs. The basic procedures should not necessarily be expected to identify the relative contributions of all the factors that affect the data base, such as audiometric methods, testing environment, effectiveness of hearing protection devices, off-the-job noise exposures, and population characteristics. However, ADBA should determine the total effect resulting from all these influential parameters plus occupational noise exposure.

ADBA techniques should be usable by general industry, which may have only limited computing and analytical abilities. Therefore, proposed ADBA statistics should be easy to apply and the results should be easy to understand. If an audiometric data base has been computerized, it is possible to utilize many different types of statistical techniques; however, general industry usually lacks the resources to apply or interpret complex statistics. The authors are currently evaluating simplified ADBA statistics which could provide a reliable indication of HCP effectiveness.

ADBA procedures can detect the specific areas in which the HCP needs improvement so that specific corrective steps can be implemented. Certain departments within the plant may need special attention, or the results may show a more general deficiency in one of the phases of the HCP. Patterns of ADBA results help the professional differentiate between group threshold changes due to inconsistent audiometric testing, audiometer calibration shifts, inadequate hearing protection utilization, and other sources (L.H. Royster & Royster, 1984b).

ADBA results provide tangible evidence of problems which can be used to demonstrate to managers the need for improvement. With the support of the ADBA findings, the responsible professional is more likely to obtain approval for the resource allocations or policy changes necessary to correct problems.

ADBA statistics can also show managers evidence of the benefits the company derives when the HCP is improved. Potential compensation costs for occupational hearing loss will decline after an effective HCP is established (L.H. Royster & Royster, 1984b), and management is often most impressed by the resulting savings.

The motivational benefits of ADBA are invaluable in gaining support for the HCP among employees and supervisors. Simple bar charts prepared from the ADBA statistics for various departments can convince supervisors that enforcement of hearing protection utilization is necessary. The same information can be discussed with employees in safety meetings or presented in poster format to show that hearing trends are truly better for groups who consistently wear their hearing protection. Employees can also learn to understand their own threshold changes better if they compare themselves to the normal age-effect reference data used in ADBA evaluations.

In short, the information provided by ADBA can be applied to give meaningful feedback to all levels of personnel involved in the HCP in order to stimulate interest and support for a stronger hearing conservation effort.

ADBA Options

The lower portion of Figure 9.1 illustrates the component ADBA processes utilized to date in evaluating the effectiveness of HCPs. The procedures are divided into two groups: population comparisons and evaluation criteria.

POPULATION COMPARISONS

One approach to evaluating HCP effectiveness is to compare the changes in the population's mean hearing threshold levels with expected changes over time. Conceptually this is a simple procedure; however, a problem arises in selecting the reference or control population. To illustrate the variety of options available we will assume that the ISO 1999 draft standard (ISO, 1982) is an appropriate model for predicting the threshold levels for a noise-exposed population. This model includes two elements: an estimate of the noise-induced permanent threshold shift (NIPTS) component of the population's threshold level and an estimate of the age-related component, referred to as the age-related threshold level (ARTL). For estimating the ARTL component, the proposed standard allows the user to employ either the threshold levels for a highly screened "pure presbycusis" population of males or females, or the threshold levels for a more appropriate custom reference group if one is available.

When evaluating the hearing trends for an industrial noise-exposed population (INEP) over time, we assume that the population's hearing level is composed of two additive components: (1) a component resulting from on-the-job noise exposures, and (2) a component resulting from nonoccupational factors such as recreational gun exposures, military service history, pathology, and presbycusis. We use the term *age effect* to designate this second component, which includes all hearing influences except on-the-job noise exposures. Reference data are available for non-industrial-noise-exposed populations (NINEPs), which provide an estimate of the age effect component (L.H. Royster & Thomas, 1979; L.H. Royster, Driscoll, Thomas & Royster, 1980). The only screening criterion utilized to eliminate subjects from the NINEPs was more than two weeks of exposure to significant on-the-job noise during the subject's prior work experience. Therefore, we assume that the NINEPs exhibit the same effects of the aging process, pathology and harmful noise — *except* industrial noise — as would an industrial population. Because sex and race are factors associated with hearing ability, separate NINEPs are available for black and white females and males. Until a data base is developed which scientifically matches the industrial population in terms of every age-effect hearing influence, we feel that the NINEPs provide the best available control data base against which to compare industrial audiometric data in estimating the amount of on-the-job hearing change.

To illustrate the consequences of selecting an inappropriate ARTL population as a control, we will present an example showing the expected changes in a population's threshold levels predicted using the draft version of ISO 1999. The population, which will be represented by the .5 fractile (median group), is assumed to have been exposed to a daily time-weighted average (TWA) of 95 dBA since age 20. The NIPTS component at the 4-kHz test frequency as a function of age is shown in Figure 9.2, as predicted using the N.A.N. computer program (L.H. Royster, Royster, & Royster, 1983). Also presented in Figure 9.2 are the threshold levels predicted for this population by combining the NIPTS component with the ARTL data for different potential reference populations. Note that the ISO 1999 draft model predicts the population's hearing level (HL) as follows:

$$\text{HL} = \text{NIPTS} + \text{ARTL} - ((\text{NIPTS} \times \text{ARTL})/120) \qquad \textbf{(9.1)}$$

A direct summation of NIPTS and ARTL would, in extreme cases, predict HLs greater than 100 dB. Therefore, the correction factor (NIPTS × ARTL)/120) is subtracted as an attempt to predict extremes in the population's threshold levels more accurately when the values for either NIPTS or ARTL, or both, become very large.

In Figure 9.2 the four ARTL populations used are the NIOSH male (NIOSH,M) reference population (NIOSH, 1972), which is also referenced in the Hearing Conservation Amendment, the ISO male (ISO,M) population defined in the ISO 1999 draft document, a white male NINEP (NINEP, WM), and a black female NINEP (NINEP, BF). If the assumed exposed population was free of NIPTS at age 20 and was not protected from noise thereafter, then the expected changes in the population's threshold level over time would be as shown in Figure 9.2 using the different ARTL populations. The predicted thresholds differ significantly depending upon the ARTL population selected, due to factors including age effects and sex and race characteristics.

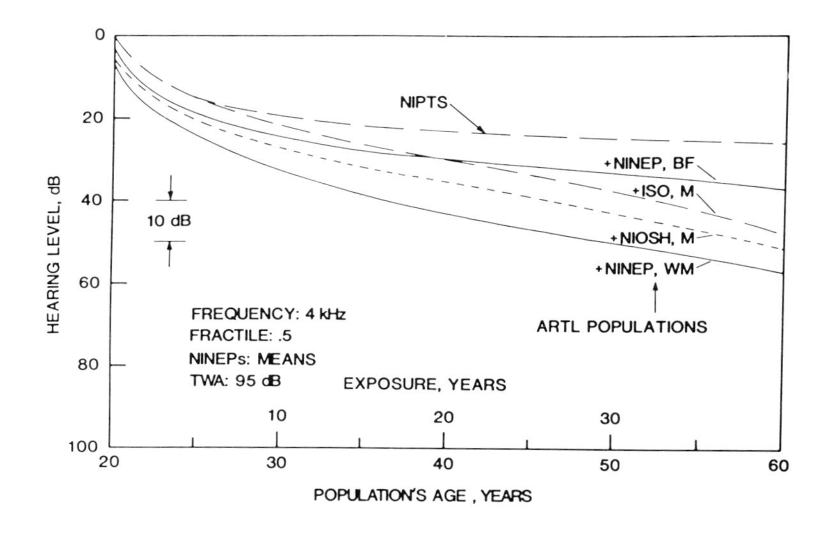

Figure 9.2. Predicted population hearing levels at 4 kHz as a function of population mean age and years of exposure to a TWA of 95 dB for the .5 fractile, calculated by adding the predicted NIPTS component to the age-related components for each of four potential ARTL populations.

In contrast, if the population were properly protected from on-the-job noise since initial exposure at age 20, then the expected threshold levels over time would be as shown in the top portion of Figure 9.3, depending upon the ARTL population selected. These curves represent the rates of change in thresholds with age in the absence of industrial noise exposure for these reference populations. Also plotted in the top portion of Figure 9.3 are the measured mean threshold levels at 4 kHz for a white male INEP in which each employee has been given ten consecutive annual audiometric evaluations. The TWA for this population has remained relatively constant at 87 dBA over the ten years represented by the test results, and the wearing of hearing protection is a condition of employment. Extensive ADBA procedures have also confirmed adequate protection from on-the-job noise exposures for this low-noise reference INEP.

Presented in the lower half of Figure 9.3 are the age-corrected threshold levels for the same white male INEP shown in the upper portion. The age correction was accomplished by subtracting from each employee's threshold levels the white male NINEP threshold levels predicted for his age.

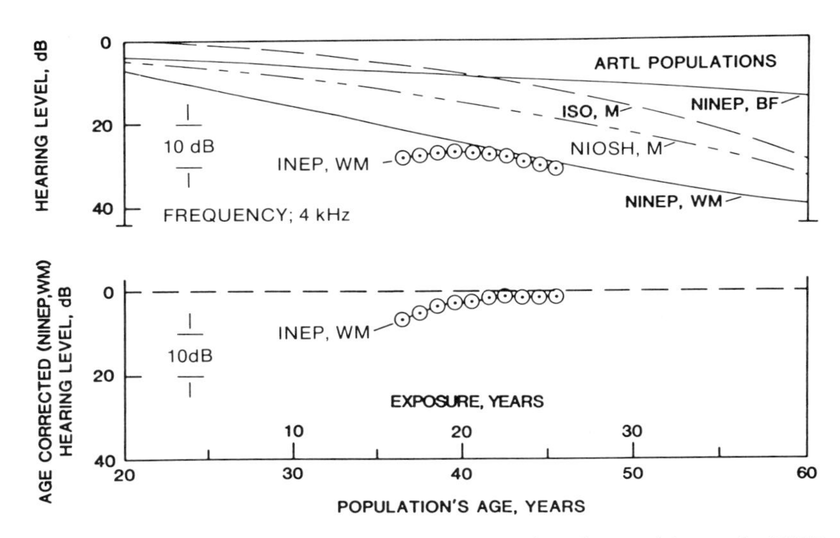

Figure 9.3. The hearing levels at 4 kHz over time for a white male INEP compared to the expected hearing levels for different ARTL populations (upper panel), as well as the effects of age-correcting the white male INEP data using the white male NINEP data (lower panel).

The horizontal line represents the expected age-effect hearing change for the white male NINEP. This portion of Figure 9.3 will be discussed in a later section with respect to the learning effect.

Several important observations are apparent from the data shown in Figure 9.3. First, the amount of pre-existing on-the-job NIPTS assumed for the white male INEP would differ significantly depending upon the male population selected as the ARTL reference. Because the learning effect results in improved thresholds over several years of testing for the INEP, and the NINEP age-effect curve does not include learning improvements, the best estimate of pre-existing NIPTS is the difference between the INEP mean threshold on the first test and the NINEP age-effect hearing level at the same age. (It is assumed that the white male INEP did not have significant prior audiometric test experience). The amount of pre-existing NIPTS for this white male INEP would be overestimated if a screened presbycusis ARTL population such as the white male ISO data were used, or if a mixed-race population such as the male NIOSH data base were used.

A second observation from Figure 9.3 is that the predicted rate of change, or slope (in dB/year), of the ARTL curves could differ significantly depending on the mean age of the white male INEP and on the particular ARTL population selected. A third observation is that the white male

NINEP seems to be a proper control population for this white male INEP, at least at the 4-kHz test frequency, because the white male INEP data become parallel to the white male NINEP age-effect curve after the learning-effect improvement has diminished.

As shown in Figure 9.1 there are at least three different types of control populations that can be utilized as comparison groups for evaluating the changes in an INEP: matching NINEPs or other similar ARTL populations (as discussed above), industrial audiometric data bases that are known to be representative of effective HCPs, and appropriately selected internal control populations of non-noise-exposed employees. Internal control groups are especially useful in attempting to account for possible audiometer calibration shifts in the data and in establishing the minimum test-to-test variability achievable for the audiometric testing environment and techniques used in a particular plant facility.

The preceding discussion has briefly pointed out the importance of considering several parameters when attempting to evaluate the effectiveness of a HCP by using ADBA procedures. These parameters include age, sex, race, the learning effect, the choice of an appropriate ARTL population, and the amount of pre-existing NIPTS at various test frequencies. If an INEP's initial threshold levels were extremely poor, less shifting would be anticipated with continued exposure to the same noise environment. However, for the ADBA evaluations conducted to date, only rarely have population mean threshold levels exhibited extreme hearing loss. In such a case the population can be divided by hearing level in order to evaluate trends for the groups with more sensitive hearing.

EVALUATION CRITERIA

In Figure 9.1 are presented the evaluation criteria that we have been studying as part of our efforts to identify a simple statistic that would reliably predict whether an HCP provides satisfactory protection. Our past efforts have concentrated on two types of statistics: those that can be used during the first few years of the HCP, when the affected employees are new to the program, and statistics that can be used for older HCPs in which employees have more audiometric test experience. The test criteria investigated to date had their origin in two observations made while studying industrial audiometric data bases (Berger, 1976; Lilley, 1980; L.H. Royster, Lilley, & Thomas, 1980; L.H. Royster & Royster, 1980; L.H. Royster & Thomas, 1979). The first observation was that low-noise-exposed populations and populations subjectively judged to be properly wearing hearing protection always exhibited an improvement in indicated mean threshold levels as a function of repeated testing, if not already audiometrically experienced.

The second observation was that the variability in the test-retest data, either with respect to the baseline audiogram or between sequential audiograms, seemed to be inversely correlated with the level of protection afforded the noise-exposed populations.

The Learning Effect

The observed improvement in the population's mean hearing levels was named the *learning effect*. This learning effect was illustrated by the changes in the age-corrected mean threshold data for the white male INEP presented in the bottom portion of Figure 9.3, where the age-corrected mean thresholds approach the horizontal line representing the expected NINEP age-effect trend, rather than remaining parallel to the line. The same effect is also visible in the top portion of Figure 9.3, where the mean thresholds for the white male INEP do not follow the worsening trend which would be predicted by any of the ARTL populations.

A clearer understanding of the degree of learning expected is demonstrated in Figure 9.4. The mean threshold levels at 6 kHz are shown for a

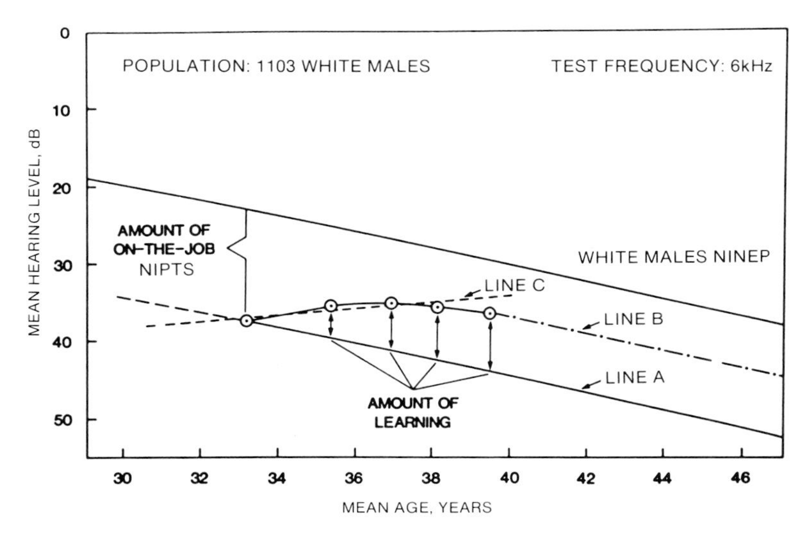

Figure 9.4. Mean HTLs as a function of age on the first five audiograms for a white male INEP, shown with the age curve for the white male NINEP. Line A shows the HTLs expected in the absence of learning. Line B shows the HTLs predicted when learning is occurring. Line C is the linear regression through mean HTLs for tests 1-4. Pre-existing occupational NIPTS and the degree of learning are labeled.

white male INEP restricted to those employees with five consecutive approximately annual audiometric tests. Also shown are the age-effect curve for the white male NINEP, and a parallel curve (line A) that passes through the value for the INEP's mean threshold for the first test. Intuitively one would expect that, in the absence of significant on-the-job noise exposure, the INEP would exhibit age effects approximately equal to the NINEP. However, as shown, the white male INEP's mean thresholds improve relative to line A up to approximately the fifth or later test, after which they are projected to follow line B. The observed level of learning is indicated by the vertical difference between lines A and B. Since the white male NINEP curve was obtained from a single test of a population without significant prior audiometric experience, it is assumed that the NINEP age-effect hearing curve also would have been improved by learning if this population had been given several audiometric evaluations (L.H. Royster & Thomas, 1979; L.H. Royster, Driscoll, Thomas, & Royster, 1980).

Another way of indicating the degree of learning is to age-correct the data for the INEP, as illustrated for the white male INEP data presented in the bottom half of Figure 9.3. The magnitude of the learning effect for this population would be the difference between the mean threshold value for the first test and the minimum later value, or 7 – 2 = 5 dB.

A similar procedure was utilized to establish the maximum magnitude of the learning effect for different segments of this same white male INEP. The threshold data were age corrected using the white male NINEP reference values, and the population was divided based on their mean thresholds at the .5, 1 & 2 kHz frequencies and at the 3, 4 & 6 kHz frequencies. The results are presented in Table 9.1 for the total population and for three 50% segments of the population as indicated. The findings indicate a maximum learning effect in the range of 5-6 dB, which is consistent with

TABLE 9.1
Estimated Maximum Level of Learning for One White Male INEP Exhibiting 10 Consecutive Annual Hearing Tests.

Population Segment	N	Mean HTLs, dB		Estimated Maximum Learning, dB	
		Frequency Range, kHz			
		.5,1,2	3,4,6	.5,1,2	3,4,6
Total	61	9.5	27.3	5.0	4.8
0 - 50 %	31	6.1	16.1	5.2	4.9
25 - 75 %	29	8.7	23.5	5.6	5.1
50 - 100 %	29	13.7	41.3	5.0	5.7

previous observations for properly protected or low-noise-exposed populations (Berger, 1976; Lilley, 1980; L.H. Royster, Lilley, & Thomas 1980; L.H. Royster, Driscoll, Thomas, & Royster, 1980; L.H. Royster & Royster, 1982). As indicated by the last column in Table 9.1 and in previous research, the degree of learning tends to increase for segments of the population with poorer thresholds at the higher test frequencies (J.D. Royster & Royster, 1981a).

Estimating the amount of on-the-job NIPTS depends not only upon selecting an appropriate ARTL population but also, as noted earlier, upon the relative audiometric experience of the two populations compared. Therefore, if we assume that the NINEP is an appropriate ARTL population, the best indicator of pre-existing on-the-job NIPTS would be a comparison between the initial mean thresholds of the INEP and the NINEP hearing level at the age corresponding to the mean age of the INEP at the time of the first test, as indicated in Figure 9.4. Comparisons of later tests to the NINEP curve would underestimate NIPTS because of the learning effect.

Obviously, learning can be partly or completely overshadowed by on-the-job NIPTS. As an example, if an audiometrically inexperienced INEP exhibited NIPTS exactly equal in magnitude over time to the degree of learning exhibited, then the mean thresholds would follow Line A in Figure 9.4 as age-effect hearing decline occurred. If the amount of NIPTS incurred was greater than the degree of learning, the mean thresholds would decline faster than line A, indicating an even less effective HCP. However, if the population already had enough audiometric experience to have exhausted the learning effect, then the interpretation would be altered: thresholds following line A would indicate adequate protection.

Since learning diminishes significantly after the first three to four audiometric tests, a simple approximation of the degree of learning can be derived by fitting a linear regression line through the measured mean thresholds at 4 kHz for the first four tests, as indicated by line C in Figure 9.4. The magnitude of the slope of line C, in dB/year, is then used as an indicator of program effectiveness. Typically, acceptable HCPs will exhibit slope magnitudes of slightly less than 0 dB/year to −1.0 dB/year over the first four years of annual audiometric evaluations (L.H. Royster & Royster, 1982; J.D. Royster & Royster, 1982a). As shown in Figure 9.4, line A exhibits a positive slope because the line represents increasing threshold levels with increasing population age. However, line C represents decreasing threshold levels with increasing age and therefore exhibits a negative slope. If line C were horizontal (parallel to the mean age axis), then line C would exhibit a slope of 0 dB/year, indicating no change in threshold level with increasing age.

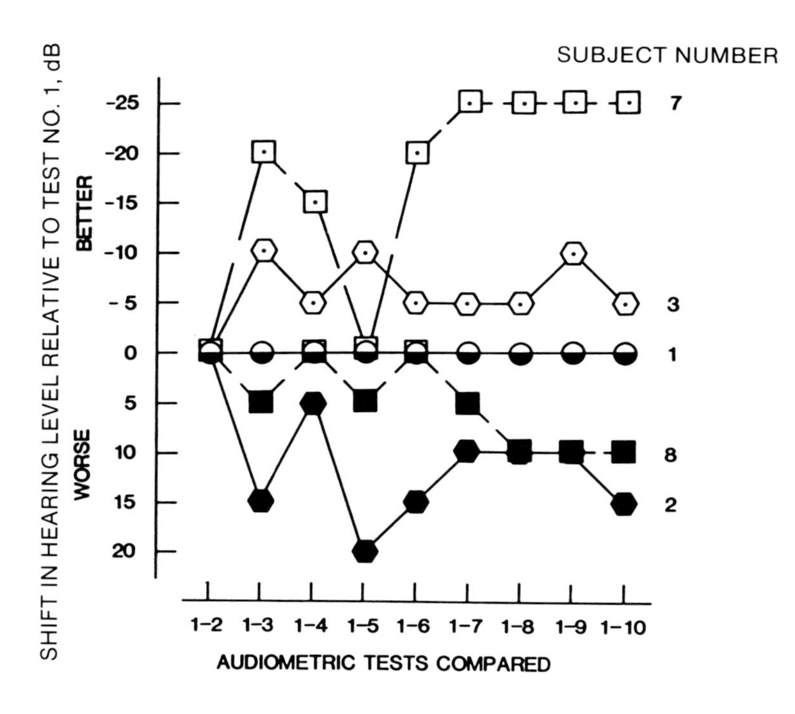

Figure 9.5. Examples of the degree of shifting in individuals' hearing levels at 4 kHz with respect to the initial test results taken from one industrial audiometric data base where the employees had received 10 annual audiograms.

Test-Retest Statistics

Records for randomly selected employees who have been audiometrically evaluated over several years display patterns of change such as those shown in Table 9.2 for the white male INEP data base under discussion. By studying individual threshold levels from Table 9.2, one observes that some employees show an improvement in the measured threshold level over the ten years of testing, some employees' thresholds show no net change, and other employees exhibit generally declining thresholds. A clearer picture of this variability in measured thresholds is obtained if samples of the data presented in Table 9.2 are plotted as shown in Figure 9.5. To obtain Figure 9.5, the individual employee threshold levels for each test were subtracted from the levels for the employee's initial test.

Mean threshold changes for large audiometric data bases reflect patterns similar to those in the sample individual records presented in Figure 9.5. At

TABLE 9.2
Examples of Observed Test Variability at 4 kHz for a White Male INEP Exhibiting 10 Consecutive Annual Tests.

Test Number	Subject Number Measured Hearing Level, dB									
	1	**2**	**3**	**4**	**5**	**6**	**7**	**8**	**9**	**10**
1	0	45	20	15	15	10	65	15	30	5
2	0	45	20	15	15	10	65	15	30	10
3	0	60	10	10	15	5	45	20	30	5
4	0	50	15	10	15	5	40	15	25	5
5	0	65	10	15	15	0	65	20	25	10
6	0	60	15	15	20	5	45	15	30	10
7	0	55	15	15	20	5	40	20	30	5
8	0	55	15	15	15	5	40	25	25	0
9	0	55	10	15	15	5	40	25	30	5
10	0	60	15	15	30	5	40	25	25	5
HTL (10-1)	0	15	-5	0	15	-5	-25	10	-5	0

least three basic trends in audiometric patterns may occur with respect to the baseline threshold levels at any one test frequency: oscillation about the initial threshold, general improvement from baseline values, or general decline in hearing ability. Typical employees may show various patterns of threshold change at different test frequencies.

Based on these patterns of changing threshold levels, the following four shift statistics were developed for use in attempting to judge the effectiveness of HCPs:

(1) Percent better baseline statistic (%Bb) - a shift of 15 dB toward better hearing (lower threshold levels) at any test frequency in either ear with respect to the initial baseline test (test 1 to 2, 1 to 3, 1 to 4, etc.),

(2) Percent worse baseline statistic (%Wb) - a shift of 15 dB toward worse hearing at any test frequency in either ear with respect to the initial baseline test (test 1 to 2, 1 to 3, 1 to 4, etc.),

(3) Ratio of percent better baseline to percent worse baseline (%Bb/%Wb) - a ratio formed from the %Bb and %Wb values for any selected year of testing.

(4) Percent better or worse sequential (%BWs) - a shift of 15 dB toward either better or worse hearing at any test frequency in either ear in a sequential comparison of one test to the preceding test (test 1 to 2, 2 to 3, 3 to 4, etc.).

Different shift magnitudes were considered in trial evaluations of audiometric data bases from low-noise-exposed industrial control populations, and populations in HCPs judged as effective and as ineffective in order to select an appropriate size shift to trigger the shift statistics (L.H. Royster, Lilley, & Thomas, 1980). If too small a shift is required by the criterion statistic, then very high percentages of employees will be included even for the low-noise-exposed populations. Conversely, if the level of shift required to flag an individual employee is too high, then the statistic would not be sensitive enough to distinguish between good and poor HCPs. The final shift magnitude selected, 15 dB, seemed to be a satisfactory compromise.

These shift statistics are based on the assumption that the general distribution of the employees' threshold shift patterns with respect to their initial tests is one indication of HCP effectiveness. The %Bb is obviously related to the learning effect, while the %Wb is related to hearing decline. The ratio %Bb/%Wb indicates whether threshold improvement or threshold decline is the dominant factor affecting mean thresholds. Factors that affect this statistic include whether the baseline mean threshold is higher or lower than normal, the learning effect, the degree of actual protection achieved from wearing hearing protection devices as typically worn, pathology, age effects, normal test variability associated with threshold determination, and exposures to damaging noise environments. Since the %Bb/%Wb statistic is based on comparisons of later tests to the initial baseline thresholds, after several years of audiograms this statistic would approach zero because normal age effects would tend to reduce the %Bb to zero. Therefore, this statistic is appropriate for use only over the first few years of testing for populations that did not exhibit significant audiometric test experience at the time of their first evaluation. Since a significant level of learning would be expected, as indicated in Figure 9.3, a higher percentage of the employees would be expected to meet or exceed the %Bb statistic than the percentage that meet or exceed the %Wb statistic. Therefore one would expect the %Bb/%Wb ratio to be generally greater than unity over the first four years of testing for adequately protected INEPs.

The %BWs statistic described as (4) above indicates the overall variability in the shifting audiometric data. One would expect the measured thresholds for a non-noise-exposed population to exhibit variability around the mean value, and this degree of spread in the data can be used as an indicator of overall program effectiveness. As an example, assume that a previously non-noise-exposed population is introduced to a significant noise hazard and provided hearing protection. If the hearing protection is not adequate for some employees and the employees are tested during the normal working hours, the resulting random temporary threshold shifts (TTSs) exhibited by some employees would increase the level of variability in the data base, which would increase the percent of the population counted in the %BWs statistic. The magnitude of the %BWs statistic, compared to the normally expected value for a non-noise-exposed population, is a useful indicator of HCP effectiveness.

Since the %BWs statistic is based on comparisons between sequential audiometric tests, normally administered once per year, this statistic is not as sensitive to age effects as the statistics based on baseline comparisons. However, the %BWs statistic *is* sensitive to the learning effect, resulting in high %BWs values during the first four to six years of audiometric testing if learning is present. Therefore, this statistic is more appropriate for evaluating a HCP's overall effectiveness after the first four to six years of test results are obtained, unless the population already has extensive audiometric experience. For a properly protected population the %BWs statistic will be less than 30% after the learning effect has diminished.

The rationale for using the %BWs statistic as an indicator of HCP effectiveness is an assumption that it is not practical for industry to produce an industrial audiometric data base that exhibits low test-retest variability (as indicated by a %BWs value of less than 30%) without implementing program elements that are characteristic of effective HCPs (L.H. Royster, Royster, & Berger, 1982). Examples include: (1) taking the time necessary to obtain an acceptable audiometric test, (2) providing effective feedback to employees at the time of the yearly audiometric evaluation, and (3) properly fitting hearing protection and instructing the employee in its use, checking for proper utilization, and replacing worn-out hearing protection devices.

Significant Threshold Shift Criteria

There is a growing tendency for industry to utilize the percentage of a population exceeding the standard threshold shift (STS) criterion as defined by OSHA in the 1983 Hearing Conservation Amendment as an indication of HCP effectiveness. The limited available data on the reliability of this criterion in judging HCP effectiveness do not support its use for this purpose.

We have studied the relative effectiveness of various threshold shift criteria, including the OSHA STS criterion, in differentiating between better and poorer HCPs (J.D. Royster & Royster, 1981b; J.D. Royster & Royster, 1982b). A summary of these findings is presented as Figure 9.6 for four of the shift criteria evaluated. The various STS criteria were applied to the audiometric data bases for several industrial HCPs that had been judged poorer or better HCPs on the basis of ADBA procedures in terms of protecting employees from on-the-job noise. The *OSHA STS* criterion is a shift of 10 dB or greater between the most recent test and the baseline in the averaged hearing threshold levels at 2, 3 and 4 kHz for either ear. The

AAO-HNS criterion involves averaging the hearing threshold levels at .5, 1, and 2 kHz and at 3, 4 and 6 kHz, then comparing the two threshold averages to the corresponding baseline values in each ear to see if either average value meets or exceeds a shift criterion of 10 dB. The *20 dB* shift criterion involves making the comparison between the present test and baseline to see if a shift of 20 dB or greater has occurred at any test frequency (.5 to 6 kHz) in either ear. The *15 dB twice* shift criterion (J.D. Royster & Royster, 1981c) is unique in that it requires a shift of 15 dB or greater from baseline to be present in two consecutive tests at any test frequency in either ear, in order to eliminate tags due to normal variability. That is, a shift of 15 dB or greater in either ear at any test frequency between the baseline and a later test must *still* be present in the next test for the criterion to be triggered.

The data in Figure 9.6 show the relative ability of each of these shift criteria to differentiate between the poorer and better HCPs, based on the percentage of employees tagged for the first time in each baseline comparison (new tags). The higher percentages indicated for the 1-2 test comparison are mainly due to invalid baseline threshold determinations. Note that for the

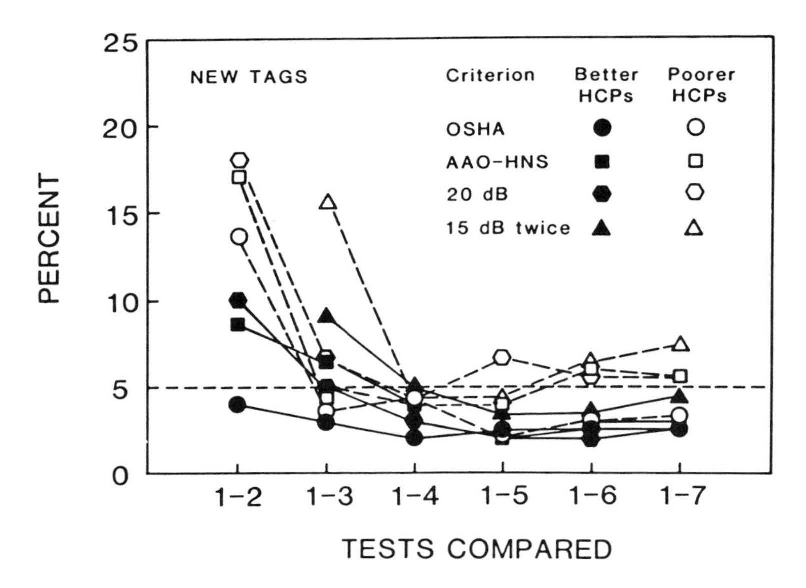

Figure 9.6. Mean percentages of employees in two better HCPs and two poorer HCPs who were tagged for the first time as showing a shift, in comparisons of tests 2 through 7 to baseline test 1, by each of four shift criteria.

later test comparisons only the OSHA STS criterion fails to differentiate significantly between the better and poorer HCPs. The data presented in Figure 9.6 have not been age-corrected; consequently there is a gradual rise in the percentages of new tags for all programs after the 1-5 test comparison. Age-correcting the populations' hearing data would have reduced the percentages of new tags indicated.

The failure of the OSHA STS criterion to differentiate significantly between the poorer and better HCPs is within itself justification for not using this shift criterion as even a relative indicator of HCP effectiveness between plant sites within a corporation or between HCPs in different industries. The limited data presented in Figure 9.6 indicate that regardless of which of the defined shift criteria is used, the percentage of new tags after approximately the third test should be less than 5%. However, it is strongly emphasized that data from additional HCPs will have to be analyzed using various shift criteria before definite ranges of acceptability for individual threshold shift percentages can be proposed for judging HCP quality.

Application of ADBA Concepts

EXAMPLE 1: DETERMINATION OF HCP EFFECTIVENESS

In order to explain the application of the ADBA procedures we will summarize the findings for the white male INEP introduced earlier by the mean threshold data in Figure 9.3.

Population Characteristics

This population consists of 61 employees who had been given ten annual audiometric evaluations. Their typical TWA of 87 dBA was determined by combining the sound survey results across all work stations and across all ten years of sound survey findings. The characteristics of the noise environment for this industrial facility did not change significantly over the ten-year period. The population was extensively trained in the use of hearing protection at the beginning of the HCP and has satisfactorily worn hearing protection for the duration of the HCP. Therefore, the effective exposure level for this population should be less than 80 dBA.

The total population at this industrial facility included 35 black males, but in this example we will consider only the findings for the larger group of 61 white males. At the time of their first audiogram this population's mean age was 36.6 years and their mean length of service was 5.7 years. The average thresholds over the .5, 1 and 2 kHz and the 3, 4 and 6 kHz frequencies are indicated in Table 9.1.

The Baseline Test-Retest and Learning-Related Statistics

The learning-related statistics for this INEP are shown in Figure 9.7. The ranges of acceptable, marginal, and unacceptable values shown in Figure 9.7 were established using ADBA results for low-noise-exposed populations and populations from acceptable and unacceptable HCPs (L.H. Royster, Lilley, & Thomas, 1980).

The statistic shown on the vertical axis is derived from the yearly %Bb and %Wb percentages over tests 1-4, which are shown in Table 9.3. For example, in the test 1-2 comparison, 19.6% of the 61 white males exhibited a shift toward worse hearing of 15 dB or more in at least one test frequency in either ear, and 14.8% of the population exhibited a shift of 15 dB or more

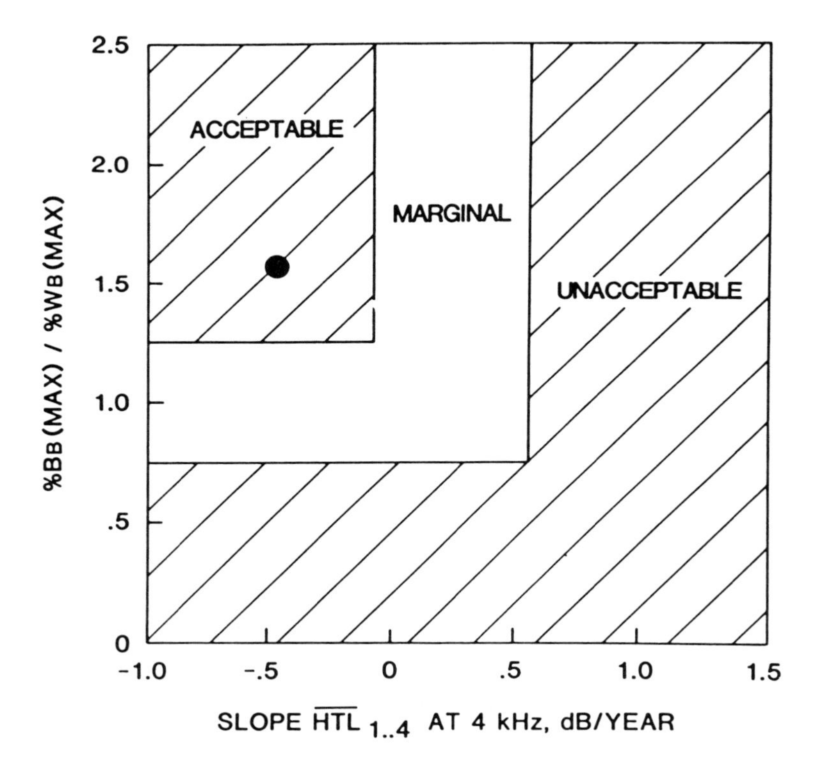

Figure 9.7. Effectiveness of the HCP for 61 white males as judged by two learning-related statistics applied over the first four years of testing: 1) the ratio of the maximum %Bb to the maximum %Wb, and 2) the slope of the linear regression line through mean hearing thresholds at 4 kHz.

TABLE 9.3
Example Calculation for Determining the Statistic %Bb(max)/%Wb(max)

Tests Compared	%Wb (1)	%Bb (1)
1 - 2	19.6	14.8
1 - 3	*26.2*	14.8
1 - 4	18.0	*40.9*

Therefore the %Bb(max)/%Wb(max) ratio =40.9/26.2 = 1.56

(1) Based on a shift criterion of 15 dB from baseline at any test frequency in either ear.

toward better hearing for at least one test frequency in either ear. Once the percentages for the three test comparisons have been established, then the maximum values obtained for the %Wb and %Bb statistics are determined as indicated by the italicized values of 26.2% and 40.9%. The ratio of the maximum %Bb value to the maximum %Wb value forms the %Bb (max)/%Wb(max) statistic which is labeled along the vertical axis of Figure 9.7. In this example the value of %Bb(max)/%Wb(max) is 40.9/26.2 = 1.6.

The statistic defined by the horizontal axis as shown in Figure 9.7 is established by determining the slope of a linear regression line fitted through the mean hearing levels at 4 kHz for the first four audiometric tests. For the population being studied the mean hearing levels at 4 kHz and corresponding ages are shown in Table 9.4. The resulting fitted linear regression line through these four points is shown plotted in Figure 9.8. The linear regression line has a slope of –.42 dB/year. This calculated slope is the statistic that is labeled along the horizontal axis of Figure 9.7.

TABLE 9.4
The White Male INEPs Mean HTL at 4 kHz and Corresponding Mean Age for the First Four Audiometric Tests.

Test Number	Mean HTL At 4 kHz, dB	Mean Age, Yrs.
1	28.0	36.6
2	27.6	37.6
3	27.1	38.6
4	26.6	39.6

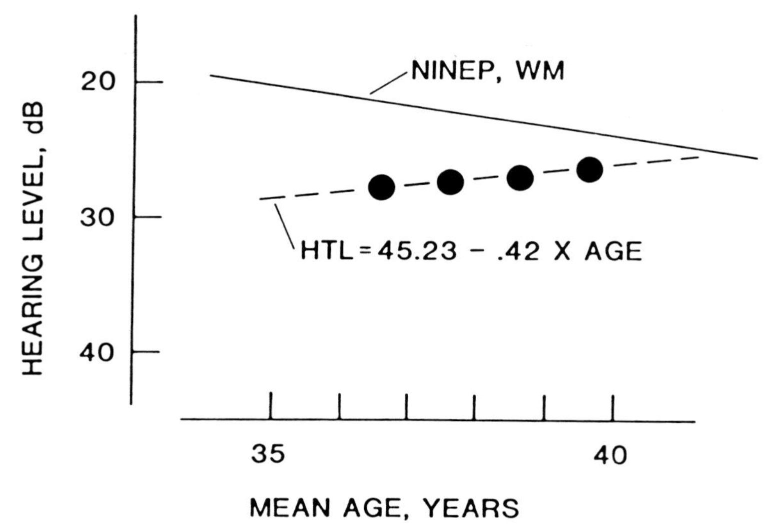

Figure 9.8. Linear regression line fitted to the mean hearing levels over the first four annual audiograms for 61 white males, plotted as a function of mean age. The learning-effect improvement for this INEP contrasts with the expected age-effect hearing decline for the white male NINEP shown over the same age range.

The point of intersection for the values of the two statistics is plotted on the criterion range graph, as indicated in Figure 9.7 by the darkened circle. For this population the HCP would be judged acceptable in terms of the degree of overall protection that has been provided the noise-exposed population, based on this single indicator. This finding should be expected because the employees' TWA is very low, 87 dBA, and the employees wear hearing protection properly.

The Variability-Related Statistic

The ADBA statistic which reflects the overall variability in the population's threshold level data is the %BWs statistic. Recall that this statistic is established by determining the percentage of the population that exhibits a shift of 15 dB or more toward better or toward worse hearing at any test frequency in either ear when a sequential comparison is made between approximately annual hearing evaluations (that is, a comparison of test 1 to test 2, 2 to 3, 3 to 4, etc.). For the sample white male INEP being evaluated, the values of the %BWs statistic are plotted in Figure 9.9 As pointed out earlier, a strong learning effect, which is of course desirable, can inflate the value of this statistic during the first few years of testing. However, after the first three to four years of audiometric test experience, the value for this statistic should be less than 30%. For the population evaluated, the established values for the last five years are all less than 10% except for the 13% value indicated for test 8-9 comparison. Based on previous experience gained from applying this statistic to data bases from low-noise-exposed populations and programs judged as acceptable and unacceptable, we have selected acceptable and unacceptable ranges for this statistic as indicated in Figure 9.9.

This example of the application of ADBA statistics to a low-noise-exposed INEP has confirmed that the population was well protected from on-the-job noise. The following second example involves a population in a more hazardous environment.

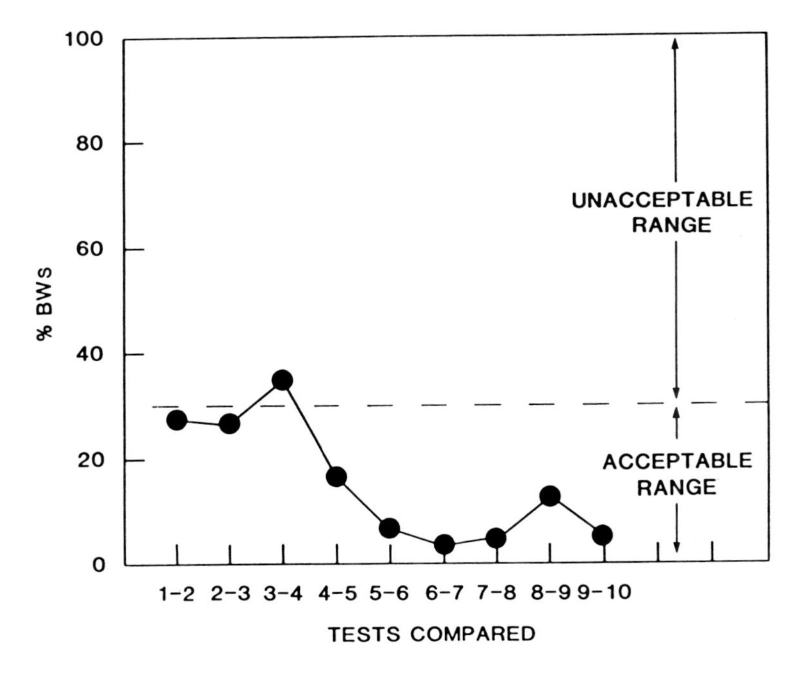

Figure 9.9. Percentages of the white male INEP exhibiting HTLs either better or worse by 15 dB in sequential test-retest comparisons (%BWs) over the first 10 annual audiograms.

EXAMPLE 2: EVALUATION OF HEARING PROTECTOR EFFECTIVENESS

ADBA procedures were used to evaluate the HCP for an industry where the daily TWA in the main production facility was 107 dBA, The values of the %Bb(max)/%Wb(max) statistic in conjunction with the slope of the mean threshold levels at 4 kHz (the learning-related statistic), and of the %BWs statistic, indicated that the HCP was unacceptable in terms of the level of protection being provided the noise-exposed workforce (L.H. Royster, Royster, & Cecich, 1984). Because the HCP appeared to exhibit the administrative characteristics of effective programs (L.H. Royster, Royster, & Berger, 1982), it was suspected that the available hearing protection devices were inadequate to protect the noise-exposed population at such a high TWA, causing excessive variability in the audiometric data base as indicated by the ADBA findings. The three hearing protectors used by the employees over the most recent four years of the HCP were the North Com-fit earplug, Flents Silenta Model 080 earmuff, and the E-A-R foam earplug.

The real-world effectiveness of these three hearing protection devices was investigated in two ways. First, the prevention of temporary threshold shift (TTS) was assessed by measuring the employees' hearing levels at the beginning and end of a work shift. The mean threshold shifts (pretest minus posttest) across all test frequencies were: Com-fit earplug, - 2.49 dB; Silenta earmuff, - .50 dB; and E-A-R earplug, .37 dB (negative values indicate the presence of TTS). In summary, the measured mean TTS was the greatest for the Com-fit earplug hearing protector, while the Silenta earmuff protector exhibited the next highest mean TTS, and the E-A-R foam earplug wearers actually showed a slight improvement in hearing from pretest to posttest, indicating the highest relative level of protection of the three devices compared.

The second method of investigating relative protection by the three devices was to apply the %BWs statistic to the most recent three years of audiometric data for the three wearer groups. The mean percentages of employees exceeding the %BWs statistic, averaged over the last three sequential audiogram comparisons, are presented in Figure 9.10. The relative ranking of real-world protection obtained from the %BWs findings is identical to that obtained from the temporary threshold shift study. That is, the E-A-R foam plug wearers exhibited the lowest level of variability between annual audiometric tests as evaluated using the %BWs statistic, while the Silenta earmuff wearers exhibited intermediate variability, and the Com-fit wearers exhibited the highest level of variability. *A very important finding was that the %BWs statistic accurately evaluated the*

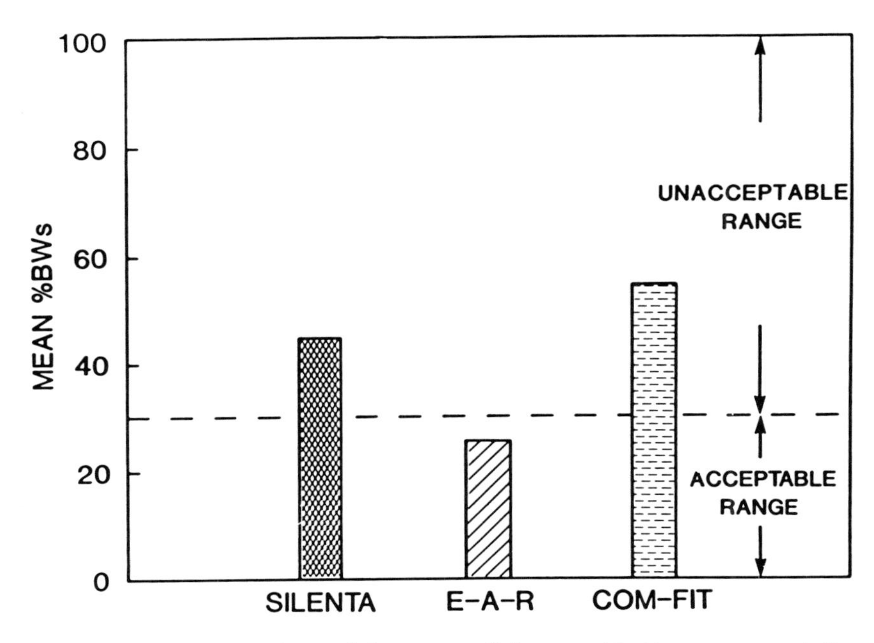

Figure 9.10. Mean values of the sequential percent better or worse statistic (%BWs) for each HPD-wearer group over the most recent three annual audiograms.

relative effectiveness of the two hearing protectors for every sequential comparison for the two groups which included at least 30 ears (E-A-R and Com-fit). The observation, if confirmed by future ADBA studies, would suggest that the %BWs statistic exhibits the potential to predict the relative effectiveness of different hearing protectors, using the results of only two annual audiometric tests.

Concluding Remarks

This chapter has briefly reviewed several concepts for audiometric data base analysis (ADBA) and demonstrated the application of three of the more promising techniques. Space does not permit discussion of other important considerations such as the effect of audiometer calibration adjustments on ADBA findings, or specific methods of using ADBA results to motivate increased support for the HCP. For more detailed information the reader is encouraged to study the referenced papers.

References

Berger, E.H. (1976). "Analysis of the Hearing Levels of an Industrial and Nonindustrial Noise Exposed Population," M.S. thesis, North Carolina State University, Raleigh, NC.

International Organization for Standardization (1982). Draft International Standard ISO/DIS 1999, "Acoustics - Determination of Noise Exposure and Estimation of Noise-Induced Hearing Impairment," Denmark; Secretariat for ISO, TC 43, SCI.

Lilley, D.T., Jr. (1980). "Analysis Techniques for Evaluating the Effectiveness of Industrial Hearing Conservation Programs," M.S. thesis, North Carolina State University, Raleigh, NC.

NIOSH (1972). Criteria for a Recommended Standard ... Occupational Exposure to Noise, U.S. HEW, Report No. HSM 73-11001, Cincinnati, OH.

OSHA (1981). "Occupational Noise Exposure; Hearing Conservation Amendment," *Fed. Reg. 46(11)*, 4078-4181, January 16, 1981.

OSHA (1983). "Occupational Noise Exposure; Hearing Conservation Amendment," *Fed. Reg. 48(46)*, 9738-9783, March 8, 1983.

OSHA (1984). Field Operations Manual, Instructions CPL 2.45A, Ch-3.

Royster, J.D., and Royster, L.H. (1981a). "Judging the Effectiveness of Hearing Protection Devices by Analyzing the Audiometric Database," in *Proceedings of Noise-Con 81*, edited by L.H. Royster, F.D. Hart, and N.D. Stewart, Noise Control Foundation, Poughkeepsie, NY 157-160.

Royster, J.D., and Royster, L.H. (1981b). "Judging the Effectiveness of Significant Threshold Shift Criteria for Industrial Use," Report submitted to OSHA, U.S. Department of Labor, submission No. 366 to docket OSH-11.

Royster, J.D., and Royster, L.H. (1981c). "Response to the U.S. Department of Labor's Request for Comments Concerning the Noise Regulation," Final report prepared for the N.C. Department of Labor, Raleigh, NC, submission no. 501-2 to docket OSH-11.

Royster, J.D., and Royster, L.H. (1982a). "Evaluating the Effectiveness of Hearing Conservation Programs by Analyzing Group Audiometric Data," Unpublished paper presented at the annual convention of the American Speech-Language-Hearing Association, Toronto, Ontario.

Royster, J.D., and Royster, L.H. (1982b). "Comparing Different Threshold Shift Criteria for Industrial Hearing Conservation," Unpublished paper presented at the annual convention of the American Speech-Language-Hearing Association, Toronto, Ontario.

Royster, L.H., Driscoll, D.P., Thomas, W.G., and Royster, J.D. (1980). "Age Effect Hearing Levels for a Black Nonindustrial Noise Exposed Population," *Am. Ind. Hyg. Assoc. J. 41*, 113-119.

Royster, L.H., Lilley, D.T., Jr., and Thomas, W.G. (1980). "Recommended Criteria for Evaluating the Effectiveness of Hearing Conservation Programs," *Am. Ind. Hyg. Assoc. J. 41*, 40-48.

Royster, L.H., and Royster, J.D. (1980). "Industrial Hearing Conservation -- New Considerations," *Hearing Instruments 31(3)*, 4-5.

Royster, L.H., and Royster, J.D. (1982). "Methods of Evaluating Hearing Conservation Program Audiometric Data Bases," in *Personal Hearing Protection In Industry*, edited by P.W. Alberti, Raven Press, New York, NY, 511-540.

Royster, L.H., and Royster, J.D. (1984a). "Audiometric Data Base Analysis Concepts," Invited address presented at the NIOSH Conference on Medical Screening and Biological Monitoring for the Effects of Exposure in the Workplace, July 10-13, Cincinnati, OH.

Royster, L.H., and Royster, J.D. (1984b). "Making the Most out of the Audiometric Data Base," *Sound and Vibration 18(5)*, 18-24.

Royster, L.H., Royster, J.D., and Berger, E.H. (1982). "Guidelines for Developing an Effective Hearing Conservation Program," *Sound and Vibration 16(5)*, 22-25.

Royster, L.H., Royster, J.D., and Cecich, T.F. (1984). "An Evaluation of the Effectiveness of Three Hearing Protection Devices at an Industrial Facility with a TWA of 107 dB," *J. Acous. Soc. Am. 76*, 485-497.

Royster, L.H., Royster, W.K., and Royster, J.D. (1983). "The N.A.N. Computer Program for Estimating a Population's Hearing Threshold Characteristics Version (2.0)," Environmental Noise Consultants, Inc. Cary, NC.

Royster, L.H., and Thomas, W.G. (1979). "Age Effect Hearing Levels for A White Nonindustrial Noise Exposed Population (NINEP) and Their Use in Evaluating Industrial Hearing Conservation Programs," *Am. Ind. Hyg, Assoc. J. 40*, 504-511.

Ward, W.D. (1980). "Noise-Induced Hearing Loss: Research Since 1973," in *Noise As A Public Health Problem: Proceedings of the Third International Congress,* ASHA Report , No. 10. American Speech-Language Hearing Association, Rockville, MD.

Noise and Hearing Conservation Manual, edited by
E.H. Berger, W.D. Ward, J.C. Morrill and L.H. Royster

10 Hearing Protection Devices

Elliott H. Berger

Contents

	Page
Introduction	320
Physics of Hearing Protector Performance	321
Sound Paths to the Occluded Ear	321
Occlusion Effect	324
Estimating Protection	325
Laboratory Test Methods	325
Long-Method Calculation of HPD Noise Reduction	326
Single-Number Calculation of HPD Noise Reduction	328
Inter- and Intralaboratory Variability	331
Selection, Fitting, Use and Care of HPDs	332
Earmuffs	333
Earplugs	338
Premolded Earplugs	340
Formable Earplugs	343
Custom Molded Earplugs	346
Semi-Aural Devices	350
Special Types and Combinations of HPDs	352
Double Hearing Protection	352
Amplitude Sensitive Devices	355
Communication Headsets	356
Recreational Earphones	356
The Initial Ear Examination and HPD Hygiene	358
Real-World Performance of HPDs	359
Summary of Existing Field Data	360
Suggestions for Derating Laboratory Data	362
A Comment Concerning the Long Method	363
OSHA's Calculational Procedure	364
Evaluation of the Audiometric Data Base	365

Effects of HPDs on Auditory Communication 365
Understanding Speech 365
Responding to Warning and Indicator Sounds 368
Localization and Depth Perception 369
General Remarks 369
Hearing Protector Standards and Regulations 370
Standards and Related Documents 370
EPA Labeling Regulation 371
Concluding Remarks 372
Recommendations 372
Final Considerations 373
References 374

Introduction

As the number and variety of available hearing protection devices (HPDs) increased dramatically in the past 30 years, their quality also improved. The crucial variable that controls hearing protector effectiveness, however, has remained the same — the wearer, and how he or she fits and uses the device. Thus much of this chapter is focused upon increasing the awareness of professional hearing conservationists regarding practical aspects of hearing protector selection and utilization and the details of real-world performance. Nevertheless, the more technical elements of hearing protection such as the physics of HPD performance, the results of laboratory attenuation tests, the effects of HPDs on auditory communications, and the complexities of standards and regulations will not be overlooked. The purpose is to provide a comprehensive yet pragmatic reference source for the occupational hearing conservationist.

While reviewing this chapter it is well to remember that, regardless of the cost or laboratory-measured attenuation of a particular device, there is little benefit in distributing it to a noise-exposed population without first providing adequate indoctrination concerning its fitting, use, and care. Additionally, a means of motivating the workers to properly implement their training also must be included. Therefore, education and motivation, which are reviewed in Chapter 11, are important collateral topics to the discussion at hand.

Physics of Hearing Protector Performance

The acoustical performance of an HPD involves not only the physics of the protector and its interface with the ear, but the anatomical/physiological limitations of the human auditory system as well.

SOUND PATHS TO THE OCCLUDED EAR

The occluded ear receives sound energy along four primary transmission paths which are illustrated schematically in Figures 10.1A, 10.1B, and 10.1C. These are:

1) Air Leaks — For maximum protection the device must make virtually an airtight seal with the canal or the side of the head. Inserts must precisely fit the contours of the ear canal and earmuff cushions must accurately fit the areas surrounding the external ear (pinna). Air leaks can typically reduce attenuation by 5 - 15 dB over a broad frequency range (Nixon, 1979).
2) Hearing Protector Vibration — Due to the flexibility of the ear canal flesh, earplugs can vibrate in a piston-like manner within the ear canal. This limits their low-frequency attenuation. Likewise, an earmuff cannot be attached to the head in a totally rigid manner. Its cup will vibrate against the head as a mass/spring system, with an effective stiffness determined by the flexibility of the muff cushion and the flesh surrounding the ear, as well as the air volume entrapped under the cup. For earmuffs, premolded inserts and foam inserts, these limits of attenuation at 125 Hz are approximately 25 dB, 30 dB and 40 dB, respectively.
3) Material Transmission — The exterior surfaces of an HPD will undergo some deformation of shape (vibration) in response to the forces applied by impinging sound waves. These vibrations are

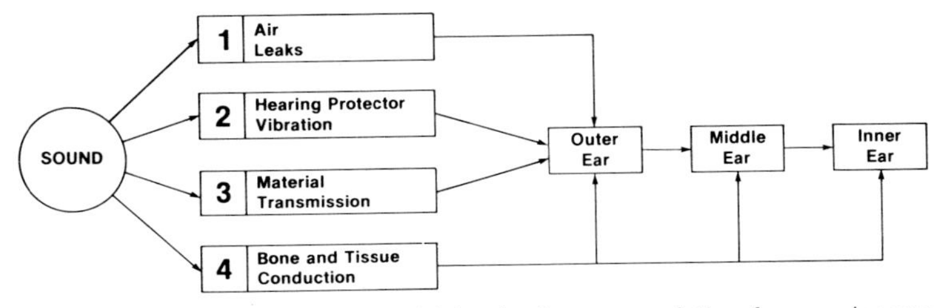

Figure 10.1A — Generalized block diagram of the four primary sound pathways to the occluded ear.

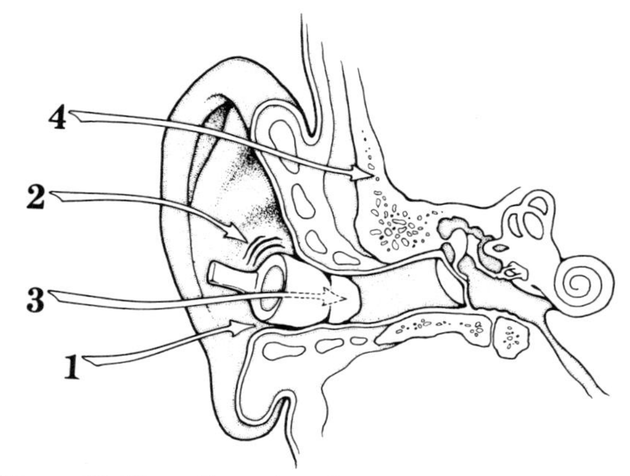

Figure 10.1B — Sound pathways for an earplug.

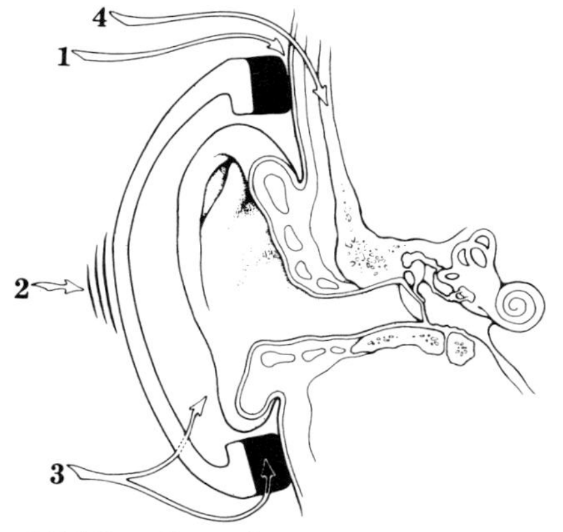

Figure 10.1C — Sound pathways for an earmuff.

transmitted through the material of the protector to its inner surface where the resultant motion radiates sound, of diminished intensity, into the enclosed volume between the HPD and the wearer's eardrum. The amount of sound reduction is dependent upon the mass, stiffness, and internal damping of the hearing protector material.

Material transmission through the cup and cushion components of earmuffs is significant, normally providing a limitation to attenuation at frequencies above 1 kHz. This path is generally less important with insert HPDs due to the fact that their exposed surface areas are much smaller. However, a special case exists for certain fibrous materials, such as cotton, which are easily permeated by air. In this latter instance, attenuation will be very low since sound will pass through the substance of the device in a manner analogous to transmission path 1), *i.e.* as though there were many tiny air leaks.

4) Bone and Tissue Conduction — Even if the HPD were perfectly effective in blocking the preceding three sound paths, sound energy would still reach the inner ear (see Figure 10.1A) via bone and tissue conduction (BC). Energy transmitted in this manner is said to flank or bypass the protector. It imposes a limit on the real-ear attenuation that any HPD can provide. However, the level of the sound reaching the ear by such means is approximately 50 dB below the level of air-conducted sound through the open ear canal. This is illustrated in Figure 10.2. The early estimates are from Zwislocki (1957) and Nixon and von Gierke (1959) and the recent estimates from Berger (1983b).

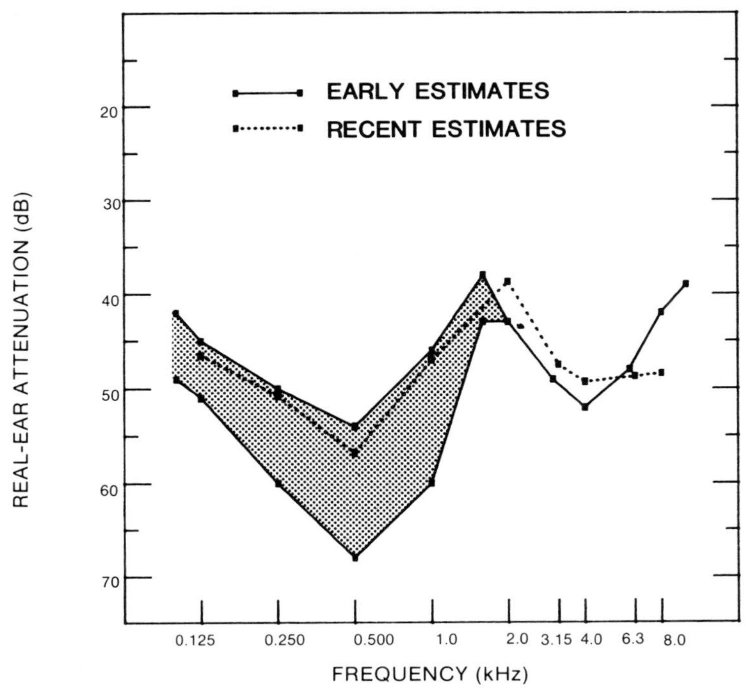

Figure 10.2 — Estimates of the bone conduction limitations to hearing protector attenuation (from Berger, 1983b).

Evidence suggests that the BC limits to attenuation are constant regardless of sound level. This indicates that the relative importance of the energy transmitted via the BC pathways is independent of the noise level in which the hearing protector is worn. Sound transmitted via the BC paths is normally not of primary concern unless an HPD's attenuation approaches 40 - 50 dB (depending on frequency), which is rarely the case due to air leaks, protector vibration, and material transmission.

Since the regions of the head around the external ear are only a small portion of the total skull area exposed to sound, covering them with an earmuff is of little significance with respect to BC, perhaps 3 - 4 dB in the 1 - 2 kHz region (Berger and Kerivan, 1983). Thus, the relative performance of plugs compared with that of muffs is not, in practice, determined by the BC paths, but by factors inherent in the design of the HPDs and their interface to the head. Increasing the limits to attenuation shown in Figure 10.2 would require completely enclosing the head in a rigid helmet with face mask. Therefore use of earmuffs in conjunction with a hard hat, which covers only part of the head and has many gaps through which the sound energy can penetrate, provides no more protection than the muffs alone.

OCCLUSION EFFECT

The efficiency of the BC paths in conducting energy to the inner ear is affected by the state of occlusion of the ear canal. When the ear is occluded with an HPD, the efficiency of transmission of bone-conducted energy for frequencies below 2 kHz is enhanced relative to the unoccluded ear. At those frequencies, vibratory displacements of the canal walls, concha, and/or pinna develop significantly greater pressure at the eardrum in the occluded condition since the sound is unable to escape freely from the ear as it can in the unoccluded case. This is called the earplug effect (Zwislocki, 1957), or more commonly the occlusion effect (Tonndorf, 1972).

An occlusion effect is observed whenever the ear is plugged or covered, either with an insert, semi-aural, or circumaural HPD. However, the magnitude of the occlusion effect is a function of how the occluding device is fitted (Berger and Kerivan, 1983). The occlusion effect is maximized when the canal is sealed at its entrance with a semi-aural or canal cap type of device and diminishes either as the occluder is more deeply inserted (as is the case with earplugs instead of semi-aurals and more deeply fitted earplugs vs. less deeply fitted ones) or as the occluder begins to contain a large internal volume (as is the case with earmuffs instead of semi-aurals and larger volume earmuffs vs. smaller volume earmuffs). This variation of the occlusion effect as a function of the method of occlusion is why a range of BC limits, instead of just a single value, is plotted for the frequencies below 2 kHz in Figure 10.2.

The occlusion effect can be easily demonstrated by capping one's ear canals while speaking aloud. When the canals are properly sealed or covered one's own voice takes on a bassy, resonant quality due to the enhancement of the BC paths by which talkers partially hear their own speech. The occlusion effect is often cited by users as one of the objectionable features of wearing HPDs. For those who are particularly sensitive to this effect it can be minimized by inserting the earplugs more deeply or using larger volume earmuffs. The occlusion effect is also a useful hearing protector fitting aid, as is later discussed in the section entitled Selection, Fitting, Use and Care of HPDs.

Estimating Protection

LABORATORY TEST METHODS

More than a dozen different laboratory methods for evaluating HPD attenuation have been reported in the literature (Berger, 1986). The tests have included measurements of threshold shift, loudness balance, masking, and lateralization procedures using human subjects, and also various instrumented measurements using human subjects, artificial heads, and cadavers as acoustical test fixtures. The most common, and reputedly one of the most accurate tests (Berger and Kerivan, 1983), is that of real-ear attenuation at threshold (REAT), which is the subject of one international and three American Standards (ISO 4869-1981; ANSI Z24.22-1957; ASA STD-1 1975; ANSI S12.6-1984) as well as documents in many other countries.[1] Virtually all available manufacturers' reported data have been derived via this method.

The test is based upon the determination of the difference between the minimum level of sound that a subject can hear without wearing an HPD (open threshold) and the level needed when the subject is wearing the HPD (occluded threshold). The difference between these two thresholds, the threshold shift, is a measure of the REAT afforded by the device. Since the test is conducted at a relatively low sound pressure level (SPL), it cannot accurately characterize the performance of HPDs that claim to offer attenuation that increases with sound level. However, for linear HPDs

[1] As of this writing, the revised American REAT standard, ANSI S12.6, has just been issued. It is too new for any data acquired in conformance with it to have yet been published. For that reason, and due to its strong similarity to ASA STD 1, the discussion in this chapter focuses on the 1975 standard.

(those not containing valves, orifices, diaphragms, or active electronic circuitry) the attenuation measured by an REAT evaluation will accurately represent the performance of the device regardless of sound level (Martin, 1979; Berger and Kerivan, 1983; Humes, 1983).

According to ASA STD-1, the measurements are to be conducted in a diffuse sound field using 1/3-octave bands of noise and a minimum of 10 subjects whose open and occluded thresholds are measured three times each. This results in 30 data points at each frequency, from which a mean attenuation ($\bar{x}$) as well as a standard deviation (σ) are computed, the latter parameter providing an indication of the variability in attenuation across subjects and replications. The mean attenuation represents protection that approximately 50% of the test subjects meet or exceed. When it is appropriate to estimate the protection that a greater percentage of the subjects attain, adjustments to the mean may be computed by subtracting one or more standard deviations.

Numerous data are still available that have been measured in accordance with the older test standard, ANSI Z24.22. That standard required the subject to be seated in a directional sound field, usually achieved by testing in an anechoic chamber with a loudspeaker in front of and facing the subject. The test sounds were pure tones. These conditions were less representative of typical industrial exposures than are the noise bands and nondirectional sound field specified in the newer standard, ASA STD-1. In practice, the electroacoustic differences between the two standards have little effect on the measured attenuation of earplugs, but earmuff attenuation tends to be lower, especially in the 1 and 4 kHz regions, using the newer standard (Bolka, 1972; Martin, 1977). Additionally, σs tend to be lower for both types of HPDs when tests are conducted using noise bands in a diffuse sound field.

REAT data are normally limited to the frequency range of 125 Hz to 8 kHz. Some laboratories routinely test to frequencies as low as 63 or 75 Hz (Martin, 1977; Camp, 1979), but few data are available at frequencies above 8 kHz. In a recent study (Berger, 1983d) the attenuation of five earplugs, one semi-aural device, four earmuffs, and one earplug-plus-earmuff combination were examined at frequency extremes. The results suggested that the attenuation values of HPDs at 80 Hz, and at 12.5 and 16 kHz, could be approximated by assuming that they were equivalent to the performance at the 125-Hz and 8-kHz 1/3-octave bands respectively.

LONG-METHOD CALCULATION OF HPD NOISE REDUCTION

If one can be assured that the user of an HPD is wearing it in the same way as did the laboratory test subjects (see Real-World Performance of HPDs), then the most accurate method of applying laboratory test data to estimate a user's protected exposure is the long method calculation illustrated in Table 10.1 (Method 1 in NIOSH, 1975). At each frequency, the HPD's $\bar{x}$ minus a σ correction (2σ is used in Table 10.1), is subtracted from the measured A-weighted octave band SPL. The protected levels are then logarithmically summed to determine the A-weighted sound level under the protector. This computation requires that the user conduct or have available an octave band analysis (see Chapter 3) and perform the appropriate calculations for each individual noise spectrum; *i.e.*, the amount of protection afforded cannot be calculated independently of the noise spectrum in which the HPD will be worn.

TABLE 10.1
Long Method Calculation of HPD Noise Reduction

Octave Band Center Frequency (Hz)	125	250	500	1000	2000	4000	8000	dBA[A]
1. Measured sound pressure levels	85.0	87.0	90.0	90.0	85.0	82.0	80.0	
2. A-weighting correction	-16.1	-8.6	-3.2	0.0	+1.2	+1.0	-1.1	
3. A-weighted sound levels [step 1 - step 2]	68.9	78.4	86.8	90.0	86.2	83.0	78.9	93.5
4. Typical premolded earplug attenuation	27.4	26.6	27.5	27.0	32.0	46.0[B]	44.2[C]	
5. Standard deviation × 2	7.8	8.4	9.4	6.8	8.8	7.3[B]	12.8[C]	
6. Estimated protected A-weighted sound levels [step 3 - step 4 + step 5]	49.3	60.2	68.7	69.8	63.0	44.3	47.5	73.0

The estimated protection for 98% of the users in the noise environment, assuming they wear the device in the same manner as did the test subjects, and assuming they are accurately represented by the test subjects is: **93.5–73.0 = 20.5 dBA.**

[A]Logarithmic sum of 7 octave band levels in the row.

[B]Arithmetic average of 3150 and 4000 Hz data.

[C]Arithmetic average of 6300 and 8000 Hz data.

It is important at this point to understand clearly the meaning of the σ correction. It is *not* a method of correcting the laboratory data to estimate performance in the real world; rather it adjusts the mean laboratory data to reflect the attenuation achieved by 84% (for a 1 σ correction) or 98% (for a 2 σ correction) of the laboratory subjects. The correction also applies to actual users to the extent that it can be assumed that they are accurately represented by the test subjects and are wearing the device in the same manner as did the laboratory subjects. However, such assumptions, especially the latter one, are not ordinarily justified.

SINGLE-NUMBER CALCULATION OF HPD NOISE REDUCTION

The primary advantage of a single number rating is that it can be precalculated by the manufacturer and supplied to the user with the device. This permits estimation of wearer noise exposures using only simple mathematics and a single noise measurement.

One of the more accurate and also one of the more common single number ratings is the Noise Reduction Rating (NRR), which was adapted by EPA (1979) from earlier NIOSH (1975) work. It is calculated in a manner similar to the long method, except that a pink noise spectrum (equal energy in each octave band from 125 Hz to 8 kHz) is used instead of the particular noise spectrum, and a negative 3-dB spectral uncertainty-correction is included in the computational procedures. The 3-dB correction is a safety factor to protect against overestimates of HPD noise reduction that might arise from differences between the shape of the assumed (pink) and particular noise spectra. In computing the NRR, the $\overline{x}$ values are reduced by subtracting 2σ so the resultant NRR should theoretically estimate the degree of protection at the 98th percentile. The meaning of this 2σ correction is the same as discussed above for the long method calculation. The NRR calculated in this manner will be denoted by the abbreviation NRR_{98}.

A representative NRR calculation is presented in Table 10.2. This table is organized in a manner similar to Table 10.1 so that the computational procedures may be compared. The NRR, which is finally computed in step 9, is found by subtracting the protected (interior) A-weighted sound levels from the unprotected (exterior) C-weighted sound levels, less a 3-dB spectral uncertainty correction. The NRR, using the same earplug data as are found in Table 10.1, is 20.8 dB.

The NRR is used to estimate wearer noise exposures by subtracting it from the C-weighted sound level as shown in Equation (1).

$$\text{Estimated Exposure (dBA)} = \text{Workplace Noise Level (dBC)} - \text{NRR} \quad \textbf{(1)}$$

TABLE 10.2
Method of Computation of the NRR

Octave Band Center Frequency (Hz)	125	250	500	1000	2000	4000	8000	dB(X)[A]
1. Assumed sound pressure levels	100.0	100.0	100.0	100.0	100.0	100.0	100.0	
2. C-weighting correction	-0.2	0.0	0.0	0.0	-0.2	-0.8	-3.0	
3. C-weighted sound levels [step 1 - step 2]	99.8	100.0	100.0	100.0	99.8	99.2	97.0	108.0 [dBC]
4. A-weighting correction	-16.1	-8.6	-3.2	0.0	+1.2	+1.0	-1.1	
5. A-weighted sound levels [step 1 - step 4]	83.9	91.4	96.8	100.0	101.2	101.0	98.9	
6. Typical premolded earplug attenuation	27.4	26.6	27.5	27.0	32.0	46.0[B]	44.2[C]	
7. Standard deviation × 2	7.8	8.4	9.4	6.8	8.8	7.3[B]	12.8[C]	
8. Estimated protected A-weighted sound levels [step 5 - step 6 + step 7]	64.3	73.2	78.7	79.8	78.0	62.3	67.5	84.2 [dBA]

9. NRR = step 3 - step 8 - 3[D]
NRR = 108.0 - 84.2 - 3 = 20.8 dB

The NRR represents the attenuation that will be obtained by 98% of the users in typical industrial noise environments, assuming they wear the device in the same manner as did the test subjects, and assuming they are accurately represented by the test subjects.

[A]Logarithmic sum of 7 octave band levels in the row. This is a C-weighted sound level for step 3 and an A-weighted sound level for step 8. See Chapter 2 for method of computation.

[B]Arithmetic average of 3150 and 4000 Hz data.

[C]Arithmetic average of 6300 and 8000 Hz data.

[D]The 3-dB spectral uncertainty factor is to protect against overestimates of the HPD's noise reduction that could arise from potential differences between the assumed spectrum and that of the user's actual exposure.

To compare the long method calculation to the NRR, the C-weighted sound level for the spectrum shown in Table 10.1 must be computed. It is 95.2 dBC. When the NRR is subtracted from this C-weighted sound level as indicated in Equation (1), the predicted exposure is 74.4 dBA versus the 73.0 dBA estimated using the long method. The more conservative estimate of protection that is provided by the NRR is expected since it includes the 3-dB spectral correction that is not found in the long method. This correction is intended to assure that in most instances errors arising from spectral variability will lead to under- instead of overestimates of protection.

The practice of using the NRR with C-weighted sound levels may seem peculiar since one would expect that estimation of an employee's protected A-weighted exposure should be accomplished by simply subtracting a single number rating from the A-weighted workplace noise level. However, a number of authors have examined this question, and the weight of empirical evidence leads to the conclusion that "the best single-number method, representing a practical compromise between accuracy and convenience, is one which generates a rating number to be subtracted from the C-weighted SPL of the noise" (Sutton and Robinson, 1981). This relationship exists because the attenuation of most HPDs decreases as frequency decreases, falling off at approximately the same rate as the increase in difference between the C- and A-weighting curves. Alternatively, the relationship may be explained by observing that the A-weighted noise reduction of a hearing protector decreases as the proportion of low-frequency energy in the noise spectrum increases, and this may be accounted for by taking the difference between the C- and A-weighted sound levels of the noise spectrum.

If one chooses to subtract the NRR from A-weighted sound levels, considerable accuracy is lost, and an additional 7-dB safety factor must be included in the computations as shown in Equation (2) (NIOSH, 1975; OSHA, 1983).

$$\text{Estimated Exposure (dBA)} = \text{Workplace Noise Level (dBA)} - (\text{NRR} - 7) \quad \textbf{(2)}$$

The 7-dB safety factor in Equation (2) accounts for the largest differences typically expected between the C- and A-weighted sound levels of industrial noises. Since it is a "worst-case" correction, in most cases it will overestimate actual C-A differences. As an alternative, one can correct the A-weighted TWA (obtained from dosimeters which are usually only capable of A-weighted measurements) by using a sound level meter to develop a C-A value for typical processes, areas, or job descriptions (Berger, 1984; Royster and Royster, 1985). This C-A value is then added to the A-weighted TWA to calculate an estimated C-weighted TWA or workplace noise level, from which the NRR can be subtracted using the procedure of Equation (1). To the extent that an accurate C-A value can be estimated, this method will provide enhanced accuracy over use of Equation (2) for those situations in which C-weighted sound levels are unavailable.

Berger (1983a) has shown that the primary source of error in applying the NRR to estimate user noise exposures is not in the computation of the NRR from the basic laboratory data, but in the fact that laboratory data are not representative of the values attained by actual users. Later in the chapter (see Real-World Performance of HPDs) the divergence between laboratory and field data will be discussed, and suggestions for derating laboratory data will be presented.

INTER- AND INTRALABORATORY VARIABILITY

Another consideration regarding laboratory attenuation data is the reproducibility of the data from different test facilities. Berger *et al.* (1982) reported the results from a round robin test program initiated by the U.S. EPA. Four HPDs representing a wide range of commercially available products were tested. Seven laboratories participated directly; data obtained separately from an eighth laboratory also were included in the evaluation. The data from laboratory eight were included since that facility was responsible for greater than 80% of manufacturers' reported data on file with EPA at the time of the study.

The results showed significant variation among the different laboratories in both $\bar{x}$ and σ, leading to substantial differences in reported NRRs (see Figure 10.3). One source of the variability appeared to be the uncertainty of obtaining the proper fit to avoid acoustic leaks. Other sources included subject selection and training, as well as data reduction techniques.

The results demonstrated that the task of rank ordering the performance of a group of HPDs was not possible using presently available data unless the user could be assured that all of the data were from one laboratory. Therefore all of the subsequent laboratory data that are illustrated in this chapter are from only one facility, the E-A-R Division Acoustical Laboratory.

Even if one analyzes data from just one laboratory when comparing HPD attenuation values and/or NRRs, accuracy and repeatability are of concern. Over a period of years or even months, a facility's subject selection, fitting, and sizing techniques may vary, the characteristics of subsequent samples of test devices may differ from those tested initially, and other uncontrolled parameters may change and give rise to divergent results. Berger (1983c) has provided representative data illustrating many

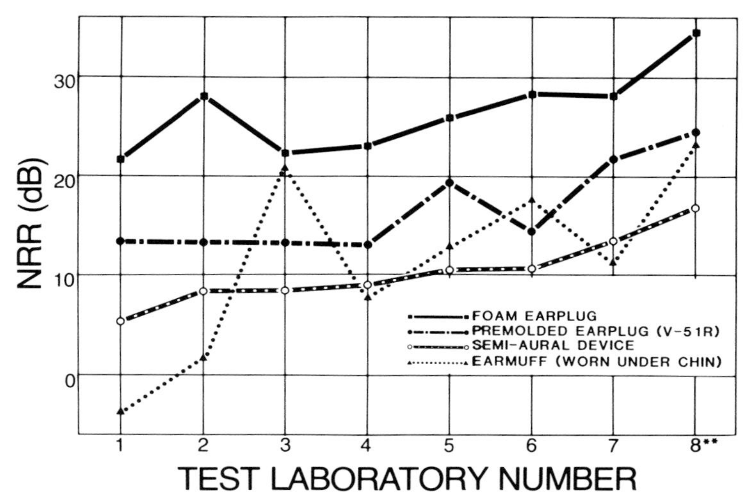

Figure 10.3 — NRRs from eight U.S. laboratories.*

of these effects, and based upon his results and those of others, concluded that changes in the NRR of less than approximately 3 dB should not be regarded as having any practical importance.

Selection, Fitting, Use and Care of HPDs

Hearing protection devices may be broadly categorized into earmuffs, which fit over and around the ears (circumaural) to provide an acoustic seal against the head, earplugs, which are placed into the ear canal to form a seal, and semi-aural devices (also called semi-insert, concha-seated, and canal caps), which are held against the canal entrance with a headband to provide an acoustic seal at that point. Other categories that may be considered are helmets which contain circumaural cups and are normally used only by the military or when head protection combined with communication must be provided along with hearing protection, and devices that rest upon the external ear (supra-aural) such as are commonly used for presenting stimuli for hearing testing. Due to the considerable variety among earplugs, they can be further differentiated into three types: premolded, formable, and custom molded.

---

*Data from Berger *et al.* (1982).

**Laboratory responsible for greater than 80% of the manufacturers' label verification reports that were on file with the U.S. EPA at the time of the round robin study.

As with all human attributes, there is a wide degree of variability in the anatomical characteristics of the human head and ear. Population norms for head widths have been reported as 121, 136, and 148 mm for the 5th, 50th, and 95th percentiles respectively. Similarly, the 5th to 95th percentile head heights cover a range of 116 - 143 mm with a median value of 130 mm. Ear canals also vary widely in size, shape, position, and angle of inclination among individuals and even between the ears of the same individual. Cross-sectional dimensions range from 3 to 14 mm, with an average diameter of approximately 8 mm; the average length of the ear canal is 25 mm. Typically, ear canal cross sections are elliptical in nature, but the extremes of round and slit shaped canals are also observed. This wide range in anthropometric data must be considered by the designers of HPDs and also must be accounted for when protectors are selected for particular individuals. Obviously, no one device will be the correct choice for all concerned.

EARMUFFS

Earmuffs normally consist of rigid molded plastic earcups that seal around the ear using foam or fluid-filled cushions and are held in place with metal or plastic headbands, or by a spring-loaded assembly attached to a hard hat. The cups are lined with an acoustical material, typically an open-cell polyurethane foam, to absorb high-frequency (>2 kHz) energy within the cup. The headbands may function in a single position only, or may be of the more versatile "universal" style suitable for use over the head, behind the head, or under the chin. Most manufacturers offer plastic or fabric crown straps for use in the alternate wearing positions to provide a more snug, secure, and comfortable fit. Representative earmuffs are shown in Figure 10.4.

Earmuffs are relatively easy to dispense since they are one-size devices designed to fit *nearly* all adult users. Nevertheless, earmuffs must be evaluated for fit when initially issued, since not every user can be fitted by all models. Head or ear sizes may fall outside the range that the muff band or cup openings can accommodate. Furthermore, heads with unusual anatomical features such as facial structures with prominent cheek bones (zygomatic arches) or severe depressions below the pinna and behind the jaw (posterior to the temporo-mandibular joint) are particularly hard to fit.

Earmuffs are good for intermittent exposures due to the ease with which they can be donned and removed, and they may be suitable when earplugs are contraindicated. For long-term wearing it is often reported that earmuffs feel tight, hot, bulky, and heavy, although in cold environments their warming effect is appreciated. It is easy for supervisors to monitor that earmuffs are in use, but since their attenuation can be easily compromised by a number of factors (discussed below), it is inadvisable to assume that "use" is synonymous with "effective protection."

Instructions for earmuff usage must be verbally reviewed when the devices are issued, since details of correct placement are often ignored or overlooked in practice. For example, many cups have a preferred orientation, either left/right or top/bottom, and this must be pointed out to the employee. Long hair, sideburns, and caps also can reduce attenuation. Employees should be advised to position the earmuffs with as much of their hair removed from the cushion-to-head interface as possible, and to be certain that caps or other clothing are not placed underneath the earmuff cushions.

Earmuff performance will be degraded by anything that compromises the cushion-to-circumaural-flesh seal. This includes other pieces of personal protective equipment such as eyewear, masks, face shields, and helmets. Typically, eyeglasses will degrade earmuff attenuation by 3 - 7 dB, with the losses most noticeable at the low and high frequencies, although the effect varies widely among earmuffs (Nixon and Knoblach, 1974). The losses will be minimized with thinner, closer fitting temple pieces.

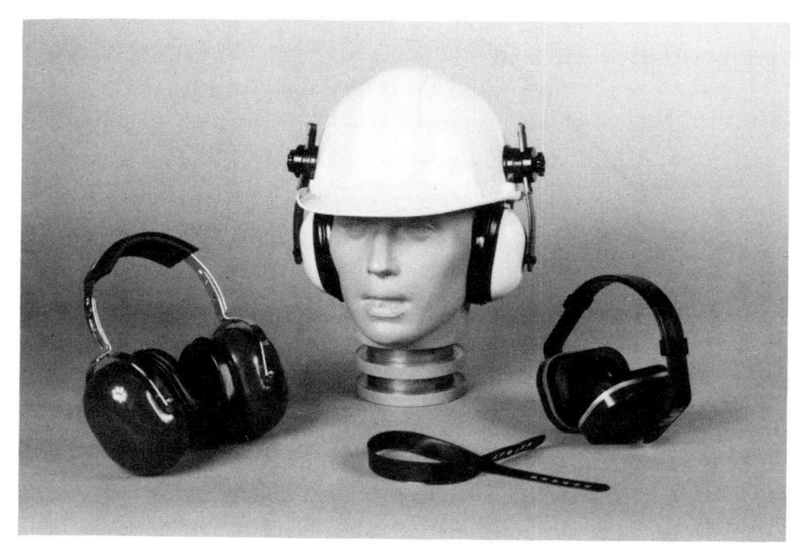

Figure 10.4 — Representative earmuffs (left to right): large volume cups with metal headband, small volume cups attached to hard hat, small volume cups with plastic headband and optional crown strap.

Foam pads are available that fit over eyeglass temples in order to relieve the pressure of the impinging cushion and also to attempt to circumvent loss of attenuation that temples may create, but they do little to minimize leaks caused by overlength temples breaking the seal behind the ear. Unfortunately the data indicate (Berger, 1982) that at least one commercially available pad device actually creates more of a loss in attenuation than simply wearing the eyeglasses alone. However, it can relieve the discomfort of temples pressing against the head and therefore should be considered for use, since comfort is crucial in motivating employees to wear their HPDs.

Acoustical leaks may also arise due to employees modifying their earmuffs by drilling holes in the cups to promote ventilation and assist the drainage of perspiration. In some instances this practice has been so extreme as to include the personalizing of earmuff cups by drilling initials in the cup wall using many closely spaced holes (Riko and Alberti, 1982). Obviously this practice must be discouraged.

When the use of protective headgear is required, hard hats with attached earmuffs provide a convenient alternative to the use of headband-attached earmuffs worn behind the head or under the chin. However, hard-hat-attached muffs are more difficult to orient and fit properly since the attachment arms can never provide as adaptable an adjustment as do the headband-attached versions, nor can they fit as wide a range of head sizes. For such devices, not only must the attachment arms be properly extended and located, but the helmet's webbing must be adjusted to locate the hat properly on the head. Compromises may need to be made between comfort and safety in adjusting either the hard hat or the earmuff cups.

A number of design parameters affect the attenuation of earmuffs, including cup volume and mass, headband force, area of opening in the cushion, and the materials from which the device is constructed. The most visible of these factors to the buyer will be the cup volume, which can vary from a minimum of slightly less than 100 cm^3 to the largest cups sold today with a volume of approximately 330 cm^3.

Volume is one of the primary factors affecting earmuff attenuation, as shown in Figure 10.5, where the attenuation of foam-cushion earmuffs (worn over the head) is illustrated. Notice that from 125 Hz to 1 kHz, earmuffs provide attenuation that increases approximately 9 dB/octave, with the larger-volume muffs providing better attenuation than the smaller ones, since in this region the extra volume and mass of their cups are the controlling physical parameters. Above 2 kHz, however, larger-volume earmuffs tend to be inferior, since in this region their increased shell surface area makes them more susceptible to developing vibrational modes within

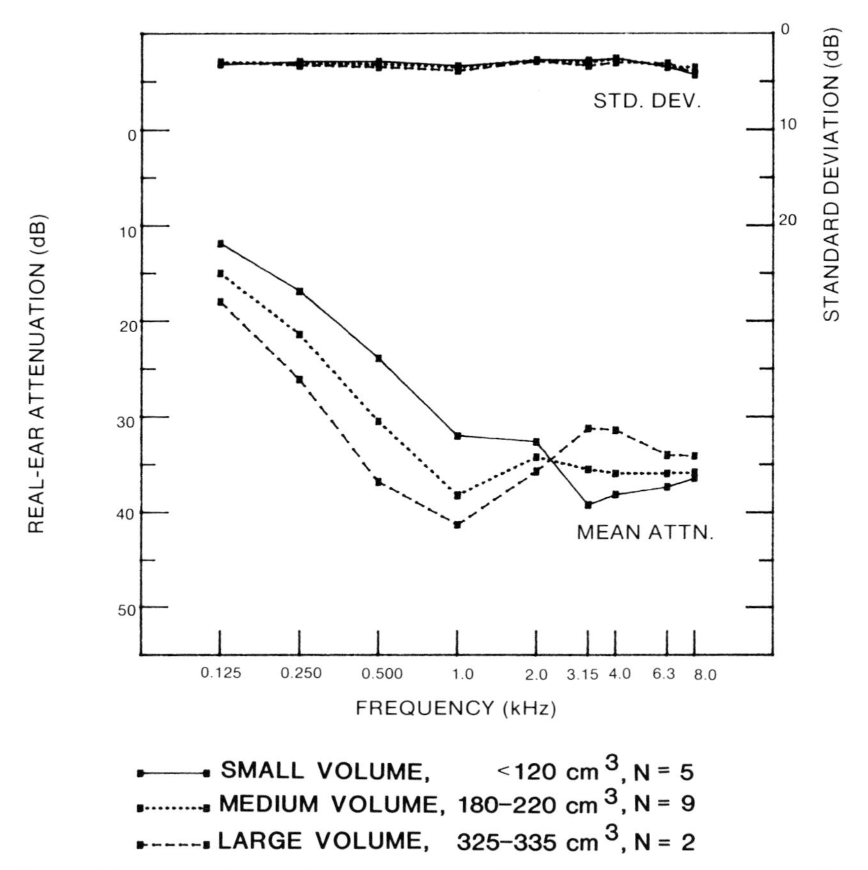

Figure 10.5 — Real-ear attenuation of 16 different foam cushion earmuffs grouped into three categories.

the cup walls and therefore less capable of blocking higher-frequency acoustical energy. At 2 kHz, the attenuation of all of the devices approaches 39 dB, the limit imposed by the BC pathways (cf. Figure 10.2).

Since the circumaural regions of the head are rarely flat or free from obstructions, sufficient force must be exerted to cause the cushion to fully contour and seal to the side of the head. Manufacturers must compromise between high forces, which yield good attenuation and poor comfort, and low forces which produce the opposite results. Band force, which can deteriorate with use and age, is also often modified by wearers if they feel a device is too tight. For many earmuffs it is easy to spring the band without breaking it, thus permanently reducing the force. Band force can be roughly checked (see Figure 10.6) by comparing the resting positions of suspect earmuff cups to those of new samples to make sure that the cup-to-cup separation is approximately the same.

The dimensions of the earmuff cup opening must also be considered when selecting a protector. Due to the transformation of sound pressure from the outside to the inside of the earmuff cup, the smaller the cup opening (other parameters being held constant), the greater will be the attenuation of the protector. Additionally, smaller diameter cup openings allow cushions to make circumaural contact nearer to the base of the pinna, which in turn tends to increase attenuation since facial contours and jaw and neck motion are minimized in those regions (Guild, 1966). However, when attenuation is increased by decreasing the cup opening, it is often at the expense of comfort and ease of use.

As the cup opening is decreased in size, the difficulty of fitting it over and around the pinna increases. For example, the length and width of male pinnae at the 75th and 95th percentiles are 67 × 38 mm and 71 × 41 mm respectively, whereas some of the higher-attenuation earmuffs sold today have openings as small as 58 × 39 mm. For such devices, wearers with even average size ears must be careful to tuck their ears into the cups. Even for

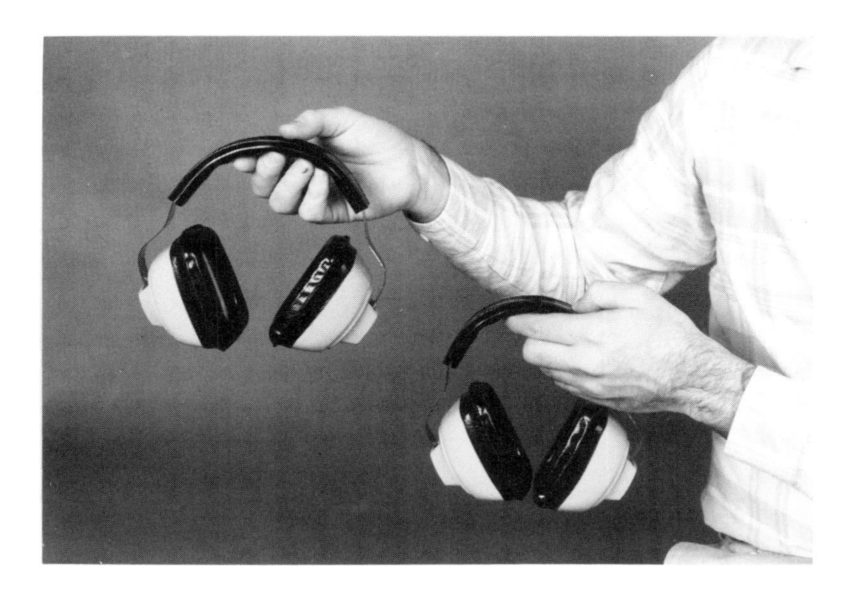

Figure 10.6 — Comparison of distorted (left) and undistorted (right) headbands of two identical new earmuffs adjusted for the same headband extension. The headband force of the earmuff on the left has been reduced by 20%.

cups with more typical openings (65 × 41 mm), employees must still be cautioned to place their pinnae fully within the earmuff cup, since some have been observed who actually rest the cushion on portions of the external ear, which not only reduces comfort but also creates a significant acoustical leak (Riko and Alberti, 1982).

The selection of whether to use foam or liquid-filled cushions is somewhat academic in today's market, since both offer similar performance, the fluid filled cushions providing slightly better protection at the low frequencies, and the foam slightly more at the high frequencies. Tests indicate that for at least one brand of earmuff, both types contour equally well around eyeglass temples (Berger, 1982). Liquid cushions always weigh significantly more, generally sell at a 10 - 20% premium, and may be punctured or split, allowing their contents to drain out.

The cover or bladder on both foam and liquid cushions is made of a plastic material that may harden and deform with time due to contact with body oils, perspiration, cosmetic preparations, and environmental contaminants. For this reason, cushions should be examined at least twice yearly and replaced when necessary. Absorbent cushion covers are available to enhance comfort, but they may reduce attenuation since they are porous and can introduce acoustical leaks.

In the selection of earmuffs for particular applications, the relatively small differences in attenuation between most popular brands and types (see small and medium volume earmuffs in Figure 10.5) suggest that except for extreme noise exposures, where the highest protection possible must be afforded, selection should be based upon other factors. Often smaller, less expensive earmuffs may provide better comfort and therefore be more readily accepted. Symmetrical designs that do not require a particular orientation will decrease the likelihood of misuse. In practice more attenuation can often be gained by assuring that properly maintained earmuffs are worn correctly and consistently than can be assured by buying heavier and perhaps more expensive "high performance" models.

EARPLUGS

Earplugs tend to be more comfortable than earmuffs for situations in which protection must be used for extended periods, especially in warm and humid environments. They can be worn easily and effectively with other safety equipment and eyeglasses, and are convenient to wear when the head must be maneuvered in close quarters. However, they are less visible than muffs, and therefore their use can be somewhat more difficult to monitor.

Earplugs come in a variety of sizes, shapes, and materials, but regardless of the particular model, care must be taken in inserting and sometimes preparing them for use. They generally require more skill and attention during application than do earmuffs. In fact, even under the best circumstances, a small percentage of users may never learn to wear them correctly, either due to their particular canal shapes, lack of manual dexterity, finger size, or missing digits.

As with earmuffs, when earplugs are initially dispensed, even the formable "one-size-fits-all" devices, the fitter must individually examine each person to be sure that a proper seal can be obtained. This can sometimes be determined by simple observation as discussed below, but almost always requires the diligent participation of the fitter and the person being fitted.

Earplug insertion will be facilitated if the wearer reaches around behind the head with the hand opposite the ear that is being fitted and pulls outwards and upwards on the pinna while inserting the plug (see Figure 10.7). This procedure helps to straighten and enlarge the canal opening. Since ear canals usually angle upwards and/or towards the front of the head, employees should be instructed to push the plugs in that direction during insertion, although due to the wide variation in human anatomy, other directions may also be appropriate. There is little likelihood of

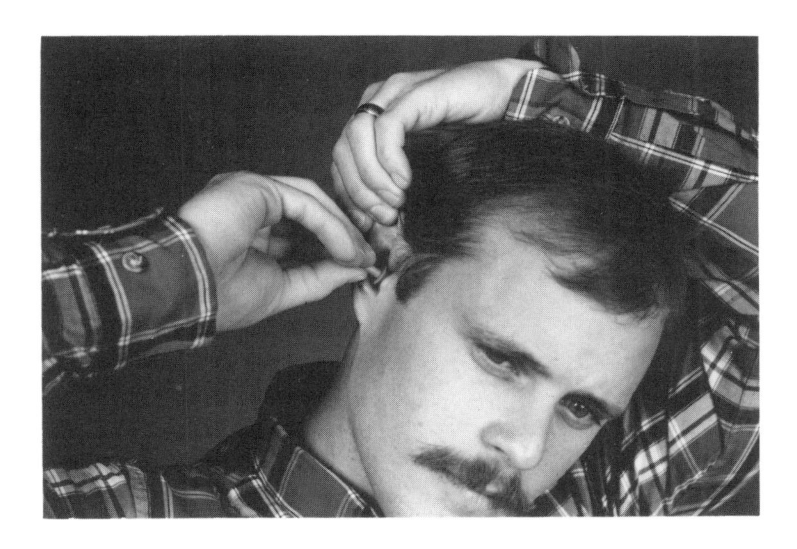

Figure 10.7 — Preferred method of pulling the pinna outward and upward while simultaneously fitting an earplug.

hurting the eardrum during insertion since the sensitivity of the adult ear canal to pressure or pain increases significantly as the eardrum is approached. The discomfort experienced due to touching these deeper portions of the canal will alert the user to stop pushing the plug before a problem can occur.

Plugs that create an airtight seal, such as premolded inserts, can be painful and potentially damaging to the eardrum if they are rapidly removed. These plugs should be withdrawn with a slow twisting motion to break the seal gently as they are extracted from the ear. With foam and fibrous plugs, which do not create a pneumatic seal (and hence cause less of a blocked-up feeling), there is little possibility of generating a sudden large pressure change upon rapid removal, and thus virtually no likelihood of damaging or rupturing the eardrum.

Finally, users must be alerted that earplugs may work loose with time and require reseating. Studies have demonstrated this effect for certain premolded and fiberglass earplugs, but at the same time have shown that foam and custom molded earplug wearers did not experience this problem (Kasden and D'Aniello, 1976; Krutt and Mazor, 1980; Berger, 1981).

Premolded Earplugs

These devices are manufactured from flexible materials such as vinyls, cured silicones, and other elastomeric formulations. Generally, the silicone formulations offer the best durability and resistance to shrinkage and hardening. Most models are available with attached cords to help prevent loss, and to improve storage and reduce contamination by permitting hanging around the neck when not in use. Typical models are depicted in Figure 10.8.

One of the oldest and most common of the premolded devices, the V-51R earplug, was developed during World War II. It is a one-flanged, PVC insert, available in five sizes, and sold under various brand names. If it is intended to use this device for an entire population, all five sizes must be stocked, in which case about 95% of adult male users can be adequately fitted (Blackstock and von Gierke, 1956). It is difficult to suggest what distribution of sizes to order since ear canal dimensions vary as a function of the racial and sexual characteristics of the population, with the black females having the smallest ear canals and the white males the largest. The black males and white females fall in between (Royster and Holder, 1982). For example, in a predominantly black female population nearly 40% would be expected to use an extra small V-51R, whereas this number would be nearer to 5% for a white male group.

The V-51R is particularly prone to shrinkage, cracking, and hardening from exposure to body oils, perspiration, and cerumen. Safe use periods for plugs of this type have been estimated at three months (Royster and Holder, 1982), but even more frequent replacement periods may be required. Another common failure mode for this plug is to split along the mold line. This is often exacerbated by the tendency of users to store the tip of one plug inside the posterior opening of the other.

Premolded plugs are also available with fewer (zero) as well as more (up to five) flanges. Generally, the greater the number of flanges, the fewer sizes are required to fit the population. Flangeless premolded plugs normally are more difficult to size correctly and more prone to work loose during use. Some two- and three-flanged varieties are molded with an entrapped air pocket to increase softness and improve comfort.

As with all HPDs, premolded earplugs are subject to user modifications to improve comfort without regard for the effects on attenuation, which will of course almost always be degraded. Common alterations include removal of flanges, punching holes through the body of the plug, and puncturing entrapped air pockets so that the plug deflates upon insertion (Gasaway, 1984).

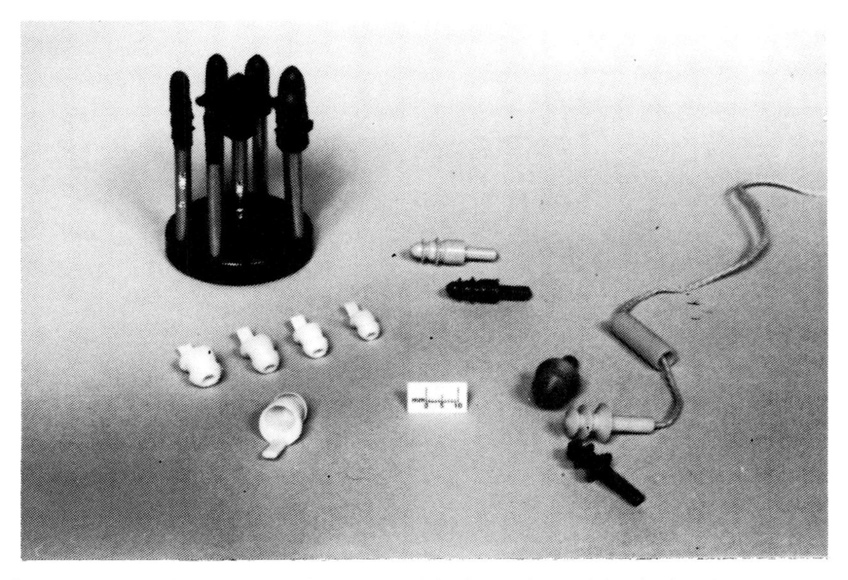

Figure 10.8 — Representative premolded earplugs (clockwise from upper left): 0-flanged 8-sized hollow plug (sizing tool shown), 2-flanged 2-sized hollow plug, 3-flanged 3-sized solid plug with insertion tool and cord, 1-flanged 5-sized plug (V-51R).

Correctly sizing and fitting premolded inserts will almost always require that a compromise be made between attenuation and comfort. The appropriate compromise can often be achieved, but only with care and skill (see Figures C 10.9, color photographs, p. 379, and C 10.10, color photographs, p. 380). Although it is often suggested that if an ear canal falls between two sizes, the larger size plug should be selected, this may be poor advice when one considers the importance of comfort. Even though the larger size may provide better attenuation, if it is not worn or not used correctly due to discomfort, the resultant protection will be nonexistent. Conversely, in the *initial* fitting care should be exercised that any errors tend towards oversizing (Royster and Royster, 1985). Then, if a second (smaller) size must be selected and tested, it will appear to be more comfortable to the wearer. Should the fitter proceed in the reverse order by initially underestimating the size of plug required, then if a second (larger) size must be selected, it is likely that it will seem less pleasant, and the employee may reject it regardless of the appropriateness of its size.

When initially inserting premolded earplugs, the fitter should easily be able to detect gross errors in sizing. Ear gauges are available from some manufacturers of premolded earplugs to aid in this process. Plugs that are much too small will tend to fall into the canal, their depth of insertion being limited only by the fitter's finger and not the plug itself. Overly large plugs will either not enter the canal at all or will not penetrate far enough to allow contact of their outermost flanges with the concha (the hollow shell-like area at the canal opening). A plug which appears to make contact with the interior wall of the canal without appreciably stretching the tissues, and which is well seated, is a good size to begin wearing (Guild, 1966). If after a couple of weeks of use the employee still experiences problems or discomfort, then another size or type of HPD should be issued.

A properly inserted premolded insert generally will create a plugged or blocked-up feeling due to the requisite airtight seal. Additionally, due to the occlusion effect, users should experience a resonant or bassy quality to their voices, as though they were talking in a drum. This will be more pronounced for males than for females, due to the lower pitch of their voices. This perception is useful as a fitting test, since if the wearer speaks aloud with only one ear correctly fitted, the voice should be more strongly heard or felt in the occluded ear (Ohlin, 1975). If this does not occur, the plug should be reseated or resized. When the second ear has been fitted correctly, the wearer should perceive his own voice as though it were emanating from the center of the head.

A record should be maintained at the fitting station of the size of the plug issued to each employee. It should be consulted when plugs are reissued to replace those that are lost or worn out. Experience suggests that in approximately 2 - 10% of the population different size premolded earplugs will be required for the left and right ears, although military data indicate that for the V-51R type earplug this number may be as high as 20% (Dept. of the Air Force, 1982). As a general rule, the more sizes a particular premolded plug is manufactured in, the greater will be the likelihood that different ones are required for each ear. The use of unmatched sizes for the two ears can pose a problem for those devices that are color coded to indicate size, since some employees may be reluctant to wear two different colored plugs.

The attenuation of a variety of premolded earplugs is plotted in Figure 10.11. The values are approximately constant up to 1 kHz clustering closely around 25 dB, and increasing to approximately 40 dB at the higher frequencies. Below 2 kHz, the lowest measured attenuation was found using a one-size flangeless plug and the highest using a three-size three-flanged device. The relatively small range in performance between devices suggests that selection would be most meaningfully based upon comfort, and the ease with which the various devices can be sized and fitted. Comparison with Figure 10.5 indicates that the average laboratory performance is better than earmuffs at 125 and 250 Hz and above 2 kHz, but is poorer at the intermediate frequencies.

Formable Earplugs

Earplugs of this variety may be manufactured from cotton and wax, spun fiberglass (often called fiberglass down, mineral wool, or Swedish wool), silicone putty (exposed, or encased in a bladder), and slow-recovery foams. Life expectancies vary, from single-use products such as some of the fiberglass down products, to multiple-use products such as the foam plugs, which may be washed, and relatively permanent items such as the encased putties. Except for the bladder-encased putties and some brands of foam plugs, formable plugs are generally not available with attached cords.

The primary advantage of formable earplugs is comfort, some of the products in this category being among the most comfortable and user-accepted of devices sold today. Additionally, formable plugs are generally sold in only one size that usually fits most, but not all, ear canals. This simplifies dispensing, record keeping, and inventory problems, but when such products are used, special attention must be given to wearers with extra small and extra large ear canals to make sure that the plugs are not too tight or too loose, respectively. Since these plugs usually require manipulation by the user prior to insertion, during which time the hands should be relatively clean, they may not be the best choice for environments in which HPDs have to be removed or reinserted many times during a work shift by employees whose hands are contaminated with caustic or irritating sub-

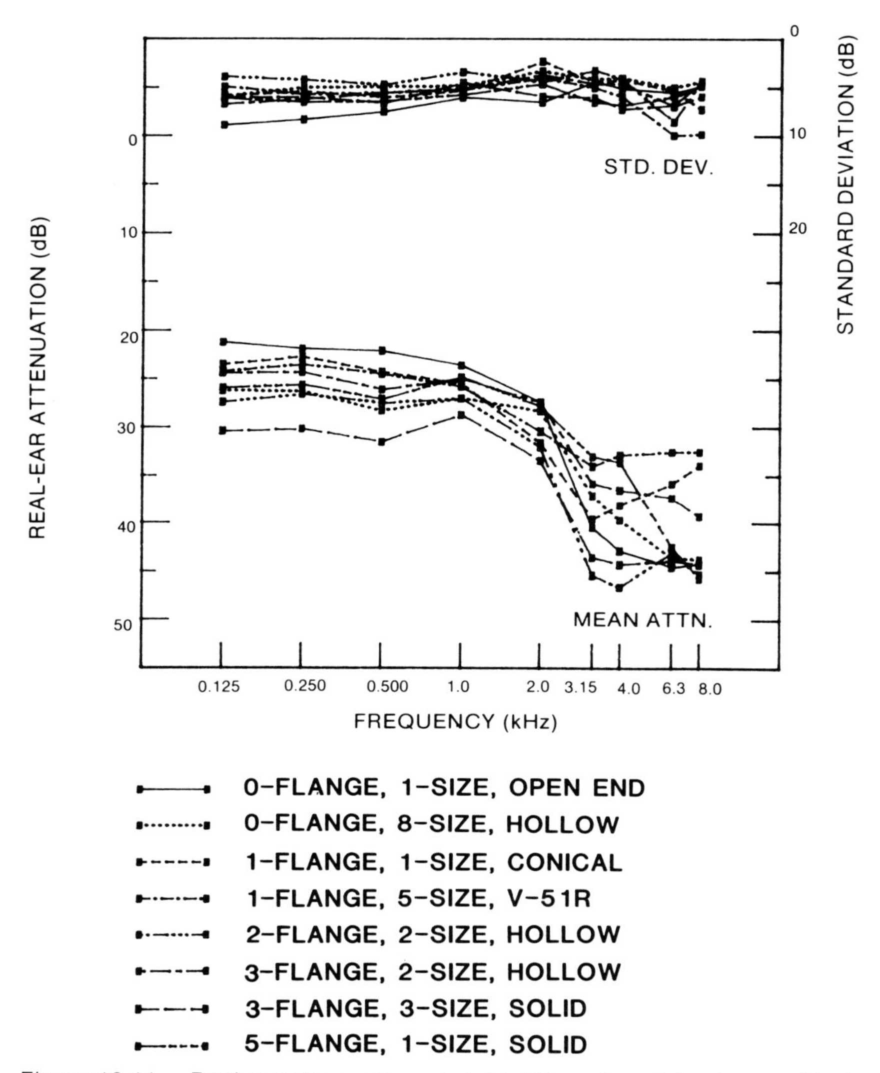

Figure 10.11 — Real-ear attenuation of eight different models of premolded earplugs.

stances or sharp or abrasive matter. User modifications and abuse typically consist of reducing the amount of formable materials used for each plug, or in the case of foam earplugs, cutting their length or diameter. Representative formable earplugs are shown in Figure 10.12.

Although cotton alone is a very poor hearing protector due to its low density and high porosity, when it is combined with wax good protection can be obtained. The wax tends to make the devices somewhat messy to prepare and use and also softens considerably in higher temperatures. These plugs are not commonly found in industrial hearing conservation programs, but they have attained popularity in the consumer market. Insertion is accomplished by kneading the material until it softens, forming it into a ball, or preferably a cone, and then pressing it into the entrance of the ear canal. Since these materials lack elasticity they may lose their seal as a result of jaw and neck motion and require frequent reseating (Ohlin, 1975).

Fiberglass down was first available in the late 1950s as a folded strip of batting which required tearing and rolling into a cone for insertion. More recently, it has evolved into versions which are partially encased in either

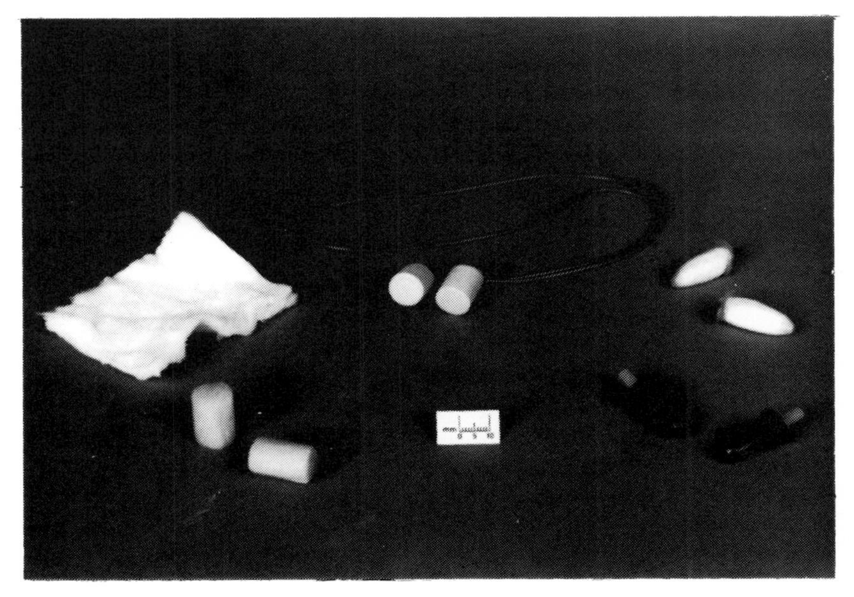

Figure 10.12 — Representative formable earplugs (clockwise from upper left): fiberglass down, slow recovery foam with attached cord, fiberglass down with taut polyethylene sheath, encased silicone putty and cotton wax.

loose or taut polyethylene sheaths. These modifications have improved the likelihood of properly fitting these types of plugs and have also reduced the possibility that small sections or pieces of the fiberglass will break off from the main body of the plug and remain in the ear canal. Sheathed fiberglass plugs are inserted by placing them into the ear canal with a slight rocking and twisting motion while using the opposite hand to pull the outer ear (see Figure 10.7).

Slow-recovery foam earplugs were first introduced in the early 1970s. They usually have a cylindrical shape and are normally about 14 mm in diameter and 18 - 22 mm in length. Insertion is accomplished by rolling the plug into a thin, tightly compressed cylinder, which (while still fully compressed) is then placed into the ear canal and gently held in position for a few seconds until expansion begins. As with other earplugs, pulling the pinna eases insertion. The plug will be properly inserted when it is actually situated in the ear canal as opposed to simply capping its entrance (see Figure C 10.13, color photographs, p. 380). This can be verified for most wearers if approximately one half of the plug is in the shape of the ear canal immediately after removal (see Figure C 10.14, color photographs, p. 381).

Unlike other formable or premolded earplugs, foam earplugs should not be readjusted while in the ear. If the initial fit is unacceptable, they should be removed, recompressed, and reinserted. And since the depth of insertion of foam earplugs can be comfortably varied, a maximal occlusion effect does not signify best fit. In fact, the deeper the insertion and the better the fit and the attenuation, the less noticeable (and annoying) will be the occlusion effect (see Occlusion Effect).

The attenuation of a number of formable earplugs is plotted in Figure 10.15. A fairly wide range of laboratory attenuation values is available, especially below 2 kHz, with the minimum values being attained by one type of sheathed fiberglass down and the highest values by the foam earplugs[2], which can offer the best low frequency attenuation of any type of HPD.

Custom Molded Earplugs

Custom earmolds are most often manufactured from two-part curable silicone putties although some are available in vinyl. The silicones are either cured by a catalyst at the time the impression is taken by the fitter, or

[2]Although recently (circa 1981) foam earplugs have become increasingly available from more than one manufacturer, the laboratory and real-world data in this chapter pertain only to the E-A-R and Decidamp brands. To my knowledge, there are no currently available real-world performance data for other brands of foam earplugs.

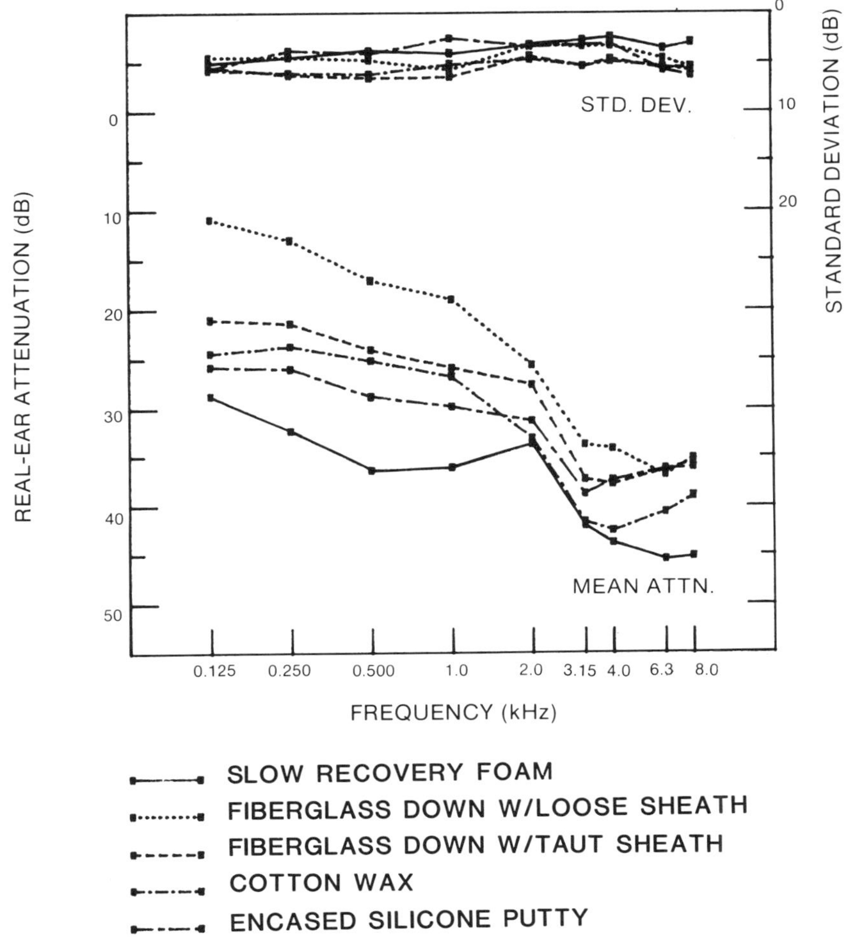

Figure 10.15 — Real-ear attenuation of five different formable earplugs.

returned to the supplier for manufacturing. Some custom earmolds are finished by inserting small handles into their exterior surface to facilitate handling, whereas others are available with attached cords. Representative devices are pictured in Figure 10.16.

Most earmolds fill a portion of the ear canal as well as the concha and pinna. The canal portion of the mold is what makes the acoustical seal to block the noise whereas the concha/pinna portion principally functions as a kind of snap-lock to maintain the HPD in position. Incorrect insertions

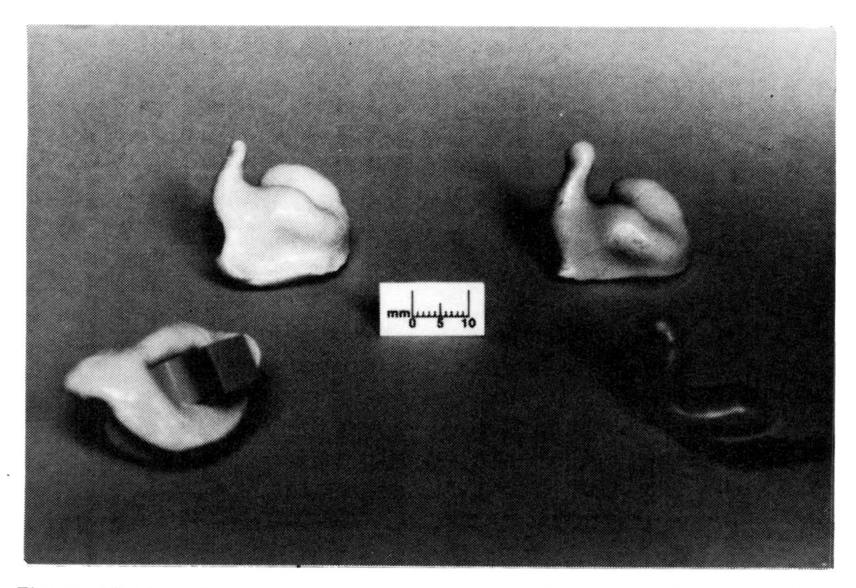

Figure 10.16 — Representative custom molded earplugs: lower left illustrates handle, lower right is mold without concha portion. All four earmolds are impressions of one subject's right ear canal. Note the significant variability in the size and shape of the canal portion of the molds.

are easily detected and the molds have little chance of working loose with time. Some molds are manufactured with only a canal impression present, on the supposition that leaving a greater area of the pinna uncovered makes them cooler and more comfortable. Unfortunately it may also impair the retention of these molds in the canal, allowing them to lose their seal during use.

Considerable skill and time are required to take individual impressions for each employee. Contrary to popular beliefs, the fact that custom earmolds are user-specific and intended to fit only the canal for which they are manufactured does not assure that they will provide better protection than other well-fitted earplugs. In fact, often the opposite is the case as has been reported in the literature (Guild, 1966) and as can be noted by comparing the laboratory attenuation for the various types of insert HPDs illustrated in Figures 10.11, 10.15, and 10.17. This observation may be partially explained by reference to Figure 10.16 which depicts four different earmolds, all of which are impressions of one subject's right ear canal, and all of which were manufactured by "experienced" fitters. Note the significant differences in the canal portion of the impressions of the three molds for which that feature is visible.

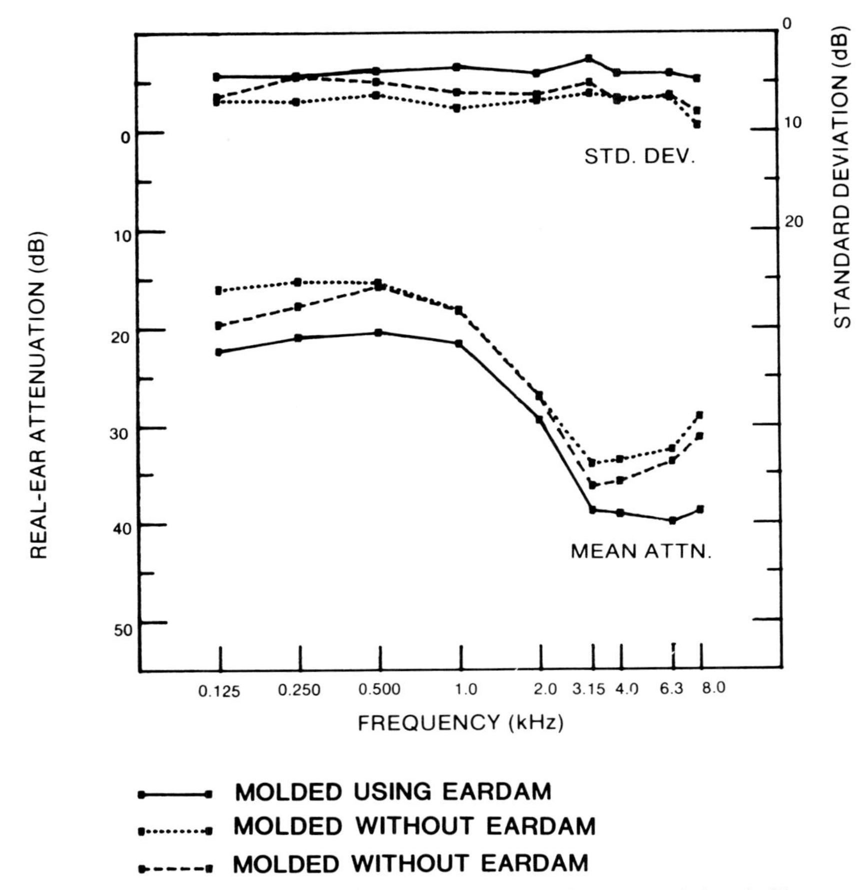

Figure 10.17 — Real-ear attenuation of three different models of silicone custom earmolds.

A worthwhile aspect of individually molded earplugs is that they are manufactured personally, for only one employee, so that they are "customized." This customization can be effectively utilized as an incentive for motivating employees to wear their HPDs. Another positive aspect of custom molds (those with both a canal and concha impression) is that due to the way in which they fit into the ear, they are less subject to misinsertion under field conditions.

Custom earmolds can be very comfortable, but experience has shown that as the molds more fully and tightly fill the canal and therefore more effectively attenuate sound, comfort deteriorates. Thus, in practice, there is a limit to how snugly a custom mold can be fitted.

In spite of the longevity claims made by some manufacturers for their "permanent" earmolds, these devices, like other earplugs, are susceptible to shrinkage, hardening, and cracking with time, and must be periodically reexamined to assure they are still soft, flexible, and wearable. They are also subject to user modifications such as whittling or total removal of the canal portion of the mold. Additionally, since they may be lost or misplaced as may all earplug-type HPDs, their use may create administrative problems since time must be allowed to remanufacture them on an as-needed basis.

Custom earmold impressions are made using a viscous material with a consistency varying from that of thick syrup to soft putty. It is mixed with a curing agent and then either formed into a cone and pressed into the canal, or placed in a large syringe and injected into the ear canal. The silicone must then cure in the ear approximately 15 minutes. When cured, it is itself the HPD. In some cases it may be dipped into a coating before use. In other cases the impressions are returned to the supplier who then uses them to make subsequent negative and positive molds in order to create the final custom earplug.

Some manufacturers recommend placing a cotton block or eardam inside the canal prior to making the impression to assure that no silicone is forced too deeply into the canal or reaches the eardrum. Additionally, the eardam helps to ensure that a better fitting impression will be taken. The reason is that, as the impression material is pressed into the canal, the dam forces it radially outwards against the canal walls to ensure a tighter fit and more effective seal. Without the dam, the viscous mold material is permitted to simply flow further into the canal without ever being forced into contact with the canal walls (see Figure 10.16, upper left impression).

The attenuation of three custom earmolds is shown in Figure 10.17. The variation between properly manufactured molds is minimal, with the average attenuation falling below that of some of the premolded earplugs, and the best protection and lowest variability being afforded by impressions taken using cotton eardams. Selection of a particular product should be based primarily upon the users' judgment of comfort and the fitters' evaluation of the ease of manufacture of the earmolds, this latter consideration being important since the final performance of the molds is so closely linked to the fitters' skill in making the initial impression.

SEMI-AURAL DEVICES

Semi-aural devices, which consist of pods or flexible tips attached to a lightweight headband, provide a compromise between earmuffs and earplugs. They can be worn in close quarters, easily removed and replaced, and conveniently carried when not in use. One size fits the majority of users. Their fit is not compromised by safety glasses or hard hats. These devices are usually available with dual position (under-the-chin and behind-the-head) or universal headbands made from either metal or plastic. The tips can be made from vinyl, silicone, or composites such as foam encased in a silicone bladder, and may cap, or in some cases enter the canal. The tips normally have a bullet, mushroom, or conical shape. Representative devices are shown in Figure 10.18.

Semi-aural devices are principally intended for intermittent use conditions where they must be removed and replaced on a repeated basis. Examples include: ground crews servicing commercial aircraft, periodic equipment inspection by personnel normally located in sound treated booths, and supervisor walk-throughs. During longer use periods, the force of the caps pressing against the canal entrance may be uncomfortable, but for those who do prefer this type of device for extended use the better ones can offer very adequate protection. Since semi-aural devices generally cap the canal at or near its entrance, they tend to create the most noticeable occlusion effect and consequently distort the wearers' perception of their own speech more than other types of HPDs. This may be objectionable to

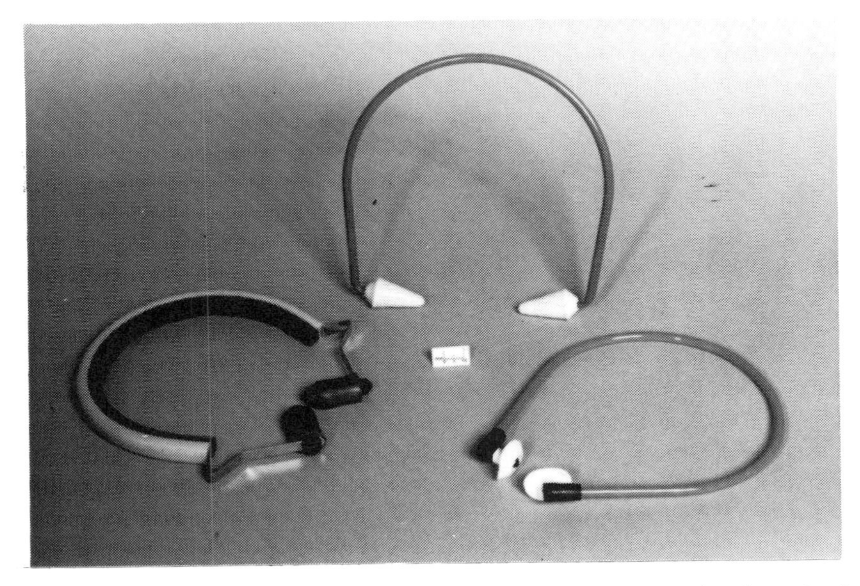

Figure 10.18 — Representative semi-aural devices (clockwise from top): conical (foam in silicone bladder), mushroom, hollow bullet.

some users. Field modifications generally involve springing the band, with a consequent reduction in protection.

Semi-aural devices are generally pushed into place at the entrance to the ear canal, although the particulars vary for each product. For example, one type requires rolling and stretching of the pods prior to insertion, another comes with the suggestion that pulling the pinna may be helpful, and others have specific orientations (left/right, top/bottom) that must be attended to if proper protection is to be attained. Certain semi-aural devices, particularly those that do not enter the canal, are prone to losing their seal periodically, especially if the band is bumped and caused to skew on the head.

The attenuation of five semi-aural devices is presented in Figure 10.19, where it can be seen that a wide range in performance is available at all frequencies, with the better protection being provided by those devices that not only cap, but also partially enter the ear canal.

SPECIAL TYPES AND COMBINATIONS OF HPDs

Double Hearing Protection

For very high level noise exposures, especially when 8-hour TWAs are greater than 105 dBA, the attenuation of a single HPD may be inadequate. For such exposures, double hearing protection, *i.e.* earmuffs plus earplugs, may be warranted. It is well recognized that double hearing protection does not simply yield overall attenuation equal to the sum of the individual attenuation of each device. This is primarily due to the BC flanking paths and the acoustical-mechanical interaction between two such closely spaced devices.

An extensive empirical study of the incremental performance to be gained by double protection was reported by Berger (1983b), who examined five inserts and three earmuffs both singly and in combination. In almost all cases a combination of plug plus muff outperformed either device individually. At individual frequencies the incremental gain in performance varied from approximately 0 to 15 dB over the better of the individual devices, except at 2 kHz where no combination exhibited a gain greater than 3 dB. The gain in the NRR for the double protection combinations was 7 - 17 dB when compared with the plugs alone, 3 - 14 dB when compared with the muffs alone, and 3 - 10 dB when compared with the better of the two individual devices.

The data suggested that although the attenuation provided by an insert HPD could be improved by wearing an earmuff over it, the choice of earmuff was relatively unimportant. However, when the situation was

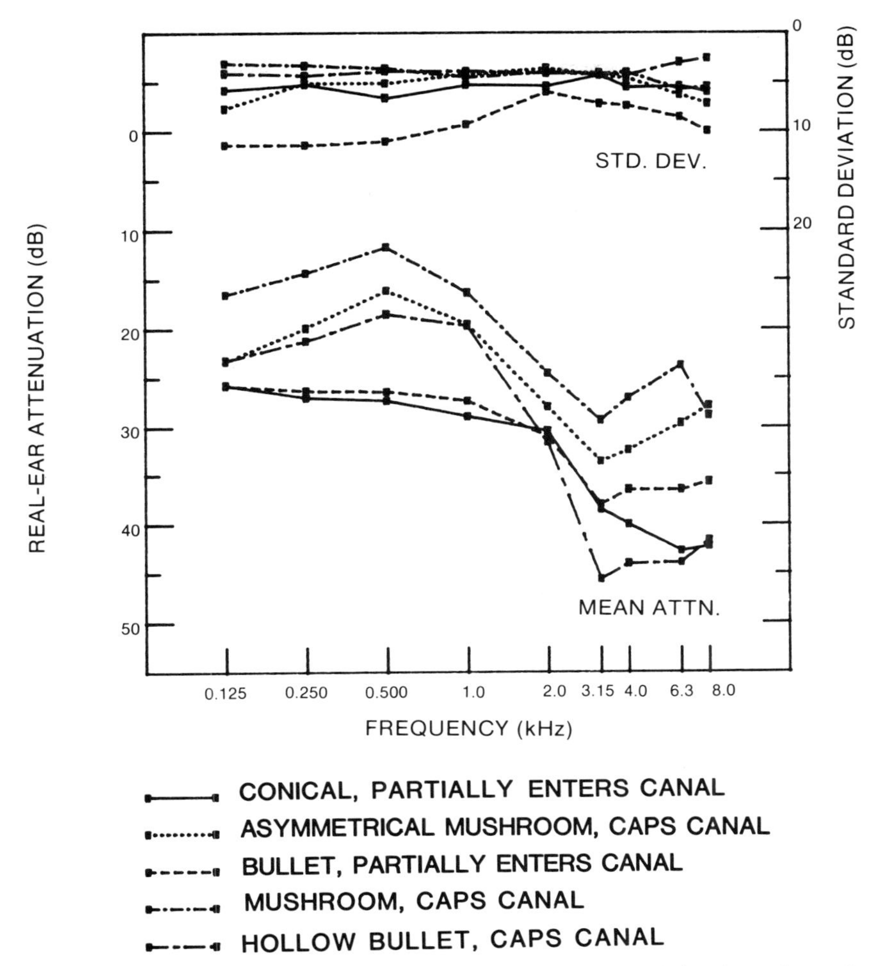

Figure 10.19 — Real-ear attenuation of five different models of semi-aural devices.

reversed and the earplug was added to the earmuff, then regardless of the earmuff worn, the choice of insert was critical at the frequencies below 2 kHz, as illustrated in Figure 10.20. At and above 2 kHz all plug-plus-muff combinations that were studied provided attenuation that was approximately equal to that of the human skull; *i.e.*, the combined attenuation was limited only by the flanking BC paths to the inner ear (cf. Figure 10.2).

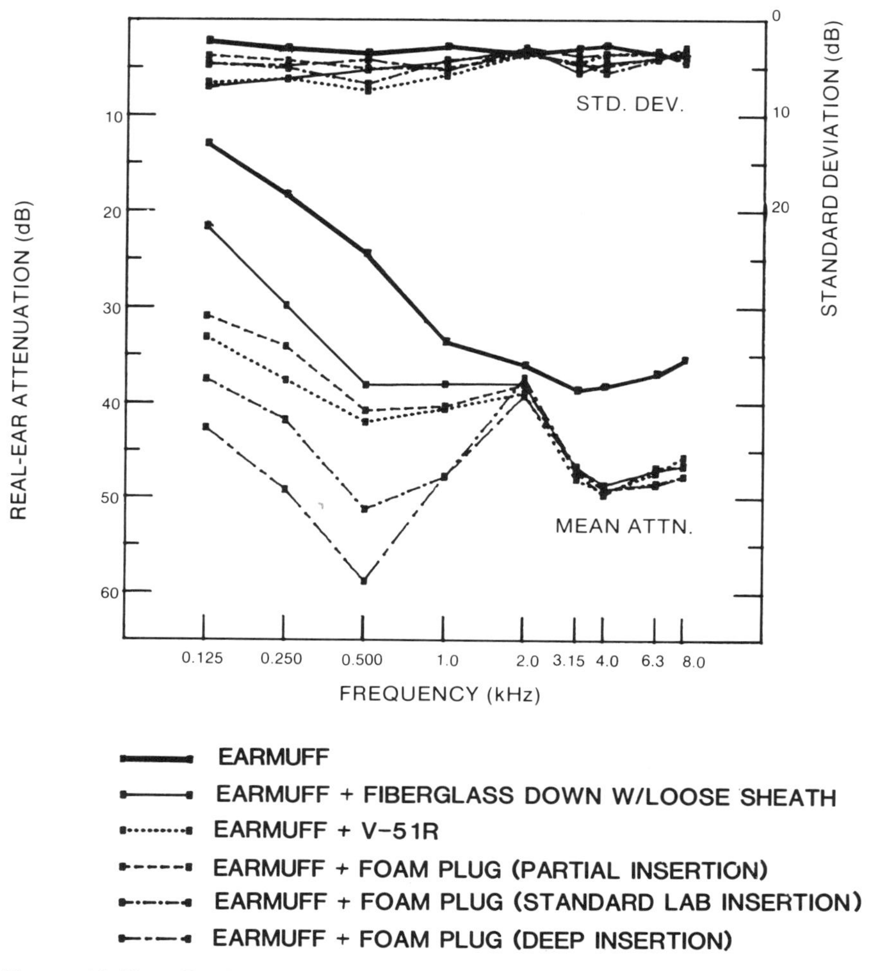

Figure 10.20 — Real-ear attenuation of a small volume (100 cm^3) earmuff alone, and in combination with each of five different earplugs (from Berger, 1983b).

Amplitude Sensitive Devices

Amplitude sensitive devices, or nonlinear HPDs as they are sometimes called, are designed to provide little or no attenuation at low sound levels, with protection increasing as sound level increases. Communication is unimpaired during quiet periods (see Effects of HPDs on Auditory Communication), but when high noise levels are present, additional protection is available. This can be accomplished to varying degrees with either passive devices that utilize orifices, valves, or diaphragms, or active devices containing electronic circuitry. Orifices and valves have been incorporated into earplugs and at least one manufacturer has offered a diaphragmatic circumaural HPD. Active electronics are currently available in earmuffs, and efforts are underway to include them in an earplug type device as well.

Both theoretical and empirical research indicates that at sound levels below 110 - 120 dB, orifice and valve type passive devices simply behave as a vented earmold with almost no attenuation below 1 kHz and attenuation increasing to as much as 30 dB at higher frequencies (Forrest, 1969; Martin, 1979; Berger, 1982). At higher sound levels, steady-state or impulsive sound waves generate turbulent airflow in the orifice which impedes the passage of sound and provides an increase of attenuation of about 1 dB for each 2 - 4 dB increase in sound level. These devices are primarily suited for gunfire exposures where peak SPLs range from approximately 135 to 175 dB. They are more effective in less reverberant conditions such as outdoor ranges (Coles and Rice, 1966; Mosko and Fletcher, 1971). These devices are of little value for most occupational and recreational exposures because the noises are rarely of the appropriate type or at a sufficient level for the nonlinear characteristics to become functional (Martin, 1976; Harris, 1979).

Active HPDs may be of the peak-limited sound transmission or noise cancelling types. The sound transmission type contains a microphone external to the circumaural cups that is used to monitor the ambient noise. When the sound levels are low, it broadcasts information into the earmuff via electronic circuits and earphones. At high sound levels the circuits do not pass the signal, permitting the physical attenuation of the cup to protect the ear. The noise-cancelling approach places both the microphone and earphone inside the HPD so that electronics can sense the sound and generate an antiphase signal through the earphone to cancel it. This latter method is generally successful only at the lower frequencies.

Active HPDs can improve speech communications in certain noisy environments, especially during intermittent quiet intervals, but at the current state of the art the cost is upwards of $150 per device. This makes them prohibitively expensive for general industrial use and all but the most specialized of environmental conditions.

Communication Headsets

In order to transmit signals to the ear, earphones may be built into both supra-aural and circumaural devices, or attached to canal inserts. Usually only the circumaural devices will provide sufficient attenuation to be suitable for use in noisy environments. They are available with both wireless (FM or infrared) and wired systems, suitable for one- and two-way communications and/or music transmission. The better devices provide specialized electronic circuits to limit the delivered SPLs so that the earphones themselves cannot present signals that could be hazardous to the wearer. The prices for communication headsets range from $100 to $1000 per unit.

In extremely high noise levels, when circumaural communication headsets may not provide sufficient attenuation of ambient noise to permit clear communications, intelligibility can be improved by wearing earplugs under the cups. This will reduce the environmental noise and the desired signal equally, but as long as the headset has sufficient distortionless gain so that its output can be increased in order to overcome the insertion loss of the earplug, the signal-to-noise ratio in the listener's ear canal can be significantly improved.

Recreational Earphones

Often employees request to use stereo earphones for protection against noise while they enjoy the music. The inappropriateness of relying on these devices for hearing protection is illustrated in Figure 10.21 in which the attenuation of a circumaural radio headset and that of a more popular set of lightweight foam supra-aural stereo earphones are compared to an industrial earmuff. The foam earphones offer almost no protection. Even the circumaural device provides no more than approximately 20 dB of attenuation at high frequencies, and actually amplifies sounds at others. This protection is inferior to that of a well designed, properly fitted HPD.

Recreational earphones alone can generate equivalent noise levels up to approximately 100 dBA (Berger, 1982). Since they offer so little attenuation, a greater concern is that employees might turn up the music to mask (*i.e.*, "drown out") the factory noise, with a consequent reduction in their ability to communicate or hear warning sounds and a significant increase in their effective noise exposures. Thus the use of recreational earphones should be prohibited when sound levels equal or exceed 90 dBA. At sound levels between 85 and 90 dBA, their use is problematical. The music they provide may alleviate boredom and increase productivity (Fox, 1971), but they offer little or no protection and can actually increase noise exposures as cited above.

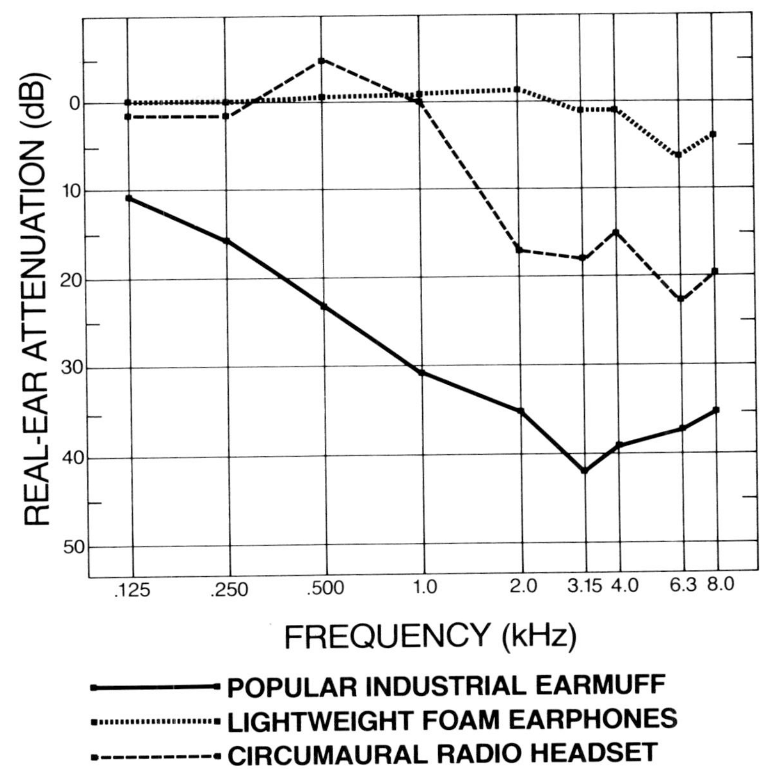

Figure 10.21 — Attenuation of recreational earphones vs. a popular industrial earmuff.

In one industrial study (Royster *et al.,* 1984) with a noise environment characterized by a variation of sound level between 80 and 90 dBA (with approximately equal time periods at each level), and a TWA of 87 dBA, employees on the average increased their equivalent exposure by only 2 dB as a result of using recreational earphones. However, about 20% of the workers were observed to play their radios at levels of 90 dBA or higher. The authors recommended that employees continue to be allowed use of such devices, with the stipulation that a significant educational effort be directed at the proper use of personal radios, that employees exhibiting permanent threshold shifts of 20 dB or more be prohibited from further use, and that the overall audiometric data base be annually analyzed for relative changes between the hearing of the wearers and nonwearers of personal radios.

THE INITIAL EAR EXAMINATION AND HPD HYGIENE

Prior to issuing HPDs, the fitter should examine the external ear to identify any medical or anatomical conditions which might interfere with or be aggravated by the use of the protector in question (also see Chapter 7). If such conditions are present, HPDs should not be worn until medical consultation and/or corrective treatment can be obtained, or the suspected condition has been shown not to constitute a problem. Areas of concern include extreme tenderness, redness or inflammation (either in or around the ears), sores, discharge, congenital or surgical ear malformations, and additionally in the case of earplugs, canal obstructions and/or impacted or excessive cerumen. The latter condition, however, is difficult to judge since few data are available on the effects of earplugs on the formation, buildup, and possible impaction of wax.

As with all clothing and equipment that comes in repeated and intimate contact with the body and the work environment, the cleanliness of HPDs must be considered. HPDs should be cleaned regularly in accordance with manufacturers' instructions, and extra care is warranted in environments in which employees handle potentially irritating substances. Normally, warm water and soap are recommended as cleansing agents. Solvents and disinfectants should generally be avoided.

Earplugs should be washed in their entirety and allowed to dry thoroughly before reuse or storage in their carrying containers. Earmuff cushions should be periodically wiped or washed clean. Their foam liners can also be removed for washing but must be replaced since they do affect attenuation. Earplugs and earmuff cushions should be discarded when they cannot be adequately cleaned or no longer retain their original appearance or resiliency.

Stressing hygiene beyond practical limits, however, can compromise the credibility of the HPD issuer/fitter. It is often difficult enough to get employees to replace or repair worn-out HPDs, let alone clean them routinely. In any event, information from authorities in the field of audiology and hearing conservation (Ohlin, 1981; Gasaway, 1985), as well as the available epidemiological data (Berger, 1985) suggests that the likelihood of HPDs increasing the prevalence of outer ear infections is minimal.

If an ear irritation or infection is reported, the exact extent and etiology of the problem should be investigated firsthand by medically trained personnel to determine whether the causative agent is an HPD or another predisposing factor. Such factors include excessive cleaning of the ear, recreational water sports, habitual scratching and digging at the ears with fingernails or other objects, environmental contaminants, and systemic conditions such as stress, anemia, vitamin deficiencies, endocrine disorders, and various forms of dermatitis (Caruso and Meyerhoff, 1980). When HPDs are implicated, a common cause has been found to be earplugs or even earmuffs that are contaminated with caustic or irritating substances, or sharp or abrasive matter. In one reported case of earplug contamination (Royster and Royster, 1985), more careful hygiene practices, combined with the use of corded plugs to allow removal without touching the protector, eliminated the problem.

Canal irritations can also arise due to the use of missized or inappropriate HPDs, omission of a "break-in period" for new users, or the use of worn out HPDs whose once resilient parts are no longer soft and flexible. In rare instances individuals may develop circumaural or canal inflammation as a result of allergic reactions to the materials of which earmuff cushions or earplugs are composed. Rectification of the above problems involves resizing or issuing alternative HPDs, retraining of users, and periodic replacement of worn-out devices.

If occurrences of external ear problems develop, it is important to determine if they are limited to a particular department or operation, to one or more brands or types of HPDs, to a change in the HPDs being utilized, to a particular time of year, or if they are perhaps due to some other policies or procedures that may have been modified within the work environment. This will allow a reasoned approach and help to avoid an overreaction which could compromise the HCP without necessarily resolving the problem at hand.

Real-World Performance of HPDs

It is invariably found that the real-world performance of HPDs is significantly less than estimated by laboratory measurement methods. This effect is often exaggerated by the particular choice of laboratory data (cf. Laboratory #8 in Figure 10.3), and by observance of the "Experimenter Fit" protocol of ASA STD 1-1975. The experimenter fit procedures are intended to develop "optimum performance" data, but this can be misleading. Optimum performance for a laboratory test wherein a trained and motivated subject sits immobile for five minutes, utilizing a test protocol in which the word "comfort" is never mentioned, and with equipment and expertise available to assure proper fitting of the HPD, is very different from the "optimum performance" that can be attained in the real world where active workers, who may consider HPDs to be an inconvenience or a major burden, must wear them for extended periods of time on a daily basis.

More explicitly, the factors that are often overlooked in hearing conservation programs and that can compromise HPD performance in the real world may be summarized as follows:

1) Comfort — This is ignored in laboratory tests but is crucial in the real world.
2) Utilization — Due to poor comfort, poor motivation or poor training, or other user problems, earplugs may be incorrectly inserted and earmuffs may be improperly adjusted.
3) Fit — Fitting and sizing of earplugs must be carefully accomplished for *each* ear. If they are not, performance will be degraded.
4) Compatibility — Since not all HPDs are equally suited for all ear canal and head shapes, the proper device must be matched to each user.
5) Readjustment — Since HPDs can work loose or be jarred out of position, employees must be advised of the need for readjustment.
6) Deterioration — No HPDs are permanent or maintenance-free. They must be inspected at least twice yearly, and replaced or repaired as necessary.
7) Abuse — Employees often modify HPDs to improve comfort at the expense of protection. This must be avoided.
8) Removal — When devices become uncomfortable they are often removed to give the ears a "break." This can dramatically reduce the effective protection (Else, 1973) as illustrated in Figure 10.22. For example, if the HPD has a nominal NRR of 25, then its effective, or time corrected, NRR would be only 20 dB if it is not worn for just 15 minutes during each 8-hour noise exposure.

SUMMARY OF EXISTING FIELD DATA

The extant literature contains 10 field studies including data from over 50 different industrial plants with a total of 1551 subjects. Various protocols have been implemented, but they all used actual employees who were participants in ongoing hearing conservation programs. The employees were tested either at their plant sites (both on the job and in special test rooms) or at remote hearing testing clinics. The studies have been summarized and critically reviewed by Berger (1983a), who amassed sufficient data to characterize the real-world performance of earmuff, premolded, fiberglass, foam, and custom molded HPDs as shown in Figure 10.23.

The data in Figure 10.23, which may be compared with the data in Figures 10.5, 10.11, 10.15, and 10.17, illustrate the significant differences

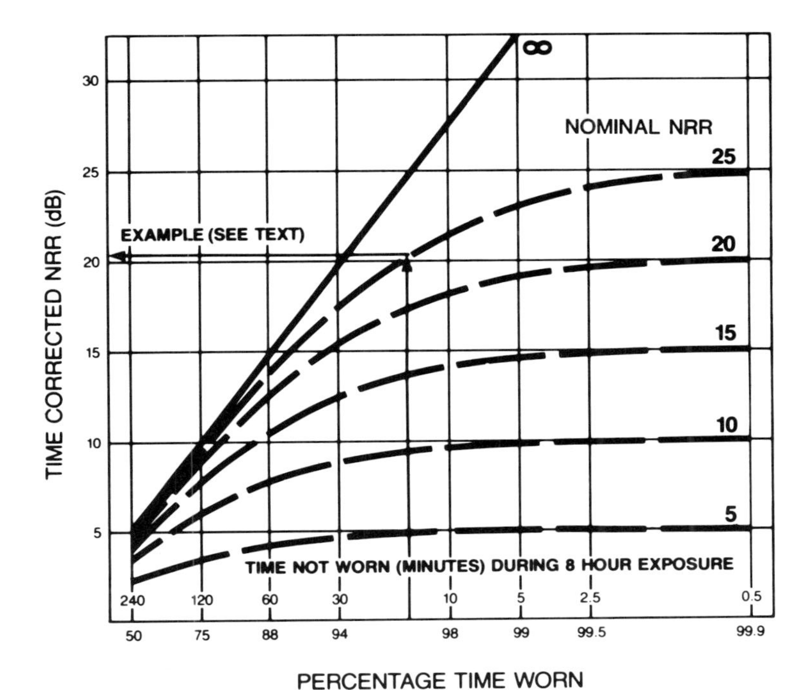

Figure 10.22 — Time corrected NRR as a function of wearing time (using OSHA 5-dB trading relationship).

that are found between laboratory and real-world performance of HPDs. The results are summarized in Figure 10.24, where the manufacturers' labeled NRR_{98}s are compared with NRRs computed from the field data using only a 1-σ correction (NRR_{84}). Thus the attenuation that at least 98% of the laboratory subjects achieved is being compared to that which was achieved by at least 84% of the real-world users. It has been argued that computations based upon real-world data should utilize a one- instead of a two-standard-deviation correction (Berger, 1983a).

Consideration of the protection afforded to 84% of the real-world users indicates that earmuffs can offer around 10 - 12 dB of protection, whereas earplugs, with the exception of the foam plugs[2], offer less than 10 dB. The average difference between the labeled NRR_{98} and the real-world NRR_{84} across all devices was 13 dB. If the labeled NRR_{98} had been compared to a real-world NRR_{98}, the average difference would have been greater than 20 dB.

[2]See bottom page 346.

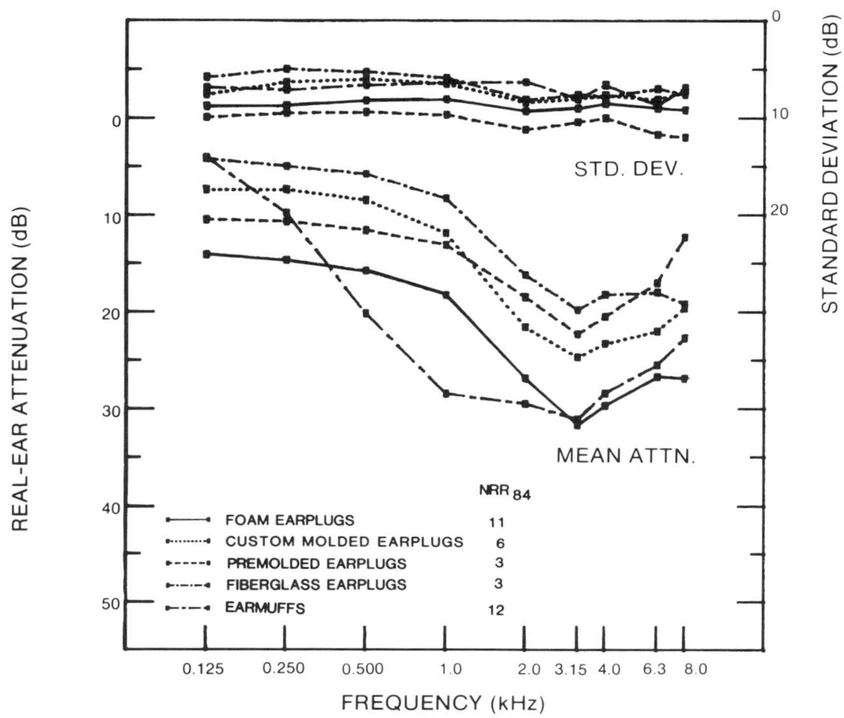

Figure 10.23 — Real-world attenuation for five types of hearing protectors (from Berger, 1983a).

SUGGESTIONS FOR DERATING LABORATORY DATA

The average NRR on devices sold in North America today is approximately 24 dB. This number clearly overstates the protection that most buyers can expect their employees to achieve, and misleads the buyer into believing that almost any HPD will reduce almost every noise exposure to a safe level. (After all, how many 8-hour exposures does one find that are greater than 85 dBA + 24 = 109 dBC?) In turn, this belief fosters programs in which HPDs are handed out indiscriminately with little or no attempt made to train or motivate the employees.

In order for users to have a more realistic perspective concerning the probable efficacy of HPDs, NRRs must be derated. An approximate real-world correction that has been suggested is to reduce the NRR by 10 dB before subtracting it from the measured C-weighted sound level. This correction is smaller than the average labeled vs. real-world differences in Figure 10.24. It is the minimum correction that is necessary to reduce laboratory generated NRRs to potentially achievable real-world values. In many instances, especially in existing hearing conservation programs, larger corrections would be warranted.

Whatever real-world adjustment is utilized for the NRR, the same one should be applied with a long method computation. For example, if 10 dB is subtracted from the NRR for use in estimating employee noise exposures, 10 dB should also be subtracted from the HPD's octave-band attenuation values prior to entering them in step 4 of Table 10.1.

A COMMENT CONCERNING THE LONG METHOD

An alternative approach to estimating protected exposure levels is to use the long method. Certainly, if one can assure the similarity of labeled and real-world attenuation for the particular user or group of users in question, and if one has the octave band analysis of the noise environment available, then the long method is preferred. In most cases, however, it is unlikely that either or both of these "ifs" will be satisfied. Most users complain of the

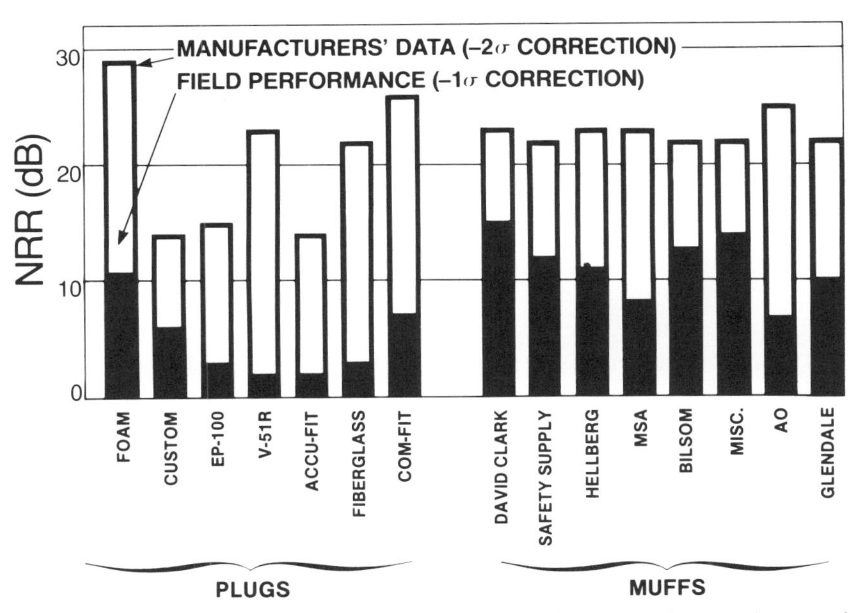

Figure 10.24 — Labeled NRRs vs. real-world performance for earplugs and earmuffs (from Berger, 1983a).

need to measure C-weighted sound levels for use with the NRR, let alone considering conducting an octave band analysis.

The primary utility of examining the mean attenuation of a hearing protector at the individual octave bands is to provide the ability to make a gross match between the device and the environment. For example, both the laboratory and real-world data show that if significant low-frequency energy is present (125 - 250 Hz), then an earmuff is a poor choice and a foam or perhaps a premolded earplug would be better. Conversely, if significant midband energy is present (primarily around 1 kHz), then an earmuff is preferred. The desire to perform this type of protector/noise spectrum matching must be tempered by the realization that assigning particular HPDs to particular jobs in a plant (based on spectrum shapes) is often impractical. It is difficult enough to assure that HPDs are worn, and worn correctly, without also trying to keep devices from being shifted among work areas with different spectral characteristics and having to assure that certain devices are worn only by employees exposed to particular spectra.

OSHA's CALCULATIONAL PROCEDURE

The Hearing Conservation Amendment specifies that employers shall evaluate hearing protector attenuation according to the methods in Appendix B of the document. (See Appendix I of this manual.) Although a note is included in Appendix B that states, "the employer must remember that calculated attenuation values reflect realistic values only to the extent that the protectors are properly fitted and worn," and although it is clear that calculated attenuation values based upon manufacturers' labeled data are unrealistic, OSHA chose to permit use of such data for compliance with the regulation. However, in a more recent administrative guideline, OSHA (1984) requires reducing published NRRs by 50%, but only for the purpose of evaluating the relative efficacy of HPDs and engineering noise controls.

The Amendment specifies that employers are to utilize the labeled data to calculate protection using either NIOSH Methods #1, #2, or #3 (NIOSH, 1975), or the NRR with either a C- or A-weighted noise measurement. NIOSH Method #1 is equivalent to the long method (described previously), Method #2 is equivalent to use of the NRR with a C-weighted noise measurement (Berger, 1980), and Method #3 is equivalent to the use of the NRR with an A-weighted noise measurement. (For additional details see Berger, 1984.)

The use of the NRR with an A-weighted noise measurement entails the subtraction of an additional 7 dB safety factor from the NRR (see Single Number Calculation of HPD Noise Reduction). This is not to be confused with the 10 dB real-world correction factor that was discussed above, since the 7 dB factor derives from the computational procedure itself and is due to the reduced accuracy of using the NRR with A- instead of C-weighted sound levels. The accuracy lost by this approach dictates that when it is used, it will add considerably more imprecision to an already rough estimate.

EVALUATION OF THE AUDIOMETRIC DATA BASE

The final arbiter of HPD efficacy will be the shifting or lack thereof in the hearing levels of the noise-exposed employees. This may be evaluated using any of a number of techniques that have been proposed for the evaluation of audiometric data bases (see Chapter 9). Regardless of the accuracy of any predictive scheme that attempts to match HPDs to employee noise exposures, if the hearing levels of individuals or groups of employees are deteriorating then additional measures must be instituted. Conversely, if the hearing levels are not changing any more rapidly than those of an appropriately selected nonindustrial noise-exposed reference population, then this is an indication of a job well done.

Effects of HPDs on Auditory Communication

An important consideration that arises when one recommends the use of HPDs is what effects, if any, they will have on the wearers' abilities to communicate verbally, listen to operating machinery, and respond to warning sounds. On the one hand, many hearing protector manufacturers claim in their literature that their devices will block out harmful high frequency noise and yet let speech through, and on the other hand, employees often complain that they can't talk to their fellow workers or hear their machinery operate when they are using HPDs. Both of these statements contain elements of truth as will become apparent after an examination of the available data.

UNDERSTANDING SPEECH

The level to which a particular sound will be attenuated by an HPD is dependent only upon its frequency and initial intensity. HPDs that do not contain active electronic circuitry cannot differentiate desired signals such as speech from useless information such as noise; both will be attenuated equally. At any one frequency the resultant signal-to-noise ratio will be unaffected by use of the HPD. But hearing protector attenuation does vary with frequency, generally increasing as frequency increases, so that the frequency balance of the attenuated signal-plus-noise spectrum will differ from the unattenuated condition. Since the predominant speech energy is located at or below 2 kHz, if the noise energy is primarily above that frequency, most HPDs will reduce the noise level more than they will diminish the overall level of the speech, thus apparently "letting the speech through and cutting out the noise." However, this is normally not the case.

The principal reason that HPDs improve the ability of normal hearing listeners to discriminate speech in high noise level environments is that, by reducing the overall level of the signal-plus-noise, the HPD permits the cochlea to respond without distortion, a characteristic which can only be assured at sound levels well below 90 dBA (Lawrence and Yantis, 1956). The effect is similar to wearing sunglasses on a very bright day. Since the total illumination of the scene is reduced, the eye is allowed to function more effectively and in a more relaxed manner. Metaphorically speaking, HPDs reduce the "acoustical glare" of high level sounds.

Speech discrimination (SD) is a measure of one's ability to understand speech. It is greatly affected by such factors as a person's hearing acuity, the signal-to-noise ratio, the absolute signal levels, visual cues (lip and hand motion), and the context of the message set. SD is measured by verbally presenting to subjects one of a number of prepared standardized word lists and determining the percentage of correct responses they achieve. The effects of HPDs on SD can be evaluated by establishing a set of test conditions and measuring SD with and without HPDs on the subjects. The results of such tests on normal hearing subjects may be summarized as follows:

1) HPDs have little or no effect on the SD ability of normal hearing listeners in moderate background noise, approximately 80 dBA, but will decrease SD as the noise is reduced below that level (Kryter, 1946; Howell and Martin, 1975; Rink, 1979).
2) At high noise levels, greater than approximately 85 dBA, HPDs actually improve SD as is demonstrated in Figure 10.25. For the pairs of curves shown, whenever the dashed line is above the solid line, the earplugs provide improved SD.

When the listener has impaired hearing, the situation is considerably more complex, and the answer is not as well defined. However, it is clear that HPDs will decrease SD for hearing-impaired listeners in low to moderate noise situations, with the effects being minimized as both the noise levels and signal-to-noise ratios are increased (Coles and Rice, 1965; Chung and Gannon, 1979; Rink, 1979; Froehlich, 1981; Abel *et al.*, 1982). The difficulty for hearing impaired listeners arises from the fact that HPDs may reduce the level of the speech signals below their threshold of audibility,

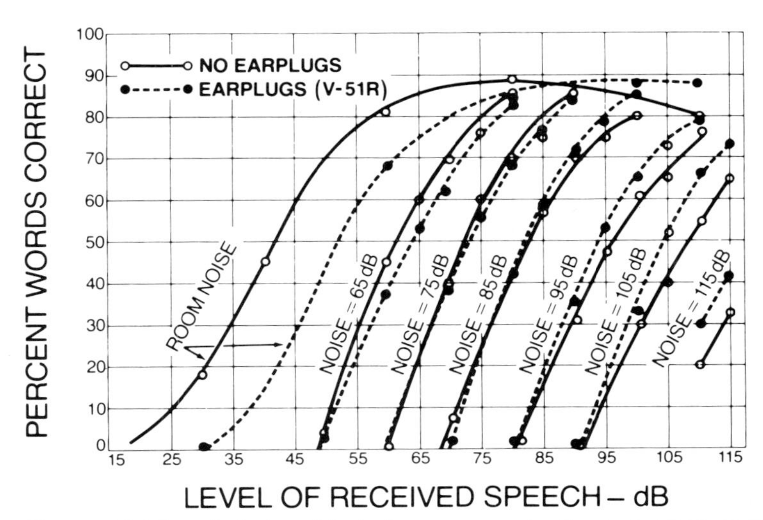

Figure 10.25 — Relationship between speech discrimination and speech level with noise level as a parameter and hearing protection as a variable. Each point represents an average of the % correct responses for eight subjects to a list of 200 words read over a speaker system in a reverberant room (from Kryter, 1946).

especially the important higher-frequency consonant sounds. None of the published studies have been able to unequivocally define the level of hearing loss that is required before hearing protection will degrade instead of improve SD in noisy environments; however, a rough estimate based upon the work of Lindeman (1976) would be a hearing threshold level of greater than about 40 dB when averaged across the frequencies of 2, 3, and 4 kHz.

The preceding generalizations can be modified in practice by additional important factors. For example, in real-world environments, communications may be either limited in scope and/or accompanied by visual cues, allowing missed words to be "filled in" and intelligibility maintained. Rink (1979) illustrated this fact for subjects listening in a 90-dBA background noise. Both his normal and hearing impaired listeners were able to maintain SD scores regardless of the use of hearing protection, as long as visual cues were presented along with the auditory stimuli. And Acton (1970) has demonstrated that employees become accustomed to listening in noise, thus performing better with respect to SD than do laboratory subjects with equivalent hearing levels. Conversely, Howell and Martin (1975) have shown that when the person speaking wears HPDs, speech quality is degraded, adversely affecting communications.

Howell and Martin's observation is at least partially explained by examining the effects of an HPD upon a talker's perception of his own voice. The HPD significantly attenuates the airborne energy, but has little effect on the BC portion, except in the lower frequencies where the perceived voice levels are actually amplified as a result of the occlusion effect (see Occlusion Effect). This alters the frequency balance of the information monitored by the talker and causes the loudness of his own voice to increase or remain the same, while existing environmental noise levels are significantly reduced. It appears as though the talker's own voice is louder compared to the noise than actually is the case, and speech levels tend to be lowered accordingly, typically by 2 - 4 dB (Kryter, 1946; Howell and Martin, 1975). As a result, employees must be taught that while wearing HPDs in noisy environments, good communications can be assured only by talking at what seems to be a louder than necessary level (Guild, 1966).

RESPONDING TO WARNING AND INDICATOR SOUNDS

The effects of HPDs on the ability of normal and hearing-impaired users to detect warning sounds is very similar to the findings with regard to speech intelligibility. However, when employees are engrossed in alternative tasks and not specifically attending to sounds that may warn them of danger or indicate a machine malfunction, then not only is the question of detection and discrimination of importance, but one must also inquire whether or not the sound will get their attention. Inattention may result in elevation of effective thresholds for particular test stimuli by 6 - 9 dB, or even more for certain individuals (Wilkins and Martin, 1978).

One field study was conducted that assessed the effectiveness of intentional (warning horn) and incidental (clinking of metal components spilling from their container) warning sounds under actual factory conditions, using subjects who were employees at the test site (Wilkins, 1980). It was found that there was no significant effect of wearing hearing protection on the ability of either normal or hearing-impaired subjects to detect the horn or the clinking sound when they were specifically awaiting its occurrence. However, when the subjects were distracted by performing their normal job duties, the clink sound was less well perceived as a warning by those with substantial hearing loss, and HPDs adversely affected its perception independent of the hearing sensitivity of the wearers. By contrast, even when distracted, there were no significant variations in the response rates for the intentional warning sound (the horn) due to either the hearing ability of the listeners or their ear condition (protected or unprotected).

Since warning sounds may be adjusted in pitch and loudness to achieve optimum perceptibility, results similar to those reported above should be achievable in most conditions. Warning sounds will be most effective when their primary acoustic output is located below 2 kHz, since hearing impaired employees will exhibit their largest losses above that frequency and many HPDs will deliver less attenuation below that frequency. Additional evidence regarding any possible hazards associated with the use of hearing protection while working in noisy environments is provided by two field studies (Cohen, 1976; Schmidt *et al.,* 1980). Those authors demonstrated that implementation of a hearing conservation program that required utilization of HPDs reduced rather than increased the number of industrial mishaps.

LOCALIZATION AND DEPTH PERCEPTION

Another effect that HPDs can have is to confuse one's ability to locate the direction of origin of sounds (Atherly and Noble, 1970; Noble and Russell, 1972). The indications are that earmuffs, which necessarily cover the entire ear, can interfere with localization accuracy to a greater extent than inserts which leave the outer ear exposed, although one recent study that measured the minimum detectable azimuthal changes of a frontally located sound source (typically $<10°$) found the opposite to be the case for one of the higher-attenuation earplugs they studied (Lin, 1981). Furthermore, experiments indicate that subjects cannot learn to compensate for the adverse effects of the earmuffs (Russell, 1977). There is also some suggestion, but no direct evidence, that HPDs may impair the ability of wearers to judge the distance to a sound source (Wilkins and Martin, 1978).

GENERAL REMARKS

The preceding data indicate that HPDs can be effectively utilized for the preservation of hearing in high noise level environments with minimal negative effects on auditory communications and demonstrated advantages in certain conditions. For hearing-impaired persons, the utilization of HPDs in lower noise levels should be carefully considered. It sometimes becomes a decision between preserving the employee's remaining hearing and creating additional communications disadvantages while at work. Often warning and indicator sounds can be augmented or replaced by visual or tactile signals to assist the hearing impaired.

Intermittent noise poses a significant problem since HPDs will cause degradation of communications during the intervening quiet intervals. Consequently, earmuffs or semi-aural devices are the preferred type of

HPDs for such conditions, since they are easily removed and replaced as the intermittence of the noise may warrant. Unfortunately, passive amplitude sensitive HPDs cannot normally be recommended, since when the noise is on, it will not generally be of sufficient level to activate their level sensitive characteristics. Thus the HPD may provide inadequate protection, especially at the lower and middle frequencies. Additionally, although amplitude sensitive HPDs can improve SD during the quiet conditions, when the noise is present they may actually provide worse SD than do standard linear protectors (Coles and Rice, 1966).

Hearing Protector Standards and Regulations

At this time there are no federal or state agencies or U.S. standards writing organizations (see Table 10.3) that approve or disapprove of particular HPDs, although there is an existing EPA regulation requiring the labeling of all HPD packaging.

STANDARDS AND RELATED DOCUMENTS

Of specific interest is ANSI S3.19 - 1974 (ASA STD 1-1975) and its recent revision ANSI S12.6 - 1984. Users often mistakenly perceive these documents to contain criteria for judging the acceptability of HPDs. In fact, these standards only describe a particular experimental method for determining hearing protector attenuation (see Laboratory Test Methods). Testing an HPD by these methods in no way confers approval or attributes any particular degree of quality to the device. It simply characterizes the laboratory attenuation of the protector, however good or bad that may be.

An agency that many program administrators mistakenly believe stipulates acceptable HPDs is MSHA. MSHA does publish a "Hearing Protector R & D Factor List," but the only criterion for having a device placed on the list is that it be tested according to ASA STD 1, and that the data be available in sales literature or on specification sheets (Marraccini, 1983).

Occasionally users will inquire about FDA approval of HPDs, especially when they are manufacturing or using other products with which the FDA is involved. At this time hearing protectors for industrial use are not considered by the FDA to be medical devices, and therefore are not subject to FDA approval (Link, 1980). Neither does the FDA approve the materials from which HPDs are manufactured. They only approve finished products, which of course in the case of those approved products, implies that the constituent materials are accepted for the intended application(s).

The only other agency with a hearing protection related regulation is OSHA, whose Hearing Conservation Amendment (OSHA, 1983) requires

TABLE 10.3
Selected Federal and National Agencies Responsible for Documents Relating to Noise

ANSI	American National Standards Institute
ASTM	American Society for Testing and Materials
EPA	Environmental Protection Agency
FDA	Food and Drug Administration
MSHA	Mine and Safety Health Administration
NIOSH	National Institute for Occupational Safety and Health
OSHA	Occupational Safety and Health Administration

that the HPDs that are used reduce an employee's TWA to 90 dBA or less, and in the case of employees demonstrating standard threshold shifts, to 85 dBA or less. No particular types or brands of HPDs are specifically recommended or proscribed. The method for assessing the amount of reduction to be expected is contained in Appendix B of the amendment (see OSHA's Calculational Procedure).

A proposed standard that has been discussed for over a decade and has passed from committee to committee is American National Standard Z137.1. This document was intended to evaluate the physical characteristics of HPDs by specifying mechanical tests that they should be subjected to, such as temperature cycling, vibration, headband extension, drop testing, *etc.* Subsequent to these tests, the device's attenuation would be measured in conformance with ASA STD 1. Currently, work on this standard has recommenced, although it is likely that final passage of such a standard will not occur until the late 1980's, at the earliest. One of the major problems that has plagued this document over the years is exactly what physical tests to specify and what, if any, pass/fail criteria can be objectively applied to the results.

EPA LABELING REGULATION

Section 8 of the Noise Control Act of 1972 empowered the EPA to label all noise-producing and noise-reducing devices. The first and only standard that was promulgated for noise-reducing devices was the hearing protector labeling regulation (EPA, 1979). As with the preceding regulations that have been discussed, it did not specify criteria by which HPDs were to be deemed acceptable or unacceptable. The regulation did however specify that the attenuation of any device or material capable of being worn on the head or in the ear canal, and which is sold wholly or in part on the basis of its ability to reduce the level of sound entering the ear, was to be evaluated according to ASA STD 1. An NRR was to be computed and placed on a label whose size and configuration was specified. The regulation also included requirements for submission of label verification data to EPA and contained enforcement and compliance audit testing provisions.

In the early 1980s, budget cuts at EPA led to elimination of the noise enforcement division. In recognition of their inability to enforce the regulation, EPA then revoked the product verification testing and the attendant reporting and recordkeeping requirements for the standard (EPA, 1983). However, the remainder of the regulation remains law unless and until Congress deletes Section 8 from the Noise Control Act. This leaves the future of the regulation in limbo. Pending Congressional action, it is possible that a voluntary labeling regulation embodying an NRR or NRR-like number and strict quality control requirements may be developed by the Safety Equipment Institute of the Industrial Safety Equipment Association.

Concluding Remarks

RECOMMENDATIONS

The factors most often considered by purchasers of HPDs are attenuation, comfort, human engineering, cost, durability, styling, and availability. It is not possible to rank order these items in a manner suitable for all applications, but in this author's opinion attenuation and comfort should be considered foremost and given equal importance. If the laboratory attenuation for a particular HPD was found to be very good, but the comfort of the device was poor, then actual in-use protection might be reduced or even nonexistent. Conversely, a lower attenuation HPD that is comfortable and can be worn regularly and consistently will provide greater effective protection to the using population.

Attenuation may be estimated by using manufacturers' published data, although a better predictor of in-field performance will be the available measured real-world attenuation data, and the best indicator of actual on-the-job effectiveness can be provided by analysis of the annually updated audiometric data base. The natural inclination to precisely match attenuation data to a particular noise exposure spectrum must be tempered by an awareness of the large differences between rated and real-world performance and of the large inter-employee variability in HPD effectiveness. Comfort, and also human engineering, should be evaluated personally by prospective buyers and their employees. Program administrators will gain a much improved appreciation of the devices under consideration if they wear them eight hours per day for a few days, in order to conduct their own subjective evaluation.

The program administrator should be able to narrow down the types and brands of HPDs that will be offered within any one program. If this is accomplished with care and perspicacity following the guidelines provided in this chapter, then the users will be offered a selection of the best available devices. At the same time, it is important to limit the selection because the problems of preparing uniform high-caliber training for a large variety of products can easily become unwieldy. Furthermore, a limited group of devices simplifies inventory and spare parts control and may allow buyers to negotiate higher volume, lower-cost contracts with their suppliers.

The selection of devices to be offered in a hearing conservation program should include a minimum of three devices representing at least two types. Generally this will consist of an earmuff and a couple of models of earplugs, but a semi-aural device or additional brands of each protector type may be warranted. The employee should be involved in the final choice of the HPD that he or she will wear. HPDs are a personal piece of protective equipment that may not adapt equally well to all head and ear canal shapes, and individual preference may vary widely. Involving the employees in the selection process will increase the likelihood of maintaining their participation in the entire program. If, after a couple of weeks of daily use, an employee is still experiencing difficulties or discomfort, the protector should be resized and/or refitted, or another hearing protector should be issued.

And finally, it is worth stressing a widely quoted but often unappreciated axiom — the best hearing protector is the one that is worn and worn correctly. And that protector will be the one that is matched to the environment, the noise, and the person in need of protection.

FINAL CONSIDERATIONS

The contents of this chapter provide a comprehensive review and analysis of the measurement, performance, selection, and fitting of HPDs. And of equal importance is the information that is provided to hearing conservationists to assist them in developing educational materials and programs to train and motivate their workforce. Hearing protectors are not a panacea and cannot be dispensed indiscriminately, but they can and do work when utilized within the context of a well defined and properly implemented hearing conservation program.

Evaluation of a large number of existing programs suggests that many protectors can provide at least 10 dB of noise reduction for 84% of the workforce. According to the best government estimates (OSHA, 1981) 92% of industrial noise exposures represent eight hour equivalent levels of less than or equal to 95 dBA (and 97% ≤ 100 dBA). Therefore 10 dB of attenuation is often all that is needed to reduce these exposures to acceptable levels. Furthermore, a number of studies have examined both short- and long-term changes in employees' hearing threshold levels and have shown that certain HPDs can and do provide adequate protection from high level industrial noise exposures (Royster, 1979; Hager *et al.*, 1982; Royster and Royster, 1982). Thus, HPDs remain one of the most important and potentially effective tools available today for the hearing conservationist to utilize in the ongoing struggle to protect workers from the hazards of occupational noise.

References

Abel, S.M., Alberti, P.W., Haythornthwaite, and Riko, K. (1982). "Speech Intelligibility in Noise: Effects of Fluency and Hearing Protector Type," *J. Acoust. Soc. Am. 71(3)*, 708-715.

Acoustical Society of America (1975). "Method for the Measurement of Real-Ear Protection of Hearing Protectors and Physical Attenuation of Earmuffs," Standard ASA STD 1-1975 (ANSI S3.19-1974), New York, NY.

Acton, W.I. (1970). "Speech Intelligibility in a Background Noise and Noise-Induced Hearing Loss," *Ergonomics 13(5)*, 546-554.

American National Standard (1981). "Personal Hearing Protective Devices for Use in Noise Environments," draft standard Z137.198X.

American National Standards Institute (1957). "Method for the Measurement of Real-Ear Attenuation of Ear Protectors at Threshold," Standard Z24.22-1957 (R1971), New York, NY.

American National Standards Institute (1984). "Method for the Measurement of the Real-Ear Attenuation of Hearing Protectors," Standard S12.6-1984, New York, NY.

Atherley, G.R.C. and Noble, W.G. (1970). "Effect of Ear-defenders (Ear-muffs) on the Localization of Sound," *Br. J. Ind. Med. 27*, 260-265.

Berger, E.H. (1980). "Suggestions for Calculating Hearing Protector Performance," *Sound and Vibration 14(1)*, 6-7.

Berger, E.H. (1981). "Details of Real World Hearing Protector Performance as Measured in the Laboratory," in *Proceedings of Noise-Con 81*, edited by L.H. Royster, F.D. Hart, and N.D. Stewart, Noise Control Foundation, New York, NY, 147-152.

Berger, E.H. (1982). Unpublished Data.

Berger, E.H. (1983a). "Using the NRR to Estimate the Real World Performance of Hearing Protectors," *Sound and Vibration 17(1)*, 12-18.

Berger, E.H. (1983b). "Laboratory Attenuation of Earmuffs and Earplugs Both Singly and in Combination," *Am. Ind. Hyg. Assoc. J. 44(5)*, 321-329.

Berger, E.H. (1983c). "Considerations Regarding the Laboratory Measurement of Hearing Protector Attenuation, in *Proceedings of Inter-Noise 83* edited by R. Lawrence, Institute of Acoustics, Edinburgh, 379-382.

Berger, E.H. (1983d). "Attenuation of Hearing Protectors at the Frequency Extremes," in 11th Int. Congr. on Acoustics, Paris, France, Vol. 3, 289-292.

Berger, E.H. (1984). "EARLog #12 — The Hearing Conservation Amendment (Part II)," *Am. Ind. Hyg. Assoc. J. 45(1),* B22-B23.

Berger, E.H. (1985). "EARLog #17 — Ear Infection and the Use of Hearing Protection," *J. Occup. Med. 27(9),* 620-623.

Berger, E.H. (1986). "Review and Tutorial — Methods of Measuring the Attenuation of Hearing Protection Devices," *J. Acoust. Soc. Am. 79(6),* 1655-1687.

Berger, E.H. and Kerivan, J.E. (1983). "Influence of Physiological Noise and the Occlusion Effect on the Measurement of Real-Ear Attenuation at Threshold," *J. Acoust. Soc. Am. 74(1),* 81-94.

Berger, E.H., Kerivan, J.E., and Mintz, F. (1982). "Inter-Laboratory Variability in the Measurement of Hearing Protector Attenuation," *Sound and Vibration 16(1),* 14-19.

Blackstock, D.T. and von Gierke, H.E. (1956). "Development of an Extra Small and Extra Large Size of the V-51R Earplug," Wright Air Development Center, Wright-Patterson AFB, OH.

Bolka, D.F. (1972). "Methods of Evaluating the Noise and Pure Tone Attenuation of Hearing Protectors," Doctoral Thesis; Penn State University.

Camp, R.T. (1979). "Hearing Protectors," *The Otolaryngological Clinics of North Am. 12(3),* 569-584.

Caruso, V.G. and Meyerhoff, W.L. (1980). "Trauma and Infections of the External Ear," in *Otolaryngology,* edited by M.M. Paparella and D.A. Shumrick, W.B. Saunders Co., Philadelphia, PA.

Chung, D.Y. and Gannon, R.P. (1979). "The Effect of Ear Protectors on Word Discrimination in Subjects with Normal Hearing and Subjects with Noise-Induced Hearing Loss," *J. Am. Aud. Soc. 5(1),* 11-16.

Cohen, A. (1976). "The Influence of a Company Hearing Conservation Program on Extra-Auditory Problems in Workers," *J. Saf. Res. 8(4),* 148-162.

Coles, R.R.A. and Rice, C.G. (1965). "Letter to the Editor: Earplugs and Impaired Hearing," *J. Sound Vib. 3(3),* 521-523.

Coles, R.R.A., and Rice, C.G. (1966). "Speech Communications Effects and TTS Reduction Provided by V51R and Selectone-K Earplugs Under Conditions of High Intensity and Impulsive Noise," *J. Sound Vib. 4(2),* 156-171.

Dept. of the Air Force (1982). "Hazardous Noise Exposure," AF Regulation 161-35, Washington, DC.

Else, D. (1973). "A Note on the Protection Afforded by Hearing Protectors — Implications of the Energy Principle," *Ann. Occup. Hyg. 16,* 81-83.

Environmental Protection Agency (1979). "Noise Labeling Requirements for Hearing Protectors," *Fed. Regist. 42(190),* 40CFR Part 211, 56139-56147.

Environmental Protection Agency (1983). "Noise Emission Standards for . . ., and Noise Labeling Requirements for Hearing Protectors; Final Rule; Revocation for

Product Verification Testing, Reporting, and Recordkeeping Requirements," *Fed. Regist. 47(249),* 40 CFR Parts 204, 205 and 211, 57709-57717.

Forrest, M.R. (1969). "Laboratory Development of an Amplitude-Sensitive Ear Plug," Royal Naval Personnel Research Committee, Report HeS 133, Med. Res. Council, London, England.

Fox, J.G. (1971). "Background Music and Industrial Efficiency — A Review," *Appl. Ergon. 2(2),* 70-73.

Froehlich, G.R. (1981). "The Effects of Ear Protectors and Hearing Loss on Sentence Intelligibility in Aircraft Noise," AGARD Conf. 311, Soesterberg, Netherlands, paper 16.

Gasaway, D.C. (1984). "'Sabotage' Can Wreck Hearing Conservation Programs," *Natl. Saf. News 129(5),* 56-63.

Gasaway, D.C. (1985). *Hearing Conservation — A Practical Manual and Guide,* Prentice-Hall, Englewood Cliffs, NJ, p. 173.

Guild, E. (1966). "Personal Protection," in *Industrial Noise Manual, Second Edition,* Am. Ind. Hyg. Assoc., 84-109.

Hager, W.L., Hoyle, E.R., and Hermann, E.R. (1982). "Efficacy of Enforcement in an Industrial Hearing Conservation Program," *Am. Ind. Hyg. Assoc. J. 43(6),* 455-465.

Harris, H.D. (1979). Personal Communication.

Howell, K. and Martin, A.M. (1975). "An Investigation of the Effects of Hearing Protectors on Vocal Communication in Noise," *J. Sound Vib. 41(2),* 181-196.

Humes, L.E. (1983). "A Psychophysical Evaluation of the Dependence of Hearing Protector Attenuation on Noise Level," *J. Acoust. Soc. Am. 73(1),* 297-311.

International Organization for Standardization (1981). "Acoustics — Measurement of Sound Attenuation of Hearing Protectors — Subjective Method," ISO 4869, Switzerland.

Kasden, S.D. and D'Aniello, A. (1976). "Changes in Attenuation of Hearing Protectors During Use," in *Proceeding of Noisexpo,* edited by J.K. Mowry, 28-29.

Krutt, J. and Mazor, M. (1980-81). "Attenuation Changes During the Use of Mineral Down and Polymer Foam Insert-Type Hearing Protectors," *Audiology & Hearing Educ.,* Winter, 13-14.

Kryter, K.D. (1946). "Effects of Ear Protective Devices on the Intelligibility of Speech in Noise," *J. Acoust. Soc. Am. 18(2),* 413-417.

Lawrence, M. and Yantis, P.A. (1956). "Onset and Growth of Aural Harmonics in the Overloaded Ear," *J. Acoust. Soc. Am. 28,* 852-858.

Lin, L. (1981). "Auditory Localization Under Conditions of High Ambient Noise Levels With and Without the Use of Hearing Protectors," Masters Thesis at North Carolina State Univ., Raleigh, NC.

Lindeman, H.E. (1976). "Speech Intelligibility and the Use of Hearing Protectors," *Audiology 15,* 348-356.

Link, D.M. (1980). Correspondence with F.E. Willcher, Jr., Executive Director of the Industrial Safety Equipment Assoc., April.

Marraccini, L.C. (1983). Personal Communication.

Martin, A.M. (1976). "Industrial Hearing Conservation 1: Personal Hearing Protection," *Noise Control Vib. and Insul. 7(2),* 42-50.

Martin, A.M. (1977). "The Acoustic Attenuation Characteristics of 26 Hearing Protectors Evaluated Following the British Standard Procedure," *Ann. Occup. Hyg. 20,* 229-246.

Martin, A.M. (1979). "Dependence of Acoustic Attenuation of Hearing Protectors on Incident Sound Level," *Br. J. Ind. Med. 36,* 1-14.

Mosko, J.D., and Fletcher, J.L. (1971). "Evaluation of the Gundefender Earplug: Temporary Threshold Shift and Speech Intelligibility," *J. Acoust. Soc. Am. 49(6 Part 1),* 1732-1733.

National Institute for Occupational Safety and Health (1975). "List of Personal Hearing Protectors and Attenuation Data," U.S. Dept. of HEW, Report No. 76-120, Cincinnati, OH.

Nixon, C.W. (1979). "Hearing Protector Devices: Ear Protectors," in *Handbook of Noise Control* (2nd Edition), edited by C.M. Harris, McGraw-Hill, New York, NY, p. 12-1 to 12-13.

Nixon, C.W. and Knoblach, W.C. (1974). "Hearing Protection of Earmuffs Worn Over Eyeglasses," Aerospace Medical Research Laboratory, Report No. AMRL-TR-74-61, Wright-Patterson AFB, OH.

Nixon, C.W. and von Gierke, H.E. (1959). "Experiments on the Bone-Conduction Threshold in a Free Sound Field," *J. Acoust. Soc. Am. 31(8),* 1121-1125.

Noble, W.G. and Russell, G. (1972). "Theoretical and Practical Implications of the Effects of Hearing Protection Devices on Localization Ability," *Acta Otolaryngol. 74,* 29-36.

OSHA (1981). "Occupational Noise Exposure; Hearing Conservation Amendment," *Fed. Regist. 46(11),* 4078-4181.

OSHA (1983). "Occupational Noise Exposure; Hearing Conservation Amendment," *Fed. Regist. 48(46),* 9738-9783.

OSHA (1984). Instruction CPL 2.45A CH-3, dated Jan. 27; an update to the Field Operations Manual.

Ohlin, D. (1975). "Personal Hearing Protective Devices Fitting, Care, and Use," U.S. Army Environmental Hygiene Agency, Report No. AD-A021 408, Aberdeen Proving Ground, MD.

Ohlin, D. (1981). Personal Communication.

Riko, K. and Alberti, P.W. (1982). "How Ear Protectors Fail: A Practical Guide," in *Personal Hearing Protection in Industry,* edited by P.W. Alberti, Raven Press, New York, NY, 323-338.

Rink, T.L. (1979). "Hearing Protection and Speech Discrimination in Hearing-Impaired Persons," *Sound and Vibration 13(1),* 22-25.

Royster, L.H. (1979). "Effectiveness of Three Different Types of Ear Protectors in Preventing TTS," *J. Acoust. Soc. Am. 66, Supp. 1,* S62.

Royster, L.H. and Holder, S.R. (1982). "Personal Hearing Protection: Problems Associated with the Hearing Protection Phase of the Hearing Conservation Program," in *Personal Hearing Protection in Industry,* edited by P.W. Alberti, Raven Press, New York, NY, 447-470.

Royster, L.H. and Royster, J.D. (1982). "Methods of Evaluating Hearing Conservation Program Audiometric Data Bases," in *Personal Hearing Protection in Industry,* edited by P.W. Alberti, Raven Press, New York, NY, 511-540.

Royster, L.H. and Royster, J.D. (1985). "Hearing Protection Devices," in *Hearing Conservation in Industry,* edited by Alan S. Feldman and Charles T. Grimes, Williams and Wilkins, Baltimore, MD.

Royster, L.H., Royster, J.D., Berger, E.H., and Skrainar, S.F. (1984). "Personal Audio Headsets and Table Radios in Industrial Noise Environments," ASHA 26(10), 77.

Russell, G. (1977). "Limits to Behavioral Compensation for Auditory Localization in Earmuff Listening Conditions," *J. Acoust. Soc. Am. 61(1),* 219-220.

Schmidt, J.W., Royster, L.H., and Pearson, R.G. (1980). "Impact of an Industrial Hearing Conservation Program on Occupational Injuries for Males and Females," *J. Acoust. Soc. Am. 67, Suppl. 1,* S59.

Sutton, G.J. and Robinson, D.W. (1981). "An Appraisal of Methods for Estimating Effectiveness of Hearing Protectors," *J. Sound Vib. 77(1),* 79-81.

Tonndorf, J. (1972). "Bone Conduction," in *Foundations of Modern Auditory Theory, Vol. II,* edited by J.V. Tobias, Academic Press, New York, NY, 195-238.

Wilkins, P.A. (1980). "A Field Study to Assess the Effects of Wearing Hearing Protectors on the Perception of Warning Sounds in an Industrial Environment," Inst. of Sound and Vibration Research Contract Report No. 80/18, Southampton, England.

Wilkins, P.A. and Martin, A.M. (1978). "The Effect of Hearing Protectors on the Perception of Warning and Indicator Sounds — A General Review," Inst. of Sound and Vibration Research Tech. Report No. 98, Southampton, England.

Zwislocki, J. (1957). "In Search of the Bone-Conduction Threshold in a Free Sound Field," *J. Acoust. Soc. Am. 29(7),* 795-804.

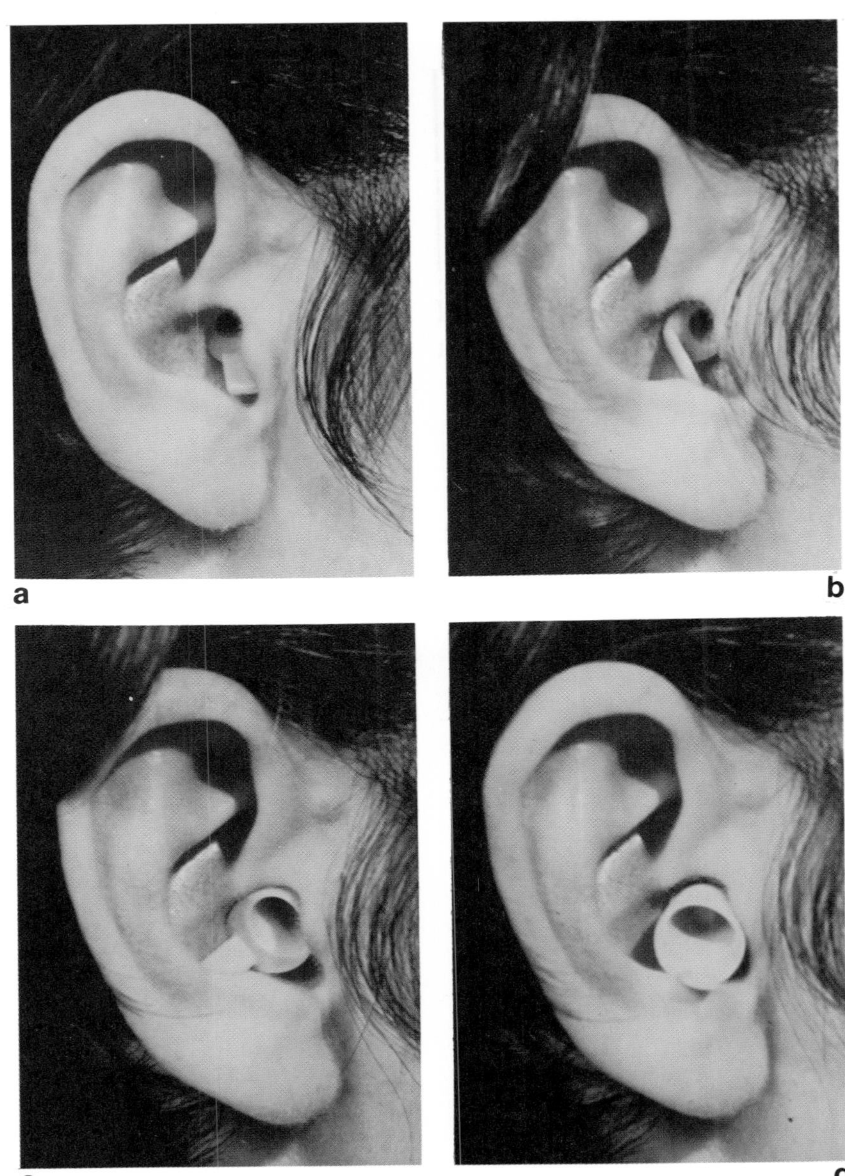

Figure C10.9 — Examples of different insertions of a V-51R earplug. a) Plug is undersized or too deeply inserted. Preferred orientation for tab is toward the rear or sometimes downward. It should never point forward or upward. b) Properly inserted. c) Properly inserted. d) Not inserted deeply enough; plug oversized.

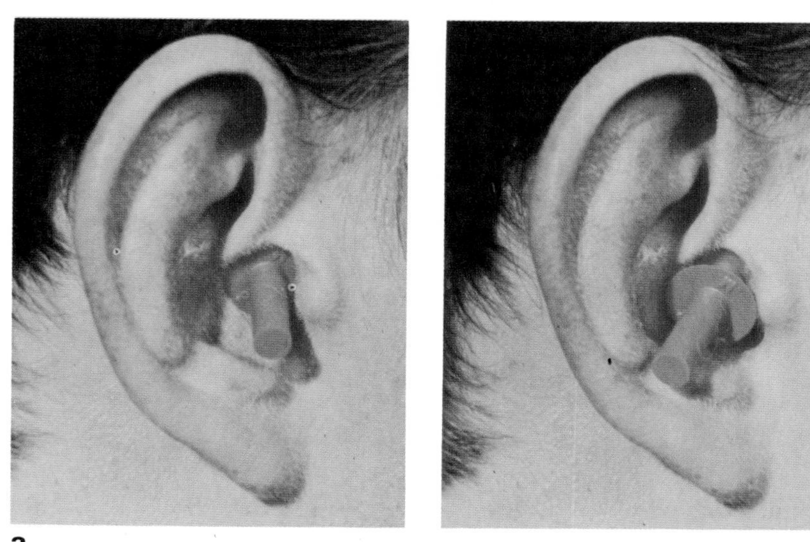

Figure C10.10 — Examples of different insertions of a three-sized three-flanged earplug. a) Properly inserted plug. Note that the outermost flange is sealing entrance to canal. b) Not inserted deeply enough and/or oversized for this ear canal.

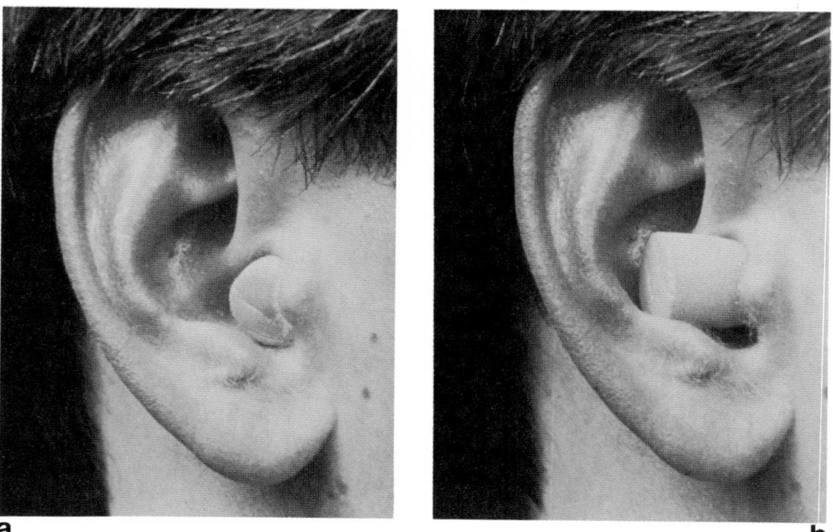

Figure C10.13 — Examples of different insertions of a foam earplug. a) Properly inserted plug. Plug should normally rest even with or slightly inside the tragus. b) Not inserted deeply enough. Principal portion of the plug is resting in the concha.

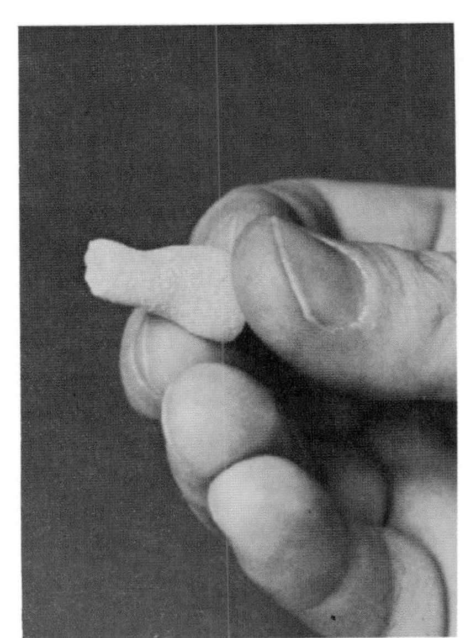

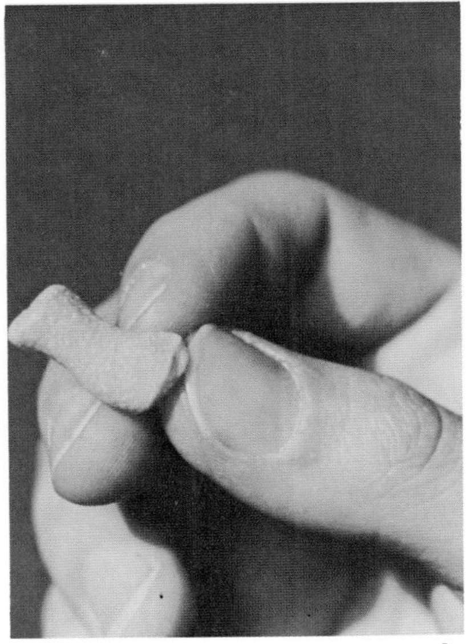

a **b**

Figure C10.14 — How to "read" a foam earplug after removal. Plug must have expanded in ear canal for about one minute prior to removal. Primarily useful for medium and smaller ear canals. a) Example of a good insertion. About ½ of plug is compressed into shape of ear canal. No creases, wrinkles, or folds are evident. b) Another example of a good insertion.

Noise and Hearing Conservation Manual, edited by
E.H. Berger, W.D. Ward, J.C. Morrill and L.H. Royster

11 Education and Motivation

Larry H. Royster
Julia Doswell Royster

Contents

	Page
Introduction	385
Real-World Needs	385
Who Is In Charge?	385
Educational Concepts	386
An Approach to Education	386
Program Guidelines	386
Timing Is Important	388
Who Needs Training?	388
Management	389
Consultants	389
Audiometric Technicians and Issuers and Reissuers of HPDs	389
Foremen or Front Line Supervisors	390
Employees	390
Special Groups	390
Formats for Education	391
One-On-One Encounters	391
Small Employee Groups	391
Regular Safety Meetings	392
Self-Education	392
Handout Materials	392
Bulletin Boards and Company Publications	393
Using Visual Aids Effectively	393
Slide Shows or Slide-Cassette Programs	393
Video Tape Programs	394
Movies	394
Visual Variety	394
Who Should Be the Educator?	394
In-House Personnel	395

External Educators 395
Hearing Conservation Consulting Firms 395
State and Federal Agencies 396
Trade Associations 396
Industrial Insurance Carriers 396
Manufacturers and Distributors of HCP Products 396
Suggested Content for Educational Programs 396
Programs for Management 397
Effects of Noise on Hearing and Productivity 397
Requirements for an Effective HCP 397
Compliance with Regulations 397
Reduction of Fears 397
Estimated HCP Costs 397
Estimated Compensation Costs 397
Expected and Achieved Benefits of the HCP 398
Programs for Primary HCP Personnel 398
Programs for Employees 398
Effects of Noise and Initial Motivation to Avoid Them 399
Hearing Protection Devices 399
Audiometric Evaluations 400
The Company's HCP Policies 401
Questions and Answers 401
Final Motivation 401
Motivational Concepts 402
Motivation for Wearers of Hearing Protection 402
Annual Audiometric Findings 402
Individual Evaluation of HPD Adequacy 404
Examples of Permanent and Temporary Hearing Loss 404
Annual ADBA Findings for Departments or Groups 406
Influence by Management and Peers 406
Employee Input into HCP Decisions 407
Examples of Worn and Abused HPDs 407
Reward Programs for HPD Utilization 408
Extra Hearing Screening 409
Motivation for Primary HCP Personnel 409
Reinforcement by Upper Management 409
Annual ADBA Results 409
Group Communication 410
Motivation for Management 410
Annual ADBA Results 410
Potential Costs for Compensation for Hearing Loss 410
Safer and More Acceptable Working Environment 411
HPD Utilization and Effectiveness Studies 411
Audiometric Findings for Managers 412

Regulations 412
Putting it all Together 413
References 413

Introduction

REAL-WORLD NEEDS

Of the five phases of an effective HCP (L.H. Royster, Royster, & Berger, 1982) the education phase (including motivational efforts) is the area least likely to receive the attention it deserves. Education is sometimes viewed as an introductory task to be completed during orientation for new workers, or as an annual nuisance in which canned presentations are repeated to disinterested employees. Poorly planned and carelessly implemented educational activities are of little value. In contrast, if educational materials are selected thoughtfully and presented meaningfully, the resulting impact on participation in the HCP can be significant.

In this chapter we have drawn on our own experiences and those of others in assisting industry in education and motivation, plus observations from our recent study (L.H. Royster and Royster, 1984a) in which we interviewed HCP personnel at industrial sites throughout the USA. Due to the variety of industrial environments, hearing conservationists must seek information from numerous knowledgeable sources and critically assess the applicability of information to their own sites before basing decisions upon it. This chapter is presented to the reader as one source of potentially useful information concerning the education and motivation of management and employees. The reader is strongly encouraged to seek out and to use several sources of materials and information when developing and modifying programs for use in the local plant facility.

WHO IS IN CHARGE?

Effective HCPs share common characteristics (L.H. Royster, Royster, & Berger, 1982), including *strict enforcement* of proper HPD utilization (Hager, Hoyle, & Hermann, 1982), the presence of a *key individual* who is personally responsible for all aspects of the HCP (Spindler, Olson, & Fishbeck, 1979), *active communications* among all levels of personnel involved in the HCP, and the availability of *potentially effective HPDs* for the existing work environments.

The key individual who sincerely believes in the value of hearing conservation is the critical factor in HCP success, and education and motivation are the main tools the key individual uses to develop and maintain the program. If an existing HCP lacks cohesion and support, an interested person can initiate change by educating top management to understand the benefits to be gained by improving the HCP. By showing management that a more effective HCP will give the company a return on its investment, the interested person will probably be given the authority to become a key individual — the coordinator for all five phases of the HCP.

Each HCP phase requires cooperative participation by other personnel, and the key individual must use education and motivation to elicit this involvement by stimulating participants' self-interest. If a hearing conservation program is administered on the basis of adhering to company policy, then supervisors and employees will thwart all efforts to save their hearing against their will. Most of us are suspicious when somebody in authority tells us that something is for our own good, and hearing conservation is no exception to this human tendency. Managers, supervisors, and employees need educational/motivational information which will show them why the daily aggravations of the HCP are worth their trouble because of the resulting benefit to themselves.

The involvement of the sincere key individual in education, motivation, and each other phase of the HCP will go a long way toward changing employees' attitudes and behaviors regarding hearing conservation. It is especially important for the in-house key individual to be involved in each phase if outside consultants are used for audiometric testing or other HCP services. Employees will not believe in the sincerity of the company if HCP phases are farmed out to consultants without the active supervision of the key individual. Whether the key individual runs the entire HCP in a small company or supervises in-house or external personnel to administer the HCP for a large firm, the unifying drive provided by the key individual can make the program succeed.

Educational Concepts

AN APPROACH TO EDUCATION

Program Guidelines

An ongoing educational process will include regular activities to present and review HCP information throughout the year. Although the formats used for the information will vary from group meetings to one-on-one conversations and from commercial films to plant newsletter articles, certain principles apply to each technique (Stapleton & Royster, 1981a):

Keep it simple.
Keep it short.
Keep it meaningful.
Keep it motivating.

Never forget that industrial employees are not eagerly awaiting new knowledge about hearing conservation. Taking a break from the job may be the only reason workers perceive for attending a safety meeting. The key individual must design the content of educational activities to capture employees' attention and interest.

The most successful educational efforts are *simple* in content and presentation. Terminology should be easy to understand, and details that are not directly applicable to the employee's daily life should be omitted. The anatomy and physiology of the cochlea are commonly overemphasized, and discussion of unseen internal areas of the ear only makes the risk of noise-induced hearing loss seem more remote. Similarly, the details of noise monitoring and TWAs are unnecessarily confusing.

Educational efforts should be *short* in length. For group presentations an ideal time is 15-25 minutes, with 30 minutes as an upper limit. If an otherwise acceptable film is too long, employees will be dozing in the darkness before the end. If a newsletter article exceeds a few paragraphs, potential readers may stop at the headline.

In order to keep the message *meaningful* the key individual must pare the content down to the core of ideas which are relevant to employees' daily functioning. Stress only the facts employees need to know: the risk of noise-induced hearing loss, the elements of hearing conservation, and the employer's HCP policies.

Beyond summarizing the facts about the HCP, the educational message must focus on *motivating* the employee to participate fully in the program. Although some employers actually offer rewards for HCP participation or penalties for nonparticipation, the ultimate motivator is avoidance of progressive hearing loss. The educational program must make the handicaps associated with hearing loss so clear to employees that they are willing to work to avoid these problems, which include impaired spoken communication ability, social isolation from friends and family, and inability to enjoy leisure activities.

Timing Is Important

In order to maintain the good will of the foremen, supervisors, and managers whose support is needed for an effective HCP, always remember to (Stapleton & Royster, 1985):

Minimize the effect on production.
Maintain flexible program schedules.

Production managers need advance notice of educational activities so they can schedule substitute workers if necessary. The key individual should work with department supervisors to arrange audiograms and HPD fitting sessions at times which will not disrupt production. In scheduling group meetings the educator must be flexible enough to meet production needs by holding programs whenever needed — for example, after second shift.

Another critical aspect of timing involves the regularity of educational activities:

Keep education going continuously.

Never let employees forget the HCP by staging all audiograms and educational sessions during one month out of the year. Spread out the range of activities throughout the entire year to remind workers about hearing conservation over and over in different ways (Karmy & Martin, 1982).

WHO NEEDS TRAINING?

The personnel who need education and motivation are: first, top management and consultants; next, nurses or other issuers and reissuers of HPDs; then foremen and front line supervisors; and finally the employees who wear HPDs and others included in the HCP (Stapleton & Royster, 1981; Else, 1982). By involving each level of the hierarchy in sequence when an HCP is initiated or revised, the company's program can be organized and implemented down through the administrative channels. Feedback at each stage helps to keep the HCP realistic in its goals and clear in its enforcement policies. Although a variety of educational activities are recommended in the following section, the basic method for organizing and reorganizing an HCP involves group meetings to allow communication among the personnel involved.

Management

Even though top management may delegate responsibility for details of the HCP, it is essential that managers have basic knowledge about the requirements for an effective HCP and the policies needed to handle administrative problems which might develop. In our experience managers typically ask "what if . . .?" questions concerning hypothetical situations which involve potential company liability (L.H. Royster & Royster, 1981). Education for management should stress the effects of noise on employee health and productivity, the firm's legal obligations for the HCP, the cost-benefit analysis for an effective HCP, and why strict guidelines for HCP procedures will minimize problem incidence. Above all, managers must understand that their *active* support (rather than nominal policy approval) is needed to establish an effective program. For example, HPD enforcement will be taken seriously by supervisors, foremen, and employees only if managers wear HPDs when they visit production areas, however briefly (Else, 1981). To keep management involved, annual HCP update meetings should be held to show the progress of the program and discuss any needs for improvement or policy change.

Consultants

Any consultants who assist in aspects of the HCP should be educated concerning company health and safety rules to avoid potential conflicts with company policy. For example, before establishing a referral relationship with physicians or audiologists, the key individual should meet with them to outline company policies and describe the operation of the HCP, the physical work environment, the degree of noise hazard, and HPD utilization requirements. By educating the consultants, the key individual will avoid problems such as receiving a doctor's note that a worker should discontinue wearing HPDs due to discomfort, since the doctor would understand from the company's training that the employee must wear HPDs to keep his job.

Audiometric Technicians and Issuers and Reissuers of HPDs

Typically the nurses or other personnel who serve as audiometric technicians attend a 2.5-day CAOHC training course which familiarizes them with all phases of the HCP, although the focus is on audiometric evaluations. It is most effective if the same staff members handle both audiometric testing and HPD issuing, though in large plants HPD reissuing may need to be performed by additional staff in several locations throughout the plant. HPD utilization and audiometric monitoring are the most critical phases of the HCP, and these contacts with employees provide the best opportunities to motivate them and answer their questions and concerns. All personnel involved in these two phases should meet together to review the company's HCP policies, to develop uniform and mutually agreeable HCP procedures, and to receive extra educational background not provided in whatever formal training courses they have attended. Some sites offer no education for HPD issuers and reissuers unless they also perform audiometry, but neglecting training for these personnel almost guarantees that HPD utilization will be ineffective. The key individual should stress to the staff involved with audiograms and HPDs just how important their roles are to the success of the HCP so they will appreciate their responsibility and take it seriously (J.D. Royster and Royster, 1984). Annual refresher sessions are needed to ensure continued uniformity in the audiometric and hearing protection phases.

Foremen or Front Line Supervisors

Because foremen and/or supervisors receive questions from workers about hearing conservation, they need more detailed information than their employees in order to handle concerns and complaints. If they lack answers for workers' questions, they will probably adopt an "anything goes" attitude to eliminate further embarrassing questions. Educational sessions for supervision also provide an opportunity for them to discuss HCP policies and clarify their roles in implementing the program. The duty which falls most heavily on foremen is daily enforcement of HPD utilization. They especially need to understand why poorly fitted, improperly worn, or inconsistently worn HPDs are ineffective. Management should take account of supervisors' HCP efforts in their performance evaluations in order to emphasize the supervisors' responsibilities in the HCP.

Employees

Finally, educational programs should be held for all employees included in the HCP. Face-to-face education is cost-effective because it reduces the occurrence of administrative problems by familiarizing workers with HCP phases and policies and engendering the beliefs and attitudes that employees need in order to take an active part in protecting their own hearing by wearing HPDs properly. Though other types of educational activities can be used effectively to remind workers of hearing conservation needs, annual group meetings are essential to provide question-and-answer periods and interaction with managers who can represent the company's commitment to the HCP.

Special Groups

Separate programs for target groups may help to accomplish specific goals. For example, if a new HCP is being established or an older HCP modified, a special program for union officials at the affected plant will smooth the transition. Educational sessions may be needed for engineering staff members to begin a new emphasis on noise controls. Similarly, vendors of equipment could be trained as a group if the company established requirements for noise levels on new equipment purchases.

FORMATS FOR EDUCATION

Depending on the size of the company and the nature of the production process, educational efforts will take different forms. A combination of educational activities scattered throughout the year will be most effective (Harford, 1978). Several options are described in this section.

One-On-One Encounters

The educational opportunities which are most effective in reaching the employee to change attitudes and behaviors occur during personal contacts when the worker receives the annual audiogram and is fitted or refitted with HPDs (Esler, 1978). After taking the audiogram the employee is usually interested in his own health status and especially receptive to constructive feedback. While the audiogram may be reviewed later by a professional, delayed written feedback after the review will have much less impact than a few immediate words from the technician. The verbal feedback should describe the amount and direction of hearing change and relate the results to HPD fit and utilization on and off the job. Praise for the conscientious HPD wearer will reinforce good habits, and concerned warnings for employees with threshold shifts will stimulate improvement. The employee is also most likely to ask questions in an individual setting.

Uninformed management often forbids feedback to employees about their hearing because they fear that employees will be more likely to file for compensation if they realize their hearing is poor. However, when management attempts to hide information from the employees, they will perceive that the HCP is a farce. Potential compensation costs can best be reduced by establishing an effective HCP, and employee feedback is one of the best ways to improve the program.

Small Employee Groups

After one-on-one encounters, the next most effective educational format involves meetings with small groups of ten workers or less. Many industries encourage group meetings of the foreman and his working group to handle personnel difficulties, production problems, and health and safety issues. Such small meetings of a production group are ideal for HCP education because the members share a common work experience and normally feel free to ask questions.

Companies which contract with mobile audiometric testing services may use the time period between audiometric evaluations to present information to the small group of employees who are waiting. This is acceptable as a part of the educational effort, but the employees would still need another opportunity later to ask questions in a less restricted time frame.

Regular Safety Meetings

Many industries hold periodic safety meetings to inform workers about safety and health issues, reinforce company policies, and obtain employee feedback about procedures. If properly structured, these meetings can be excellent educational experiences. The group size should not exceed 35 people, and the meeting should be run as a two-way discussion, not a one-way lecture. Only if the educator has charismatic communicative abilities can member interaction be achieved in a large group. Safety meetings are an especially good way to foster employee involvement by distributing samples of new HPDs for trial use, then discussing workers' reactions to the product at the next meeting (Sadler, 1982).

Self-Education

Various slide-cassette programs on hearing conservation are available for use on special video machines which can be set up for employees to view at will during their breaks, while waiting for their audiometric tests, or during other free periods. This idea is a useful way to present basic information, but interactive educational opportunities are always necessary to project management's involvement in the HCP and to answer workers' questions. Another danger of relying on self-scheduled audiovisuals is that management may start expecting employees to watch them during their own unpaid time; education necessary to protect the safety and health of the worker should always occur during company time, even if the work schedule must be extended to accommodate the training.

Handout Materials

The least effective employee educational technique is the distribution of pamphlets or similar handout materials. It is impractical to create an easily understood short booklet which would cover all the needed topics and would be so interesting that employees would take time to read it. If given to the general worker population without explanation, pamphlets are usually discarded unread. Therefore, we strongly discourage written materials as the primary educational method. However, booklets can be useful supplements to live educational presentations, especially if given as a reference source to workers who ask technical questions. Written materials also provide a review for personnel such as foremen or HPD reissuers, who need more detailed knowledge.

Bulletin Boards and Company Publications

One way to reinforce HCP educational efforts is to run hearing conservation stories in the plant newsletter and to post HCP items on bulletin boards. These posters and articles serve as reminders, not as primary educational efforts. The material should convey a positive message about how HPD utilization and other HCP efforts relate to the worker's greater enjoyment of home life, social occasions, and hobby activities. Another approach is to engender competition among production departments by praising the department with the best record for HPD utilization and/or the best ADBA results (see chapter 9).

USING VISUAL AIDS EFFECTIVELY

Carefully chosen visual aids are invaluable in stimulating employee interest during group educational presentations. Straight lecture formats are undesirable because the educator will invariably have trouble keeping employees' attention in a classroom atmosphere. The best approach is to use visual aids to illustrate basic concepts and rely on live presentations for company-specific information, discussions, and questions. Fortunately, today there is an abundance of visual aids available for consideration (Appendix II).

Slide Shows or Slide-Cassette Programs

Slide formats allow the educator more freedom than movies or video tapes because the content can be modified more easily and the presentation can be interrupted more readily. Individual slides can be replaced with shots of the company's own work environments, audiometric facilities, sound survey procedures, and personnel. The educator can easily pause to insert extra information or emphasize specific topics. Commercial cassettes can be spliced to add sections on the company's own selection of HPDs and procedures for issuing and reissuing.

Commercial slide-cassettes can be found to suit the needs of most companies, but custom programs can also be developed if care is taken in selecting the best local talent for production purposes (Stapleton & Royster, 1981b). The key individual should resist political pressure to feature powerful managers in a custom program since their degree of popularity may detract from the message being presented. Management can best express interest in the HCP by attending educational sessions to say a few words of support.

Video Tape Programs

Video equipment is increasingly available at reasonable cost, offering the same custom options as slide-cassettes with the added feature of movement on the screen. Moving pictures do project situations more realistically, but greater technical expertise is needed to produce a video tape which would compare favorably in quality with commercial films to which audiences are accustomed (Stapleton & Royster, 1981a). However, if a firm is willing to invest the time and has the resource personnel, a video is an excellent choice.

Movies

There are numerous commercial films available about hearing conservation which are suitable for use in educational programs (Appendix II). Before purchasing a film, the key individual should review several choices to select one which emphasizes the topics and point of view most compatible with the company's needs. For example, although only about 12% of the US workforce who are presently wearing HPDs choose to wear earmuffs instead of plugs or semi-aurals (L.H. Royster & Royster, 1984a), some films show nearly all employees wearing muffs because they are obvious to the camera. Older movies tended to overemphasize the anatomy of the hearing mechanism. To ensure that the general tone of the film will be appealing to employees, it is wise to ask representative workers to help in the selection process.

Visual Variety

No matter how terrific a visual aid may be, repeated viewing will dull its impact. It is fine to use the same material year after year in orientations for new employees, but the annual programs for current workers should be varied to minimize boredom and inattention. If the company cannot afford to maintain several visual aid choices to alternate among plant locations, local industries are wise to loan and borrow materials from each other to provide variety.

WHO SHOULD BE THE EDUCATOR?

From our experience, the best person to facilitate educational sessions is the one whom employees recognize as being sincerely interested in their well-being, familiar with the local plant environment, and available to help with any problems they may encounter concerning the HCP. A strong background of knowledge about all aspects of hearing conservation is also desirable, but not at the expense of the former characteristics.

In-House Personnel

Obviously, the key individual is the person displaying the greatest number of the desired qualities listed above. However, other in-house personnel are also good choices as educators to help the key individual with sessions and/or hold programs at times when the key individual is not available. The effective educators we have observed include nurses, audiometric technicians, safety directors, personnel directors, and other personnel associated with the HCP. Although in-house personnel may not be polished speakers who are comfortable in front of groups, their actual involvement in implementing the HCP makes up for any lack of speaking skills *if* they are committed to protecting employees' hearing. Their everyday involvement in audiometric evaluations, HPD issuing and reissuing, HPD utilization checks, noise monitoring, *etc.* gives them a credibility which external consultants normally do not exhibit.

External Educators

Many companies have found it necessary and/or beneficial to use outside consultive services to fulfill some of the requirements for an effective HCP. In considering potential external educators for the work force, company personnel must be careful to select someone whose knowledge and opinions are consistent with the company's own particular situations. An educator's credibility is quickly destroyed if he makes statements which are contrary to workers' own experience. For example, a lecturer once stressed the need to use white earplugs which could be easily checked for cleanliness and cautioned against using plugs with attached cords, since cords could be a safety hazard if they became caught in machinery. These opinions had some merit, but *not* for the specific audience being addressed: dairy employees who had to wear brightly colored plugs on cords to facilitate finding HPDs that might fall into the milk supply! As this anecdote indicates, company personnel may need to educate the educator about the particular requirements of their production processes.

One sensible way to evaluate potential outside educational sources is to contact other industries which have used their services. If similar industries were satisfied with the service they received, this feedback can help make the choice. Several types of sources should be considered in choosing an outside educator, as described below.

Hearing conservation consulting firms offer the services of professionals such as audiologists, industrial hygienists, and acoustic engineers. These firms may not wish to be involved in the education phase for a company's HCP unless they are involved in other phases of the program as well.

State and federal agencies such as OSHA, MSHA, Departments of Health, Industrial Commissions, etc. often have personnel who do hearing conservation educational programs free of charge with the goal of helping companies develop their own in-house capabilities. Our past experience with these services has generally been favorable, especially with basic presentations. Often management hesitates to utilize government agencies' services because of fear that the educator will tip the enforcement branch of the agency to visit if any problems are observed. Although this potential exists, we have never been told of such an occurrence. If other industries in the area have used the agency's services successfully, we believe there is no reason to avoid these free resources.

Trade associations are increasingly involved in making educational presentations and even developing custom audio-visual aids appropriate to the needs of their particular client industries.

Industrial insurance carriers often have extensive health and safety expertise to share with client industries. Assess carefully the advice of carriers. Although some firms have truly outstanding records of developing educational materials, some agents have attempted to prevent the flow of information to employees out of fear of increasing the number of compensation claims filed.

Manufacturers and distributors of HCP products often have effective reference materials available and may have knowledgeable personnel who could serve as educators. However, the company must evaluate the depth of expertise and the prejudices of such consultants before accepting their services. A few manufacturers periodically offer detailed free seminars on hearing conservation topics which can be extremely beneficial for HCP personnel such as the key individual, audiometric technicians, HPD issuers and reissuers, and those involved in noise monitoring.

SUGGESTED CONTENT FOR EDUCATIONAL PROGRAMS

In planning program content for managers, the primary HCP implementation personnel, and employees in the HCP, the key individual must take account of the perspective of each group. The educational and motivational impact can be maximized by modifying the presentation to appeal to their specific interests and needs. In selecting material for each type of presentation, identify the behaviors desired for the target group and choose materials which will provide the information and motivation to stimulate those particular behaviors, *not* general background information (Feeney & Nyberg, 1976).

Programs for Management

The goal of education for top management is to develop their committed support behind the HCP. Too often HCPs are established as token efforts to satisfy regulatory requirements; such programs are typically ineffective because there is no driving power behind the motions. To improve a poor HCP or maintain a good one, education must start at the top each year. Even if management support has already been achieved, the key individual must renew it by combating potentially adverse influences that also affect management. Legal and medical advisors, trade or business associations, and internal company officials may downplay the importance of hearing conservation or suggest changes which would hurt the program (L.H. Royster & Royster, 1981). Once a year the key individual needs to present a 1-2 hour program for managers to update them on the status of the HCP and show areas which deserve praise or need improvement.

Because the profit incentive drives management, educational programs for management must show how an effective HCP benefits the company without costing significantly more than an ineffective program. Suggested topics (L.H. Royster & Royster, 1981) for management education include:

(1) **Effects of noise on hearing and productivity,** stressing the increase in production efficiency due to easier communication between employees with normal hearing and the reduction in administrative problems when the HCP policies are clearly established and consistently followed,

(2) **Requirements for an effective HCP,** focusing on characteristics of good programs in implementing the five phases and any areas in which the company's HCP currently falls short — especially regarding the leadership role of management in rewarding the primary HCP personnel for proper program implementation and the modeling role of management in setting an example of HPD utilization for employees,

(3) **Compliance with regulations,** including the elements of the noise regulations established by OSHA and other regulatory agencies and the proper procedures and documentation to be followed for future use in potential workers' compensation cases,

(4) **Reduction of fears** by answering "What if . . .?" questions and formulating solutions to problems previously identified as concerns of management by interviews or questionnaires prior to the educational session,

(5) **Estimated HCP costs,** or actual costs if available,

(6) **Estimated compensation costs** with and without an effective HCP as determined from the company's audiometric data base,

(7) **Expected and achieved benefits of the HCP,** including a progress update on various departments.

Due to the 1-2 hour time limit, each topic should be presented concisely with a written executive summary available for managers' later reference. If questions arise which are not resolved to the educator's satisfaction, these problem topics should receive prompt followup and special attention in the next year's meeting.

Programs for Primary HCP Personnel

Included in this group are the nurses, safety/health personnel, audiometric technicians, HPD issuers/reissuers, and supervisors who actually carry out the phases of the HCP on a daily basis. The target behavior for this group is competent and consistent performance of their assigned functions in making the HCP go. Although some small group meetings might be necessary to cover particular aspects such as audiometric testing methods, it is important for *all* these personnel to meet face-to-face and communicate about their cooperation in achieving an effective program. Each person needs to understand the importance of his/her own responsibilities in the overall success of the HCP and to appreciate the purpose of the other individuals' responsibilities. We have observed too many cases, especially in larger firms, where the HCP failed because of a lack of communication and cooperation among resource personnel responsible for separate parts of a program that lacked the unifying supervision of a key individual.

Primary HCP personnel need some reward for good performance beyond the simple satisfaction of knowing they are helping employees (who may not always appreciate their efforts). Therefore, it is desirable to have a management representative attend the meeting to express the company's commitment to the HCP and to outline how the performance evaluations of the primary HCP personnel will include their HCP responsibilities.

Suggested content for the educational program for primary HCP personnel is similar to that for employees (next subhead section), but with greater detail of coverage so that these personnel will feel confident to answer workers' questions. They should be provided with a written compilation of common questions and associated responses, as well as written materials for review and reference (Berger, 1982a, 1982b, 1983a).

Programs for Employees

The content for a general educational program for employees, such as the orientation for new hires or the annual group meeting, should include at least the following areas, with emphasis distributed according to the needs of the particular industry and the conditions under which training is done. Within each topic, focus on the points which are directly related to the target behaviors desired from employees: wearing HPDs properly on the job and for leisure activities, and participating in the company's audiometric evaluations.

(1) **Effects of noise and initial motivation to avoid them**

Coverage of the physical damage to the cochlea should be minimized in order to stress the resulting social and psychological handicaps which accompany hearing loss. We all value the activities which good hearing allows us to enjoy, so the educational program should emphasize our dependence on the sense of hearing in everyday life. Concentrate on the effects of hearing loss on interpersonal communication, enjoyment of formal group situations such as meetings and religious services, and participation in recreation activities. Cochlear hair cells themselves are abstract and unappreciated, but they are worth saving because of their value in terms of these everyday activities. Although the payback for protecting one's hearing is normally longterm rather than immediate, there are also some current benefits of wearing HPDs, so the educational program should mention the reduction of fatigue, headaches, and tension as side benefits. The noise hazard areas of the production facility should be identified, as well as off-the-job hearing hazards such as gunfire, chain saws, snowmobiles, etc.

(2) **Hearing protection devices**

In most industries HPDs are the primary method of protecting the worker from noise; therefore, the educational program should spend at least one third of its allotted time covering HPD utilization. Topics must include the purpose of HPDs, how they work, the necessity for a good fit to achieve protection, the ineffectiveness of substitute materials such as cotton or user-modified HPDs with ventilation holes, etc. The types and styles of HPDs available within the company for its various TWA levels should be reviewed, emphasizing the company's willingness to work with the employee to find a good personal choice. It is important that the typical reasons given for not wearing HPDs be addressed during the group educational session (Forrest, 1982; J.D. Royster & Royster, 1984).

The educator should keep in mind that the oldest employees are usually most resistive to wearing HPDs, with the youngest workers also showing more resistance than middle-aged employees (Chung, Gannon, Roberts, & Mason, 1982; Foster, 1983; L.H. Royster & Royster, 1985).

Even though fitting and use instructions should be covered on an individual basis at audiometric test time, review information is needed concerning the correct methods for HPD utilization and care. If the group is small enough, HPD fit should be checked by the educator and assistants.

Finally, the company's procedures for obtaining replacement HPDs should be outlined, stressing the need to return to the fitter if a different style or size is desired for better comfort or greater convenience, and showing examples of deteriorated or worn HPDs to demonstrate the need for regular replacements. In developing the specific content of the educational program concerning HPDs, the educator should review Chapter 10 of this manual.

(3) **Audiometric evaluations**

Employees' attitudes toward their annual audiometric tests vary, and the educational program should attempt to counteract workers' fears, prepare them for the test (Morrill, 1984), and present the audiometric evaluations as a benefit. Employees may fear audiometric testing and even refuse to be tested if they believe that their jobs, pay rate, or chance for advancement will be jeopardized if they show hearing loss. The educator must address these concerns by showing how the audiogram results are evaluated and how they are used — to trigger followup action to provide greater protection for employees with threshold shifts, but *not* to terminate or demote workers. Audiometric testing also serves as a health benefit to employees by detecting non-occupational hearing problems for which medical attention is needed. Although a detailed explanation of audiometric test frequencies and threshold determination methods is not necessary, workers do need an understanding of the normal hearing changes with age, how noise-induced hearing loss adds to age-effect changes, and how increasing degrees of loss and progression of loss to lower frequencies mean increased difficulty in communication and activity enjoyment. It is useful to display posters showing the audiogram format and examples of age-effect loss and noise-induced loss for employees to refer to as needed after their annual audiograms.

(4) **The company's HCP policies**

The five phases of the firm's HCP should be summarized, with an assurance that the company is meeting or exceeding its legal responsibilities to the workers through the scope of the program. A representative of management should be present to voice the company's commitment to an effective HCP and to outline the respective responsibilities of the company and the employees. Although positive consequences for wearing HPDs are needed (see motivation section), we feel that the absolute enforcement of proper HPD utilization is also necessary for an effective HCP. Therefore, the negative consequences of failure to wear HPDs must be presented, with management's assurance that the enforcement policies will be followed.

(5) **Questions and answers**

Employees should be encouraged to ask questions freely during the educational program, with a special portion set aside only for this purpose. It is strongly recommended that prior to conducting an educational session, the educator should study the most commonly asked questions and develop appropriate responses (Mellard, Thomas, & Miller, 1978; L.H. Royster & Royster, 1981; Berger, 1982a; Berger, 1982b; Berger, 1983). For questions which arise after the educational session is over, employees should be told to consult their supervisors for simple concerns, with supervisors referring larger problems to the key individual.

(6) **Final motivation**

The educational program should close on a positive note. In summary statements, the educator should re-emphasize the company's interest in protecting the workers' hearing and solicit their full participation and cooperation in the HCP so that they will avoid the handicapping results of noise-induced hearing loss. If the educator does not have a true belief in the value of the HCP, employees will quickly sense this insincerity and will discount the educational/motivational message. Therefore, it is important for the educator to retain enthusiasm and self-motivation to be effective in motivating employees. Varying the structure of the educational sessions and using visual aids will make the job of presenting the programs less tiring for the educator.

Motivational Concepts

A variety of motivational concepts should, of course, be used within the formal educational programs, but motivation and educational reinforcement are also an on-going process throughout the HCP's yearly cycle of audiometric testing, HPD reissuing, and informal contacts between the primary HCP personnel and the employees. In fact, all the policies and procedures followed in implementing the HCP should be established with motivational principles in mind. Most performance discrepancies (such as failure to wear HPDs properly) are not due to a lack of skill or knowledge, but rather to a lack of positive consequences for the desired behavior (Mager & Pipe, 1970; Feeney & Nyberg, 1976).

In an excellent short reference book, Mager and Pipe (1970) outline steps for analyzing performance problems to identify potential solutions. Assuming employees have the skills needed to perform, solutions involve removing any punishments or obstacles for the desired behavior and arranging rewarding consequences. In hearing conservation the ultimate goal of preventing hearing loss is a rather abstract and distant avoidance of a negative situation, not a positive event. The immediate steps toward the goal, such as wearing HPDs, may be annoying and are seldom rewarded; indeed failure to wear HPDs may be rewarding in itself due to greater comfort. The HCP should be structured to make wearing HPDs as easy and comfortable as possible and to provide frequent positive reinforcement for proper utilization, plus punishment for failure when all else fails (Cluff, 1980; L.H. Royster & Royster, 1982; Hager, Hoyle, & Hermann, 1982).

The following sections present a variety of motivational ideas which can be employed during HCP activities.

MOTIVATION FOR WEARERS OF HEARING PROTECTION

Annual Audiometric Findings

The contact between the audiometric technician and the employee at the time of the annual audiogram provides the single best opportunity to influence that individual regarding the HCP. As mentioned earlier, the worker's interest is aroused regarding the audiogram results due to concern for his/her general health, worries about hearing loss as a sign of aging, anxiety that a hearing impairment would affect job security or promotion opportunities, and fear that the presence of a hearing loss of which the employee is already aware would become known to fellow workers. For these and other reasons the employee will be receptive to constructive comments about his/her hearing status after the audiogram (L.H. Royster, Royster, & Berger, 1982; Hager, Hoyle, & Hermann, 1982).

In some firms two individuals work as a team, with one person doing the testing while another person obtains auditory history updates before the audiogram and gives feedback about the results. Although it is not necessary to give the worker a copy of the audiogram at this time, it is extremely beneficial for a visual comparison of past and current threshold levels to be shown to the employee. Because decibel values are generally unfamiliar to the work force, the best method of feedback is to roughly sketch the results for the left and right ears onto an audiogram chart. A wipe-clean overlay sheet may be placed over a large audiogram chart as a handy method to show workers their hearing changes, and the pictorial display makes the amount of hearing improvement or decline much more concrete to the employee than listed threshold values.

The content of the feedback information provided about hearing status will depend on company policy, the experience and knowledge of the technician, and the degree of change in light of the employee's medical and auditory history. If the current thresholds indicate stable or improved hearing, the employee should be commended and urged to continue to wear HPDs during noise exposures. If the worker makes comments about off-the-job situations in which he/she wears HPDs, the technician must be especially sure to praise him/her and to make a note about the comments for future use in educational/motivational activities for other employees. For example, one worker described how she wore HPDs when she accompanied her husband for his weekend target shooting, even though she herself did not shoot. Anecdotes such as this are the best source of effective program materials, and they are free for the taking if the HCP personnel will simply take time to observe and record them.

If the audiogram results show significantly declining thresholds, the technician should caution the employee about the changes, inquire about situations involving unprotected noise exposure on or off the job, check the adequacy of HPD fit and the employee's HPD placement skills (see next topic), and ask whether there are any problems with the current HPDs which discourage consistent use. If HPD difficulties are reported, different options should be tried to find the best protector for the individual. The employee may also express general concerns, such as there being no use in wearing HPDs since he/she already has a substantial hearing loss. The technician must emphasize the need to save the remaining hearing, expressing personal concern and the company's concern for the individual. Any fears about termination or demotion must be counteracted if the worker is cooperating in the HCP efforts. However, if an employee refuses to wear HPDs, the technician must emphasize that company policy requires termination or transfer to a non-noise job for safety/health infractions involving the HCP.

Individual Evaluation of HPD Adequacy

If an employee shows significant threshold shifts or expresses concern that his/her HPDs do not seem to give good protection, the audiometer can be used to give an approximate indication of the effectiveness of earplugs as worn. The difference in measured thresholds with and without the earplugs give an estimate of the amount of protection provided in the work environment (Harvey, 1981; Berger, 1983c). If the indicated attenuation is insufficient, it will be necessary to reinstruct the worker regarding proper insertion methods, change the size of the plug, or select an alternative style of HPD.

The process of determining HPD adequacy and showing the difference between a poorly fitted or incompletely inserted plug and a proper choice can be very motivating to employees because it demonstrates that HPDs can really make a difference. This demonstration may convince the dubious or inconsistent wearer that HPDs are worth the bother. On the other hand, some conscientious workers may worry that their HPDs are inadequate because of the misconception that they should not be able to hear speech or any other sounds through well-fitted HPDs, even though the educational program has covered this topic. In this case the effectiveness check will reassure them that their protectors are doing their job. The results of HPD effectiveness checks for individuals can be described in posters for bulletin boards, vu-graphs for group educational meetings, or newsletter articles to spread their impact to other workers in the HCP.

Examples of Permanent and Temporary Hearing Loss

Any typical industrial audiometric data base contains examples of permanent and temporary noise-induced hearing loss from on-the-job as well as off-the-job sources. These audiograms, plus personal comments from the affected individuals, are excellent motivational material because they relate to real-life experiences of fellow employees. Of course, the identity of the affected persons must be kept anonymous by removing all identifying information, and the permission of employees must be obtained before using the examples. Most workers are agreeable to sharing their experiences anonymously to help others, and some individuals are willing to identify themselves as victims of hearing loss and give brief motivational talks to groups of coworkers to encourage them to protect their own hearing. This type of personal testimony is the best endorsement the HCP can receive. For example, we visited a newspaper in which the press room foreman, a 30-year veteran who lost his hearing before the introduction of the HCP, personally educated new hires about the handicaps of hearing loss and spoke each year to the press crew as a group to remind them how his life was affected by his impairment, motivating them to wear their HPDs conscientiously.

When creating a poster or newsletter article from an individual's example of hearing loss, the writer should include the age and sex of the affected person, the probable noise source which caused the damage, and the effects of the loss on everyday activities. Such effects can be noted from comments individuals make to the audiometric technician, as shown by these examples:

> "I get angry at my daughters when they don't speak clearly."
>
> "I have a hard time understanding the words when the church choir sings or listening to the lyrics of a song on the radio in my car."
>
> "Sometimes I feel like people avoid talking with me because they have to repeat what they say so much."

Males frequently mention difficulty understanding female speakers, especially their daughters or grandchildren, because of the higher frequency of female voices. Discussion of these types of personal comments brings the effects of noise close to home, showing employees why their hearing matters to them enough to protect it.

Hearing aids should be discussed to counteract the impression that if a hearing impairment occurs, then an aid will make things good as new. Although modern hearing aids have improved tremendously, and many individuals with high-frequency loss can benefit more from amplification now than in the past, an aid does not improve poor hearing to the degree that eyeglasses correct poor vision. Therefore, no one can afford to be careless about HPDs. The best way to communicate the limitations of hearing aids is to present the experiences and opinions of several employees who wear them.

It is also useful to present examples of temporary threshold shifts (TTSs) from occupational noise or hobby exposures such as guns or chain saws to demonstrate that single noise exposures really do affect the ears temporarily, with permanent damage gradually accumulating. Examples of TTS are often observed during the annual audiometric test sessions, and these provide effective motivational materials, especially if the results of retesting are available to show improved thresholds after TTS recovery. Special TTS studies can be done to measure thresholds at the beginning and end of a work shift to convince workers that inadequate HPD utilization (poor fit and/or placement) will result in temporary loss (Zohar, 1980a; Zohar, Cohen, & Azar, 1980; L.H. Royster, Royster, & Cecich, 1984).

In promoting HPD utilization it is effective to tell workers about nonoccupational and even non-noise-related uses of HPDs, especially those with a twist of humor. One employee noted that he wore his earplugs more off the job than at work; asked why, he responded that they were the best thing he knew to shut out his mother-in-law. The spouses of habitual snorers also find HPDs indispensable. We have heard that closed-cell foam earplugs are good fishing tools if placed below a sinker on the line, so that the foam's buoyancy will gradually wiggle the bait up through the water. Knowing extra uses for HPDs may make employees take a fresh look at them, as well as adding interest to the educational program.

Annual ADBA Findings for Departments or Groups

It is difficult to motivate some employees on an individual basis because they do not like to be singled out and are more comfortable receiving information as part of a group. In addition, group feedback provides an extra incentive for the members to compete with other departments in HCP activities. One way to achieve group feedback is to post the findings of annual ADBA procedures (see Chapter 9) in simple bar graph formats. However, care must be exercised to present a range of acceptable results which all departments can achieve, rather than simply comparing raw numbers for different groups. Population differences in characteristics such as age and mean hearing level could affect the results regardless of the degree of protection being achieved.

Influence by Management and Peers

There is no more effective way to destroy the employee's interest in the HCP than for top managers, middle managers, primary HCP personnel, or front line supervisors to enter noise hazard areas without wearing HPDs (Else, 1981; Hager, Hoyle, & Hermann, 1982; Lofgreen, Holm, & Tengling, 1982). The reason these individuals usually give for their failure to observe HCP rules is that they will be in the area only a short time. Although they may not be overexposed on the basis of TWA, from the standpoint of showing support for the HCP *any* exposure is long enough to warrant wearing HPDs. The appearance of managers in the formal educational programs to advocate HCP policies will reinforce their support for the program, but their everyday participation will mean much more to employees. In fact, management should ask workers to ask *any* individual who enters a noise hazard area without protection to leave and obtain HPDs before returning (Sadler & Montgomery, 1982). We are familiar with successful HCPs in which employees would actually ask their supervisors and guests to put on HPDs or leave the area.

In addition to the formal company hierarchy, there is also an informal pecking order within work groups which can be tapped to the advantage of the HCP (Cluff, 1980; Carroll, 1982). If the lead man or woman within a group — the one who commands the respect of coworkers — can be motivated to set a good example by wearing HPDs properly, then others will follow. Such employees are especially appropriate to be asked to try out new types of HPDs which are under consideration for purchase; they can collect feedback from their work groups about the practicality and comfort of new styles, as described in the next subsection.

Employee Input into HCP Decisions

When employees help to select their own protective equipment they are more willing to wear it in spite of inconveniences (Else, 1982). For example, one safety director obtained special disposable respirators because employees had complained about the trouble involved in washing their current style. After a trial period with the new respirators, the employees decided that they preferred the old ones after all and were content to wear them because they had made the decision. Similarly, if an individual who complains of discomfort with HPDs is given individual attention and allowed to try each permissible option the company has to offer, then the worker will feel more satisfied even if a degree of discomfort remains (L.H. Royster & Royster, 1985).

Wise key individuals know that they can pick up good information, on an informal basis, about the practicality of various HPDs for their own work environments by spending a few minutes chatting with employees and foremen in the coffee break area. Employees are often able to offer potential solutions for problems they experience in HPD utilization.

On a more formal basis, it is good policy to introduce any changes in the HPD selection or other aspects of the HCP in safety meetings in order to ask workers for their feedback and suggestions about implementation of the change (Komaki, Heinzmann, & Lawson, 1980). For example, if operators are consulted before the construction of noise-controlling machinery enclosures, the enclosures can usually be designed to permit efficient machine operation as well as noise reduction. Some industrial managers frown upon this type of employee involvement, but it is a good way to foster worker support for HCP policies and procedures.

Examples of Worn and Abused HPDs

One of the more difficult daily challenges for the hearing conservationist is to motivate employees to replace HPDs which are worn out or deteriorated, and not to abuse their protectors by making creative modifications (Ohlin, 1981; L.H. Royster & Holder, 1982; Gasaway, 1984a; L.H. Royster

& Royster, 1985). Attitudes regarding HPD replacement vary dramatically from plant to plant; in some sites workers throw away reusable HPDs on a daily basis because they are unconcerned about company costs, while in other environments employees hang onto HPDs designed for short-term use for weeks or months because they are cost conscious. Individual employees may become attached to their HPDs, especially custom molded types or rubbery plugs which have shrunk to a very comfortable size, and keep them for years (examples we know of include 7-year-old custom molds and 8-year-old V-51Rs). Over-extended use of most HPDs normally reduces their effectiveness due to material deterioration. In addition to inevitable wear, intentional HPD abuse is a perennial problem, especially when HPD utilization is enforced without any display of sincere company interest in the HCP.

Obviously HPD wearers need training and motivation regarding HPD replacement and proper HPD care. The audiometric technician should always inspect the worker's HPDs at audiogram time to detect wear or abuse (Else, 1982; L.H. Royster & Royster, 1985). Deteriorated or modified HPDs should be saved so that they can be used as examples in educational programs. If possible, it is useful to do protection level checks of the old HPDs versus new ones to document the poorer effectiveness of the old HPDs, then make motivational posters from the results. Often abuse occurs when employees are attempting to reduce discomfort or relieve difficulties in communicating with coworkers or detecting machinery noises (Acton, 1977); in these cases selection of a more appropriate HPD for the individuals can increase their willingness to wear HPDs.

In addition to one-to-one counseling with identified offenders, departmental utilization compliance records should be compiled by making unannounced walk-through surveys, then posting the results. Replacement HPDs should be issued on the spot to any employee found wearing deteriorated HPDs (Guild, 1966). Periodically the HCP personnel should set up a replacement station at the plant entrance at shift change time to distribute new HPDs to those who need them. HCP procedures should also be designed to ensure that it is easy and convenient for employees to obtain new HPDs throughout the year by making them available at numerous locations within the plant.

Reward Programs for HPD Utilization

Behavioral positive reinforcement techniques have been reported as successful in increasing HPD acceptance. Zohar (1980a) summarized two studies in which token economies were established in industrial settings to reward HPD wearers. In each case utilization rose to 90% after the reward program began and remained at that level after token distribution was discontinued, indicating that the technique had fostered a new group norm. Others have proposed that once employees are induced to try HPDs through token economies, the extra-auditory benefits of regular wear will begin to motivate HPD utilization (Lofgreen, Holm, & Tengling, 1982).

Extra Hearing Screening

As a special health benefit to promote the HCP, it is easy to utilize the audiometric technicians to screen the hearing of non-noise-exposed plant personnel and especially the children and spouses of employees. Screenings can be held on weekends or during annual employee picnics or other parties. Usually out of 100 children screened, 4-5 will be identified with temporary or permanent hearing loss. Referral of these children to appropriate medical, audiological, and educational resources for needed treatment can result in a degree of appreciation for the company's HCP which can be achieved in no other way. Of course, it is recommended that such screening programs be implemented under the guidance of a local audiologist, otolaryngologist, or company physician.

MOTIVATION FOR PRIMARY HCP PERSONNEL

The audiometric technicians, safety/health personnel, and supervisors who implement the day-to-day procedures of the HCP carry the main burden of the program. The motivation of these personnel is critical to the HCP's success. If they do not perform their tasks consistently and with sincerity, then the program's credibility is destroyed.

Reinforcement by Upper Management

Management must be convinced to make HCP implementation an integral part of the job descriptions upon which the primary HCP personnel are evaluated during annual performance appraisals. On a more frequent basis, managers should inquire about the status of the HCP and any implementation difficulties being encountered, and should praise the primary HCP personnel for their efforts.

Annual ADBA Results

The findings from annual ADBA procedures can provide a concrete reward to HCP personnel. If the HCP is effective, mean hearing levels will improve for several years due to the learning effect. If results differ by department, the personnel will see where they have succeeded and where they need to concentrate additional effort. Knowledge of the results of their efforts will help these personnel keep going.

Group Communication

Primary HCP personnel should be assembled for progress updates often enough to ensure that there is good communication and cooperation among the individuals who carry out various phases of the program (L.H. Royster, Royster, & Berger, 1982). If the key individual facilitates a process of expressing appreciation for tasks well done and offering helpful suggestions for areas which need improvement, the personnel will see why their performance matters and will develop a team spirit of cooperation.

MOTIVATION FOR MANAGEMENT

Managers do not need to know many details about the operation of the HCP, but they do need summary information about the status of the program and its cost-effectiveness for the company.

Annual ADBA Results

Presentation of summary graphs illustrating ADBA findings for various plants and departments will help the key individual obtain continued or increased resources and support for the HCP. Demonstration that the results are actually poorer for the section in which the supervisor is less committed to the HCP will stimulate the responsible manager to re-emphasize the importance of HCP duties to that subordinate. Suppose, for example, that HPDs with greater attenuation are needed in a certain area, but they are more expensive. Differential ADBA results for that area can generate approval of the extra budget allocation to offer the needed protectors. If the ADBA results look good, then management can be encouraged to praise the primary personnel responsible, and the key individual can thank the general managers and production manager for their part in supporting the successful program.

Potential Costs for Compensation for Hearing Loss

The one topic that always gets managers' attention is the company's liability for workmen's compensation claims. Although occupational hearing loss has accounted for only a fraction of one percent of the compensation industry currently pays (Berger, 1985), the expense could be much greater if more eligible workers filed claims. Establishment of an effective HCP will result in declining potential compensation costs for several years due to the learning effect (L.H. Royster & Royster, 1984b) and to reduced long-term costs as hearing loss in younger workers is prevented. The immediate learning-related decline in potential costs can be several orders of magnitude greater than the annual cost of administering the HCP, demonstrating to management the concrete benefit of the program.

Safer and More Acceptable Working Environment

Increased safety and job satisfaction mean increased profit to the company, and an effective HCP can help accomplish these ends (Staples, 1981; Lofgreen, 1982). The available data which attempt to relate injury rates with the implementation of an effective HCP suggest that workers who are adequately protected from noise will exhibit significantly fewer injuries than workers who are not properly protected or who have not been sufficiently motivated to wear the available HPDs (Cohen, 1976; Komaki, Heinzmann, & Lawson, 1980; Schmidt, Royster, & Pearson, 1982). This relationship may well be an expression that employees who have been motivated to wear HPDs are also more likely to be safety-conscious in other ways. Regardless of the underlying reason for this observation, the relationship between effective HCPs and reduced injury rates can be used to generate managerial support for the HCP. Injury data are typically compiled in most companies, so it is usually not a major effort for the key individual to correlate ADBA findings with department or plant injury rates.

HPD Utilization and Effectiveness Studies

Key individuals often ask us how to convince management to make decisions related to HPD utilization, such as allowing use of a more comfortable, lightweight earmuff which has a lower NRR (Noise Reduction Rating) than the heavyweight clunker approved by the outside consultant, removing from inventory a HPD which seldom gives an acceptable fit, or strengthening lax HPD enforcement. If the audiometric data are computerized, ADBA procedures can provide the data needed to accomplish these purposes. However, if ADBA is not a feasible option, other approaches can be used.

To motivate management to improve HPD enforcement, first try to determine the reason for laxity. If upper managers simply do not believe that enforcement is poor, HPD utilization surveys can provide the needed evidence that a problem exists. If management allows laxity because of fear of confronting employees about HPD utilization, then ADBA results would probably be needed to demonstrate to management the real costs they are incurring through lack of enforcement.

We have observed that managers typically place unwarranted emphasis on small differences in NRRs between protectors when deciding which types will be allowed, when actually the real-world protection offered by different types varies less than the NRRs would suggest (Berger, 1983b). Regardless of the NRR, it is the employee's willingness to wear the HPD consistently during noise exposures that determines the actual protection received. Therefore, management may actually achieve greater real-world protection by offering a lower NRR device which is more comfortable. Temporary threshold shift studies of groups of employees wearing the two HPDs can provide the needed data (L.H. Royster, Royster, & Cecich, 1984). If the mean TTSs for the wearer groups do not differ, then the more comfortable HPD is just as effective as the higher NRR device.

By evaluating the problem situation, the key individual can usually decide what type of data it would be useful to collect in order to demonstrate the relevant need to management.

Audiometric Findings for Managers

One indirect motivational tool for influencing management is to include them in the HCP by administering annual audiograms and providing HPDs for hobby use as well as plant visits. Typically a significant percentage of managers will exhibit mild to severe high-frequency hearing loss, often from military noise exposure or other nonoccupational exposures. Counseling from the key individual about how wearing HPDs could have prevented such a loss will help managers see the benefit of the program. Unfortunately, some managers may be reluctant to take part in audiometric testing because of embarrassment about hearing loss. However, if top level management participates in the program, their example will encourage others to follow.

REGULATIONS

There is no doubt that the dominant force behind the implementation of HCPs in US industry is federal and state safety and health regulations. Workmen's compensation regulations also influence management, but not nearly to the same degree as the OSHA regulations. Should regulations be used to motivate management to act or employees to participate more fully in the company's HCP? The answer is both no and yes.

No, federal regulations should not be used as the main source of motivation to get employees to wear hearing protection or to force management to establish a HCP or modify HCP policies. This approach would probably result in failure on both fronts. The employee will not use HPDs because government agencies think he/she should. Management will develop a HCP on paper to comply with the regulations but will probably not be willing to allocate the necessary financial or administrative commitment necessary to establish an effective HCP.

Yes, both employees and management need to be familiar with relevant federal and state regulations that relate to workplace safety and health, including the minimum requirements for compliance. Knowledge of each party about the responsibilities placed on workers and on management will provide a degree of motivation for both. However, the main emphasis should be on the background information that resulted in the creation of the regulation: the effects of noise and what is necessary for adequate protection of the employee.

Although management needs to understand details of the regulation to evaluate the HCP for compliance with it, the educator must make certain not to convey the idea that the primary reason for HCP implementation is meeting regulatory requirements. When this happens, the educator has failed to motivate management properly.

Putting it all Together

At the beginning of this chapter we referred to the multiplicity of types of work environments across the USA where workers and management are involved in industrial HCPs to protect the hearing of noise-exposed populations. As a consequence of this diversity of conditions we do not feel that it is possible to promote a fixed educational program format for workers or management that would adequately satisfy the needs of general industry. Likewise we do not feel that any single motivational strategy exists that would be appropriate for all situations.

Because of this variety of needs in industrial HCPs with respect to education and motivation we have attempted to cover these two subjects broadly, leaving the formulation of the particular educational and motivational activities for a plant facility up to the responsible key individual.

Our goal in writing this chapter was to provide the educator with general guidelines and ideas for developing ongoing programs to educate employees and managers and motivate them to be interested in and involved in the HCP. However, in the final analysis the level of success achieved by you, the educator, will be determined by factors beyond our range of influence — mainly by your commitment to hearing conservation and the health and safety of the work force for which you are responsible.

References

Acton, W.I. (1977). "Problems Associated with the Use of Hearing Protection," *Ann. Occup. Hyg. 20,* 387-395.

Barker, D.M., Driscoll, D.P., and Florin, Janice D. (1982). "Programming Hearing Conservation for a Large Corporation," *National Safety News.* November, 54-57.

Berger, E.H. (1981a). "EARLog #6 — Extra-Auditory Benefits of a Hearing Conservation Program," *Occup. Health Saf. 50(4),* 28-29.

Berger, E.H. (1981b). "EARLog #7 — Motivating Employees to Wear Hearing Protection Devices," *Sound and Vibration 15(6),* 10-11.

Berger, E.H. (1982a). "EARLog #8 — Responses to Questions and Complaints Regarding Hearing and Hearing Protection (Part I)," *Occup. Health Saf. 51(1),* 28-29.

Berger, E.H. (1982b). "EARLog #9 — Responses to Questions and Complaints Regarding Hearing and Hearing Protection (Part II)," *J. Occup. Med. 24(9),* 646-647.

Berger, E.H. (1983a). "EARLog #10 — Responses to Questions and Complaints Regarding Hearing and Hearing Protection (Part III)," *Prof. Saf. 28(3),* 14-15.

Berger, E.H. (1983b). "Using the NRR to Estimate the Real World Performance of Hearing Protectors," *Sound and Vibration 17(1),* 12-18.

Berger, E.H. (1983c). "Assessment of the Performance of Hearing Protectors for Hearing Conservation Purposes," *J. Acoust. Soc. Am. Suppl. 1, 74,* S94.

Berger, E.H. (1985). "EARLog #15 — Compensation for a Noise Induced Hearing Loss", E-A-R Div. of Cabot Corp., Indianapolis, IN.

Caroll, B.J. (1982). "An Effective Safety Program Without Top Management Support," *Prof. Saf. 27(7),* 20-24.

Carroll, C., Crolley, N., and Holder, S.R. (1980). "A Panel Discussion of Observed Problems Associated with the Wearing of Hearing Protection Devices by Employees in Industrial Environments," in *Proceedings of a Special Session on the Evaluation and Utilization of Hearing Protection Devices (HPDs) in Industry,* edited by L.H. Royster, D.H. Hill Library, North Carolina State Univ., Raleigh, NC.

Chung, D.Y., Gannon, R.P., Roberts, M.E., and Mason, K. (1982). "Hearing Conservation Based on Hearing Protectors: A Provincial Project," in *Personal Hearing Protection in Industry,* edited by P.W. Alberti, Raven Press, New York, NY 559-568.

Cluff, G.L. (1980). "Limitations of Ear Protection for Hearing Conservation Programs," *Sound and Vibration 14(9),* 19-20.

Cohen, A. (1976). "The Influence of a Company Hearing Conservation Program on Extra-Auditory Problems in Workers," *J. Saf. Res. 8(4),* 148-162.

Else, D. (1981). "Hearing Protection: Who Needs Training?," *Occup. Health 33(9),* 451-453.

Else, D. (1982). "Hearing Protection Programme Establishment," in *Personal Hearing Protection In Industry,* edited by P.W. Alberti, Raven Press, New York, NY 471-484.

Esler, A. (1978). "Attitude Change in an Industrial Hearing Conservation Program: Comparative Effects of Directives, Educational Presentations and Individual Explanations as Persuasive Communications," *Occup. Health Nursing 26(12),* 15-20.

Feeney, R.H., and Nyberg, J.P. (1976). "Selling the Worker on Hearing Conservation," *Occup. Health Saf. 435(4),* 56-59.

Forrest, M.R. (1982). "Protecting Hearing in a Military Environment," Scand. Audiol., Suppl. 16, *Proceedings of the Oslo International Symposium on Effects of Noise on Hearing,* edited by H.M. Borchgrevink, Norway 7-12.

Foster, A. (1983). "Hearing Protection and the Role of Health Education," *Occup. Health 35(4),* 155-158.

Gasaway, D.C. (1984a). "'Sabotage' Can Wreck Hearing Conservation Programs," *Natl. Saf. News 129(5),* 56-63.

Gasaway, D.C. (1984b). "Motivating Employees to Comply with Hearing Conservation Policy," *Occup. Health Saf. 53(6),* 62-67.

Guild, E. (1966). "Personal Protection," in *Industrial Noise Manual, Second Edition,* Am. Ind. Hyg. Assoc., 84-109.

Hager, W.L., Hoyle, E.R., and Hermann, E.R. (1982). "Efficacy of Enforcement in an Industrial Hearing Conservation Program," *Am. Ind. Hyg. Assoc. J. 43(6),* 455-465.

Harford, E.R. (1978). "Industrial Audiology," in *Noise and Audiology,* edited by D.M. Lipscomb, University Park Press, Baltimore, MD 299-327.

Harris, D.A. (1980). "Combating Hearing Loss Through Worker Motivation," *Occup. Health and Saf. 49(3),* 38-40.

Harvey, D.G. (1981). "A Method to Increase the Effectiveness of Ear Protection," *Sound and Vibration 15(5),* 24-27.

Karmy, S.J. and Martin, A.M. (1982). "Employee Attitudes Towards Hearing Protection as Affected by Serial Audiometry," in *Personal Hearing Protection in Industry,* edited by P.W. Alberti, Raven Press, New York, NY 491-509.

Komaki, J., Heinzmann, A.T., and Lawson, L. (1980). "Effect of Training and Feedback: Component Analysis of a Behavioral Safety Program," *J. Appl. Psycol. 65,* 261-270.

Lofgreen, H. (1982). "The Human and Economic Benefits of Hearing Protection in the Plant," *Canadian Occup. Saf., 20(6),* 2-3, 9.

Lofgreen, H., Holm, M., and Tengling, R. (1982). "How to Motivate People in the Use of Their Hearing Protectors," in *Personal Hearing Protection in Industry,* edited by P.W. Alberti, Raven Press, New York, NY, 485-490.

Lutz, G.A., Decatur, R.A., and Thompson, R.L. (1973). "Psychological Factors Related to the Voluntary Use of Hearing Protection in Hazardous Noise Environments." U.S. Army Med. Res. Laboratory Report No. 1,006. Fort Knox, KY (NTIS-AD-777520).

Maas, R. (1970). "The Challenge of Hearing Protection," *Indus. Med. 39(3),* 29-33.

Maas, R. (1971). "Compliance with OSHA on Hearing Conservation," *J. Environ. Control and Saf. Manage.,* December, 11-13.

Maas, R.B. (1972). "Industrial Noise and Hearing Conservation," in *Handbook of Clinical Audiology,* edited by J. Katz, Williams and Wilkins Co., Baltimore, MD 772-818.

Mager, R.F., and Pipe, P. (1970). "Analyzing Performance Problems or 'You Really Oughta Wanna'," Fearon Publishers, Belmont, CA.

Mellard, T.J., Doyle, T.J., and Miller, M.H. (1978). "Employee Education — The Key to Effective Hearing Conservation," *Sound and Vibration 12(1),* 24-29.

Morrill, J.C. (1984). "Instructions and Techniques for Reliable Audiometric Testing," *Occup. Health Saf. 53(8),* 64-68.

Ohlin, D. (1981). "User Training and Problems," in *Proceedings of Noise-Con 81,* edited by L.H. Royster, N.D. Stewart, and F.D. Hart, Noise Control Foundation, New York, NY 131-136.

Royster, J.D., and Royster, L.H. (1984). "Hearing Protection Practices, Problems and Solutions," *ASHA 26,* 77.

Royster, L.H., and Holder, S.R. (1982). "Personal Hearing Protection: Problems Associated with the Hearing Protection Phase of the Hearing Conservation Program," in *Personal Hearing Protection in Industry,* edited by P.W. Alberti, Raven Press, New York, NY 447-470.

Royster, L.H., and Royster, J.D. (1981). "Education Programs for Management and the New OSHA-US Noise Regulations (Educational Requirements)," in *Proceedings of a Special Session on Hearing Conservation Programs — The Educational Phase,* edited by L.H. Royster, D.H. Hill Library, North Carolina State Univ., Raleigh, NC.

Royster, L.H., and Royster, J.D. (1984a). "Hearing Protection Utilization Survey Results Across the USA." *J. Acoust. Soc. Am. 26, Suppl. 1,* S 43.

Royster, L.H., and Royster, J.D. (1984b). "Making the Most Out of the Audiometric Data Base," *Sound and Vibration 18(5),* 18-24.

Royster, L.H., and Royster, J.D. (1985). "Hearing Protection Devices," in *Hearing Conservation in Industry,* edited by Alan S. Feldman and Charles T. Grimes, Williams and Wilkins, Baltimore, MD.

Royster, L.H., Royster, J.D., and Berger, E.H. (1982). "Guidelines for Developing an Effective Hearing Conservation Program," *Sound and Vibration 16(1),* 22-25.

Royster, L.H., Royster, J.D., and Cecich, T.F. (1984). "An Evaluation of the Effectiveness of Three Hearing Protection Devices at an Industrial Facility with a TWA of 107 dB." *J. Acoust. Soc. Am. 76(2),* 485-497.

Sadler, O.W. and Montgomery, G.M. (1982). "The Application of Positive Practice Overcorrection to the Use of Hearing Protection," *Am. Ind. Hyg. Assoc. J. 43(6),* 451-454.

Schmidt, J.W., Royster, L.H., and Pearson, R.G. (1982). "Impact of an Industrial Hearing Conservation Program on Occupational Injuries," *Sound and Vibration 16(1),* 16-20.

Spindler, D.E., Olson, R.D., Fishbeck, W.A. (1979). "An Effective Hearing Conservation Program," *Am. Ind. Hyg. Assoc. J. 40(7),* 604-608.

Staples, N. (1981). "Hearing Conservation — Is Management Short Changing Those at Risk?," *Noise and Vib. Control Worldwide 12(6),* 236-238.

Stapleton, L., and Royster, L.H. (1981a). "Educational Programs for Hearing Conservation," in *Noise-Con 81 Proceedings,* edited by L.H. Royster, N.D. Stewart and F.D. Hart, Noise Control Foundation, New York, NY 153-156.

Stapleton, L., and Royster, L.H. (1981b). "Educational Programs for Hearing Conservation," *Proceedings of a Special Session on Hearing Conservation Programs — The Educational Phase,* "edited by L.H. Royster, D.H. Hill Library, North Carolina State Univ., Raleigh, NC.

Stapleton, L., and Royster, L.H. (1985). "The Education Phase of the Hearing Conservation Program," *Sound and Vibration 19(2),* 29-31.

Wilkins, P.A., and Acton, W.I. (1982). "Noise and Accidents — A Review," *Ann. Occup. Hyg. 25(3),* 249-260.

Wyman, C.W. (1969). "Industrial Hearing Conservation, Administration and Human Relations Aspects," *Natl. Saf. News 99(5),* 65-70.

Zohar, D. (1980a). "Promoting the Use of Personal Protective Equipment by Behavior Modification Techniques," *J. Saf. Res. 12(2),* 78-85.

Zohar, D. (1980b). "Safety Climate in Industrial Organizations: Theoretical and Applied Implications," *J. Appl. Psychol. 65(1),* 96-102.

Zohar, D., Cohen, A., and Azar, N. (1980c). "Promoting Increased Use of Ear Protectors in Noise Through Information Feedback," *Human Factors 22(1),* 69-79.

Noise and Hearing Conservation Manual, edited by
E.H. Berger, W.D. Ward, J.C. Morrill and L.H. Royster

12 Engineering Controls

Robert D. Bruce
Edwin H. Toothman

Contents

Page

Engineering Control 418
Plant Design 422
Replacement 430
Quieter Equipment 430
Axial *vs.* Centrifugal Fans 430
Fan Speed 430
High Speed Drive Gears *vs.* Turbines 430
Belt Drives *vs.* Gear Drives 431
Pneumatic *vs.* Mechanical Ejectors of Parts 432
Hopper Vibrators 432
Pneumatic *vs.* Electric Tools 433
Hydraulic Press *vs.* Riveting Hammer 434
Impact Tools 435
Quieter Processes 441
Quieter Materials 443
Source Modification 444
Reduce Driving Force 444
Reduce Speed 444
Maintain Dynamic Balance 445
Increase Duration of Work Cycle 446
Provide Vibration Isolation 446
Vibration 447
Natural Frequency 447
Forcing Frequency 450
Transmissibility 450
Transmissibility Curve 451
Vibration Isolators 453
Selection of Isolators 454
Reduce Response of Vibrating Surface 458
Damping 458
Reduce Area of Vibrating Surface 467
Use Directivity of Source 467

Reduce Velocity of Fluid Flow 468
Air Ejection Systems 468
Valves and Vents 471
Mufflers and Wrapping Treatments 475
Path Modification 477
Enclosures 477
Transmission Loss *vs.* Noise Reduction 478
Calculation of Required TL 480
Unlined Enclosure 483
Lined Enclosure 485
Enclosure of a Working Environment (Within a Noisy Area) 486
Simplified Design for Enclosures 494
Multiple Wall Enclosures 497
Partial Enclosures 497
Shields or Barriers 500
Room Absorption 502
Lined Ducts and Mufflers 508
Summary 520
Acknowledgments 520
References 520

Engineering Control

The engineering control of industrial noise requires some fundamental knowledge of acoustics and the development of a solution to a particular noise problem also demands a high degree of ingenuity and determination on the part of the person responsible for noise control. This individual must be able to gain the cooperation and participation of the plant operating staff and others who understand the plant's operational requirements.

The general acoustical principles of noise control are well established. However, it is not always possible to predict accurately the results of noise reduction efforts. This is due at least in part to the complex character of industrial noise sources, the different environments encountered, and the limits imposed by the operational and maintenance requirements of the equipment. However, these uncertainties should not discourage the industrial hygienist, because noise problems have been solved in many industrial plants by individuals with only a limited noise control background. Case histories of some of these problems are included in this chapter. Although apparently similar noise problems vary considerably in the amount of noise reduction required, these case histories can serve as a starting point for noise control solutions. Since there are often several different possible solutions to a noise problem, the specific noise reduction techniques finally employed should be designed to fit each individual case.

The first step toward solving any noise problem is to define it. This necessitates measuring the noise level and securing complete information on employee exposure time in the noisy environment and in any other environment to which the employee might be exposed during the working day. If the noise to which an employee is exposed varies considerably during the day, it will be necessary to measure the noise levels and the length of time the employee is exposed to the various noise levels. Then, the daily noise dose or the time-weighted average sound level (TWA) can be computed. Alternatively, it is possible to evaluate the exposure by means of a dosimeter as discussed in Chapter 4. In either case, the exposure must be compared to the specific requirements of the current noise regulation to determine whether or not the exposure is in compliance. If possible, the noise exposure should be reduced to a TWA below 85 dBA. This will better protect the employee against hearing impairment and will usually eliminate the need to place the employee in a hearing conservation program.

After determining that reduction of the employee's noise exposure should be investigated, the industrial hygienist must then consider various measures for controlling exposure:

> engineering controls
> administrative controls
> hearing protection devices

The preferred method of noise control is engineering control, *i.e.,* noise reduction at the source. It may eliminate the requirements of hearing protection, audiometric testing, and/or limitation of exposure time. It may also improve speech communication and reduce annoyance. However, noise reduction is not always feasible, in which case, the other control measures must be employed. If engineering controls are to be considered, a more detailed noise analysis will be required in order to determine the most efficient and economical method of reduction.

After engineering controls have been installed, the noise reduction that was obtained should be quantified. This evaluation may be necessary to show compliance and may be useful in solving future noise problems.

When more than one noise source is involved in an area, it is essential to reduce the noisiest source(s) if effective reduction is to be achieved. For example: Three noise sources of 90 dB, 95 dB and 101 dB combine to create a noise level of 102 dB. If the 90-dB source is completely removed, the sum of the 95- and 101-dB sources is still 102 dB. If the 90- and 95-dB sources are both removed, the 101-dB source still remains. On the other hand, if the 101-dB source is removed, the sum of the 90- and 95-dB sources is only 96 dB.

These calculations hold true as long as the noise level measurements are of the same type: for example, A-weighted, overall, octave-band, or narrow-band noise measurements. Chapter 2 presents details of how to add or subtract noise levels.

Sound from a single source can travel by multiple paths to the listener. It is important to recognize all of the alternative paths. Figure 12.1 presents examples of how the sound is radiated from five different sources. For instance, a sound source inside an enclosure can radiate sound as follows:

(1) direct radiation of sound through openings in the enclosures;
(2) sound radiation from the enclosure due to structure-borne vibration from the source; and
(3) indirect radiation from the enclosure, that is, airborne sound from the source to the inside of the enclosure and subsequent re-radiation from the outside of the enclosure.

The problem is to determine which paths carry the most sound energy and then to select appropriate methods of obtaining the desired reduction along each path.

The introduction of engineering noise controls into the industrial environment is a challenge for the industrial hygienist and plant employees. Since many employees have not worked with engineering controls, they may not appreciate their importance. Thus it is imperative that engineering noise controls be carefully designed and installed. In situations where engineering controls such as enclosures are being designed for machines, equipment, and processes, the operating, maintenance, and service personnel should have the opportunity to discuss the noise problem and the proposed controls. This involvement will reduce the employees' objections to the engineering controls. These employees, from their experience, will be able to provide valuable guidance on the proper location of access doors for operation and maintenance as well as information on acceptable locations for lubrication points and control switches. They can also provide insight into the design of the controls to ensure safe and efficient operation of the machines, equipment, and processes. Finally, the plant employees must be trained in the use and care of these controls in order to minimize the rate of deterioration and maximize the obtainable benefits.

In the design and installation of engineering noise controls, ergonomics (the application of human biological science to engineering) must be used to achieve optimum work efficiency. For example, work posture (sitting, standing, bending) as well as existing environmental factors (lighting, heating and cooling) must be taken into consideration. This is especially

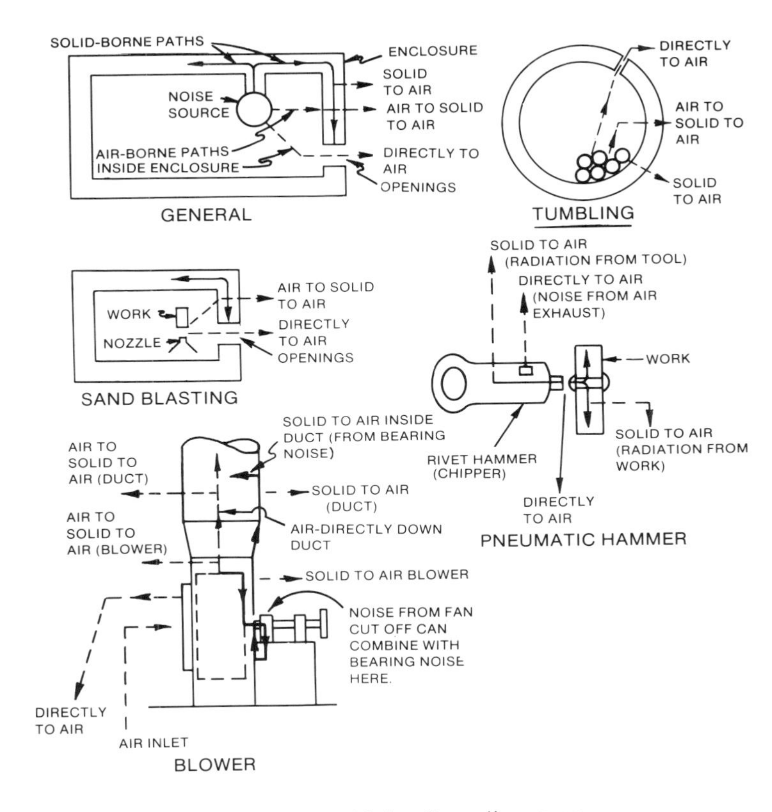

Figure 12.1 — Noise flow diagrams.

true where employee enclosures are involved. Lighting, heating and cooling must be sufficient to prevent reduced efficiency and, possibly, a deterioration in work quality. Enclosures should be adequate in size and contain sufficient window area to prevent a feeling of claustrophobia. The windows should be tilted to prevent glare and situated in the best location to enhance proper usage by the employee(s) requiring protection from noise.

When a noise problem is being investigated, it is important to consider as many potential solutions as practicable. One way to ensure this is to systematically consider all possibilities. Basic engineering control approaches to a noise problem include initial purchase of quiet machinery, replacement of noisy equipment, source modification and path modification. The outline shown below can be used in making such an analysis.

PLANT DESIGN
> noise specifications

REPLACEMENT
> quieter equipment
> quieter processes
> quieter materials

SOURCE MODIFICATION
> Reduce driving force on vibrating surface
> reduce speed
> maintain dynamic balance
> increase the duration of the work cycle
> provide vibration isolation
>Reduce response of vibrating surface
> add damping
> increase stiffness
> increase mass
> shift resonance frequencies
>Reduce area of vibrating surface
> reduce over-all dimensions
> perforate surface
>Use directivity of source
>Reduce velocity of fluid flow
>Reduce turbulence

PATH MODIFICATION
> enclosures
> shields or barriers
> room absorption
> mufflers

Examples of these control measures will be illustrated in this chapter.

Plant Design

One of the greatest opportunities for the industrial hygienist in the field of noise control is to guide the design of new plants and the modernization of existing ones. In this manner, many potential noise problems can be solved in the design stage. Successful planning for noise control involves: (a) knowledge of the noise characteristics of each machine and process and the proposed location of noise sources, operators, and maintenance personnel, (b) prediction of employee work schedules, and (c) selection of design goals based on appropriate exposure limits.

Engineering specifications for design and selection of equipment should incorporate either a limit on the acceptable noise or a requirement for the vendor to provide noise performance data. In most cases, the machinery manufacturer is in the best position to reduce the noise of the machine at its source with built-in designs. It is expected that many such designs will not significantly increase the cost of the machine. The noise specification presented in Figure 12.2 can serve as a guide for either specifying maximum acceptable noise levels or for requiring the vendors to submit data.

The vendor may provide sound power level (L_w) data or sound pressure level (L_p) data. In either case, the industrial hygienist will need to calculate L_p at the operator positions in the plant. Most frequently, vendors provide L_p data. The following equation can assist in this analysis:

$$L_{p_2} = L_{p_1} - 10 \log\left(\frac{Q_1}{4\pi r_1^2} + \frac{4}{S_1\bar{\alpha}_1}\right) + 10 \log\left(\frac{Q_2}{4\pi r_2^2} + \frac{4}{S_2\bar{\alpha}_2}\right) \quad \textbf{(12.1)}$$

where L_{p_2} is the L_p in the plant
L_{p_1} is the L_p supplied by the vendor
Q_1 is the directivity of the source in the vendor's setup
r_1 is the distance in feet from the machinery for L_{p_1}
$S_1\bar{\alpha}_1$ is the total absorption in the vendor room for measurements (see Equation 12.9).
Q_2 is the directivity of the source in the purchaser's facility
r_2 is the distance in feet from the machinery to one of the operator's positions
$S_2\bar{\alpha}_2$ is the total absorption in the space where the machinery is to be located

Alternatively, these calculations can be performed graphically using Figure 2.7. This graph shows the relation between L_p and L_w. It also shows how this relationship is affected by distance from the source (r), directivity factor (Q), and room constant (R; see Equation 12.10).

L_w is the total sound power level of the noise source and is independent of distance or environment. By definition, $L_w = 10 \log(W/W_o)$, where $W_o = 10^{-12}$ watt and W is the acoustical power of the source. L_p is the sound pressure level at some distance from the source. In order to determine L_p from L_w, one must state the environmental conditions considered in Figure 2.7 (r, R, and Q). The room constant (R) and the directivity factor (Q) are discussed in Chapter 2.

Example 12.1:

Given: The A-weighted L_w of a vendor's machine is 100 dBA. The operator will work 10′ from this machine. The machine will be installed on the floor in the center of a large room which has a room constant of 1000 sq ft.

Figure 12.2 — Equipment Noise Specification.

1. GENERAL

1.1 This Specification is a means of establishing the limiting value of noise generated by equipment to be installed in an industrial plant. It also provides a uniform method of conducting and recording noise tests to be made on such machinery.

1.2 This Specification describes limits and methods of measuring sound emission in the purchase of equipment. Tests are to be made by the vendor and witnessed by the purchaser unless otherwise specified. Confirming or additional measurements by the purchaser shall be permissible.

2. INSTRUMENTS

2.1 A type 1 or 2 Sound Level Meter, as specified in ANSI S1.4--1983, when used alone, measures overall noise levels only.

2.2 An Octave Band Analyzer, as specified in ANSI S1.11--1966 (R1971), is the preferred instrument for measuring broad-band noise by this Specification and is used in conjunction with a sound level meter.

2.3 Instruments shall be calibrated as recommended by the instrument manufacturer. Overall calibration of the instruments, including the microphone and internal calibration of the meters, shall be made before and after the test of each piece of equipment.

3. NOISE TESTS

3.1 Ordinarily, the test will be made at the factory or in a test room provided by the vendor at his expense. The test room should provide conditions free of extraneous sounds that would interfere with the noise tests.

3.2 Ambient sound levels within the test room should be 10 dB or more below the sound level that prevails when the tested equipment is in operation. If the ambient levels are not at least 10 dB below the equipment and sound levels, corrections to the data are required as presented in 5.1.

3.3 Unless otherwise specified, equipment tested should be at full load and at anticipated load(s). Loading devices may be provided or specified by the purchaser.

3.4 The placement of the microphone during the test should be such as to protect it from air currents, vibration, electric or magnetic fields, and other disturbing influences that might affect the readings obtained. Positioning of the microphones at ear level and a horizontal distance of 3 feet from the nearest major surface is usually satisfactory. The entire area surrounding the equipment should be explored to ensure that the maximum noise levels are measured.

3.5 Measurements shall be made at a minimum of 6 points approximately 60 degrees apart in the plane specified in paragraph 3.4 and at any operator position(s). Start at the position of maximum noise level. No measurement position shall be located further than 10 feet from adjacent positions. Additional readings will be specified when a directivity pattern is to be established. When multiple, similar units are to be purchased, tests on more than one unit might be requested.

3.6 If desired, noise measuring methods specified in national, international, or trade association standards can be utilized. See ASA Standards Index (1985) for a comprehensive listing.

Figure 12.2 (continued)

4. RECORDS

4.1 Records of tests for each piece of equipment shall include the information and readings called for in this specification.

4.2 Test results are to be reported to the purchaser for analysis and acceptance before equipment is shipped, unless otherwise specified.

5. SOUND LEVEL SPECIFICATIONS

5.1 The location and orientation of the microphone for measurements of total (ambient plus machinery) and ambient noise levels shall be identical. If either the machine or ambient noise levels fluctuate appreciably, maximum levels shall be recorded. Sound levels shall be measured over a long enough time to account for all fluctuations. If the difference between total and ambient readings is less than 5 dB, the ambient level is unsatisfactory for measuring the noise produced by the machine. If the difference is 10 dB or more, the higher readings are essentially the noise levels generated by the machine. For differences of 5 to 10 dB, the machinery noise levels shall be determined by applying the correction values indicated below.

Correction Values Allowed for High Ambient Sound Levels

Difference between Total and Ambient Noise Levels, dB	Correction to be Subtracted from Total Sound Level, dB
3	3
4-5	2
6-9	1
10	0

5.2 For the purpose of this specification, it is assumed that narrow-band or pure-tone noise (or both) exist when the noise level of an octave band is at least 6 dB above the noise level in adjacent octave bands. If narrow-band noise as specified here does exist, the noise is more objectionable than broad-band noise upon which criteria are based. Therefore, it should be accounted for by assuming a 5-dB lower criterion for that band.

5.3 Purchase orders should specify maximum acceptable octave band sound power levels re 10^{-12} watt or octave band or A-weighted sound pressure levels re 20 μPa. If sound pressure levels are specified, environmental conditions must also be specified, as discussed in this chapter. Maximum sound pressure levels for both slow and peak (impulse) meter responses should be specified.

6. SPECIAL REQUIREMENTS

6.1 Equipment or locations which create special noise problems not covered by these specifications (such as neighborhood noise) require special consideration. Supplementary specifications and additional descriptions will be necessary.

Figure 12.2 (continued)

EQUIPMENT NOISE SPECIFICATION

Type of Equipment ______

Manufacturer ______ Vendor ______

Vendor's Order No. ______ Serial ______ Shop ______

Purchaser's Project No. ______ Machine No. ______ Order ______

Equipment Specifications: ______ Model No. ______ Serial No. ______

Size ______ Capacity ______

Speed ______ Horsepower ______

Machine Load—% Capacity ______

Test Room—Dimensions Length ______ Width ______ Height ______

Material Floor ______ Wall ______ Ceiling ______

Noise Description Continuous Intermittent-Impact

Does Narrow Band Noise Exist Yes ______ No ______

Octave Band Analyzer Make ______ Model ______ Serial No. ______

Sound Level Meter Make ______ Model ______ Serial No. ______

Microphone Make ______ Type ______ Serial No. ______

Calibration of Instrument Before ______ After ______

Location of Microphone ______

OCTAVE-BAND RECORDING FORM

Test No.	Time Hrs.	Position	Conditions	Sound Pressure Level dB *re* 20 μPa										
						Octave Band Center Frequency: Hz								
				A-Weighted Level	Over-All Level	31.5	63	125	250	500	1k	2k	4k	8k

Comments: ______

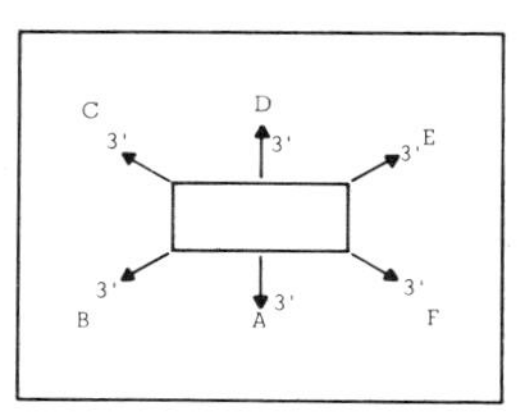

Indicate on the sketch the position of the equipment as placed in the room and orient the machine by identifying features. Note the sound level readings at appropriate locations such as A-F. Indicate total area of open windows and doors.

Question: What will be the sound level experienced by the employee?

Solution: The radiation will be hemispherical and thus Q=2. From Figure 2.7, enter the graph at $r/\sqrt{Q} = 10/\sqrt{2} = 7.1$. Then proceed vertically upward to R = 1000. Then, proceed horizontally to the right to find that L_p is 12 dB below L_w. This means that the operator would be exposed to 100 - 12 or 88 dBA.

Example 12.2:

Given: A vendor's machine produces a maximum sound level (L_p) of 104 dBA, when r = 3 ft, Q = 4 in the direction toward the location where the 104 dBA exists, and R = 2000 sq ft. You are to install it in a plant where r = 10 ft, Q = 2, and R = 2000 sq ft.

Question: What will be the worker's noise exposure?

Solution: Under conditions at the vendor's plant, the relation between L_p and L_w is determined as follows:

Enter the graph of Figure 2.7 at $r/\sqrt{Q} = 3/\sqrt{4} = 1.5$, then proceed vertically upward to R = 2000 sq ft, then horizontally to the right to find that L_p is 6 dB below L_w.

Under purchaser plant conditions, enter the graph of Figure 2.7 at $r/\sqrt{Q} = 10/\sqrt{2} = 7.1$, then proceed vertically upward to R = 2000 sq ft, then horizontally to the right and find that L_p is 14 dB below L_w.

Since the noise source is the same in both cases, that is, at the vendor's test stand and at the purchaser's plant, L_w will be the same. Therefore, the difference in L_p at the purchaser's plant would be 14 - 6 or 8 dB. This means that the machine that produced 104 dBA at 3′ under the vendor's test stand conditions will produce 96 dBA at the operator position in the purchaser's plant location.

By using Figure 2.7, noise levels can be estimated from the vendor's performance data whether expressed in L_w or L_p. It is to be emphasized, however, that if the vendor chooses to express noise performance data in terms of L_p, he must also supply data on r, Q and R, so that the purchaser can calculate the sound level in his facility. One way to remind the vendor to supply the necessary information for these calculations is to supply him with a noise specification as suggested in Figure 12.2, Section 5.3.

Upon receiving the data requested in the noise specifications, the industrial hygienist has to evaluate this information. If there are a number of suppliers of the equipment being purchased, the relative noise outputs should be considered when comparing prices and other factors that determine which product will be purchased. For example, Figure 12.3 shows the noisiest and quietest of six pneumatic screwdrivers capable of performing a required job when operating unloaded at the speed developed by 90 psi air pressure. These data formed the basis upon which a purchaser chose the quietest tool.

If all of the equipment under consideration fails to meet the specified noise level requirements and if the suppliers refuse to provide noise control in their respective products, consideration should then be given as to whether some products are more amenable to noise control treatment than others. The cost of applying noise control measures to less expensive but noisier machines should be considered at the time of purchase.

Once the equipment has been selected, and when the information reported on the specification indicates the equipment has a noise problem, noise control will have to be provided by the purchaser. In such cases, plant layout may be an important factor in controlling employee exposure. Before a location in the plant is selected for the noisy equipment, the following questions should be considered.

1. Is the noise intermittent or continuous?
2. Is this a single or multiple machine installation?

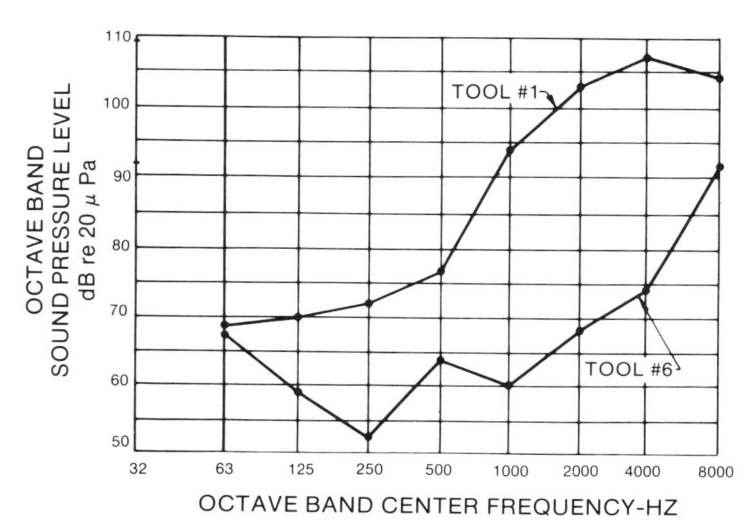

Figure 12.3 — Comparison of noise levels for two pneumatic screwdrivers.

3. Does an operator tend each piece of equipment at all times?
4. How many people other than the operator will be exposed to the noise?
5. Can the equipment be enclosed without affecting production or maintenance?
6. What are the current noise levels in the plant location?

The answers to the above questions may be helpful in placing the equipment or controlling the equipment noise before installation so as to minimize noise exposures.

A single machine producing intermittent noise may be located with much more freedom than equipment producing steady noise of the same magnitude. However, it should be remembered that equipment purchased for intermittent duty often comes into full-time operation and this can create a noise problem. Usually these problems can be solved more cheaply at the time of installation than after the equipment is in production. In the same manner, a single, noisy operation has a way of expanding into a multiple machine installation which must be handled in a far different manner. Usually, a single machine can be isolated acoustically before or after installation. Multiple machine installations require acoustical planning if individual enclosures are to be attempted or if machine spacing and room treatment are to be combined to yield minimum noise levels between machines.

In multiple installations, where each machine is tended by an operator, but where individual machines operate only a small percentage of the time, the operator's exposures may be controlled by enclosure of the individual machines. Where all machines operate continuously, or where a single operator tends a number of machines, exposure control is usually achieved through operator isolation or room treatment. A single noisy machine should not be placed in the center of a quiet, populated area. If it is necessary to locate a noisy machine in such an area, allow sufficient area for an enclosure or separate room. This is easier to do in the planning stage of a project. For equipment containing intake and exhaust openings which radiate high noise levels, space should be allowed for intake and exhaust mufflers or the machines should be located so that the noise may be ducted outside the building. It is often cheaper to collect exhaust noise and duct it outside the building than to install and maintain an exhaust muffler system. However, one must be careful not to create a noise problem in an adjacent office building or neighborhood by ducting noise to the outside.

Replacement

QUIETER EQUIPMENT

Sometimes it is possible to replace a noisy operation with a quieter machine, process or material. The existing machines in a plant were probably selected because they were the most economical and efficient means of producing the product or they provided a desired service at the time they were installed. Noise was very likely not considered as an important factor at the time of purchase. Often, it is more economical to purchase quieter equipment than to purchase noisy equipment and to reduce the noise with engineering controls. Indeed, it is sometimes more economical to retire noisy equipment earlier than planned and to replace it with quieter equipment than it is to design and install engineering controls.

Axial vs. *Centrifugal Fans*

As shown in Figure 12.4, propeller and axial flow fans produce more high-frequency noise than centrifugal type blowers of the same capacity. Since hearing conservation criteria allow higher noise levels at low frequencies, a squirrel cage type blower may be more desirable in some installations. On the other hand, since high-frequency noise control for fans is less costly than low-frequency noise control, there may be situations where the axial fans are more desirable.

Fan Speed

Fans or blowers running at high speeds are considerably noisier than when running at lower speeds; in fact, the sound power output will vary as the 5th power of the speed. Figure 12.5 illustrates the noise reduction achieved by using a larger, lower-speed blower. The blower for a vapor collection system produced excessive noise while moving 3600 cfm at 2.8″ static pressure. It ran at 3450 rpm and had a 12.5″ diameter wheel. It discharged into a cylindrical filter consisting of 1.5″ thick glass fiber compressed to 0.75″. A quieter fan was selected and the noise reduction achieved is shown on the graph. The new blower ran at 900 rpm and had a 32.625″ diameter wheel.

High Speed Drive Gears vs. *Turbines*

In general, steam turbine drives are usually less noisy than motor drives with gear increasers where high speeds are involved.

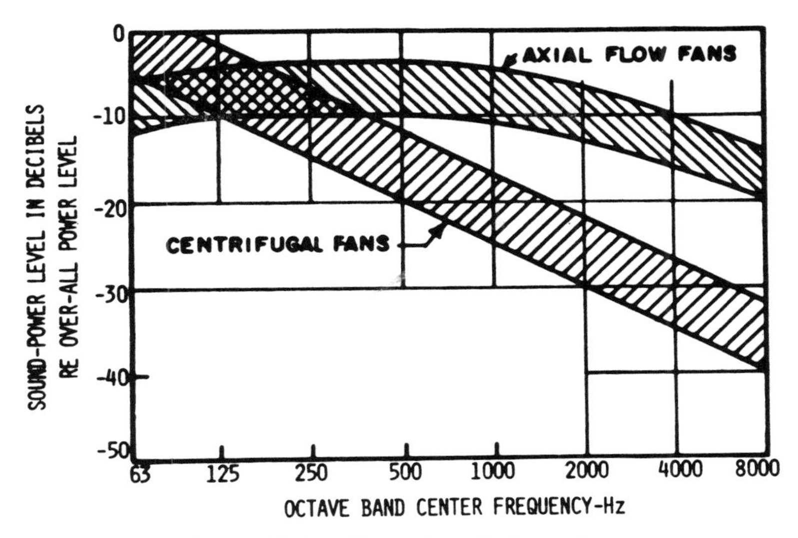

Figure 12.4 — Characteristic fan noise.

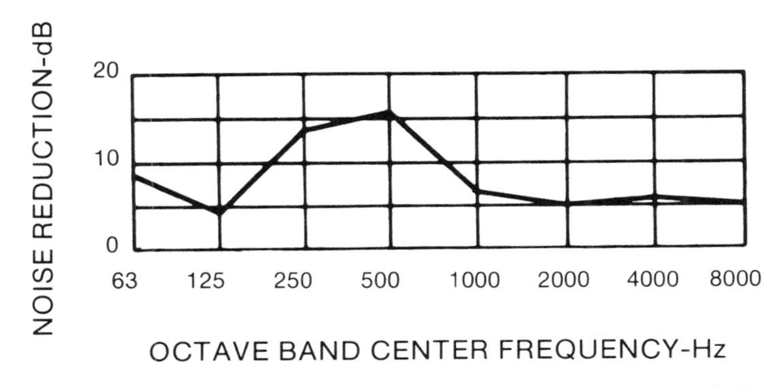

Figure 12.5 — Noise reduction obtained by replacing a fast, small diameter fan with a slower, larger fan.

Belt Drives vs. Gear Drives

Quieter operation can usually be assured by using belt drives instead of gear drives. However, rubber-tooth belt drives can produce excessive noise when running at high speeds.

Pneumatic* vs. *Mechanical Ejectors of Parts

Compressed air is commonly used for part ejection on punch presses, drill presses, and other high-speed production machines. Figure 12.6 shows the noise reduction achieved by substituting a mechanical part ejector for a 0.25″ dia. air nozzle with 85 psi line pressure. Mechanical ejectors of this type are readily adapted to many such operations and perform with little or no trouble.

Hopper Vibrators

Mechanical hopper vibrators cause noise by part-to-part impact and, more importantly, part-to-hopper impact. Basically, the goal should be to shake only the part of the material-handling system which impedes the flow, such as the bottom cone of hoppers. Figure 12.7 shows an 8′ dia. hopper with an electric solenoid type vibrator. A live bottom bin was installed. The following values of noise reduction were accomplished:

OB Ctr. Freq. (Hz)	63	125	250	500	1000	2000	4000	8000
Noise Reduction (dB)	7	6	20	22	16	12	12	9

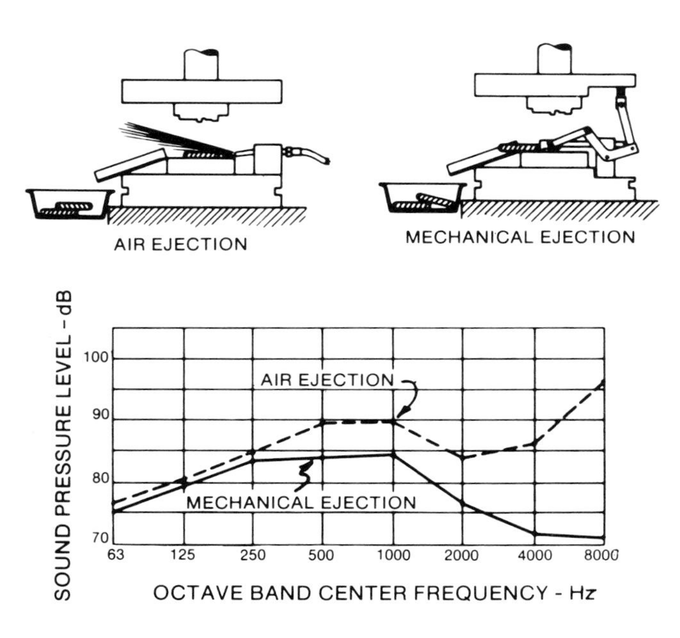

Figure 12.6 — Comparison of sound pressure levels for air ejection with those for mechanical ejection.

Figure 12.7 — 8′ diameter hopper.

This noise reduction was achieved because only the cone is vibrated. As a result, there is much less vibratory power required and there are no metal to metal impacts. Vibrators employing metal to metal impacts should be avoided.

Pneumatic vs. *Electric Tools*

In general, pneumatic portable tools are noisier than electric tools, because of the exhaust air. Figure 12.8 shows how one manufacturer has incorporated a muffler in a portable pneumatic grinder. The noise reduction is also presented. Figure 12.9 presents an external muffler built for a specific tool by the manufacturer. The noise reduction shown in the figure is for a position near the operator's ear.

Figure 12.10 shows a diamond-faced core drill used for drilling holes in concrete. In comparison to an air hammer and star drill, a diamond-faced core drill with water to cool the bit and flush away cuttings is a very fast and relatively quiet operation. For small holes, carbide-tipped masonry drills provide relatively quiet operation when drilling through concrete. Operation of the carbide-tipped masonry drills is much more dust-free than operation of the air hammer.

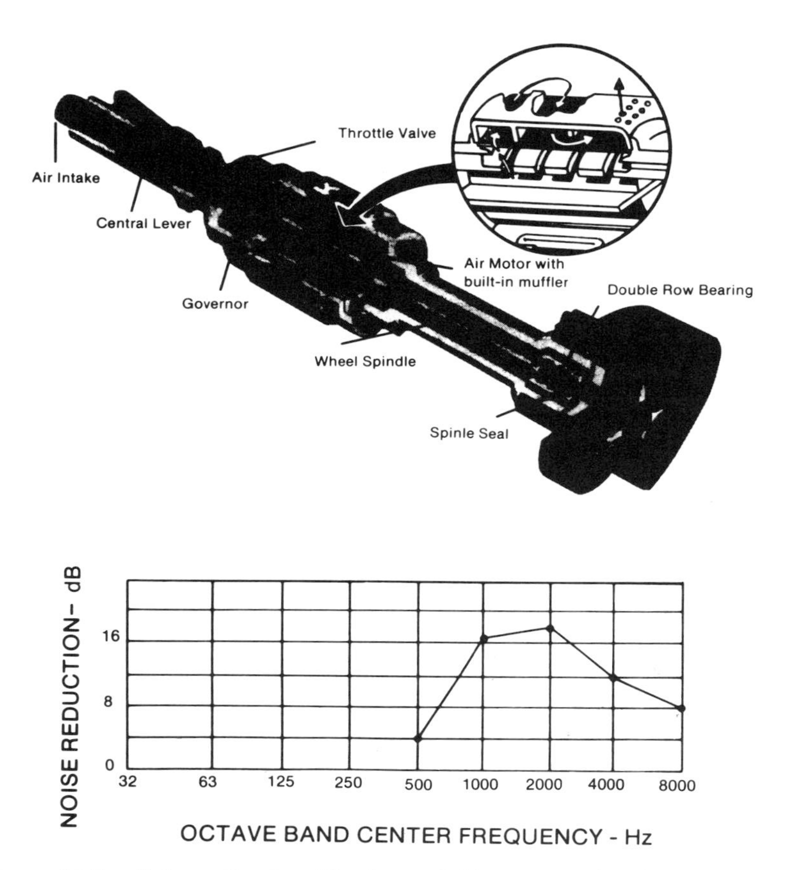

Figure 12.8 — Schematic of muffler for portable pneumatic grinder and noise reduction due to muffler.

Hydraulic Press vs. Riveting Hammer

A hydraulic press is quieter than a pneumatic riveting hammer. In fabrication shops, heated rivets are often driven by pneumatic or hydraulic presses where the configuration of the work permits. If unmuffled, the exhaust of compressed air from the operation cylinder of pneumatic presses is usually quite noisy. The noise can be diminished by reducing the velocity of the air. This can be done by distributing the air flow over a larger area by using a dispersive muffler. Figure 12.11 shows a silencer and the noise reduction achieved by installing it on the discharge of an automatic valve on an 80-psi

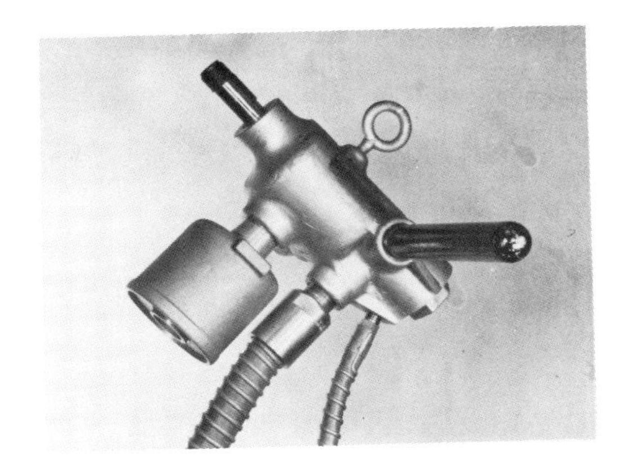

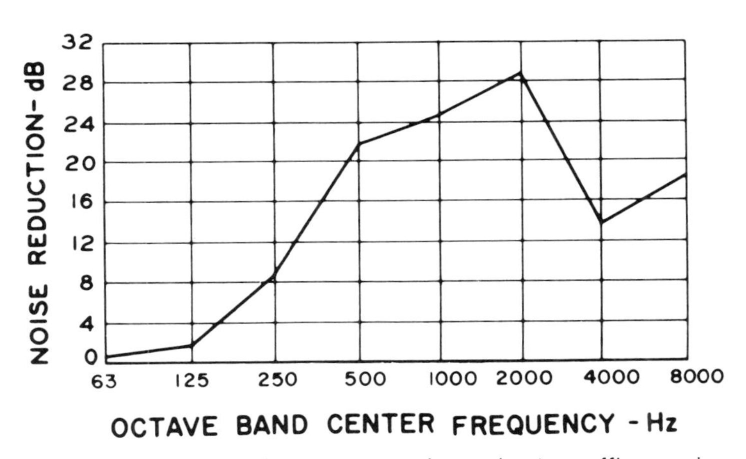

Figure 12.9 — Noise reduction at operator's ear due to muffler on air motor.

compressed-air line. Figure 12.12 shows a silencer used on the air exhaust of a 100-ton bull press. The sound pressure levels after the silencer was installed approach the shop's background noise.

Impact Tools

If an impact tool must be used, the smallest one capable of doing the job should be employed. Figure 12.13 shows the noise reduction that was accomplished by using the smallest acceptable air hammer for weld peening on a ⅝″ thick stainless steel tank 5′ in diameter and 8′ long with one open end instead of the hammer normally used by the employee. Employees need to be encouraged to use less noisy operations when those operations perform as well as the noisy ones.

Figure 12.10 — Quiet-operating masonry drills cutting through floor slab.

Large electric solenoids are used to trip clutches, valves, and various mechanisms; they can also be sources of high impact noise. A study of this problem shows that most large solenoids can be replaced with quieter electrically-controlled air cylinders. Of course, the discharge of the air cylinders must be muffled. In one case, hydraulic presses used in the manufacture of electrical relays were originally equipped with noisy electric solenoids. These solenoids actuated hydraulic valves that controlled the operation of the rams in sequence with other components of automatic transfer machines. The solenoids were replaced with pneumatic cylinders. This replacement proved to be satisfactory in every respect and superior with regard to noise, smoothness of operation, and valve wear. Resilient bumpers inside the solenoids at the point of impact can also be effective in reducing noise.

Figure 12.14 presents a photograph of an electric lock seam roller which seals seams quietly at rates of 18 to 30′ per minute depending on the gauge of metal. Noise reduction data are not given because the noise of the original impact tools varied considerably depending upon the material being worked and because the noise of the substitute tools was below the

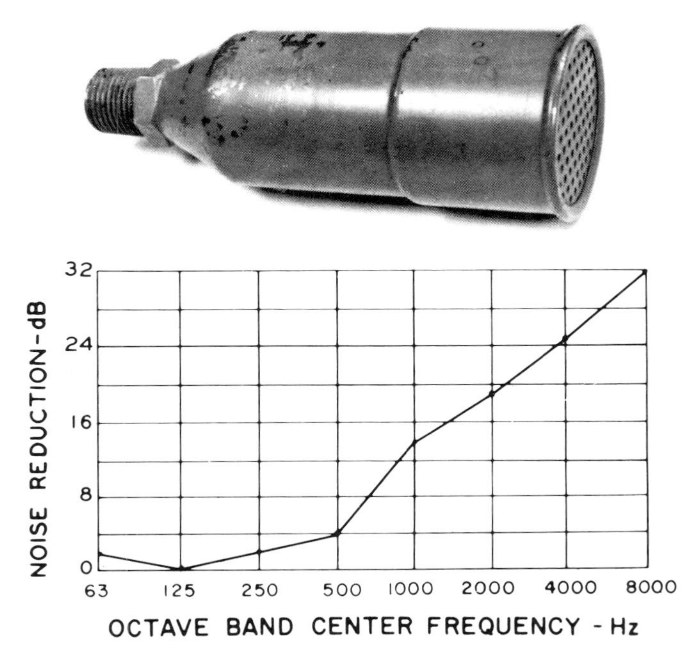

Figure 12.11 — Silencer for compressed air exhaust and noise reduction.

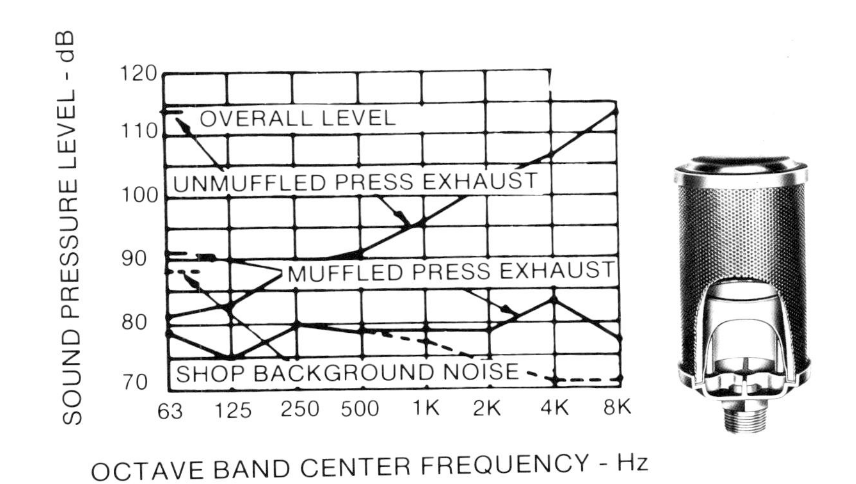

Figure 12.12 — Silencer used on air exhaust of a 100-ton bull press and sound pressure levels before and after installation.

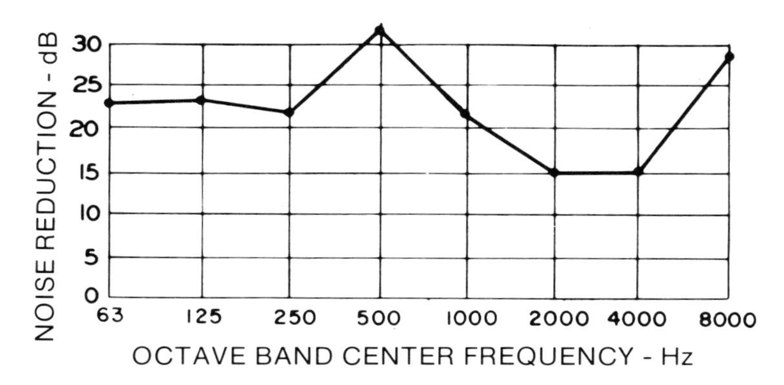

Figure 12.13 — Noise reduction achieved by using the smallest acceptable air hammer for a particular weld peening project.

Figure 12.14 — Seam roller for metal ducts.

background noise. Figure 12.15 shows a pneumatic rotary shear used in trimming metal. It replaced a pneumatic chisel. Air discharge noise from the squeeze tool and shear must be muffled to achieve maximum noise reduction. The portable air compressor shown in Figure 12.16 was designed with noise control as a specification. All functional components of this diesel-engine-powered machine, including engine, compressor, mufflers, fuel tanks, receiver separator tank and frame, are completely enclosed in an aluminum, glass-fiber, sheet-steel sandwich-panel material. Improved cooling by increased airflow through mufflers was required for this enclosed machine. The noise reduction achieved by this design, measured 3′ from the compressor, is shown below.

OB Ctr. Freq. (Hz)	63	125	250	500	1000	2000	4000	8000
Noise Reduction (dB)	8	13	24	17	10	10	10	14

Figure 12.17 shows a pavement breaker. One of its major noise sources is the air exhaust. Some measure of noise reduction can be achieved by muffling this air exhaust, which flows up through the first expansion chamber, reverses, and moves down through a second expansion chamber. At a distance of 3′ the noise reduction achieved was about 8 dB. Since pavement breakers usually operate at about 108 to 110 dB at this location, the sound is reduced to about 100 to 102 dB. This means that the operator must continue to wear hearing protection. However, at a distance of 20′, hearing protection would probably not be necessary.

A vane-type air-driven impact gun was tested to determine the noise produced when an internal muffler was used as compared to piping the noise to a remote location by means of a rubber hose. The air gun was

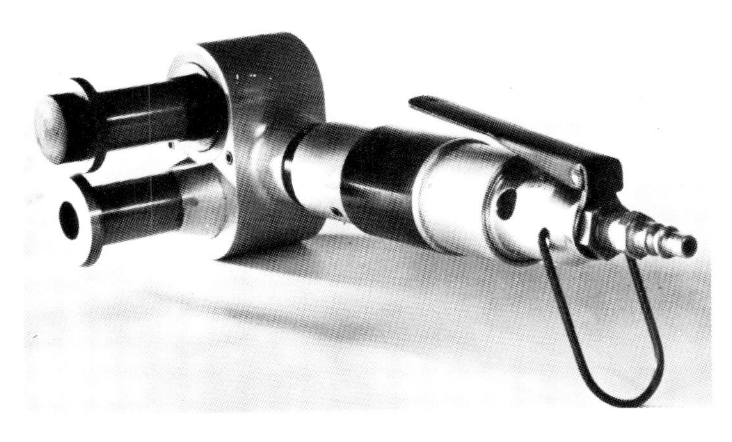

Figure 12.15 — Pneumatic rotary shear.

Figure 12.16 — Portable air compressor with noise control.

Figure 12.17 — Jack hammer with air exhaust muffler.

running free during these tests. The results of these tests and photographs of the impact guns are shown in Figure 12.18.

The airpowered disc polisher shown in Figure 12.19 has an internal muffler using a sintered stainless steel fiber metal as an absorbing medium.

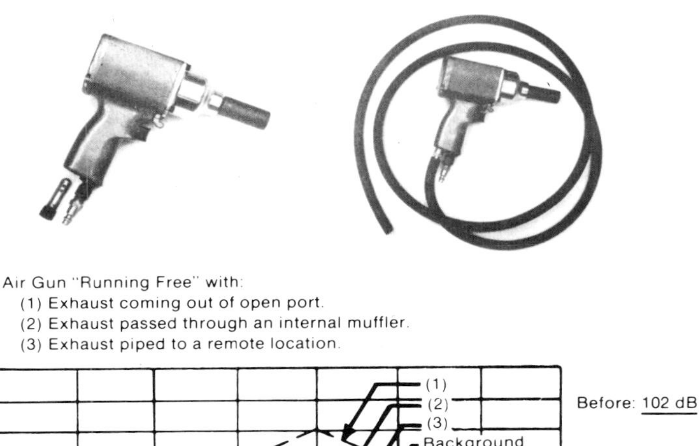

Figure 12.18 — Pneumatic impact gun noise control.

A 2-⅞″ diameter tube cleaner created excessive noise. The manufacturer's solution includes a built-in muffler for the air discharge of the turbine drive as shown by Figure 12.20. The noise reduction achieved was:

OB Ctr. Freq. (Hz)	63	125	250	500	1000	2000	4000	8000
Noise Reduction (dB)	14	30	24	9	8	18	18	14

QUIETER PROCESSES

Sometimes it is possible to replace a noisy process with a quieter one with the same or better production. For example, welding is significantly quieter than percussion riveting unless chipping is required in the weld preparation. Appreciable noise reduction is also accomplished by the substitution of high-strength bolts for percussion riveting. The impact wrench used to tighten the nuts is an undesirable noise source in itself, although not as bad as a riveting hammer. Also, the use of high-strength

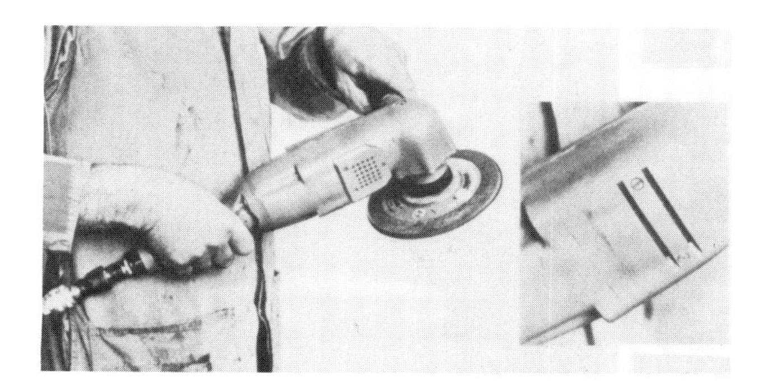

Figure 12.19 — Airpowered disc polisher with internal muffler.

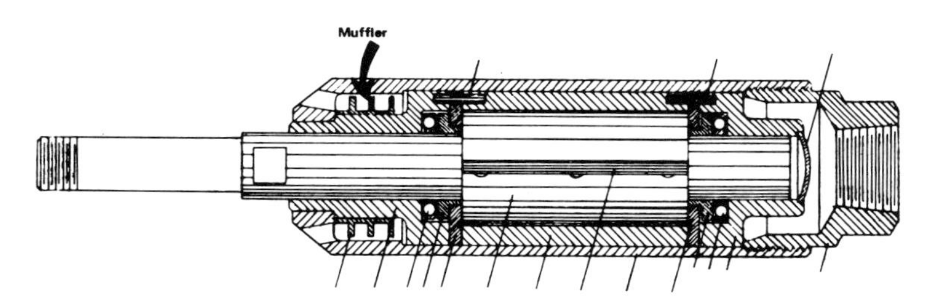

Figure 12.20 — Tube cleaner with internal muffler.

bolts provides several advantages over other fasteners, especially for erection in the field.

It is sometimes possible to reduce noise by substituting grinding for chipping. Alternatively, chipping noise can be eliminated by substituting arc or flame gouging. In the arc-air process, the metal removal is accomplished by melting the metal with an electric arc struck by a carbon electrode and blowing it away with a high-velocity air stream. The noise levels measured near the operator of arc-air equipment and a typical chipping operation are as follows:

OB Ctr. Freq. (Hz)	63	125	250	500	1000	2000	4000	8000
Chipping [126 dBA]	106	107	108	113	118	117	116	124
Arc-air metal removal [103 dBA]	76	71	74	82	96	100	95	92

It is claimed that the arc-air process may be used on any metal or alloy in all types of fabrication. The operation is effective in removing defects in welds and castings and in gouging and grooving. However, the configuration of the electrode holder makes it difficult to use the process on any but readily accessible surfaces.

Flame gouging is a process of metal removal where a cutting torch is used to remove metal in narrow strips by melting and oxidizing it. Flame gouging may be used to remove defects from welds and castings. Any standard scarfing torch may be used for removing small defects. The noise levels produced by flame gouging are generally lower than those shown for arc-air metal removal.

QUIETER MATERIALS

The materials used to construct buildings, machines, pipes and containers have a vital relationship to the noise radiated from the machine and transmitted through structure-borne connections. Some materials and structures have high internal damping; others have little internal damping and ring when struck. This noise can be reduced by applying damping to the material. An example is the parts tote box. Parts are dumped into the box by a forklift truck or a conveyor. The noise can be reduced by damping the sides and bottom of the box or by constructing the box from a wire mesh which is small enough to retain the parts. Another way to reduce ringing noise is to insert a resilient bumper between the materials banging together. Elastomers are good materials to use for bumpers. Examples are gaskets, seals, industrial truck tires, and caps for hammer heads.

Rotary dryers commonly use hammers (knockers) on the outside of the dryer shell to prevent product buildup on the inside. The metal-to-metal impact noise produced is usually objectionable. This noise can be reduced by providing resilient heads for the hammers. By providing sufficient striking area between hammer and shell, the resilient facing material can usually be made to transmit the desired vibration to the dryer shell without causing the objectionable metal-to-metal impact noise. In one case, the overall noise level was reduced 28 dB. Common materials used for the face of hammers are neoprene, nylon, Fabreeka and rawhide.

Noise produced by steel wheels on hand trucks can be reduced considerably by using rubber or plastic tires. The wheel diameter, tire width, and hardness should be considered when using resilient tires in order to assure satisfactory handling performance as well as noise reduction.

By replacing a steel gear with a fiber gear in a 15-hp high-speed film rewind machine and flooding the gears in oil, the noise levels were reduced as follows:

OB Ctr. Freq. (Hz)	63	125	250	500	1000	2000	4000	8000
Noise Reduction (dB)	10	6	5	5	8	20	16	14

Source Modification

In discussing industrial noise control, it is desirable to separate noise sources into two general categories determined by the mechanics of sound wave generation. The first category includes noise sources in which the sound waves result from surface motion of a vibrating solid. The second category consists of sound sources that result from turbulence in a fluid medium, *i.e.*, the result of interaction between high-velocity fluid flow and the surrounding air. There are also combinations of these two categories, for example, vibrating surfaces that produce turbulence in the air and turbulence that causes vibrations in solid structures.

Since noise control at the source is the best solution and often the least expensive, one of the first steps toward noise control should be the reduction of forces that ultimately result in noise-generating vibrations. In the same manner, the reduction of fluid flow velocity will have an effect upon noise generation by turbulence. As can be seen from Figure 12.1, there are usually intermediate steps between the driving forces and generation of sound waves in the air surrounding the source. In a broad sense, all of the intermediate steps shown are also part of the noise source.

Any noise source in the first category will consist of a source of vibratory forces coupled to a sound radiating surface. Control at the source may then consist of:

> reduction of driving force,
> reduction of radiating surface response to driving forces, and
> reduction in radiation efficiency of sound generating surfaces.

The driving forces commonly found in industrial equipment may be described as repetitive mechanical forces and non-repetitive impact forces.

REDUCE DRIVING FORCE

Reduce Speed

Repetitive forces, often the result of imbalance or eccentricity in a rotating member, increases with an increase in rotational speed. Repetitive unbalanced forces may also be produced by reciprocating members such as pistons or rams. Increased speed results in greater forces and, usually, higher noise levels. Therefore, no machine should be operated at an unnecessarily high speed. It may be desirable to use a larger machine that can be run at a slower speed rather than a smaller higher-speed machine operating at its maximum power. Alternatively, variable-speed drives can be used. This approach can reduce the noise levels for the length of time the equipment can be operated at the lower speed.

Maintain Dynamic Balance

Dynamic balance is of the utmost importance in minimizing the magnitude of repetitive forces. Maintenance of bearings, proper lubrication and alignment are essential to maintaining dynamic balance. Figure 12.21 illustrates an exhaust blower (705 rpm, 6″ static pressure, 13,800 cfm) that was badly out of balance. As a result, the bearings were worn. The figure also shows the noise reduction achieved by replacement of the bearings and

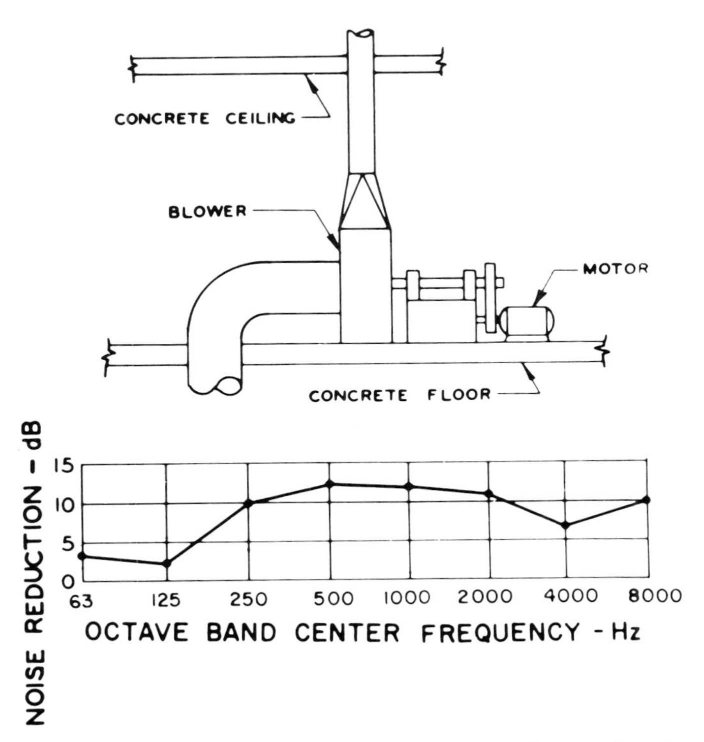

Figure 12.21 — Noise reduction by replacement of worn bearings and balancing.

balancing of the blower. High-speed equipment must be well balanced to minimize varying centrifugal forces which can cause excessive vibration and resultant noise.

Increase Duration of Work Cycle

Impact force is present in most metal fabricating operations such as punching, forging, riveting, and shearing. Because of the short duration of most impact forces, the noise is strongly dependent upon the maximum amplitude of the force. Figure 12.22 shows how the same work can sometimes be accomplished with a smaller force spread over a longer period of time. Figure 12.23 shows a 48″ film cutter which has skewed, segmented blades to provide a shear type cut.

Provide Vibration Isolation

Sometimes it is possible to reduce impact forces by the use of resilient material at the point of impact. Examples of this method include lining tumbling barrels, chutes, hoppers, stock guides, etc. with resilient material such as rubber or elastomers. Figure 12.24 shows a tumbler that has been lined with a tough rubber material. The details of the liner are shown on the

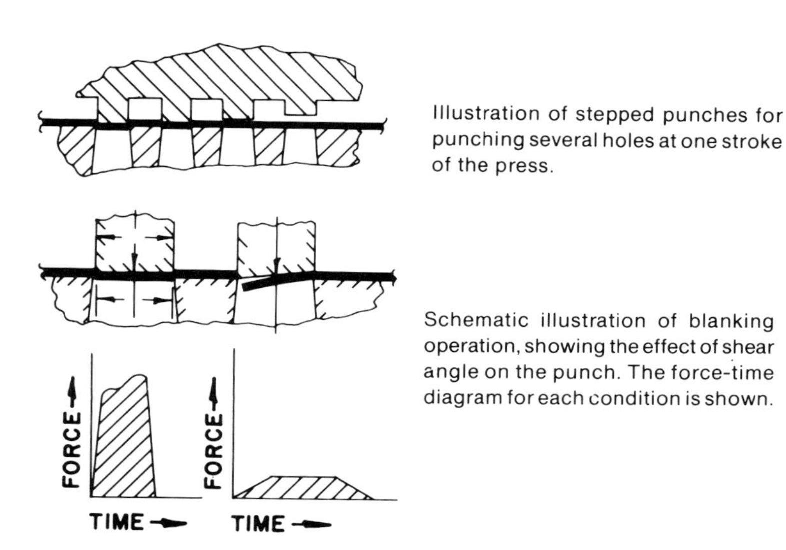

Figure 12.22 — Illustration of how to reduce the force required by increasing the duration of the work cycle.

Figure 12.23 — 48″ Film cutter with segmented skewed blades.

upper right. As can be seen, the noise levels were significantly reduced with this treatment above 250 Hz. However, the radiated noise level below 250 Hz actually increased, a result that is often observed when the noise control treatment creates better acoustic coupling to the surrounding air.

The procedure in the following sections can be used to analyze the source of vibration and determine the best method of vibration isolation.

Vibration

For a system such as a fan structure, many different modes of vibration are possible. However, if the structure is assumed to vibrate as a rigid body, the modes of vibration are limited to six degrees of freedom. Three of these occur in a vertical direction (y-axis), in a horizontal direction (x-axis), and in a transverse horizontal direction (z-axis). The other three are known as the rotational or rocking modes and they appear as rocking actions around the x-, y-, and z-axes, respectively.

Vibration is caused by forces occurring in any of the six modes or combinations thereof. Unbalanced rotating parts generate varying centrifugal forces which can cause excessive vibration and noise.

Natural Frequency

Any mechanical system with mass and stiffness, such as a mounted motor and its isolator, will have a natural frequency of vibration. This is the rate in cycles per second (Hz) at which the system will vibrate if deflected, freed, and allowed to oscillate. This natural frequency must be determined

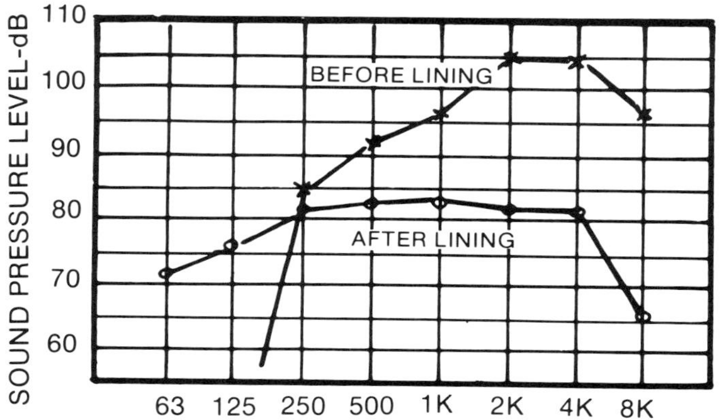

Figure 12.24 — Noise levels before and after rubber lining was applied to tumbler.

in order to predict the degree of isolation to be expected. [For a rigid system with a given mass, this frequency depends only upon the static deflection of the isolator, assuming that the supporting structure (floor) does not deflect.] The natural or resonance frequency can be calculated for linear modes from the following formula:

$$f_n = \frac{1}{2\pi}\sqrt{\frac{Kg}{W}} \quad \textbf{(12.2)}$$

where f_n = natural frequency, Hz
K = stiffness of isolator, lbf/in
g = acceleration of gravity = 386″/sec^2
W = weight of the system per mount, lbm

A second method of calculating natural frequency is known as the static deflection method. The expression for natural frequency in terms of static deflection is:

$$f_n = 3.13 \times \frac{1}{\sqrt{S_{st}}} \quad \textbf{(12.3)}$$

Where S_{st} is the static deflection of the mount, in inches, under its intended load. A curve representing this equation is shown in Figure 12.25.

The natural frequency for the rotational modes can be calculated from a similar formula as follows:

$$f_n = \frac{1}{2\pi}\sqrt{\frac{K_r}{I}} \quad \textbf{(12.4)}$$

where K_r = rotational, torsional, or rocking stiffness of mount, inch-pounds per radian of angular displacement about the axis being considered

I = mass moment of inertia of the supported load about the same axis and through the center of gravity, pound-inch-seconds squared

The natural frequency of a system may be altered by changing any of the factors used in the above three formulas (i.e., the weight of the system,

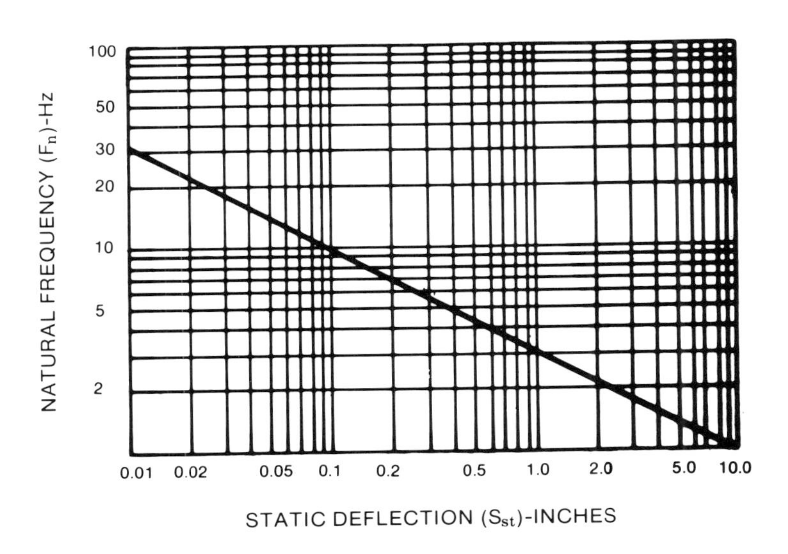

Figure 12.25 — Static deflection at natural frequency.

stiffness of the mount, moment of inertia of the supported load, or the rotational, torsional, or rocking stiffness of the mount). Such modifications will result in a change in the natural frequency. In most cases, changing the stiffness of the mount (it is often difficult to change stiffness without changing the mass) is the simplest method of altering the natural frequency of the system. However, because a system's natural frequency is proportional to the square root of the ratio of stiffness to mass, a major change in stiffness, in mass, or in the stiffness-to-mass ratio will be necessary in order to obtain a significant change in f_n.

Forcing Frequency

The forcing frequency, the primary frequency of the exciting vibration source which is being isolated, must be determined. If many frequencies exist, the lowest frequency is the one that normally is considered first.

Transmissibility

Transmissibility (TR) is the term used to express the effectiveness of an isolation system. It is the ratio of the force (or motion) experienced by the supporting structure to the force (motion) exerted.

In a force-excited system, such as that found where unbalanced rotating elements produce vibrating forces, the transmissibility of the isolation system would be the ratio of the force experienced by the supporting structure to the force exerted by the machine. For example, if an unbalanced rotor running at a speed of 1800 rpm was producing 100 lb of centrifugal force and it was desired to limit this force to 2 lb on the foundation, the transmissibility of the isolation system would have to be 0.02 (2/100); *i.e.,* 98% isolation of the force would have to be achieved.

In a motion-excited system, such as that found where an instrument is shaken by vibration of the instrument panel, the transmissibility of the isolation system for the instrument would be the ratio of the amplitude of motion of the mounted instrument to the amplitude of motion of the panel. For example, if the panel were vibrating at 1″ amplitude and it was desired to limit the instrument motion to one hundredth of an inch, the transmissibility of the isolation system would have to be 0.01.

The reduction of structure-borne noise by vibration isolation also can be expressed in terms of transmissibility. On the decibel (dB) scale, it is defined as:

$$\text{Noise Reduction (NR in dB)} = 20 \log \text{TR} = 20 \log \left[\frac{\text{Exciting force}}{\text{Transmitted force}} \right] \quad \textbf{(12.5)}$$

For example, if a beam were generating noise at 30 dB and it were desired to reduce the level to 10 dB, the transmissibility of the isolation system would have to be 0.1 to obtain the 20-dB reduction. Naturally, if the beam were not the source of noise, no noise reduction would be accomplished by isolating the beam. Great care must be taken to identify the noise source before application of the treatment.

Since transmissibility is a function of frequency, TR in the above equation will be the value of transmissibility at the frequency of the vibration. Figure 12.26 presents a family of curves of transmissibility as a function of the ratio of the forcing frequency to the natural frequency for different damping ratios.

The damping ratio (δ) is the ratio of damping constant (C) to critical damping (C_c). Critical damping is the value of damping at which the mass when displaced would slowly return to its equilibrium position without "overshoot" (Thomson, 1981). In addition to damping ratio, there are a number of other terms that can be used to describe damping at resonance. For small damping, these terms are related (Beranek, 1971).

$$\delta = \frac{C}{C_c} = \frac{\eta}{2} = \frac{2.20}{f_n T_{60}} \quad \textbf{(12.6)}$$

where δ = damping ratio, dimensionless
C = damping constant, lbf-sec/in
C_c = critical damping constant, lbf-sec/in
η = loss factor, dimensionless
f_n = natural frequency, Hz
T_{60} = reverberation time, sec

It is evident that transmissibilities below 1.0 are necessary before an isolation system becomes effective, and transmissibilities above 1.0 may occur if proper consideration is not given to the resonance frequencies of the system. The most important consideration in designing a vibration isolation system is the relationship between the forcing frequency of the exciting force and the natural frequency of the isolation system.

Transmissibility Curve

Since low transmissibilities are essential for effective vibration isolation, it is only necessary to examine a transmissibility curve, or formula, to find the desirable frequency ratio (the ratio of forcing frequency to natural frequency) which must be attained to make the system effective. The curves shown in Figure 12.26 can be applied to each of the six modes, which may be considered separately. The curves show the following principal facts:

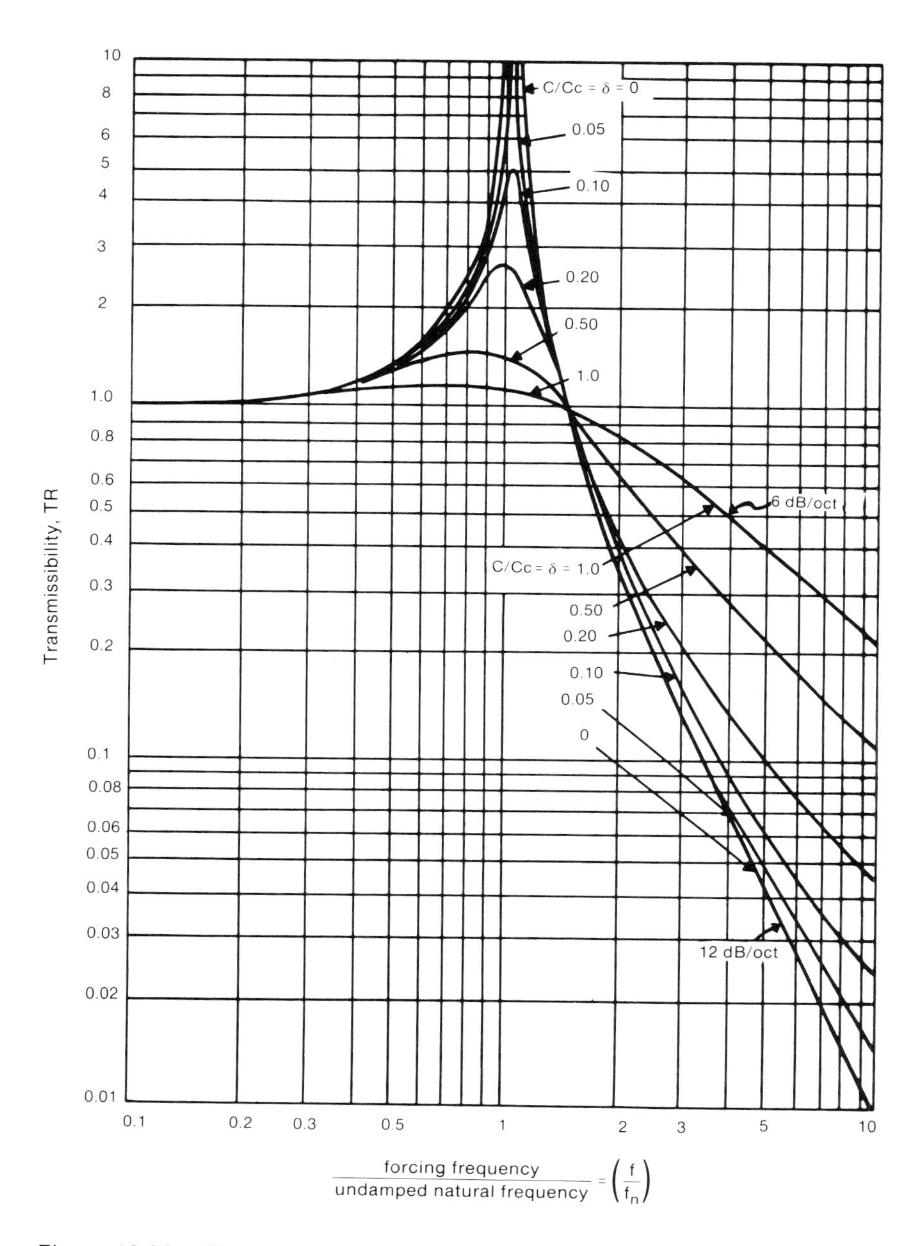

Figure 12.26 — Transmissibility curve—single degree of freedom curve with damping (assuming constant stiffness of isolator with respect to frequency).

1. When the forcing frequency equals the natural frequency of the mounted system, the transmissibility becomes much greater than 1.0 and no isolation is attained. Instead, an amplification of the exciting force, or motion, occurs and the phenomenon is known as resonance.
2. It is not until the forcing frequency is 1.4 (or more) times the natural frequency of the mounted system, that any isolation is achieved.
3. To achieve a transmissibility of 0.1 (which means 90 percent isolation efficiency), the frequency ratio must be at least 3.3 if the system is assumed to exhibit zero damping.

The transmissibility curves illustrated in Figure 12.26 are idealized cases; *i.e.,* the stiffness and damping factor of the isolators are constant with respect to the forcing frequency. This is generally true for metallic spring isolators but not necessarily so for elastomeric isolators. This effect may cause severe detrimental effects on isolation efficiencies of the elastomeric isolators.

Damping limits amplification at resonance but it also limits isolation efficiency. This means that systems which need some damping to control resonance while being brought up to speed or while being shut down will sacrifice isolation efficiency to some extent.

The following approximate values of damping ratio (C/C_c) can be used to select a transmissibility curve:

> (a) steel -- 0.005,
> (b) composition materials (Fabreeka, Micarata, etc)--0.05,
> (c) elastomer--0.10,
> (d) dash pots, shock absorbers--0.5 and higher.

If available, the vendor's values should be used whenever possible in place of these approximate ones. More detailed discussions of vibration isolation are available (Beranek, 1971; Crocker, 1975; Harris, 1979; Thomson, 1981).

Vibration Isolators

A vibration isolator is an elastic member which supports the vibrating machine and isolates it from the surrounding structure. Vibration isolators may provide a low natural frequency for a system if properly specified and installed.

Vibration isolators may be constructed of metal springs, elastomers, felts, or composition pads, or combinations thereof. However, an arbitrary elastic material or quantity of the material will not necessarily provide isolation. Generally, spring vibration isolators or air springs are used for low-frequency isolation below 30 Hz. Elastomer and composition isolators are more effective above this frequency.

Metal springs wear well and are not affected by oil or other contaminants. However, transmission of sound at frequencies at or above the "surge" frequency of a spring can impair the performance of a spring isolator. The surge frequency is a function of the diameter of the wire and the length of wire used in manufacturing the spring. Generally, surge frequencies are in the mid- or high-frequency range and neoprene elements are used in series with the metal spring to help overcome this deficiency. For large springs, however, the lowest surge frequency may be in a range where the neoprene is not highly effective (Harris and Crede, 1961).

Elastomer, felt, and composition vibration isolators have more internal damping than steel springs. This prevents excessive build-up of vibrations during slow acceleration and deceleration where the forcing frequency passes through the point of maximum transmissibility. However, it may still be necessary to utilize snubbers to limit the system's maximum motion when passing through resonance.

All bolts, pipes, *etc.* connecting the machine to surrounding bodies must be isolated to prevent direct transmission of vibration. Figure 12.27 shows one method for isolating bolts which pass through an elastomer mounting. Note that metal-to-metal contact would "short circuit" the isolator. Thus, isolating the head of the bolt without also isolating the shaft is of no value.

Selection of Isolators

In situations involving a system on a rigid foundation, the following step-by-step procedure is suggested for selecting a vibration isolator:

1. Determine the lowest forcing frequency in the machine to be isolated.
2. Establish the permissible transmissibility or desirable noise reduction level.
3. Using Figure 12.26, determine the required natural frequency.
4. Using Figure 12.25, determine the necessary static deflection to obtain the natural frequency required.
5. Determine the weight on each mounting point of the machine to be isolated.
6. From the load deflection data supplied by manufacturers of vibration isolators, determine the isolator or pad size required.

Also, consider the effect of creep, wear, and resistance to corrosion, ozone, oil or other detrimental agents.

Since the choice of an elastomeric isolator depends most critically upon the constancy of the isolator's stiffness with varying temperature and forcing frequency, isolators made of hydrocarbon rubber, natural rubber,

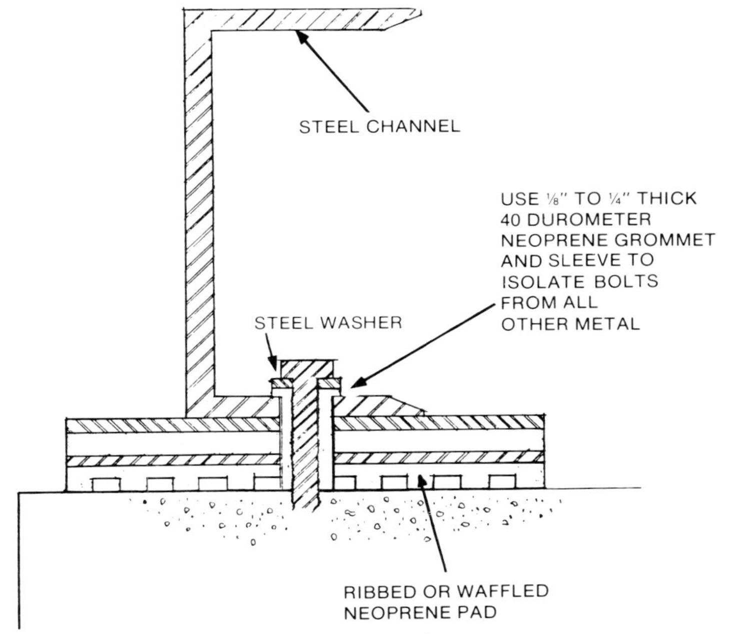

Figure 12.27 — Vibration isolation pad installation.

and silicone rubber are recommended, especially in sub-zero temperatures. Neoprene is highly desirable near room temperature and has excellent oil and ozone resistance.

Finally, if the foundation is not rigid, the design of vibration isolation becomes more complex. An example of a non-rigid foundation is a floor not directly on the ground (*i.e.*, with a basement beneath or an upper floor). The vibration isolation efficiency of an isolator can be reduced if the floor's static deflection is comparable to the isolator deflection (ASHRAE, 1981).

Figure 12.28 shows a number of commercially available vibration isolators including flexible couplings, flexible pipe connections, spring isolators, and elastomeric isolators. The following example illustrates the use of Figures 12.25 and 12.26 to solve a vibration problem by vibration isolation, assuming a rigid foundation.

Example 12.3:

Given: A 200-lb machine is transmitting vibration at 40 Hz into its support structure with a force of 100 lb. The machine is exposed to grease.

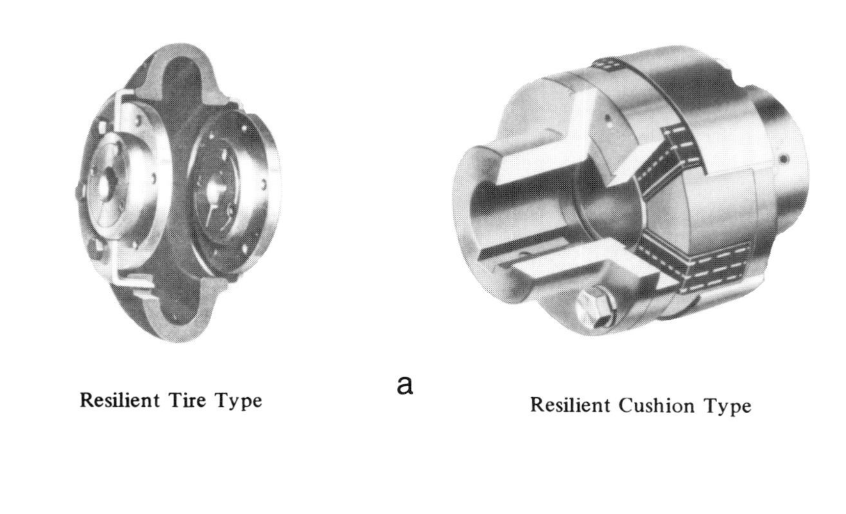

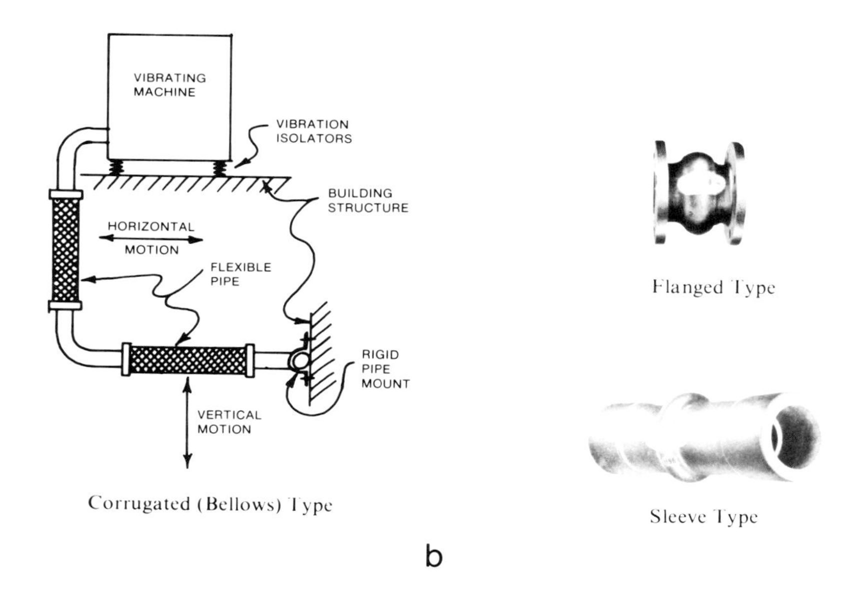

Figure 12.28 — Vibration isolators (a) flexible couplings, (b) flexible pipe connections, (c) spring-type vibration isolation mount, (d) elastomeric isolation mount.

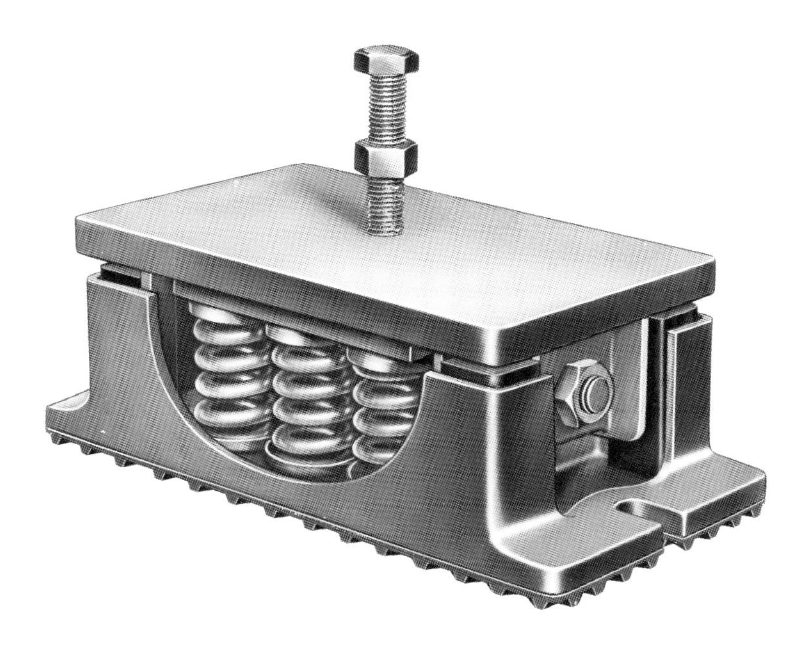

c

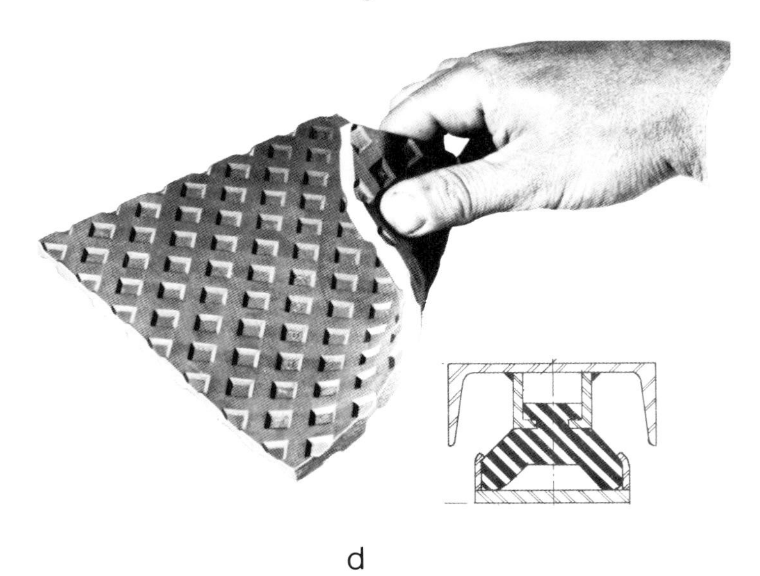

d

Figure 12.28 (continued)

Question: How would you select an isolator to limit this force to 5 lb?

Solution: This will require a transmissibility of 0.05 (5/100). Since the machine is exposed to grease, a neoprene isolator is selected. Assume the damping ratio of the isolator is 0.1. Following the 0.1 curve of Figure 12.26 to the transmissibility of 0.05, we find the ratio of the forcing frequency to natural frequency.

$$\frac{f}{f_n} = 5.5$$

$$f_n = \frac{40}{5.5} = 7.3 \text{ Hz}$$

From Figure 12.25, it can be seen that a static deflection of about 0.2″ is required for the isolator.

Assuming equal distribution of load on four points of support, the load per mount becomes 50 lb so the stiffness of the isolators must be 50/0.2 or 250 lb/in. From information in the vendor's catalog, it is now possible to select the proper isolators.

REDUCE RESPONSE OF VIBRATING SURFACE

The response of a vibrating member to a driving force can be reduced by damping the member (if the member is vibrating at a resonance frequency), by increasing its stiffness if excited below resonance, or by increasing its mass if excited above resonance. When the frequency of the driving force is equal to the natural frequency of the member being vibrated, large displacements result because the member is being driven at a resonance frequency.

Damping

Most mechanical structures have a family or series of resonances which are rather widely spaced in the low frequency range but more closely spaced at higher frequencies. Because of the large displacements achieved at resonance, there is usually increased noise radiation. Vibrations at resonance frequencies may be limited very effectively by damping. Basically, there are two forms of damping treatments. These are the free-layer or extensional damping and constrained-layer damping. Both of these types of treatments have been applied to industrial noise problems. Damping treatments reduce the vibrations at the resonance frequencies, create faster decay of vibrations in the surface, reduce noise created by repetitive impacts on the surface, and improve the high-frequency transmission loss. The performance of damping materials is often strongly temperature-dependent. Manufacturers of damping materials can provide detailed performance information about their products. Optimizing a damping treatment is usually a complicated procedure and, in cases where the cost can be justified, it is best left to experts. However, for many industrial situations, it is satisfactory to use a rough rule-of-thumb guideline, as follows.

For the free-layer or extensional damping treatment, the rule of thumb is that the viscoelastic treatment should be about the same thickness as the surfaces being treated. Viscoelastic materials are preferred because it is possible to predict the performance of a treatment using the available information from the manufacturers.

A truck manufacturer made a new aluminum-bodied cab-over-engine truck and found that the driver was exposed to sound levels that were higher than desired. In order to increase the damping at the coincidence frequency (Irwin and Graf, 1979) and thus the transmission loss of the cab floor panels, a free-layer damping treatment was applied to the panels. Both the panel and the treatment were about ⅛″ thick. The octave-band sound pressure levels measured at the operator's ear at the governed speed are presented in Figure 12.29. The A-weighted sound levels were reduced from 86 dBA to 82 dBA (Kuntz, 1985).

A coating of vibration damping compound was applied to the guards and exhaust hoods of a ten-blade gang ripsaw to damp these vibrating surfaces. Figure 12.30 shows the sound pressure levels before and after treatment. The measurements were made 3′ in front of the saw while the machine was idling. Under operating conditions, the results of such a treatment will vary considerably with the material being cut.

For the rough treatment typical of industrial environments, constrained-layer damping is often preferred to free-layer damping. This is accomplished by covering the vibrating surface with a thin sheet of damping material with the proper adhesive on both sides and then adding an outer covering of sheet metal. If absolutely necessary, the sandwich so formed can be bolted together on 6″ to 8″ centers. The rules of thumb are:

- for vibrating panels having a thickness of up to 16 ga, use an outer steel plate (restraining plate) of the same ga as the vibrating plate;
- for vibrating plates of 16 ga to ⅛″ thick, use a restraining plate of 16 ga steel;
- for vibrating plates ⅛″ to ¼″ thick, use a ⅛″ thick restraining plate;
- for vibrating plates ¼″ thick or heavier, use a ¼″ thick restraining plate.

The most common damping materials are damping felt, liquid mastics, and elastomeric damping sheeting. All materials selected should be compatible with exposure to existing temperature and environmental conditions and to chemicals and oils. Note that flat, unsupported surfaces are the ones radiating the most noise. Corners of box-like structures, reinforcing

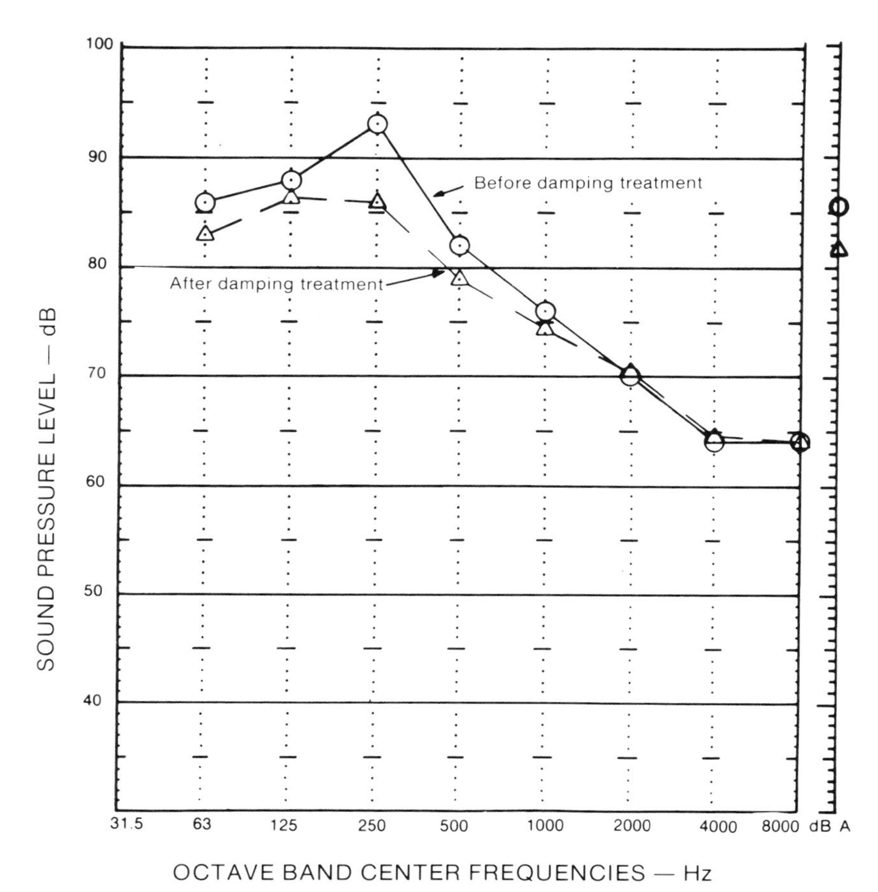

Figure 12.29 — Octave-band sound pressure levels. Stationary, engine at governed rpm, driver's ear test position. Effect of damping the bottom of the cab.

bosses, etc. are so rigid they probably do not require damping. This simplifies the damping treatment because it eliminates many double curved surfaces which would be difficult to laminate. Manufacturers of damping materials can provide advice regarding the most effective use of their materials. However, do not hesitate to use the heavy restraining plates suggested. They make the treated panels significantly stiffer and may reduce not only the resonance response but the driven response as well. The extra weight and stiffness might be the most important factor in reducing the noise. In the transfer of small parts by means of chutes, hoppers, and tote boxes, considerable noise is often generated from the impact of the

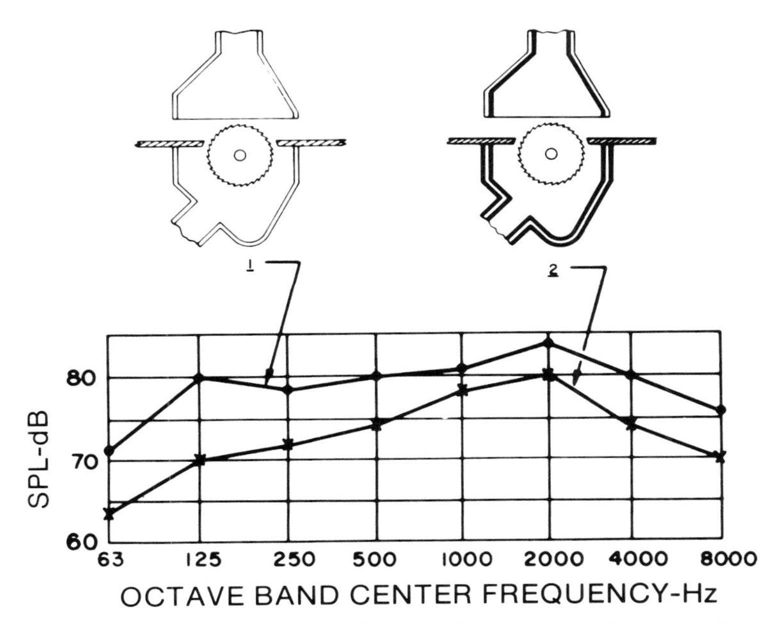

Figure 12.30 — Noise reduction for guards and exhaust hoods of saws.

parts dropping on metallic surfaces. If the weight of the parts and the distance they fall are controlled by the operation, noise reduction can best be accomplished by treatment of the surface on which the parts fall.

Assembly components being dropped into steel plate hoppers generated sound levels of 122 dBA during 2-sec interval of maximum noise. The hoppers were open-topped. Hopper panels were ¼″ steel. The operator works at a product assembly station positioned between two hoppers, each of which is about 3′ away from the operator's ear. Treatment consisted of covering the exterior of one hopper with a layer of 3/16″ damping material that, in turn, was covered with an outer layer of ⅛″ steel. An adhesive was used to bond the damping material to both steel surfaces. The outer perimeter of the steel cover plate, which slightly overhung the damping layer, was welded around the edges to the base plate. Figure 12.31 presents the noise reduction accomplished with this treatment. The sound level during the 2-sec interval of maximum noise was reduced from 122 dBA to 113 dBA (NIOSH, 1978).

The drum-type continuous miner consists of a large rotating drum fitted with steel projections, or teeth, which tear at and crush stone, coal or minerals. Dislodged fragments are then raked onto a conveyor bed which carries them some 25′ to the rear of the machine, where they are dumped into a hopper or cart. The primary source responsible for the operator's

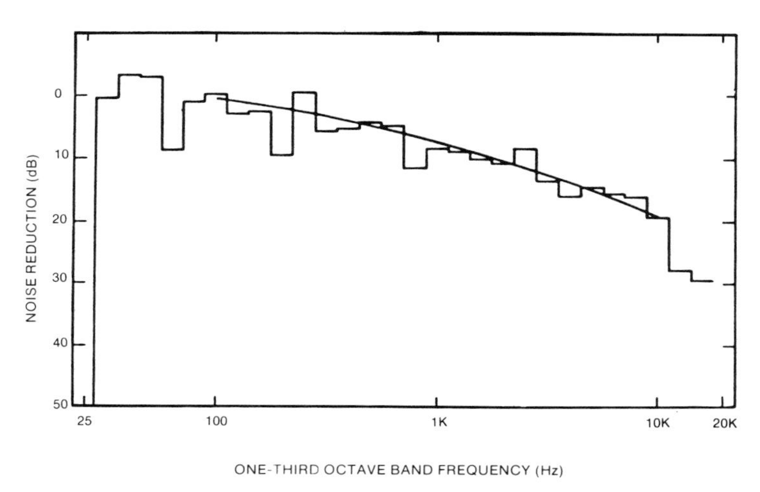

Figure 12.31 — Hopper noise reduction achieved by damping.

exposure is the noise of the chain-driven conveyor "flights" impacting the trough and sides of the bed. Figure 12.32 shows these "flights." The conveyor bed is constructed of ¾″ steel plate and the sides are typically ¼″ in thickness, also of steel. The bed was damped using a 3/16″ thick sheet of PVC-based damping material, adhesively bonded and constrained by ⅜″ steel sheet, edge welded. The damping material is also shown in the figure. The overall noise reduction at the operator's position after treatment was 7 dBA. Noise reductions at other locations near the conveyor were between 6 and 10 dBA (Lilley, 1985).

In the finishing of large bronze marine propeller castings, a loud ringing sound was produced during chipping. The vibration of the blades was found to be responsible for the noise produced. A special vibration damper was developed with which it was possible to reduce the ring of the propeller during chipping well below that of the chipping tool alone. Figure 12.33 shows several of the dampers attached to a 25,000-lb destroyer propeller. Each damper consists of ½″ × 6″ × 6″ steel plates welded to the jaws of a C-clamp. The faces of the plates that bear on the propeller are covered with 0.1″ to 0.125″ thick asphalt-impregnated felt attached with a compatible contact adhesive. The differential motion of the clamp and blade causes shear, flex, and compression in the felt and these motions convert the vibrational energy to heat.

The relative noise levels produced during chipping on the propeller under various conditions are also shown in Figure 12.33. The upper curve shows the noise levels produced by chipping on the undamped propeller and the lower solid curves show the levels with 3, 6, and 12 dampers

Figure 12.32 — Damping treatment for continuous miner. The conveyer belt beneath the "flights" is a constrained-layer damped structure. The side panels are extensionally damped.

attached to each blade. The noise levels produced when the chipping tool was operated without touching the blade are shown by the dotted curve. The chipping noise decreased as more dampers were applied to each blade. With 6 dampers attached to each blade, the chipping noise was about the same as the tool noise. The tool noise exceeded the chipping noise in some octave bands with 6 to 12 dampers per blade because the chipping tool made somewhat more noise when running free than when used to remove metal.

High-temperature damping materials (temperatures up to 600° F) are also available and have been applied to steel muffler shells, high speed blower housings, steel piping systems, gear boxes, fan housings and turbine exhausts. The material is applied as paste, sheet, or tile.

Stock tubes of screw machines are inherently noisy because there is nearly continuous impact between the tube and stock. One manufacturer has used constrained-layer damping to reduce the noise. Inner and outer steel tubes with damping material between the tubes were used. Significant noise reduction, measured one foot from the middle of the stock tube, was achieved as follows for a 4000 rpm operation with ½″ hexagonal stock.

OB Ctr. Freq.	(Hz)	63	125	250	500	1000	2000	4000	8000
Noise Reduction	(dB)	12	15	15	14	20	29	34	30

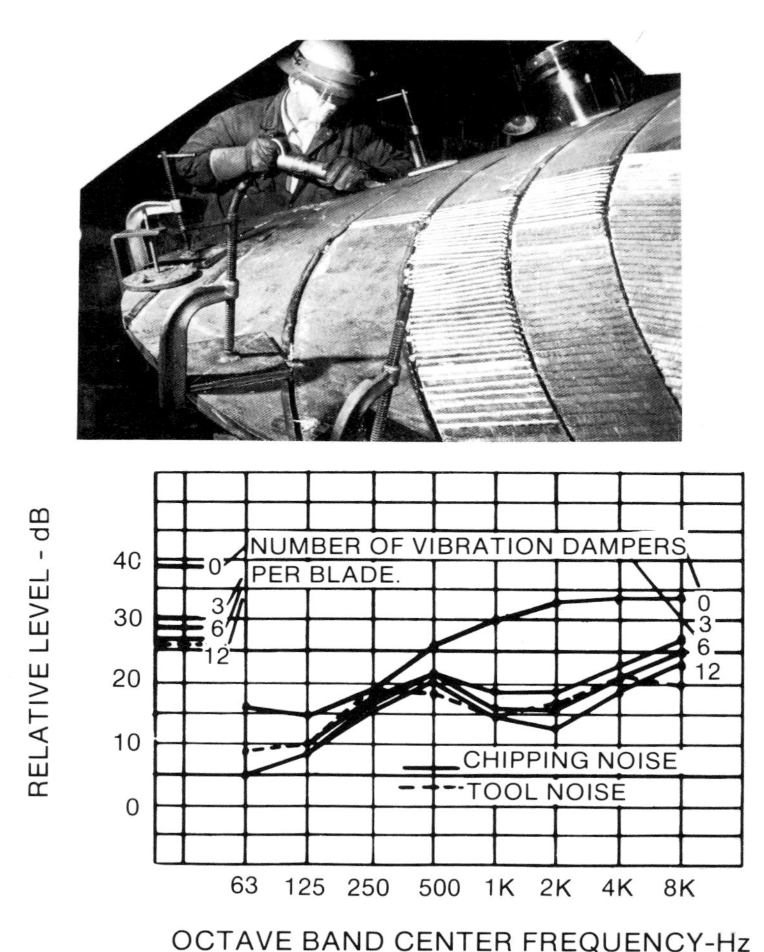

Figure 12.33 — Noise reduction by damping of propeller blades.

Another approach to noise reduction of stock tubes is to line the tubes with a polymer extrusion as shown in Figure 12.34. The polymer tubes are spiral wrapped with open cell polyurethane foam and then inserted into a steel tube. With four spindle machines running ½″ to ¾″ hexagonal stock at 700 to 800 rpm, an average noise reduction of approximately 18 dB has been reported. Reduction of stock tube noise by itself will not be effective in reducing overall room noise unless all room noises are controlled. In particular, noise from the cutter end of the screw machine must also be controlled.

A rubber compounding mill had its lower section encased in a metal shell. The shell acted like a sounding board and amplified the motor, gear and roll noise of the mill. An application of extensional damping material to the inner surface of the metal shell reduced the noise. The maximum reduction achieved was:

OB Ctr. Freq.	(Hz)	63	125	250	500	1000	2000	4000	8000
Noise Reduction	(dB)	11	8	14	10	6	7	9	11

Sometimes damping treatments are combined with enclosure techniques to provide significant noise reduction. An example of this technique is shown in Figure 12.35. The casing of a 2000 hp extruder gear was radiating excessive noise. The gear case cover was ⅜″ steel. The base was 1″ steel with 1″ thick 9″ deep ribs. Measurements with an accelerometer showed that the ⅜″ steel and the 1″ steel were vibrating at approximately the same level. The ⅜″ steel was damped with ¼″ damping felt plus an outer covering of ¼″ steel. The sandwich (⅜″ steel + ¼″ felt + ¼″ steel) was bolted together on 8″ centers.

The irregularity of the 1″ steel base made constrained layer damping (as used on the cover) impractical. Instead, ¼″ steel plate was welded to the 9″ deep ribs and the voids were filled with sand. Figure 12.35 also presents the sound pressure levels before and after treatment.

Figure 12.34 — Polymer tubes for screw machine.

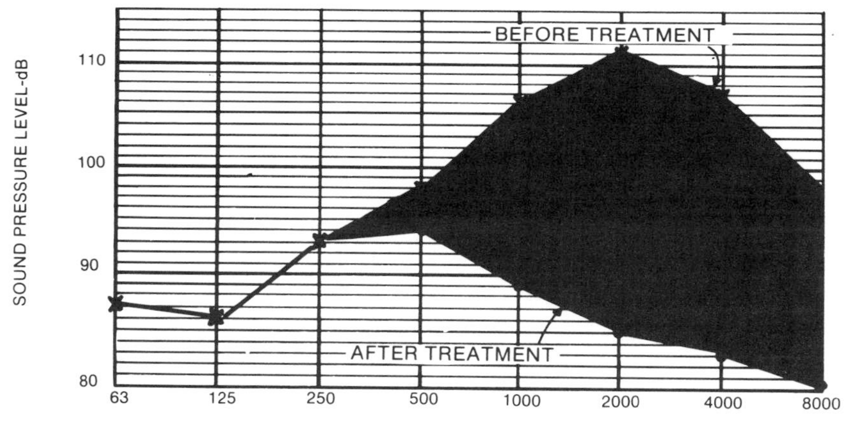

Figure 12.35 — Extruder gear treatment. Shaded area represents reduction due to treatment. Levels at 250 and 500 Hz, after treatment, are due to other equipment noise.

Figure 12.36 show a multistage high speed centrifugal compressor where noise control was considered during the design stage. The primary noise control treatment was to use heavy cast construction for the machine. To meet the environmental criteria of 90 dBA, the only parts that required acoustical coverings were the gear case cover and the steel interstage piping couplings. The heavy construction of the machine illustrates economical noise control without the inconvenience of enclosures.

REDUCE AREA OF VIBRATING SURFACE

The sound energy generated by vibrating surfaces depends not only upon the velocity of surface motion, but also upon the area of the radiating surface. In general, any regularly shaped area with one dimension greater than one-fourth wavelength can effectively radiate sound at the frequency corresponding to that wavelength in air. Therefore, the effective radiation of low-frequency sound is usually limited to large surfaces. Conversely, any surface of more than several square inches can effectively radiate sound at frequencies above 1000 Hz.

Surfaces radiating low-frequency sounds can sometimes be made less efficient radiators by dividing them into smaller segments or by reducing the total area. The use of perforated or expanded metal can often result in less efficient sound radiation from sheet metal guards and cover pieces.

USE DIRECTIVITY OF SOURCE

Many industrial sound sources are directive; that is, they radiate more sound in some directions than others. Common examples of directive sources include intake and exhaust openings, partially enclosed sources, and large sheet metal surfaces. A discussion of directivity can be found in

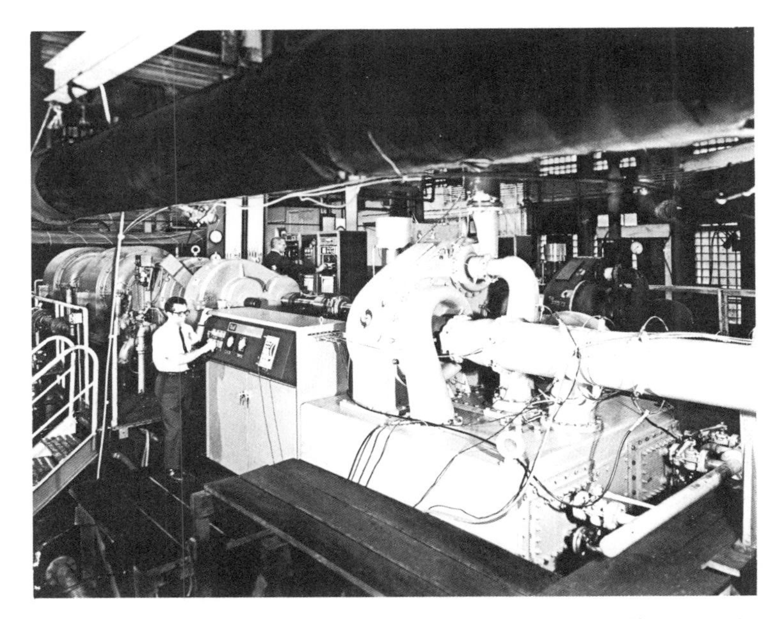

Figure 12.36 — Centrifugal compressor with increased stiffness and mass.

Chapter 2, in this chapter (Figure 12.2) and in other texts (Beranek, 1971; Peterson, 1980).

It is sometimes possible to utilize directivity of the source to provide noise control in a particular region of the sound field. This type of control is achieved by directing the source so that a minimum in the sound field occurs at the point or area of interest.

It is not possible to achieve worthwhile noise reduction by source direction when the point of interest lies in the reverberant portion of the sound field. In the reverberant field, the sound level depends only on the source strength and the amount of sound absorption in the room. For enclosed areas containing little sound absorption, the reverberant field may extend to within a few feet of the source, and direction of the source will have little effect on the sound levels throughout most of the area. However, under these conditions, there is some advantage in directing the source to an area of highly absorbing material, for this effectively reduces the source strength as far as the remainder of the room is concerned.

REDUCE VELOCITY OF FLUID FLOW

Several types of noise problems involving fluid flow are air ejection systems, valves and venting high pressure gas to the atmosphere.

Air Ejection Systems

There are three sources of noise in air ejection systems:

(1) noise generated by turbulence upstream from the opening,
(2) noise generated at and just downstream of the opening and
(3) noise generated by high velocity flow over sharp edges or cavities on the dies or target.

Since the level of the generated noise depends on the stream velocity, reduction of the supply pressure will provide some noise control. However, the reduction of stream velocity must be governed by the thrust requirements necessary to eject the part.

A typical impact pressure profile of a small jet stream is shown in Figure 12.37. It illustrates that the high-velocity portion of the stream is only two jet diameters wide at most. Therefore, by accurately aiming the jet stream at the target, maximum thrust can be obtained with minimum velocity. In air ejection systems, direction of the jet is very important if satisfactory operation is to be achieved with minimum stream velocity. It is often possible to obtain the necessary thrust by utilizing multiple nozzles, a larger jet having a lower velocity, or by moving the nozzle closer to the part being ejected. The treatment of the exit opening can provide reduction of the jet noise under some conditions. The use of multiple-opening nozzles to replace a single opening nozzle of ¼″ OD (outside diameter) copper tubing results in the reduction of exit noise. Where exit noise predominates, the sound spectra generated by small jets are largely in the higher octave bands. It is the noise in these higher octave bands that is effectively changed with the substitution of multiple opening nozzles.

Figure 12.38 presents 4 different types of nozzles. Also shown are comparisons of the sound level as a function of thrust and the air flow rate as a function of thrust. With the same pressure, the simple and multiple-jet nozzles provide approximately 6-10 times the thrust of similar size air-shroud and restrictive nozzles. The multiple-jet and air-shroud type nozzles generally give 3 to 10 dBA noise reduction for the same thrust with very little or no increase in air consumption. For ejection of parts and other applications requiring concentrated thrusts, the multiple-jet nozzles are the better choice. Where large quantities of air with lower and more uniformly distributed thrust is desirable, for example, in drying or dust cleaning, the air shroud nozzles are preferable (Huang and Rivin, 1985).

Once the velocity of the stream has been reduced to a minimum, additional reduction may be obtained by streamlining the targets as shown in Figure 12.39. Streamlining is quite important upstream from the target. An example of reducing the turbulence is shown in Figure 12.40 where the clearance between cutter blades and the bed knife was increased. This edger-planer for trimming foamed plastic generated low-frequency tones because the cutter blades were chopping the conveying air stream. By

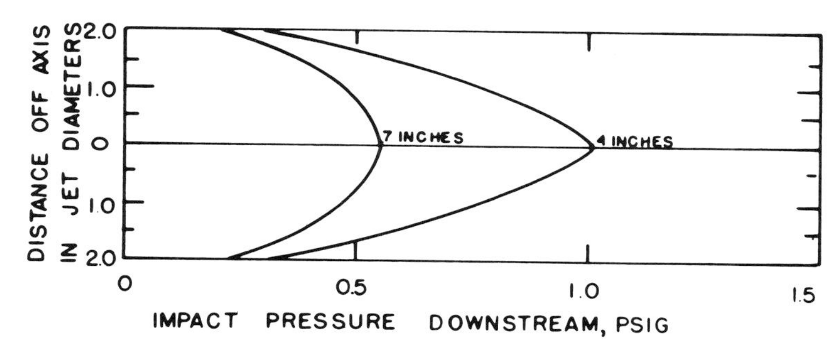

Figure 12.37 — Impact pressure profile downstream from single ¼″ diameter copper tube nozzle at 50 psig.

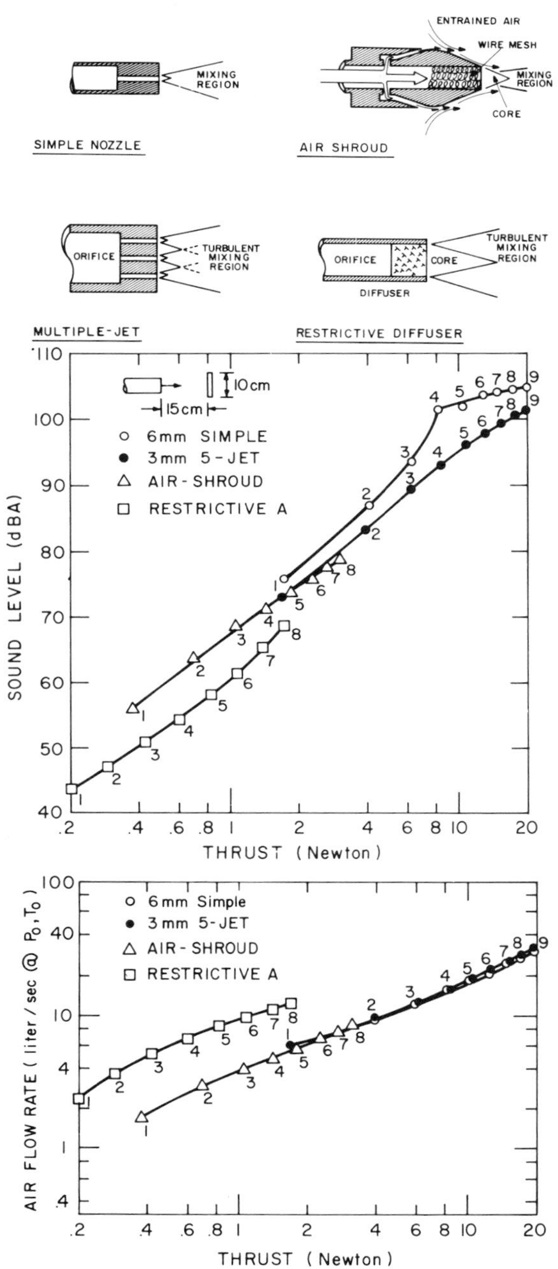

Figure 12.38 — Nozzle noise, thrust, and air consumption. From Huang and Rivin (1985).

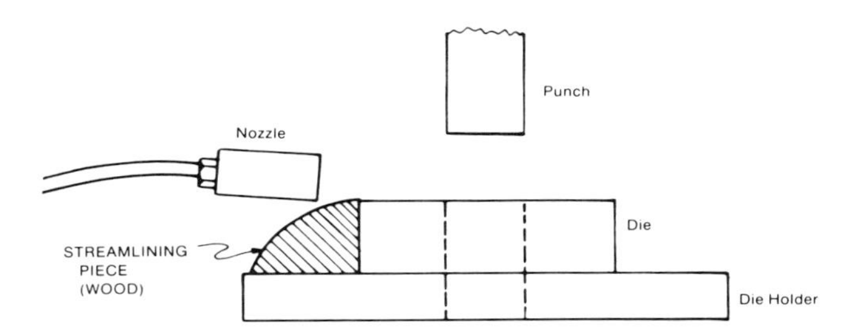

Figure 12.39 — Streamlining the jet path.

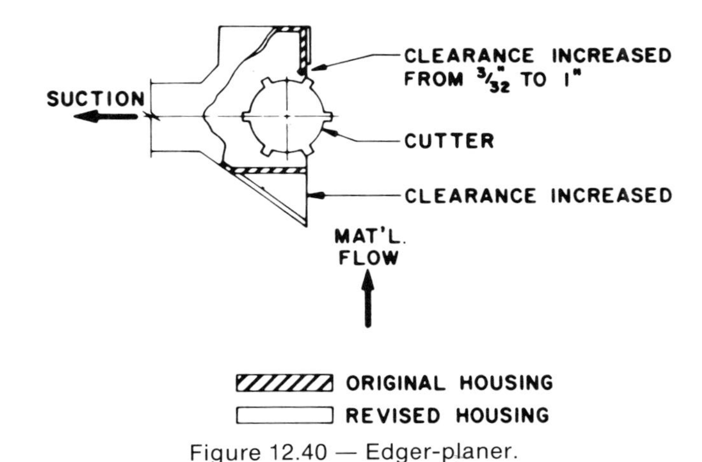

Figure 12.40 — Edger-planer.

increasing the clearance between the cutter blades and casing from 3/32″ to 1″, these low frequency levels were reduced from 102 to 84 dB.

Valves and Vents

One of the most common noise problems associated with fluid flow is that of venting high-pressure gas to atmosphere. When the ratio of absolute pressure upstream of the pressure-reducing valve to downstream pressure of the valve is 1.9 or greater, the port velocity in the valve will be sonic and L_w will vary in proportion to the mass flow or gas. The valve is then in a choked flow condition. Valve manufacturers can provide sound power and/or sound pressure levels for many of their valves. In addition, Figure 12.41 can be used to estimate the approximate aerodynamic sound power level of valves. This curve (Ingard, 1973) is based upon the pressure drop through the valve and the flow rate on the downstream side of the valve.

The L_w obtained from this graph applies to the octave bands containing the cutoff frequency of the pipe or duct downstream from the valve and at frequencies greater than the cutoff frequency. The cutoff frequency can be calculated as follows:

$$f_{co} = \frac{0.58c}{2D}$$

where f_{co} = cutoff frequency, Hz
c = speed of sound in the gas in the pipe, ft/sec
D = inside diameter of downstream pipe, ft.

The speed of sound in any gas can be obtained as follows:

$$c = 1{,}051 \sqrt{\left(\frac{t + 460}{460}\right)\frac{29}{M}} \qquad \textbf{(12.7)}$$

where c = speed of sound in the gas, ft/sec
t = gas temperature, °F
M = molecular weight of gas

Below the cutoff frequency, the sound power level will drop roughly 6 dB for each octave band.

Example 12.4:

Given: Saturated steam (M=18) at 150 psig (365° F) is to be reduced to 15 psig. The mass rate of flow is 2000 lb/hr. The line downstream from the valve is 4″ dia.

Question: What will be the L_w spectrum of the aerodynamic noise generated in the valve, assuming an insignificant change in steam temperature?

Solution:

$$\text{Mass flow} = \frac{2000 \text{ lb/hr}}{60 \text{ min/hr}} = 33.3 \text{ lb/min}$$

$$\text{psia} = \text{psig} + 14.7 \cong \text{psig} + 15$$
$$p_1 = 150 + 15 = 165 \text{ psia}$$
$$p_2 = 15 + 15 = 30 \text{ psia}$$

$$\frac{p_1}{p_2} = \frac{165}{30} = 5.5$$

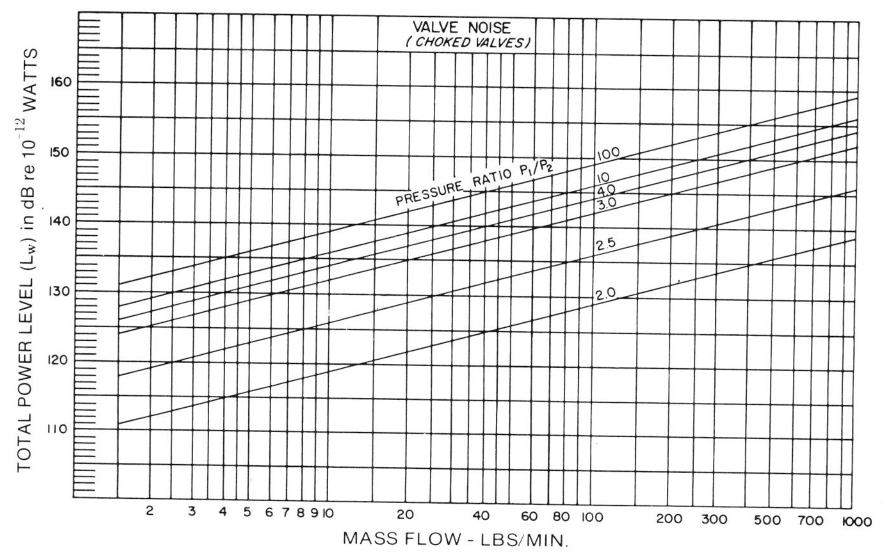

Figure 12.41 — Pressure reducing valve noise. From Ingard (1973).

From Figure 12.41, L_w = 139 dB

The velocity of sound in steam at 365° F and M = 18 is

$$c = 1{,}051\sqrt{\left(\frac{365 + 460}{460}\right)\left(\frac{29}{18}\right)} = 1787'/\text{sec}$$

The cutoff frequency for a 4″ dia. pipe is:

$$f_{CO} = \frac{0.58c}{D} = \frac{0.58 \times 1787}{1/3} = 3109 \text{ Hz}$$

The anticipated L_w is 139 dB in the octave bands at and above the cutoff frequency of 3109 Hz and in higher frequency octave bands. In this case, the L_w drops 6 dB in each octave below the 4000-Hz octave band. Thus, the anticipated octave band noise spectrum would be as shown below:

OB Ctr. Freq.	(Hz)	63	125	250	500	1000	2000	4000	8000
L_w	(dB)	103	109	115	121	127	133	139	139

Based upon these estimated sound power levels or values provided by the manufacturer, noise control may be required.

One approach to eliminating the noise of pressure-reducing valves is to design the valves with gradual pressure drop and expanding volume so that sonic velocity is never reached. One such valve that has proven quite successful is the drag valve shown in Figure 12.42. This valve consists of a stack of discs approximately 3/16″ thick. The discs have small gas flow passages that provide a tortuous path for the gas flow from the high-pressure to the low-pressure side. As the valve stem moves, more or fewer of these discs are exposed and gas volume control is provided. These valves are designed for gas velocities low enough that mufflers or acoustical lagging are not required. Typical values of noise reduction of this valve relative to one in comparable service are:

OB Ctr. Freq. (Hz)	63	125	250	500	1000	2000	4000	8000
Noise Reduction(dB)	24	19	28	17	18	25	34	34

For existing valves and for new installations where drag valves are not justified on an economic basis, mufflers and wrapping treatments can be applied.

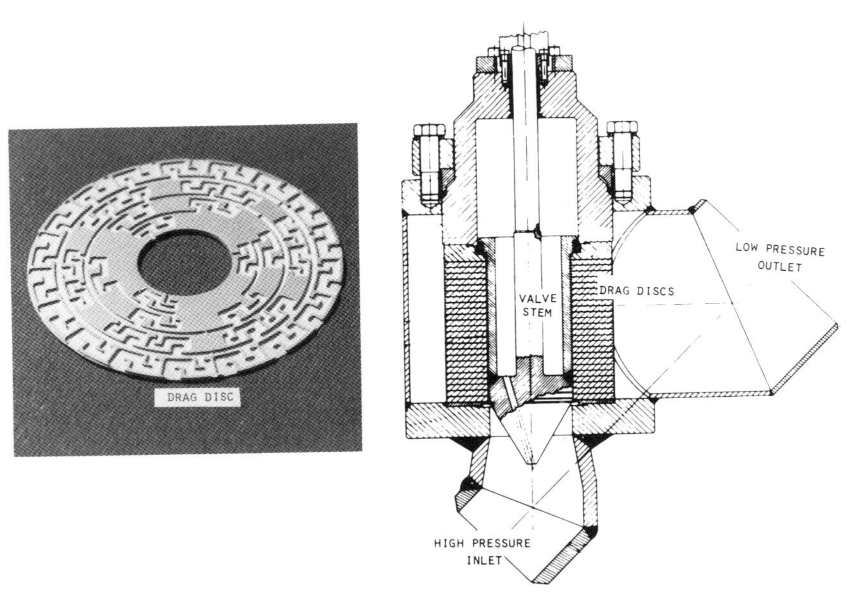

Figure 12.42 — Quiet pressure-reducing valve.

Mufflers and Wrapping Treatments

Figure 12.43 illustrates three common types of muffler for controlling noise from exhausts. The dissipative mufflers in the two upper illustrations reduce noise by absorbing it in acoustical linings. Dissipative mufflers provide broadband noise reduction. The thickness of acoustical linings should be determined by the noise characteristics of the source. Thicker linings attenuate low-frequency energy more effectively than do thinner ones. The straight-through lined duct type muffler shown in the upper illustration of Figure 12.43 adds very little pressure drop to the system. It is generally used only for diameters of 6″ or less. For diameters greater than 6″, the centered body type shown in the middle illustration is used, particularly where low-pressure drops are required.

Where acoustical linings cannot be used, the non-dissipative or reactive muffler shown in the lower illustration in Figure 12.43 may be used. These are most useful in controlling pulsating flow noise in a restricted frequency range. This type of muffler consists of cavities and restrictions or side branches designed to trap and/or reflect the sound energy at a particular set of frequencies back into the sound source. Because of the interactions between flow, adjacent sections of the muffler, and the sound source, the design of reactive mufflers is quite complicated and the final designs are often developed experimentally. It is usually expedient to purchase commercially available mufflers based on performance.

The two most common approaches for controlling the noise of an existing pressure-reducing valve are:

> muffler downstream of valve with wrapping
> extensive wrapping of valve and piping

Figure 12.44 illustrates a muffler downstream of the valve. In addition, the valve and the pipe between the valve and the muffler are covered with an acoustical lagging consisting of 2″ glass fiber (3 to 4 lbs/cu ft) with an outer covering of 16 gauge steel. Where a muffler cannot be used in the downstream piping, the valve and the downstream piping can be covered with an acoustical lagging as shown in Figure 12.45. The wrapping can be the same as mentioned above. Depending on the details of the installation, it may be necessary to wrap as much as 100′ of the downstream piping. There are many other materials that can be used for acoustical lagging, including mineral fiber and foam for the inner cover and lead-loaded vinyl and composite lead-aluminum materials for the outer cover. In addition, it is important not to excite equipment located downstream of the valve, such as thin-walled heat exchangers, separators, and spray towers.

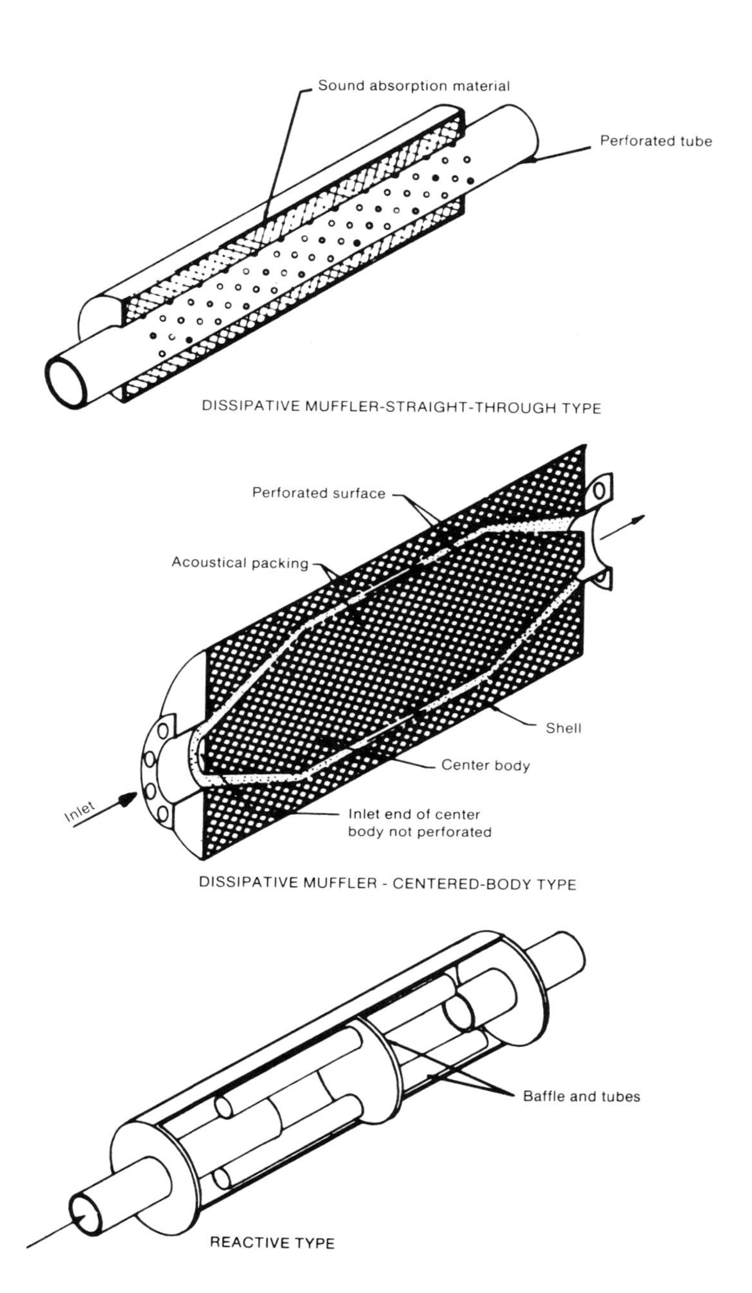

Figure 12.43 — Muffler designs.

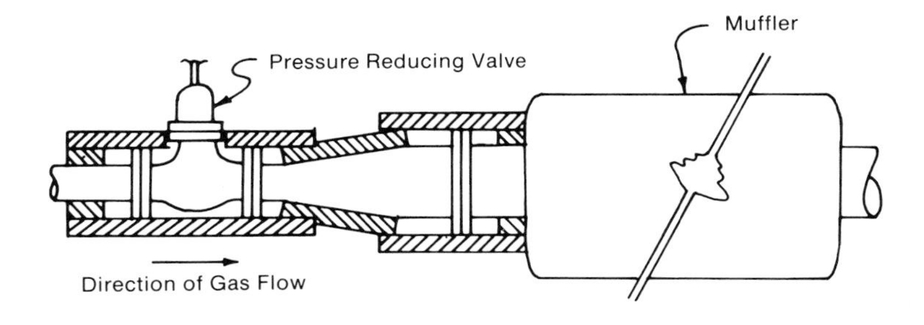

Figure 12.44 — Pressure-reducing valve noise control by muffler and lagging.

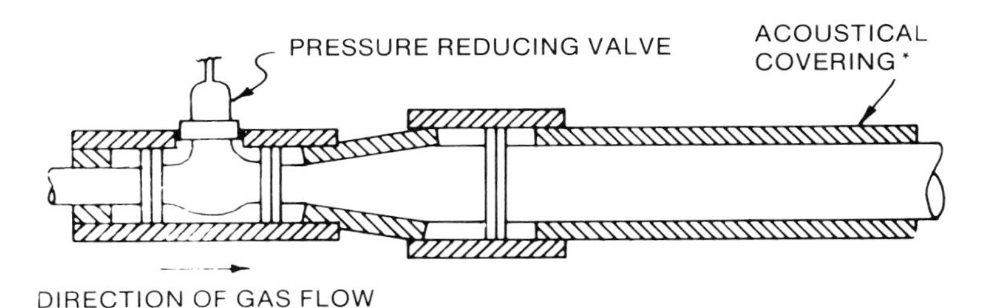

*2" THICK GLASS FIBER (AT LEAST 3LB. PER CU. FT. DENSITY) PLUS AN OUTER COVERING OF LEAD-LOADED VINYL.

Figure 12.45 — Pressure-reducing valve noise control by valve and pipe lagging.

Path Modification

ENCLOSURES

Where reduction of noise at the source is impractical, attenuation of airborne sound can be achieved by enclosing the source. Although there are many factors that govern the actual construction of an enclosure, only acoustical considerations will be included in the following design procedure. To keep the example as simple as possible, a single octave-band level of the overall noise is considered and a single material of construction is chosen for the enclosure. In practice, all significant octave bands and all types of materials should be considered.

Transmission Loss vs. Noise Reduction

To design effective acoustical enclosures, two terms must be understood: noise reduction (NR) and transmission loss (TL). In this context, NR may be defined as the difference between sound level measurements inside and outside an enclosure. NR is dependent upon the acoustical environment both inside and outside the enclosure. On the other hand, TL is independent of environment. It is the number of decibels by which sound energy incident on a partition is reduced in transmission through the partition, *i.e.*, the ratio of the incident to the transmitted energy. Table 12.1 lists the measured TL of common building materials (NIOSH, 1973; NIOSH, 1975). Measured values of TL rather than theoretical values should be used in the design of enclosures.

In designing enclosures, many factors must be considered if the enclosure is to prove satisfactory from both an acoustical and a production point of view.

1. Enclosure wall: The materials used to construct the basic shell will largely determine the NR of the enclosure. In order to achieve even a small amount of isolation (10 dB reduction), it is necessary to construct the shell from materials impervious to air flow.

TABLE 12.1
Sound Transmission Loss of General Building Materials and Structures (NIOSH, 1973; NIOSH, 1975)

Material or Structure	Weight lb/sq ft	Transmission Loss, dB Octave Band Center Frequencies, Hz					
		125	250	500	1000	2000	4000
DOORS							
1-¾", wood	4.8	23	28	27	25	30	32
1-¾", wood with 16ga. steel facing	6.8	25	27	32	35	28	30
1-¾", flush wood door	8.1	25	33	34	35	42	46
2-½", acoustical door with 12 ga. steel facing	21.9	43	49	48	55	57	44
10", acoustical door with damped metal absorbing insulation, hardware and gasketing as core	36	47	53	61	64	64	65
GLASS							
¼" plain surface structure, core of cast acrylic plastic	1.45	16	17	22	28	33	35
¼" glass with Saflex polyvinyl Butyral interlayer		29	34	35	33	41	
Two layers of ¼" glass with 0.045" Butacite core	6.2	31	32	34	35	40	47
Two layers of ¼" plate glass with ½" air space	6.2	25	20	27	33	29	36
¾" plate glass		33	35	40	43	41	

TABLE 12.1 — continued

Material or Structure	Weight lb/sq ft	Transmission Loss, dB Octave Band Center Frequencies, Hz 125	250	500	1000	2000	4000
WALLS-HOMOGENEOUS							
0.120″ flexible mastic	1.0	14	15	27	35	33	41
.008″ sheet lead	0.5	13	14	15	22	27	32
¼″ sheet steel	10	23	38	41	46	43	48
⅜″ sheet steel	15	26	39	42	47	41	51
20 gauge steel roof decking, 1.5″		30	39	46	49	50	50
6″, 2 coats of bondex cement base paint facing; core of masonry wall, lightweight concrete blocks, no paint or plaster on back	26	38	36	40	45	50	56
4″ hollow concrete block wall, dense aggregate		30	39	43	47	54	50
WALLS-NONHOMOGENEOUS							
3-¾″ wall with ½″ gyp board on both sides of 2-½″ steel studs (no insulation)		17	28	37	45	43	42
4-⅜″ wall with 2 layers of ½″ gyp board on one side and 1 layer of ½″ gyp board on the other side of 2-½″ steel studs (no insulation)		28·	31	46	51	52	47
5-⅜″ wall with ⅝″ gyp board on resilient channel on one side and ⅝″ gyp board on the other side of wood studs with 3-½″ fiberglass	6.4	29	40	51	55	48	58
10-¾″ wall with 2 layers of ⅝″ gyp board on both sides of a double row of wood studs with 3-½″ insulation		37	47	55	63	67	67
3-¼″ wall with perforated 18 ga. galvanized steel C-liner with glass fibers sealed in polyethylene bags and channel wall 20 ga. galvanized steel	4.7	18	20	29	38	40	45

2. Acoustical linings: The inside of the enclosure shell should be lined with sound absorbing material to prevent the reverberant build-up of noise. The thickness and density of lining will depend upon the frequency at which the greatest noise reduction is required. For instance, if glass fiber material is used, 1″ thickness and 3-lb density provides absorption coefficients of greater than 0.5 for frequencies of 500 Hz or above, as listed in Table 12.2. Figure 12.46 illustrates the effects of material thickness on sound absorption.

3. Seals: If a noise reduction of more than 10 dB is required of an enclosure, it must have tight-fitting joints. Figure 12.47 shows various seals that might be used. A noise enclosure must have all cracks and openings tightly sealed in order to reduce leakage of noise. The enclosure in the figure illustrates applicable principles of sealing and fastening the base, wall, door, and observation window. It should not be considered a complete enclosure design because of the possible need for ventilation, acoustical lining, *etc.*

4. Mounting: The enclosure should be isolated from any vibrating part of a machine. If the machine is mounted on a heavy concrete base or floor, it is usually satisfactory to fasten the enclosure to the floor. However, if the machine causes considerable vibration in the base or floor, either the machine or the enclosure (preferably the machine) should be vibration isolated.

5. Access for product flow or maintenance: In most cases, provisions must be made for convenient access of operation and maintenance. Usually, this can be done by providing access doors, using remote indicating gauges, providing removable panels, or by providing one permanent panel through which oil, water, and electrical lines can be run. The purpose of the permanent panel is to make it easy to remove the rest of the enclosure for access.

6. Effect on machine: Forced ventilation of the enclosure might be required if the enclosure will cause overheating of the machine. If the enclosure is ventilated, the inlet and discharge ducts must be muffled to prevent the escape of noise.

Calculation of Required TL

If an enclosure is placed over a noise source, the L_p inside the enclosure will depend upon the L_w of the source and the room constant of the enclosure, as given by the following equation:

$$L_p = L_w + 10 \log \left[\frac{Q}{4\pi r^2} + \frac{4}{R} \right] + 10.5, \text{ dB} \tag{12.8}$$

where Q = directivity factor
r = distance from the acoustic center of the source, in ft
R = room constant, in sq ft
L_w = sound power level re 10^{-12} watts

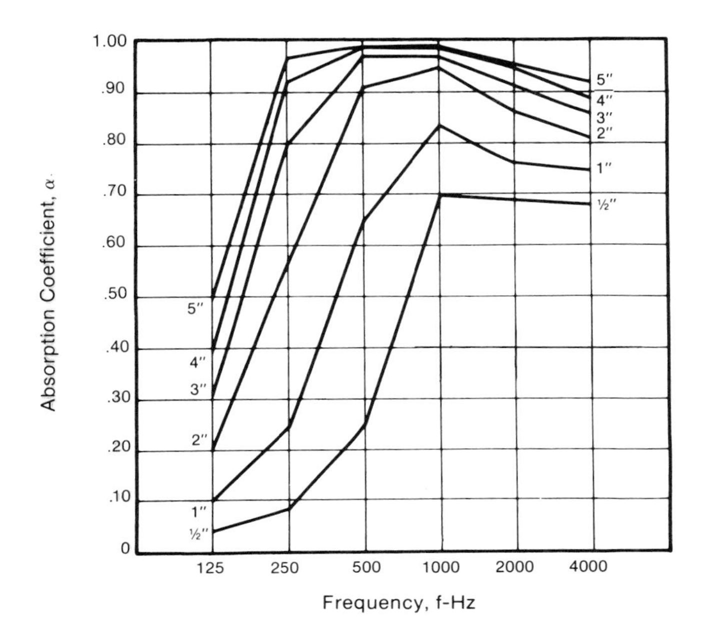

Figure 12.46 — Variation in the absorption coefficient of 6 lb/ft^3 fiberglass as a function of frequency, with material thickness as a parameter.

TABLE 12.2
Typical Sound Absorption Coefficients for Common Building Materials (AIMA, 1974; NIOSH, 1975)

MATERIAL		SOUND ABSORPTION COEFFICIENT Octave Band Center Frequencies, Hz					
		125	250	500	1,000	2,000	4,000
Ashes dumped, loose (2.5	11" thick	.90	.90	.75	.80		
lb. water per cu ft)	3" thick	.25	.55	.65	.80	.80	
Brick, unpainted		.02	.02	.03	.04	.05	.05
Brick, unglazed		.03	.03	.03	.04	.05	.07
Brick, unglazed, painted		.01	.01	.02	.02	.02	.03
Carpet, heavy, on concrete		.02	.06	.14	.37	.60	.65
Same, on 40 oz hairfelt or foam rubber		.08	.24	.57	.69	.71	.73
Same, with impermeable latex backing on 40 oz hairfelt or foam rubber		.08	.27	.39	.34	.48	.63
Concrete block, coarse		.36	.44	.31	.29	.39	.25
Concrete block, painted		.10	.05	.06	.07	.09	.08
Fabrics:							
Light velour, 10 oz per sq yd, hung straight, in contact with wall		.03	.04	.11	.17	.24	.35
Medium velour, 14 oz per sq yd, draped to half area		.07	.31	.49	.75	.70	.60
Heavy velour, 18 oz per sq yd, draped to half area		.14	.35	.55	.72	.70	.65

TABLE 12.2 — continued

MATERIAL	SOUND ABSORPTION COEFFICIENT Octave Band Center Frequencies, Hz					
	125	250	500	1,000	2,000	4,000
Floors:						
Concrete or terrazzo	.01	.01	.02	.02	.02	.02
Linoleum, asphalt, rubber or cork tile on concrete	.02	.03	.03	.03	.03	.02
Wood	.15	.11	.10	.07	.06	.07
Wood parquet in asphalt on concrete	.04	.04	.07	.06	.06	.07
Foams:						
1", 2 lb/cu ft polyester	.23	.54	.60	.98	.93	.99
2", 2 lb/cu ft polyester	.17	.38	.94	.96	.99	.91
Glass:						
Large panes of heavy plate glass	.18	.06	.04	.03	.02	.02
Ordinary window glass	.35	.25	.18	.12	.07	.04
Glass fiber:						
1", 3 lb/cu ft	.23	.50	.73	.88	.91	.97
1", 6 lb/cu ft	.26	.49	.63	.95	.87	.82
Gypsum Board, ½", nailed to 2 × 4's 16" o.c.	.29	.10	.05	.04	.07	.09
Marble or Glazed Tile	.01	.01	.01	.01	.02	.02
Openings:						
Stage, depending on furnishings			.25 to .75			
Deep balcony, upholstered seats			.50 to 1.00			
Grills, ventilating			.15 to .50			
Plaster, gypsum or lime, smooth finish on tile or brick	.01	.02	.02	.03	.04	.05
Plaster, gypsum or lime, rough finish on lath	.14	.10	.06	.05	.04	.03
Same, with smooth finish	.14	.10	.06	.04	.04	.03
Plywood paneling, ⅜" thick	.28	.22	.17	.09	.10	.11
Sprayed-on acoustical material, 1" cellulose applied to metal lath, 2.5 lb/cu. ft.	.47	.90	1.10	1.03	1.05	1.03
Water surface, as in a swimming pool	.008	.008	.013	.015	.020	.025

ABSORPTION OF SEATS, AUDIENCE AND AIR
per square foot of seating area or per unit

	125	250	500	1,000	2,000	4,000
Air, per 1000-ft^3 @ 50% RH				.9	2.3	7.2
Audience, seated in upholstered seats, per sq ft of floor area	.60	.74	.88	.96	.93	.85
Unoccupied cloth-covered upholstered seats, per sq ft of floor area	.49	.66	.80	.88	.82	.70
Unoccupied leather-covered upholstered seats, per sq ft of floor area	.44	.54	.60	.62	.58	.50
Wooden Pews, occupied, per sq ft of floor area	.57	.61	.75	.86	.91	.86
Chairs, metal or wood seats, each, unoccupied	.15	.19	.22	.39	.38	.30

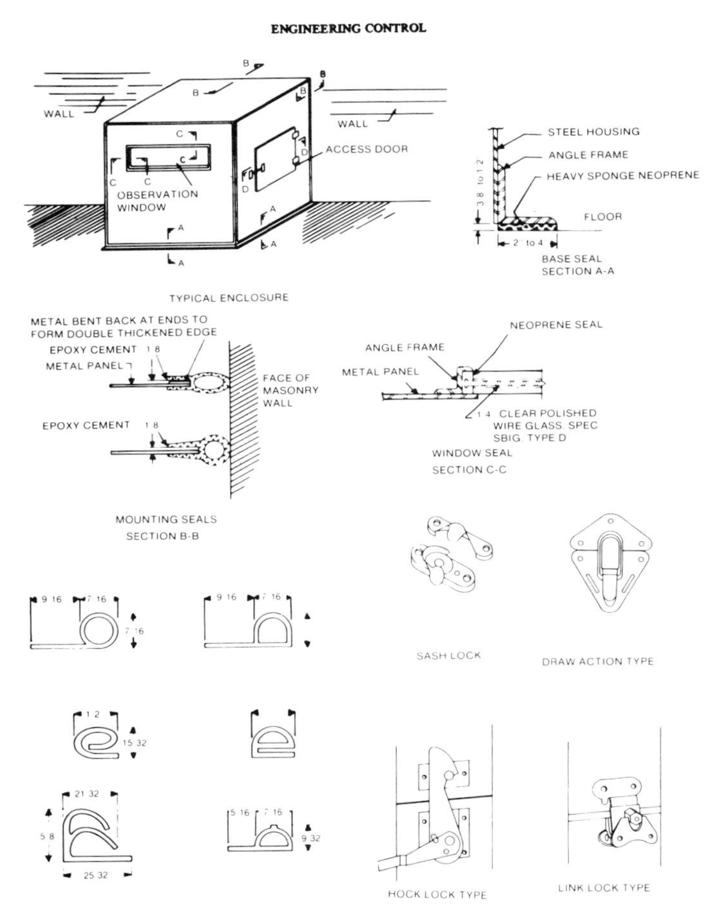

Figure 12.47 — Typical access door seals and fasteners.

Unlined Enclosure

Example 12.5:

The octave-band analysis of the noise generated by a gear reducer, at the nearest operator position, exceeds the desired noise level criterion in the 500-, 1000-, and 2000-Hz octave bands as shown in lines 1 and 2 of Table 12.3. The gear reducer is mounted on the floor near the center of a 50 × 30 × 10 ft room. The floor and ceiling of the room are constructed of unpainted poured concrete and walls of unpainted brick. It is decided that the best means of reducing the noise in this case is with a 3 × 3 × 3 ft enclosure of sheet steel.

TABLE 12.3
Enclosure of a Noise Source—Work Sheet

	Octave Band Center Frequency (Hz)	63	125	250	500	1000	2000	4000	8000
Line	**Factors**					**dB**			
1	L_p 1.5 ft from source	82	84	87	86	93	88	80	81
2	Criteria	105	98	93	85	85	85	85	85
3	Required noise reduction—NR				1	8	3		
4	Allowance for noise variation				5	5	5		
5	Allowance for reverberant buildup					15			
6	Total TL required for enclosure walls					28			

The first step is to determine the required transmission loss (TL) of the enclosure bounding walls, *i.e.,* the sheet steel.

Line 3 of Table 12.3 represents the required NR at the operator position (1.5 ft from the noise source). In this example, the analysis will be confined to the 1000-Hz octave band. Many problems in the field will require an analysis of several (or all) of the octave bands. An allowance of 5 dB is entered in line 4 of Table 12.3 as a safety factor for possible variation in the noise levels and deviation of the performance of the treatment from theoretical values. It may be noted that we are neglecting possible small-enclosure effects that should be considered for a complete noise control analysis (Irwin and Graf, 1979).

From Table 12.2, the absorption coefficient of the concrete floor and ceiling is .02. For the unpainted brick walls, the absorption coefficient is .04.

The average absorption coefficient is calculated as follows:

$$\bar{\alpha} = \frac{\alpha_1 S_1 + \alpha_2 S_2 + \alpha_3 S_3 + \ldots + \alpha_n S_n}{S_1 + S_2 + S_3 + \ldots + S_n} \quad \textbf{(12.9)}$$

$$\bar{\alpha} = \frac{(50\times30\times2\times0.02) + (30\times10\times2\times0.04) + (50\times10\times2\times0.04)}{3000 + 600 + 1000} = 0.027$$

where α_i = absorption coefficient of one of the bounding surfaces of a room,

S_i = area of the surface in sq ft

$\alpha_i S_i$ = sound absorption of surface S_i measured in *sabins* in units of sq ft; thus a "perfect" absorber with an area of 1 ft^2 has an absorption of 1 sabin.

The room constant,

$$R = \frac{S\bar{\alpha}}{1-\bar{\alpha}} = \frac{4600 \times 0.027}{1-0.027} = 128 \text{ sq ft} \quad \textbf{(12.10)}$$

r = 1.5 ft, the distance from the center of the noise source to the proposed bounding surface of the enclosure,
Q = 2, since the gear reducer is located near the floor and this produces essentially hemispherical radiation,

Substituting into Equation 12.8:

$$L_p = L_w + 10 \log \left[\frac{2}{4\pi(1.5)^2} + \frac{4}{128} \right] + 10.5, \text{ dB}$$

$$= L_w + 10 \log (0.1020) + 10.5$$
$$= L_w + 0.6$$
$$\cong L_w + 1 \text{ dB}$$

Now determine the L_p relative to the L_w within the enclosure. From equation 12.10 and using an absorption coefficient for steel of 0.02:

$$R = \frac{S\bar{\alpha}}{1-\bar{\alpha}} = \frac{54 \times 0.02}{1-0.02} = 1.10 \text{ sq ft}$$

Substituting into equation 12.8 with r=1.5 and Q=2:

$$L_p = L_w + 10 \log \left[\frac{2}{4\pi(1.5)^2} + \frac{4}{1.10} \right] + 10.5 = L_w + 16.2 \text{ dB}$$

Comparing the results obtained above, the L_p within the enclosure is approximately 16 dB above the L_w, and the L_p before the enclosure was installed was 1 dB above the L_w. This means that the enclosure will cause a 15-dB increase in L_p within the enclosure which is due to the reverberant buildup. This is entered in line 5 of Table 12.3.

The total TL required for the enclosure is the sum of lines 3, 4 and 5, and is shown in line 6 to be equal to 28 dB. The thickness of the steel sheet must be selected to give the required TL.

Lined Enclosure

The required TL of the enclosure bounding surfaces can be reduced by lining the enclosure with sound-absorbing material. If the enclosure were lined with 2″ of 6-lb/ft^3 density glass fiber blanket and if the manufacturers' literature has indicated an absorption coefficient of about 0.92 for this frequency range, the average α would be increased to 0.77 and R would become 181 sq ft. The L_p inside the enclosure would equal L_w. This means that by using a directivity factor of 2, the L_p inside the enclosure would be 1 dB less than it would be at the same location without the enclosure. The required TL would then be 8 plus 5 minus 1 or 12 dB. The enclosure lining has reduced the required TL of the enclosure from 28 dB to 12 dB, which means that a much lighter enclosure can be used.

Another point in favor of a lined enclosure is that the lower noise level inside the enclosure will make leaks or access openings in the enclosure wall less detrimental to the NR.

Enclosure of a Working Environment (Within a Noisy Area)

The NR provided by an enclosure in a noisy area (noise source outside the enclosure) depends not only upon the TL of the bounding surfaces of the enclosure, but also upon the room constant of the enclosure. These relationships are expressed in the following equation:

$$NR = TL - 10 \log\left[\frac{1}{4} + \frac{S_w}{R}\right] \qquad \textbf{(12.11)}$$

where NR = difference in L_p from outside to inside of enclosure,
TL = transmission loss of enclosure in dB,
S_w = exposed area of enclosure in sq ft, and
R = room constant of enclosure working environment in sq ft

Figure 12.48 is a nomogram of this equation for computing the amount of TL required to meet the desired NR. This equation also applies in determining the NR of a wall separating a noisy room from a relatively quiet room.

This indicates that the noise reduction provided by an enclosure with a very large absorption (a "dead" space) is limited to a maximum of:

$$NR = TL + 6 \text{ dB}$$

For a medium dead enclosure, the NR would seldom exceed the TL by 1 or 2 dB. For a live enclosure, the NR usually will be at least 5 or 6 dB smaller than the TL.

Example 12.6:

The center of a 100′ by 50′ by 14′ manufacturing area has a noise level as shown by line 1, Table 12.4. The objectives are: (1) to enclose a small office in the center of this room, and (2) to provide a noise level inside the office to meet the Preferred Noise Criteria (PNC) of line 2, Table 12.4 (Beranek *et al.*, 1971). The problem is to determine the required wall and ceiling TL to meet the criteria. The office is to have an acoustical ceiling consisting of 1″, random drilled, ceiling tile having an α of 0.65 at 500 Hz and 0.45 at 250 Hz.

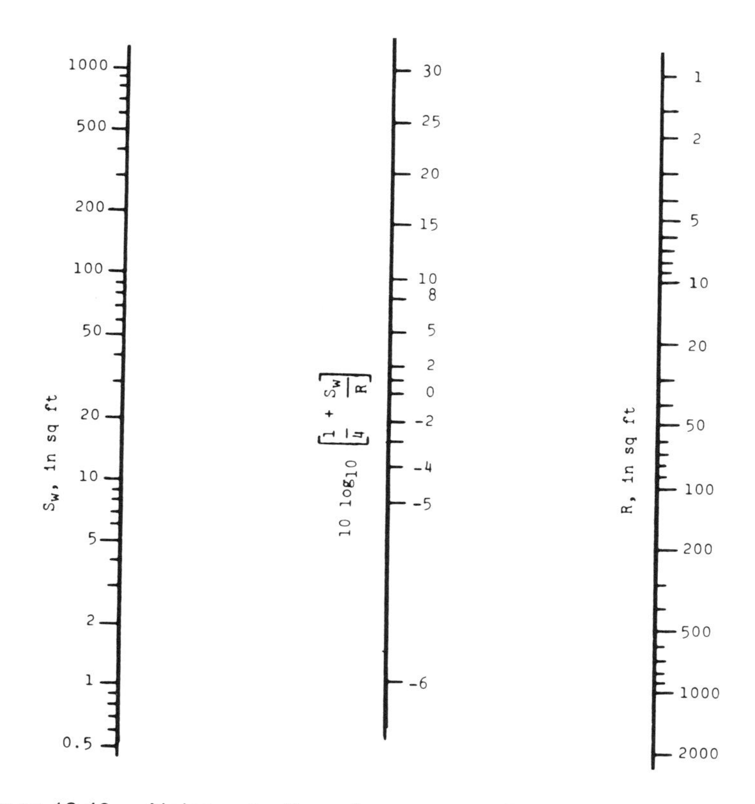

Figure 12.48 — Noise reduction of common walls between rooms. To find noise reduction: draw a line connecting S_w (area of transmitting wall in sq ft) to R (room constant of receiving room in sq ft), subtract the value at the middle bar intersection from the wall transmission loss; the remainder is the noise reduction.

The α of the floor and the walls is to be 0.02 at 500 Hz and 0.01 at 250 Hz. The office is to be 15′ by 10′ by 8′.

The average absorption coefficient of the office at 500 Hz would be:

$$\bar{\alpha} = \frac{(15\times8\times2\times0.02) + (10\times8\times2\times0.02) + (15\times10\times0.02) + (15\times10\times0.65)}{240 + 160 + 150 + 150} = \frac{108.5}{700} = 0.155$$

The room constant, R, of the office would be:

$$R = \frac{700 \times 0.155}{1\text{-}0.155} = \frac{108.5}{0.845} = 128.4 \text{ sq ft}$$

TABLE 12.4
Enclosure of a Working Environment—Work Sheet

	Oct. Band Ctr. Freq.-Hz	63	125	250	500	1000	2000	4000	8000
Line	Factors				dB				
1	Existing noise level	76	81	85	88	86	84	80	77
2	Criteria PNC-55	70	66	62	59	55	51	48	48
3	Line 1 minus Line 2	6	15	23	29	31	33	32	29
4	Allowance for noise variation	5	5	5	5	5	5	5	5
5	Required NR (line 3 plus line 4)	11	20	28	34	36	38	37	34
6	Total TL required for enclosure walls			36.5	40.6				
7	TL wall selected			39	43				
8	TL ceiling selected			39	39				

For the 500-Hz octave band from Table 12.4, the required NR is shown to be 34 (in line 5). Substituting into equation 12.11, we find:

$$34 = \text{TL} - 10 \log \left[\frac{1}{4} + \frac{550}{128.4} \right] = \text{TL} - 6.6$$

$$\text{TL} = 34 + 6.6 = 40.6 \text{ dB (enter in line 6 of Table 12.4)}$$

For the 250-Hz octave band, the average absorption coefficient would be:

$$\bar{\alpha} = \frac{(240 \times 0.01) + (160 \times 0.01) + (150 \times 0.01) + (150 \times 0.45)}{700} = 0.104$$

$$R = \frac{700 \times 0.104}{1 - 0.104} = \frac{72.8}{0.896} = 81.25 \text{ sq ft}$$

For NR = 28, we find:

$$28 = \text{TL} - 10 \log \left[\frac{1}{4} + \frac{550}{81.25} \right] = \text{TL} - 8.5$$

$$\text{TL} = 28 + 8.5 = 36.5 \text{ dB}$$

The TL of 40.6 dB at 500 Hz and the 36.5 dB at 250 Hz from line 6 of Table 12.4 can be satisfied with a 4″ hollow concrete block wall having dense aggregate with no surface treatment. The TL of this wall can be found in Table 12.1.

The ceiling TL requirements can be satisfied with two sheets of ⅜″ gypsum lath clamped tightly together and ⅞″ of sanded gypsum plaster on each side. Other panels that can meet the desired TL are presented in Table 12.1. Since the acoustical ceiling of 1″ acoustical tile was used in the example, this must be added to the interior ceiling.

A good sealant such as a dense caulking compound that maintains its resilience would be required where the wall and ceiling meet. Ceiling and wall panels should be sealed at the edges so that the noise cannot leak through a part of the panel. Any windows, doors, or vents in the building surfaces of the office should meet the TL specifications determined above. The door would require an automatic bottom door seal and an extruded rubber seal at the top and sides.

Very often it is desirable to have viewing windows and ventilation ducts in enclosures. These might have less TL than the walls of the enclosure. The TL of an enclosure with components of different TLs may be determined using Figure 12.49 to combine individual TLs two at a time. From this

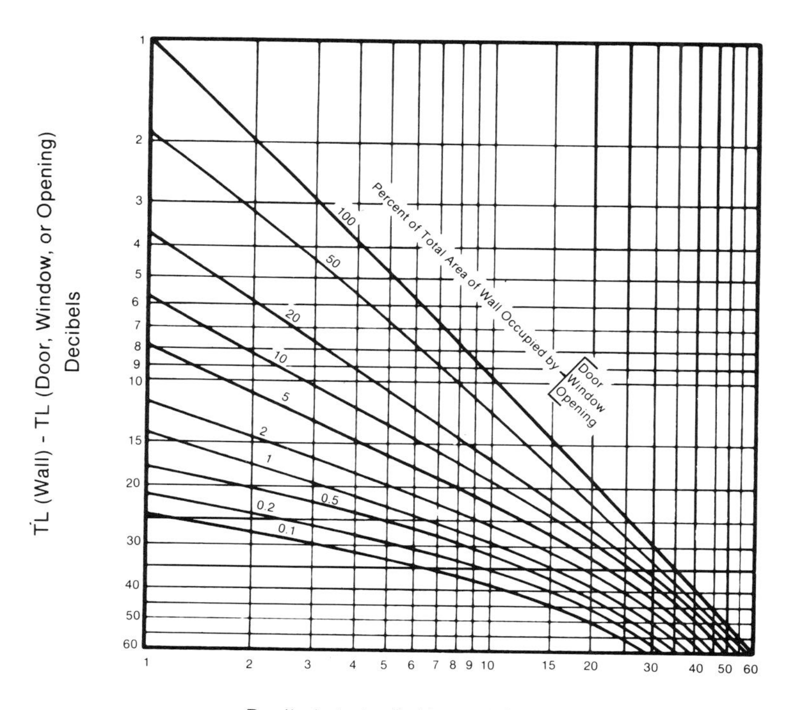

Figure 12.49 — Composite transmission loss of walls with doors, windows, and openings.

figure, it is seen that a very small area of low TL (*e.g.*, cracks) can greatly reduce the total TL. Although increasing the mass of the major portion of the enclosure will partially compensate for poorly fitted windows or doors, it is usually less expensive to adequately seal such openings. Figure 12.50 can be used to determine the effect of leaks in an enclosure.

Enclosures can be constructed for steam turbines used to drive a boiler feed pumps. In one case, turbine noise exceeded the criteria by a considerable amount and was particularly objectionable due to its narrow-band nature. To control this noise, the turbine was enclosed with 16-gauge steel lined with 1″ thick glass fiber wool having a density of 3 lb/ft^3. The noise level was reduced by 14 dB in the 4000-Hz octave band in which the narrow-band noise occurred.

Figure 12.51 shows a motor-gear assembly and the enclosure installed for noise control. Also shown is the noise reduction due to the enclosure. The enclosure is constructed of ⅛″ steel with welded joints and lined with 1″ fiberglass board. Silencers for intake and exhaust ventilation air are 12″ parallel plates of 1″ neoprene coated fiberglass board spaced 1″ apart. A solid floor separates the motor-gear area from the tank area below the floor.

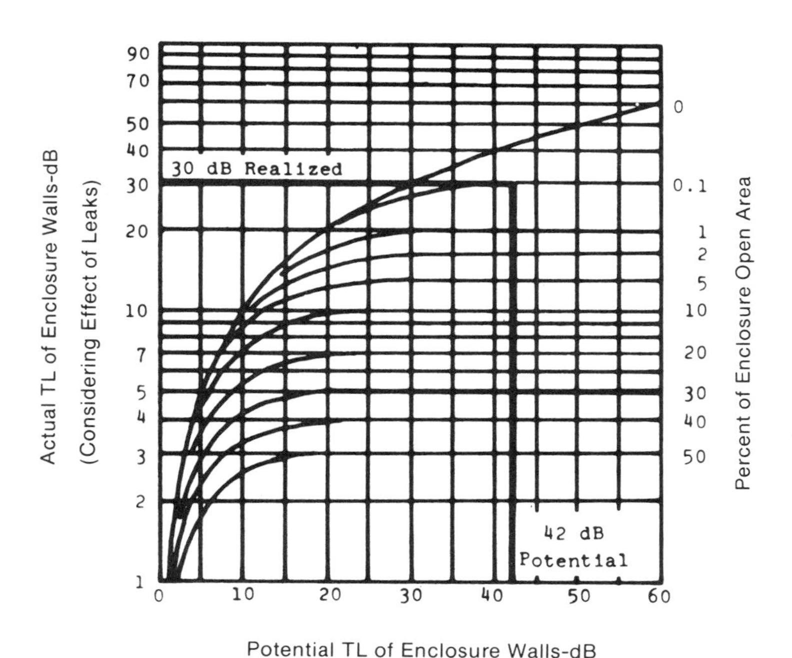

Figure 12.50 — Enclosures—effect of leaks.

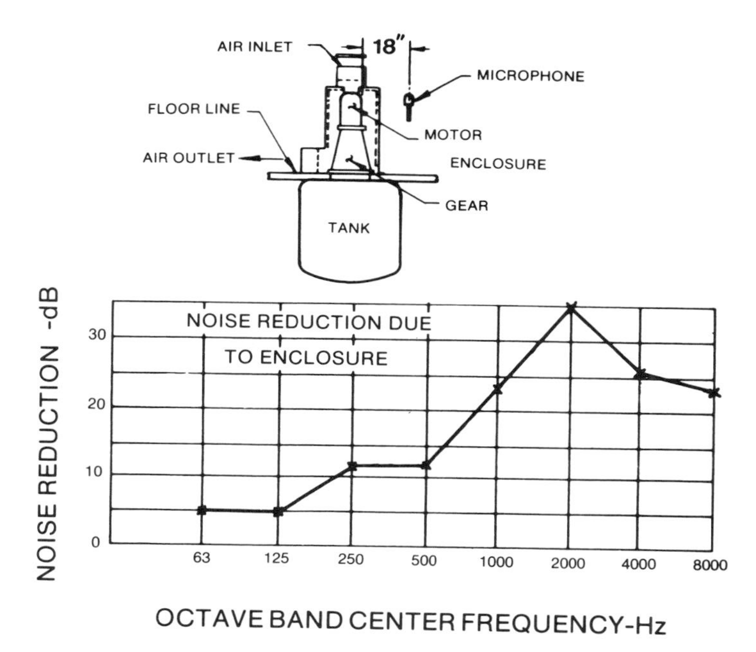

Figure 12.51 — Noise reduction of motor-gear enclosure.

Figure 12.52 shows a section of a tumbling barrel and the enclosure provided by the manufacturer. By lining the housing for the tumbling barrel with 1" thick glass fiber and coating the inside of the door with a damping compound, the sound pressure levels were reduced as shown in the figure.

Figure 12.53 shows a rubber tooth belt drive for a high-speed drill which created excessive noise in the 4000-Hz octave band. A tight fitting enclosure of 16-gauge steel lined with 1" of sound absorbing material provided the noise reduction shown in the figure. The actual reduction in belt noise was greater than shown because noise from other machines dominated after the enclosure was installed.

Figure 12.54 presents a schematic of an enclosed screen sifter used to separate metal. The figure also presents sound pressure levels measured before and after enclosing this machine. Doors were provided for the removal of the filtered material. The doors were lined with 1" thick fiberglass and isolated from the sides of the enclosure by thin strips of soft rubber. In order to minimize structure-borne noise propagation, the body of the enclosure was isolated from the exhaust hood and the floor.

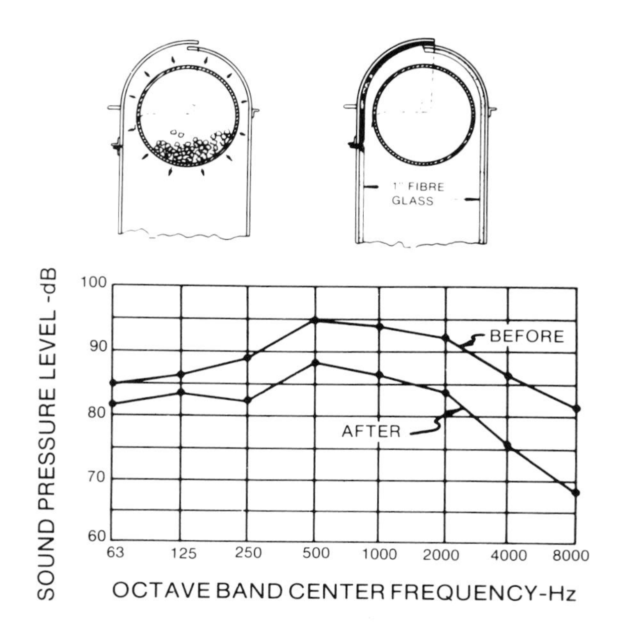

Figure 12.52 — Effect of absorption inside enclosure.

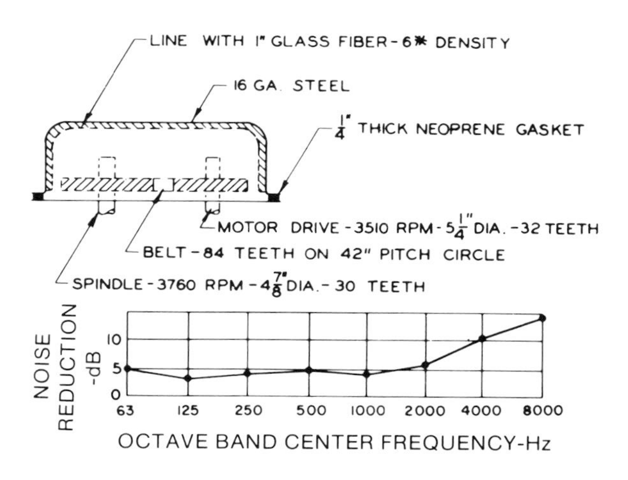

Figure 12.53 — Noise reduction of enclosure.

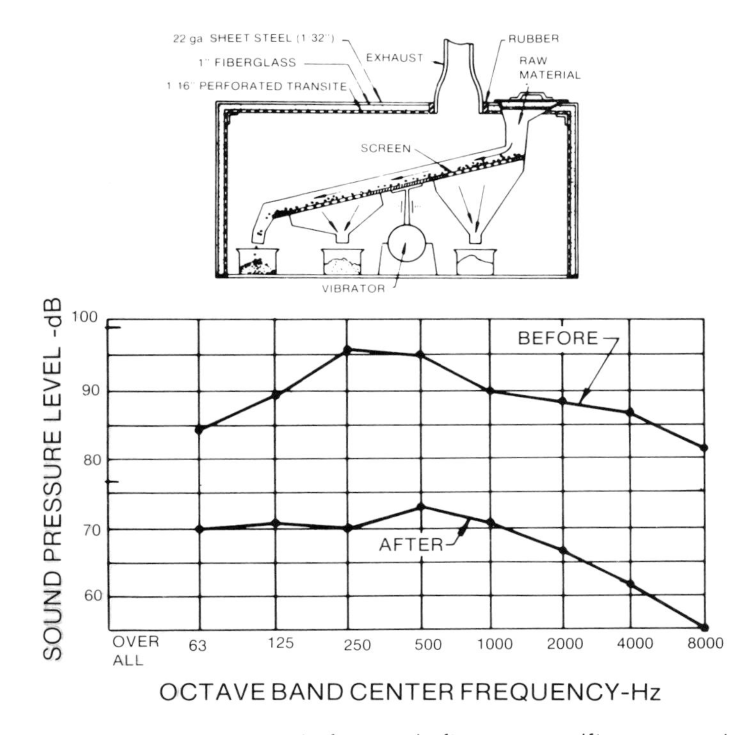

Figure 12.54 — Noise levels before and after screen sifter was enclosed.

An edger planer used for trimming foamed plastic in the fabrication of sandwich-type panels generated a noise level at the operator position of 106 dB in the 500-Hz octave band. This noise was generated by the high-speed rotary cutters planing the rigid foam. To confine this noise, a sound absorbing enclosure was constructed of plywood lined with 4″ thick flexible polyurethane foam. Particular attention was given to making this enclosure as tight as possible. The enclosure was sealed to the floor with mastic, the doors were weatherstripped, and all penetrations through the enclosure walls were sealed. The feed and discharge openings were baffled with double flaps of rubber. This treatment reduced the noise level from 106 dB to 88 dB in the 500-Hz octave band. The reduction achieved by octave bands was:

OB Ctr. Freq. (Hz)	63	125	250	500	1000	2000	4000	8000
Noise Reduction (dB)	7	5	20	18	19	10	12	11

Oil or gas burners on large boilers create excessive noise due to aspirated primary or secondary air or both. By supplying combustion air through a piped system the noise can be drastically reduced. For example, a vaporizer boiler with sealed type burners was quieter than one with open air inlets as shown below:

OB Ctr. Freq. (Hz)	63	125	250	500	1000	2000	4000	8000
Noise Reduction (dB)	11	17	20	24	13	12	7	8

Simplified Design for Enclosures

The design of enclosures for a noise source, as discussed previously, is rather complicated and time-consuming. However, it can be greatly simplified with careful selection of the enclosure lining. If the enclosure can be designed with an $\bar{\alpha}$ of 0.7 or greater, TL will equal NR for all practical purposes, and the calculations to determine this relationship can be omitted. It would still be advisable to provide 5 dB of additional TL to allow for variations from theoretical values. It is to be emphasized that $\bar{\alpha}$ must be considered for the frequency of interest. Table 12.2 shows the performance of typical acoustical absorbing materials. Note that performance varies considerably with frequency and thickness of material.

Sometimes the shape of the noise source and the space available for an enclosure dictate a close-fitting enclosure. In such situations, a "lagging" treatment, which is in contact with the vibrating surface of a noisy machine, pressure reducing valve, or piping can be effective.

A pressure-regulating valve and gas line was covered with 1-½″ of preformed 85% magnesite, in turn covered with an asphalt paper sheath for weather protection. The thermal material was used instead of glass fiber because of its availability. The sound pressure levels measured before and after the lagging was installed are shown in Figure 12.55.

Figure 12.56 illustrates how a combination of principles of noise reduction has been applied to reduce the noise of a pulp refiner. The first step was to fill the pit below the refiner with an absorptive acoustical material in order to absorb the noise within the machine housing. The outside of the refiner was covered with a 1″ coating of acoustical material and then covered with an impervious ¼″ layer of asphalt.

Figure 12.57 presents a schematic of steam vacuum jets (eductors) on a 50,000 bbl/day vacuum column which were creating excessive noise. The primary steam jet discharges into a condensing exchanger. A pair of secondary jets drawing vacuum on the primary exchanger discharge into a second condensing exchanger. Two noise control treatments were installed. The initial step was to wrap the jets from the steam chest to the inlet flange of the exchangers. The wrapping consisted of 4″, 3 lb/ft^3 glass fiber plus an

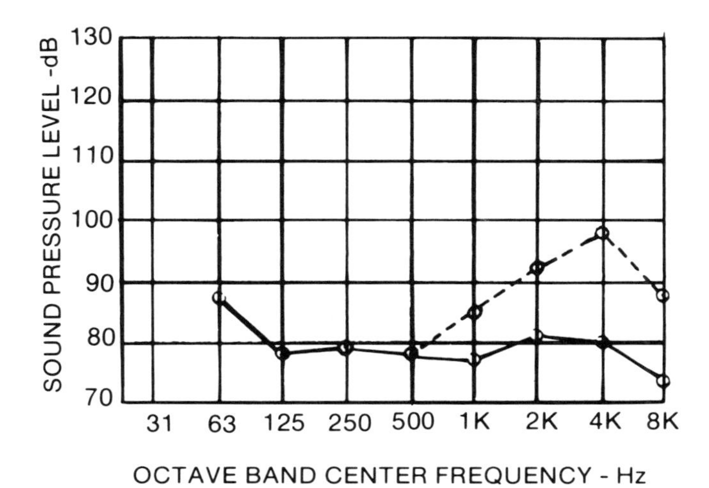

Figure 12.55 — Noise levels before and after lagging treatment.

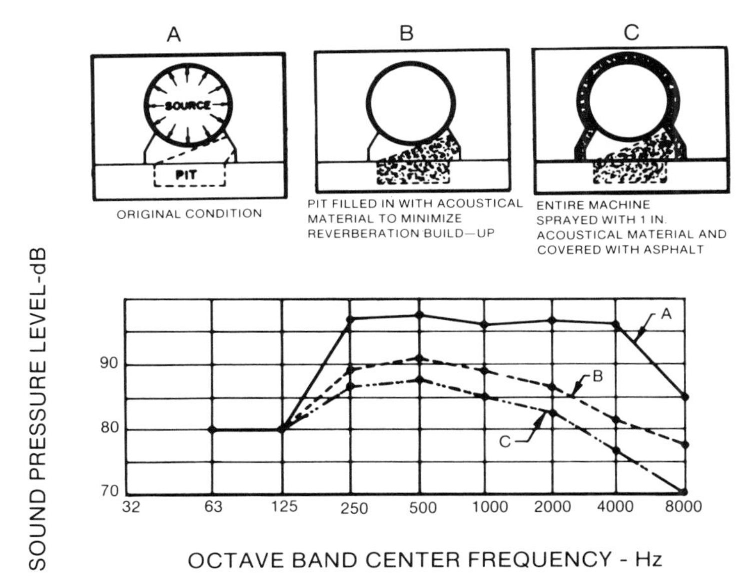

Figure 12.56 — Noise reduction for pulp refiner by acoustical absorption and lagging.

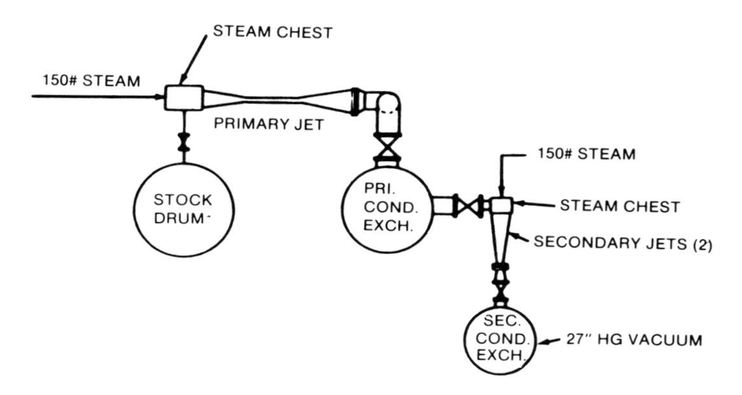

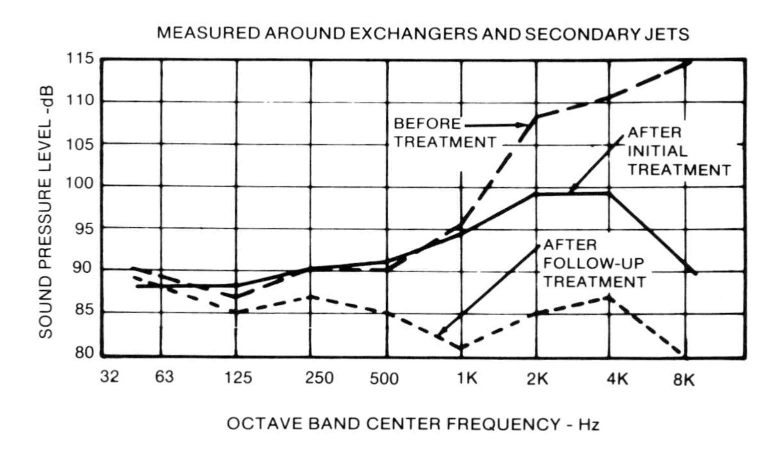

Figure 12.57 — Noise reduction of lagging treatments.

outer covering of 0.016″ aluminum. It was then obvious that too much noise was radiating from the exchanger inlets and shells (not exchanger heads). The second step was to cover these parts in a similar manner. The resulting noise levels are shown in Figure 12.57.

In all total-enclosure design, care must be taken to consider potential increases in enclosure temperature and their resultant effect on the system enclosed.

Multiple Wall Enclosures

For cases where the required TL is in excess of 50 dB, it is often more economical to use multiple wall (instead of the usual single wall) construction. When two or more impervious walls are separated by an air space with no solid tie between, the TL is usually considerably greater than would be expected on the basis of mass-law attenuation. Problems requiring such high-transmission-loss construction are usually complicated and they are beyond the scope of this chapter.

Partial Enclosures

When material flow or required access to machines prohibits complete enclosure, it is sometimes possible to control noise with a partial enclosure. The partial enclosure should separate the noise source from the employee. Partial enclosures should be lined with absorptive material to obtain maximum effectiveness. The more complete the enclosure, the greater will be the noise reduction obtained. Partial enclosures are useful mainly in giving a shadow effect for workers who would otherwise be in the high-level direct field. This shadow effect is limited to high-frequency sounds where the dimensions of the barrier are several times the wavelength of the predominant noise. The extent of the shadow zone also depends on the distance from the openings, the configuration of the enclosure and the absorption of the nearby room and machine surfaces. The reduction in level of the high-frequency direct sound will range from about 3 dBA for the partial enclosures shown in the upper part of Figure 12.58 to about 15 dBA for the more complete enclosure shown by the three lower examples.

Figure 12.59 shows a can divider, which is a device for splitting a conveyor line of cans into two lines. Cans fall into the top, drop on a bar at the bottom of the divider and roll off randomly in either direction. This device is placed in a freight car to feed cans to both ends of the car simultaneously. Due to access requirements, a complete enclosure cannot be used. However, the partial enclosure also shown in the figure is sufficient to reduce the noise to an acceptable level. The reduction achieved by such a partial enclosure will depend upon where the measurements are made. For instance, the figure shows the noise reduction that was achieved at 6″ from the side of the enclosure and at the bottom openings. Practically no reduction was achieved at the top opening.

Figure 12.60 illustrates the partial enclosures that were built around four motor-generator sets so that sound had no direct path of escape, as shown by Section A-A. This construction was chosen so that all sound waves would have to strike an absorbing surface at least once before they could

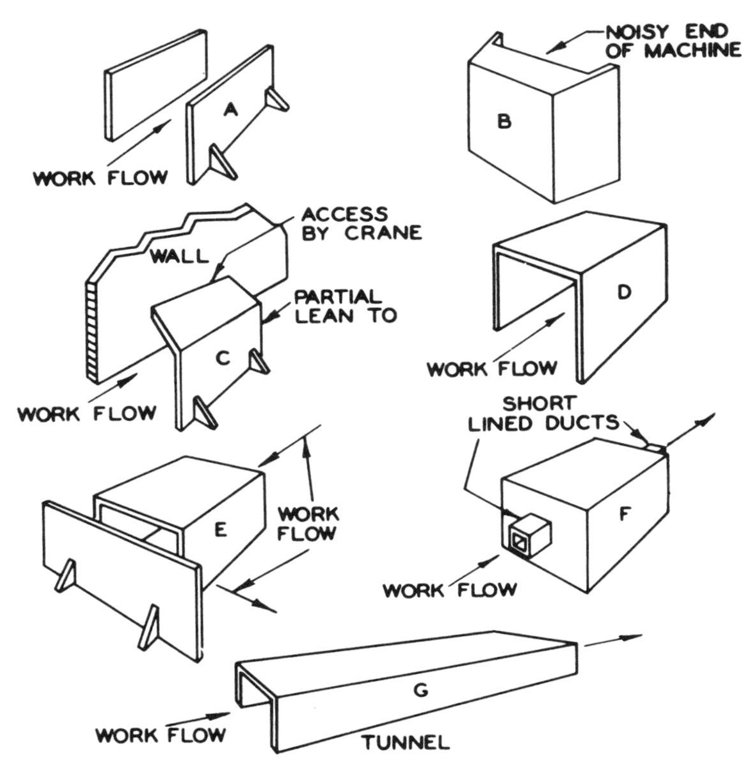

Figure 12.58 — Examples of partial enclosures.

escape into the room. The noise reduction achieved by these partial enclosures is also shown in the figure.

A river water pumping station contained four 900-hp, 20,000-gpm pumps which operated at 960 rpm. The gear reducers produced excessive noise so they were individually treated by covering with a partial enclosure such as is shown in Figure 12.61. The enclosure had a 6′ inside dia., a 9′ inside height, an outer cover of 3/16″ boiler plate, and a lining of 3″ glass wool faced with a ½″ mesh galvanized wire screen. Slots were provided for drive shafts and windows for inspection and lubrication. The performance of the enclosure is shown in the figure. The performance could be improved considerably by reducing the size of the openings.

A large friction saw is used for cutting structural steel. This saw operates in a way similar to that of a radial saw in a woodworking shop. A partial enclosure similar to a full-height telephone booth was constructed to

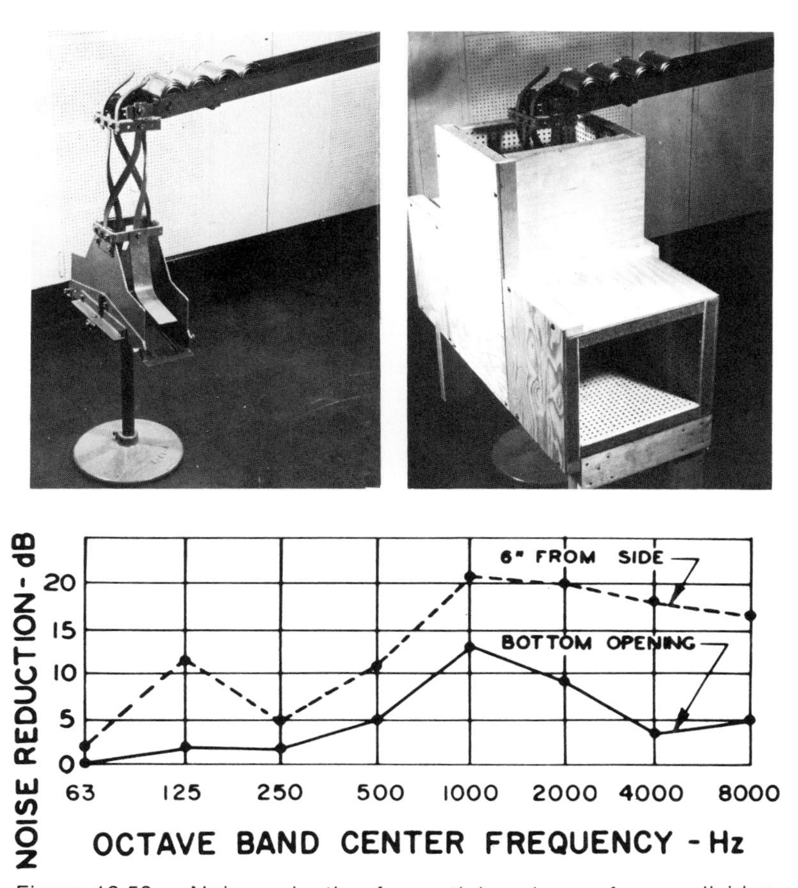

Figure 12.59 — Noise reduction for partial enclosure for can divider.

protect the operator. The walls of the booth are sheet steel outside and 4″ fiberglass plus an inner protective surface of perforated metal. It has a double-glazed safety glass observation window in one side and the opposite side is open. The noise reduction achieved was:

OB Ctr. Freq. (Hz)	63	125	250	500	1000	2000	4000	8000
Noise Reduction (dB)	6	1	10	12	14	19	20	21

A textile shuttlecock loom was located in a small room whose walls and ceiling were covered with acoustical tile. Three men normally work in this room but only one operates the loom. A partial enclosure was built around the machine. The partial enclosure was made of sheet metal lined with acoustical tile. The noise reduction achieved by the partial enclosure was:

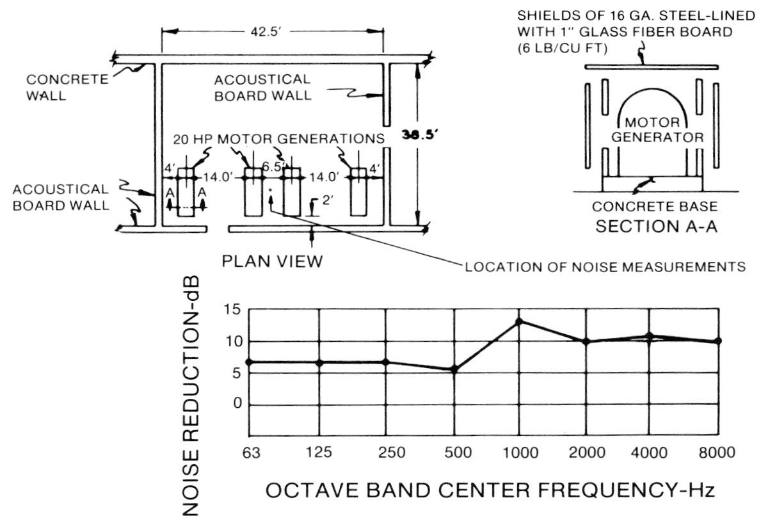

Figure 12.60 — Noise reduction of partial enclosures of motor-generators.

OB Ctr. Freq. (Hz)	63	125	250	500	1000	2000	4000	8000
Noise Reduction (dB)	2	3	2	10	10	12	9	6

Probably the best practical method for estimating the performance of a partial enclosure is to consider the percent of the radiation pattern that the partial enclosure intercepts. For instance, if 50% of the radiation area is intercepted, the reduction would be 3 dB. For 80% interception, the reduction would be 7 dB. For 90% interception, the reduction would be 10 dB. More detailed discussion of partial enclosures is given in the literature (Fehr 1951; Mariner, 1956; Bishop, 1957; Rettinger, 1957; NIOSH, 1978; Irwin and Graf, 1979).

SHIELDS OR BARRIERS

Shields or barriers can be used to control the transmission of noise when placed between the source and receiver. The attenuation is due to the incident sound waves being reflected away from the receiver by the barrier, causing a "noise shadow" behind the barrier. Such barriers can be effective for mid- and high-frequency noise but not low-frequency noise. The noise reduction of barriers is illustrated by Figure 12.62. The noise reduction depends upon the effective height of the barrier (H), the wavelength of the sound λ, and the angle of deflection θ as shown by Figure 12.63. It is apparent that the barrier should be as high as practical and as close to the

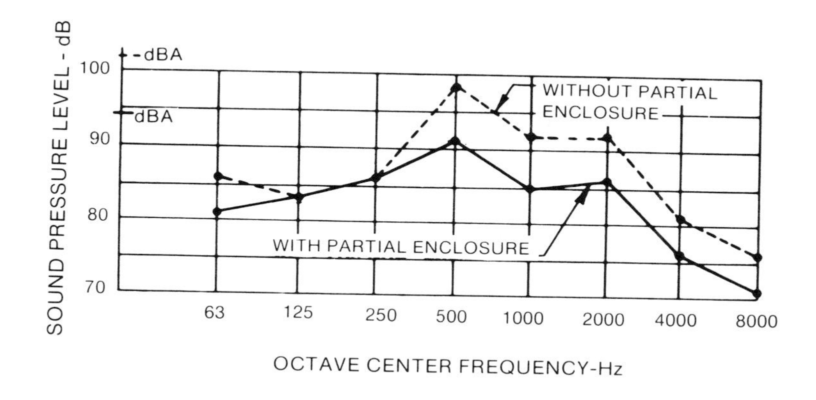

Figure 12.61 — Noise levels with and without partial enclosure for gear reducers on 900-hp water pumps.

source as possible. The barrier material should have a TL at least 9 dB greater than the expected noise reduction. The noise reduction due to the actual distance between the source and receiver (which depends on the environment) would be in addition to the effect of the barrier. Inside of buildings, barriers can be limited in their effect due to reflections off the ceiling and walls (Irwin and Graf, 1979).

Example 12.7:

> Determine the attenuation of a barrier having an H of 4′ and a θ of 30°. First, calculate the wavelength for the frequencies of interest using Equation 2.1, and record the value in Line 2 of Table 12.5. For this example, c = 1130 ft/sec was used. Then, calculate H/λ and record the value in Line 4. Finally, determine the estimated noise reduction for each frequency from Figure 12.63 and record the value in Line 6.

The use of shields between a noise source and an employee is usually quite effective when both the source and the employee are close to the shield and when the noise is predominantly high frequency. Figure 12.64 shows a shield on a punch press. Compressed air jets are used to blow foreign particles from the die. Sound pressure levels before and after a ¼″ thick safety glass shield was installed are also shown in the figure. It should be noted that with multiple units close together, this type of treatment would probably not be effective.

ROOM ABSORPTION

Small values of noise reduction (3 to 7 dB) can be achieved by room absorption under certain conditions. These conditions are: (1) the room before treatment has little sound absorption material in it, (2) the noise level of any of the sources does not exceed the design goal when operating out of doors, and (3) operators are not in the direct field. What one is really trying to accomplish is to reduce the noise to a level that would exist if the machines were running outdoors. In other words, the objective is to eliminate the reverberant buildup of sound caused by running the machines in the confines of the room. In order to do this, one must change the hard, smooth, impervious surfaces of the room (which reflect the sound) into soft, rough, porous surfaces which absorb the sound. The ability of a material to absorb sound is described by its absorption coefficient (α). This is defined as the ratio of the sound energy absorbed to the sound energy incident upon the material. The material composition and thickness, the way in which it is mounted, and the frequency of the sound all affect the value of the absorption coefficient.

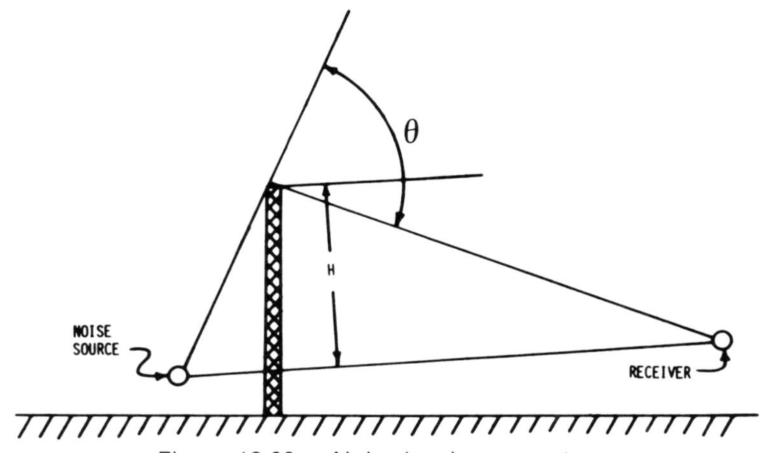

Figure 12.62 — Noise barrier geometry.

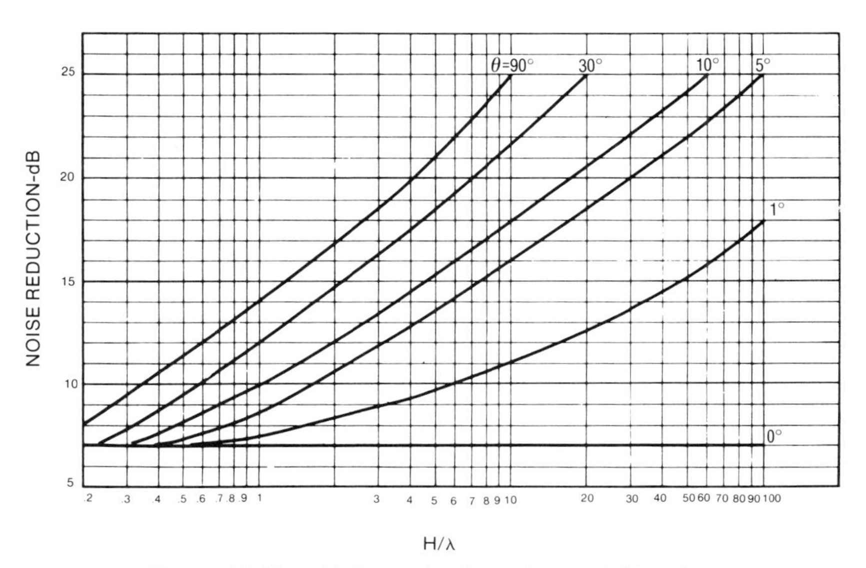

Figure 12.63 — Noise reduction of a partial barrier.

TABLE 12.5
Noise Reduction Through the Use of a Partial Barrier

Line	Octave Band Center Frequency Hz (f)	63	125	250	500	1000	2000	4000	8000
1	Speed of Sound— ft/sec (c)	1130	1130	1130	1130	1130	1130	1130	1130
2	Wave Length— feet (λ)	18	9	4.5	2.2	1.1	.56	.28	14
3	Barrier Effective Height—ft (H)	4	4	4	4	4	4	4	4
4	H/λ	0.22	0.44	0.89	1.8	3.6	7.0	14	29
5	Angle of Deflection— degrees (θ)	30	30	30	30	30	30	30	30
6	Noise Reduction (From Figure 12.63)	7	9	11	14	17	20	23	25

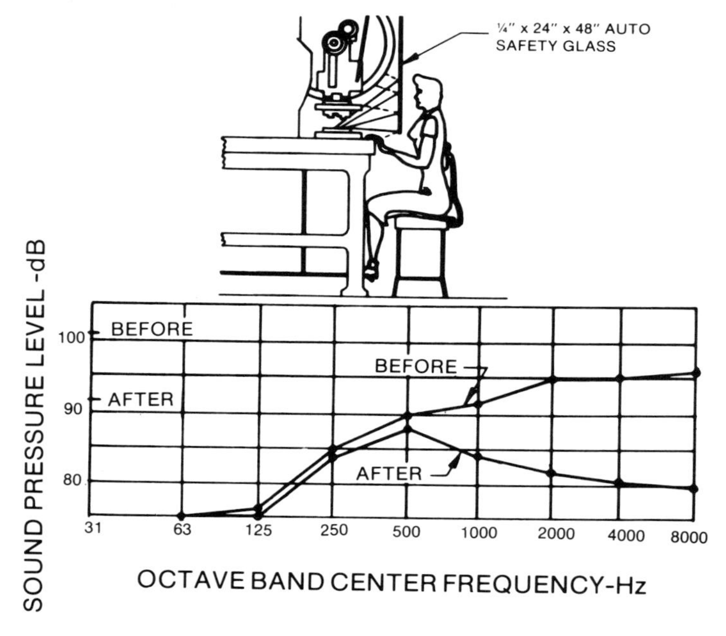

Figure 12.64 — Noise levels before and after safety glass shield was installed.

Since the absorption coefficient varies with frequency for different materials, intelligent selection requires octave-band analysis of the noise not only to choose the best material, but also to predict whether the expected results will meet the requirements. The theoretical noise reduction that can be accomplished in the reverberant field is determined by the following formula:

$$NR = 10 \log \left[\frac{A_2}{A_1} \right] \qquad \textbf{(12.12)}$$

where A_1 = total amount of absorption in the room before treatment (sabins)
A_2 = total amount of absorption in room after treatment (sabins)
NR = noise reduction (decibels)

They can be determined from the nomogram below:

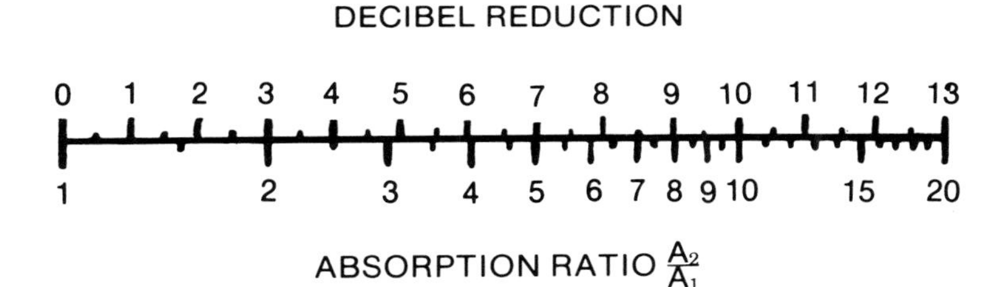

The total number of absorption units mentioned above is the sum of all the room surface areas multiplied by their respective absorption coefficients plus absorption due to other objects or people. Table 12.2 shows absorption coefficients for typical surface materials, people, and objects (AIMA, 1974; NIOSH 1975).

Example 12.8:

A typical example of the calculations involved in estimating the amount of NR that can be obtained by absorption treatment is as follows:

Room: 30′ × 60′ × 10′
Ceiling: Plaster
Floor: Concrete
Walls: Glazed tile
Steam pipes: Surface area 180 sq ft, magnesia covered

Machinery: Surface area 180 sq ft
People: Four
Frequency of interest: 1000 Hz
To calculate A_1:

	S		α for 1000 Hz (est.)		A Sabins
Ceiling	1800	×	0.02	=	36
Floor	1800	×	0.02	=	36
Walls	1800	×	0.01	=	18
Covered pipe	180	×	0.50	=	90
Machinery	180	×	0.02	=	4
Four People (4 sabins/ person)				=	16
			A_1	=	200 Total

The ceiling is to be covered with 1 inch of sound absorbing material having a coefficient of 0.80 at 1000 Hz. To calculate A_2:

Ceiling	1800	×	0.08	=	1400
Floor	1800	×	0.02	=	36
Walls	1800	×	0.01	=	18
Covered pipe	180	×	0.50	=	90
Machinery	180	×	0.02	=	4
Four People (4 sabins/ person)				=	16
			A_2	=	1564 Total

The reduction is determined simply by using the calculated values of A_1 and A_2.

$$NR = 10 \log \left[\frac{A_2}{A_1} \right] = 10 \log \left[\frac{1564}{200} \right] = 8.9 \text{ dB}$$

The results can be quite different for other frequencies; therefore, similar calculations should be repeated for all frequencies of interest. It is important to include all surfaces in making this type of calculation.

Ten automatic wire-cutting machines which have many mechanisms such as cams, gears, reciprocating parts, and metal stops were located in an alcove measuring 20′ × 60′. The operator of the machines does not have to be among them constantly. A low ceiling, only 7.5′ high, was of wood. Two end walls and one side wall were of brick. The fourth side was open to a large storage area. The floor was concrete.

Since there were multiple noise sources, a low ceiling, and very little absorption and because of the ability of the operator to be away from the nearfield of the machines, the use of acoustical absorption material could be of benefit. Perforated acoustical tile was applied to the ceiling and to a part of one end wall. A total area of 1275 square feet was covered. After installation of the acoustical tile, the employees commented that the working conditions had improved considerably. No reductions in the frequencies below 300 Hz were measured; however, in the other octave bands, noise reductions of 4 to 12 dB were obtained. At a distance of 20 feet from the center of the machines, where the operator could spend some time, the noise reduction was in the 15 to 20 dB range in the upper octave bands. The general area noise reduction was:

OB Ctr. Freq. (Hz)	500	1000	2000	4000	8000
Noise Reduction (dB)	4	4	9	12	10

A large reverberant room with many motor generator sets in it was too noisy. The room was treated with an absorptive treatment as shown in the photograph in Figure 12.65. These baffles are 6 lb/ft^3 fiberglass completely encased in Mylar film. They were hung in rows just above the level of the lights and on 3′ centers. The noise reduction achieved was:

OB Ctr. Freq. (Hz)	63	125	250	500	1000	2000	4000	8000
Noise Reduction (dB)	4	7	9	10	7	8	8	3

Figure 12.65 — Ceiling absorption with hanging baffles.

A textile weave room was treated with an absorptive ceiling treatment. The noise reduction was:

OB Ctr. Freq. (Hz)	63	125	250	500	1000	2000	4000	8000
Noise Reduction (dB)	6	9	6	6	6	11	11	12

LINED DUCTS AND MUFFLERS

Ventilation ducts and conveyor systems are often important noise transmission paths entering or leaving enclosures around noise sources or quiet areas. Ducts can be lined to attenuate sound traveling along them. The noise reduction achieved by lining ducts of regular shape with acoustical material having a given absorption coefficient may be calculated using the following formula:

$$NR = \frac{12.6\ P\alpha^{1.4}}{A}\ dB/ft \quad \textbf{(12.13)}$$

where α = absorption coefficient of the lining material at the frequency of interest
P = perimeter of duct in inches
A = cross-sectional area of duct in square inches

To simplify the use of this formula, Table 12.6 shows the value of $12.6\ \alpha^{1.4}$ for various absorption coefficients. This formula is most accurate between 250 and 2000 Hz and for $0.2 < \alpha < 0.4$ (Harris, 1979).

This formula for determining noise reduction is based on tests on ducts having cross-sectional dimensions in the ratios of 1:1 and 2:1. Even though parallel-baffle sound absorbers vary considerably from these ratios, the formula has proven to be a valuable and sufficiently accurate guide for estimating performance for most industrial noise problems. Of

TABLE 12.6
Sound Absorption Coefficient vs. $12.6\alpha^{1.4}$

α	$12.6\alpha^{1.4}$
0.50	4.77
0.55	5.46
0.60	6.16
0.65	6.89
0.70	7.65
0.75	8.43
0.80	9.22
0.85	10.02
0.90	10.87

course, actual performance data for mufflers is preferred when available. Where all of the air passages have the same sound-absorbing material, the noise reduction is calculated for one passage, not for the duct as a whole. In general, the smallest allowable distance between baffles provides the most noise reduction per foot of length. This formula becomes increasingly inaccurate as the ratio of the distance between the baffles to the wavelength increases: its use should be restricted to ratios of less than 0.1 (Beranek, 1971).

Where possible, it is desirable to avoid a line of sight noise path from the source to the receiver. Using several lined 90-degree bends or staggered absorbers in the path will greatly increase the noise reduction with little increase in cost, although the resulting increase in back pressure may need to be evaluated. More complete discussions of lined ducts and elbows are available (ASHRAE, 1981).

If the noise reduction required makes it necessary to use a longer duct than desirable, a parallel baffle absorber (sound trap) such as is shown in Figure 12.66 can be used. Figure 12.67 presents a sketch of a parallel baffle muffler and an elbow. The muffler was used to reduce the noise from the compressor intake of a 7000-hp gas turbine which was operating at 5800

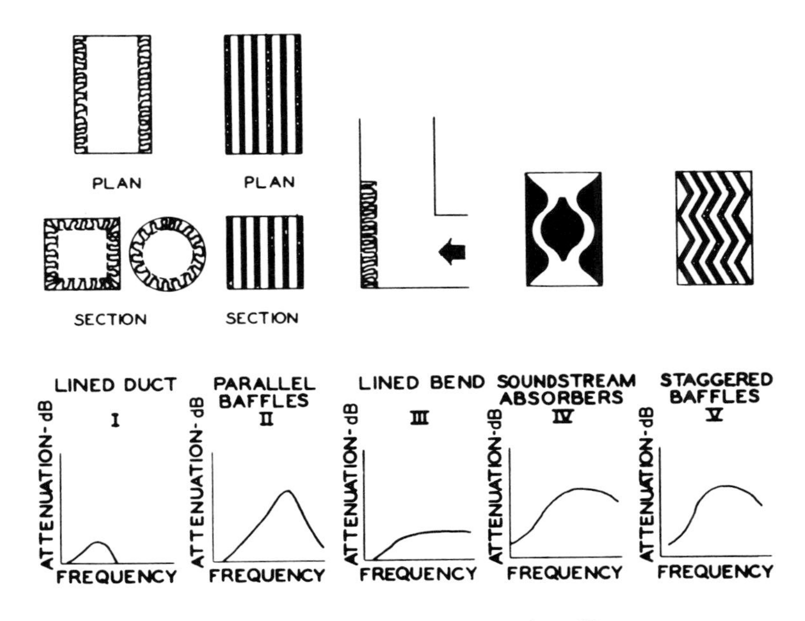

Figure 12.66 — Lined ducts and mufflers.

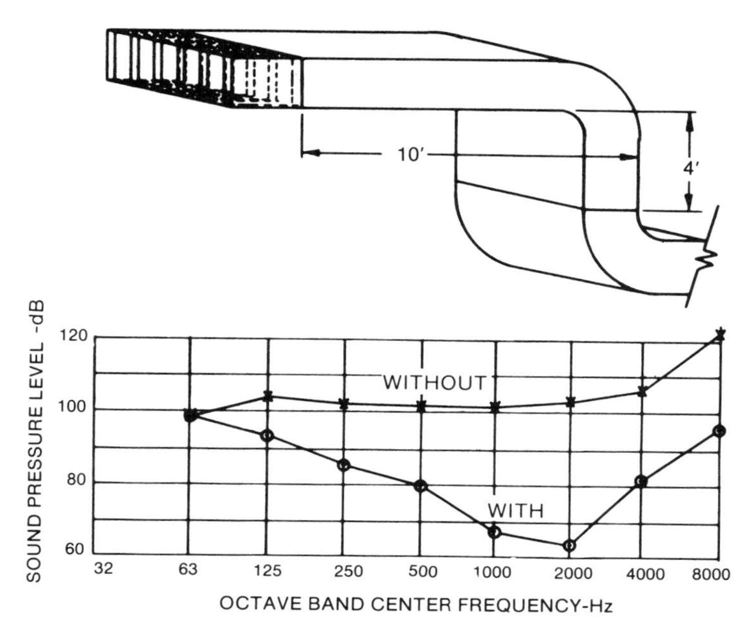

Figure 12.67 — Noise levels with and without muffler for air compressor intake for 7,000-hp gas turbine.

rpm and 6200 hp. The muffler consisted of six parallel baffles, each 3.5" wide, filled with glass or mineral fiber, and faced with 18-gauge perforated sheet steel. The cross section of the duct is 7' × 8' and the elbow is constructed of 0.25" unlined steel plate. The sound pressure levels before and after treatment are shown in the figure.

Figure 12.68 shows an air ejector used to strip waste textile fibers from perms (bobbins). The curve shows the noise reduction achieved 3 feet from the ejector by means of the dissipative muffler. The air supply line was ½" at a pressure of 100 psi; a 1" dissipative muffler was used. Notice that the noise levels in the 125- and 250-Hz octave bands were increased slightly, which is characteristic of this type of muffler. The noise is due to the pressure drop across the valve and not the velocity of air being exhausted from the pipe. This is apparent since the pipe size is the same at the inlet and discharge of the muffler.

Figure 12.69 is a photograph of a muffler installed on the discharge of an air motor. The noise reduction achieved was:

OB Ctr. Freq. (Hz)	63	125	250	500	1000	2000	4000	8000
Noise Reduction (dB)	2	7	7	9	10	23	29	23

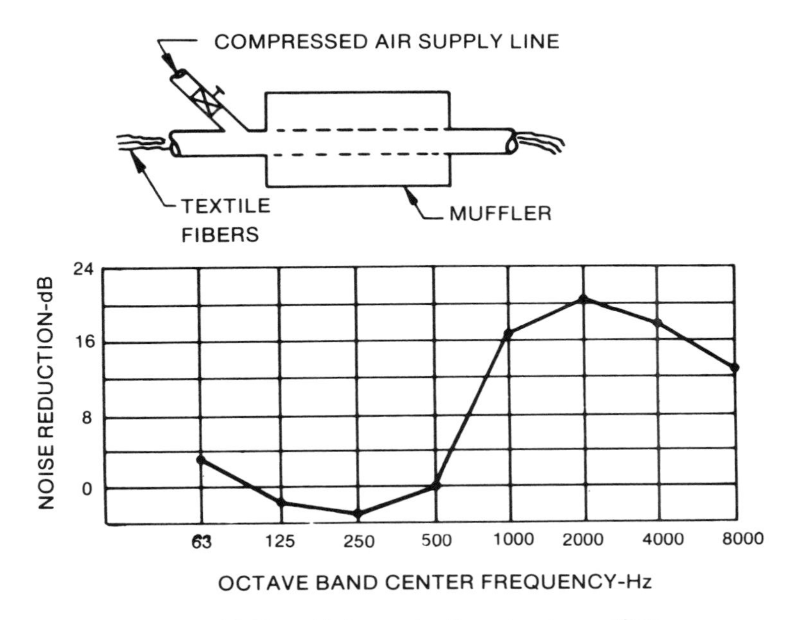

Figure 12.68 — Noise reduction due to muffler.

Air-operated cylinder-in-race type vibrators, as shown in Figure 12.70, can be operated reasonably quietly, provided that the air exhaust is muffled, the vibrator is mounted securely, and there are no loose parts on the vessel being vibrated. The noise reduction achieved by installing a muffler at the air discharge of the vibrator was:

OB Ctr. Freq. (Hz)	250	500	1000	2000	4000	8000
Noise Reduction (dB)	15	21	36	38	40	27

The vibrator should be mounted on a very heavy, rigid bracket so that the bracket itself does not become a serious noise radiator. Steel channels do not make good mounting brackets for vibrators where minimum noise radiation is desired. A preferable mount would be to weld studs to the reinforcing plate before welding the reinforcing plate to the hopper.

It is preferable to mount the muffler on a non-vibrating surface (such as the building structure) and to connect the muffler to the vibrator air discharge with rubber hose.

The intakes of reciprocating air compressors create low-frequency noise that can be objectionable to workers who work near them. An intake filter muffler shown in Figure 12.71 can be used to eliminate this problem. In this installation, the noise level was reduced by 23 dB in the 63-Hz octave band.

Figure 12.69 — Muffler installed on air motor.

Primary air intakes of 54 furnace oil burners created a noise problem in an oil refinery. Mufflers shown in Figure 12.72 were designed for the primary air intakes and installed on the burners. Primary air enters the forward bottom slot and passes over the first baffle and under the second baffle to reach the primary air openings in the burner. Noise generated by the primary air inlet must take the reverse path where it impinges on the acoustical material and is partially absorbed. The graph in Figure 12.72 shows the noise reduction achieved by the mufflers. The mufflers are constructed of an outer shell of 16-gauge metal and lined with 1″ glass wool faced with ¼″ mesh galvanized wire screen. Internal baffles, which are split for better fit around the burner venturi, are of sandwich construction with a center sheet of 16-gauge metal. The entire assembly is hinged on the top to allow the muffler sections to be opened in place for adjustment of the burners. Hinges can be slipped apart for complete separation of the muffler halves. Thumb screws are used to lock the muffler halves together and reduce the area of cracks at the joints where noise leakage might occur.

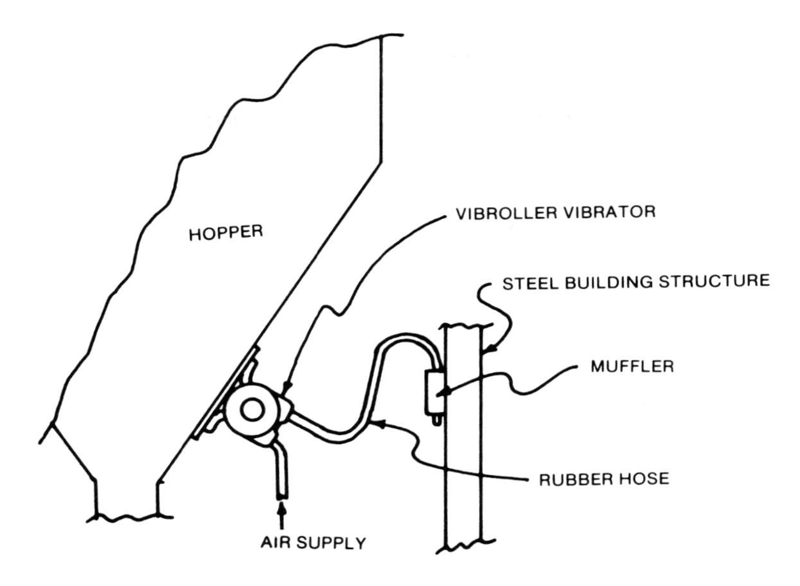

Figure 12.70 — Muffler on air discharge of vibrator.

Air hoists, especially when used on production lines, can be a serious noise source. This noise can be controlled by installing a muffler at the air discharge as shown in Figure 12.73. The noise reduction achieved was:

OB Ctr. Freq. (Hz)	500	1000	2000	4000	8000
Noise Reduction (dB)	18	31	35	38	28

Air inlet vents to a bomb room allowed noise to escape to the adjacent operating area creating a noise level of 98 dB in the 250-Hz octave band. There were two 12″ diameter vents. A muffler, as shown in Figure 12.74, was installed on each of the vents. The muffler was constructed so that the sound would strike a minimum of two sound-absorbing surfaces before escaping into the operating area. The noise reduction achieved was:

OB Ctr. Freq. (Hz)	63	125	250	500	1000	2000	4000	8000
Noise Reduction (dB)	0	0	8	10	26	23	24	26

Objectionable noise was radiating from the intake of a 2-stage centrifugal air compressor rated at 10,000 cfm at 110 psig discharge. To control this noise, a silencer was installed as shown by the photograph in Figure 12.75. Noise measurements at 3′ from the air intake are shown before and after treatment.

Aspirating gas burners on a four-burner, down-convection, end-fired heater were generating excessive noise. Twin intake aspirating air mutes

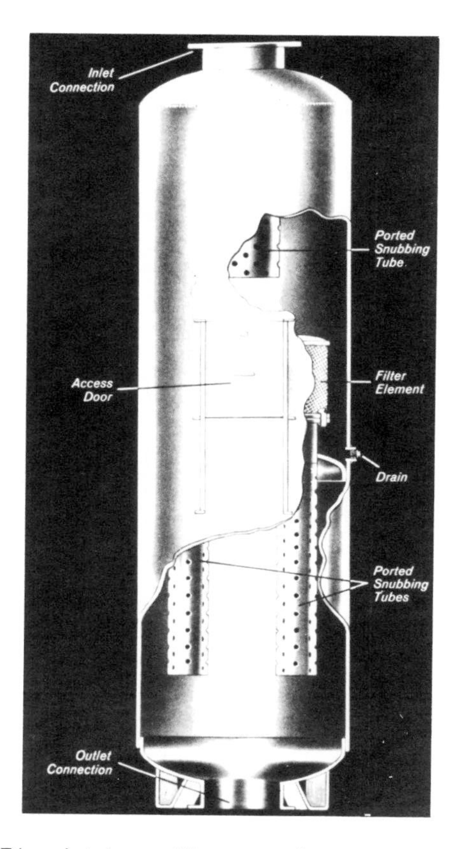

Figure 12.71 — Intake muffler on reciprocating compressor.

were installed as shown by the sketch in Figure 12.76. The noise reduction achieved is also shown. Since the mufflers had only a 1″ lining and since the low-frequency noise from the secondary air openings in the face of the burners was not muffled, noise below 250 Hz was not reduced. Measurements were made at 3′ from the aspirating air slot and between the burners.

A floor-fired cylindrical furnace having non-aspirating gas burners was generating excessive noise. Each burner was receiving 5400 scfh (standard cubic feet per hour) of natural gas at 10 psig. Special mufflers were designed to control this noise as shown by the sketch in Figure 12.77. The burners are enclosed in a common plenum chamber of galvanized steel lined with acoustic tile. The sound pressure levels at 10′ below the burners before and after treatment are also shown in the figure.

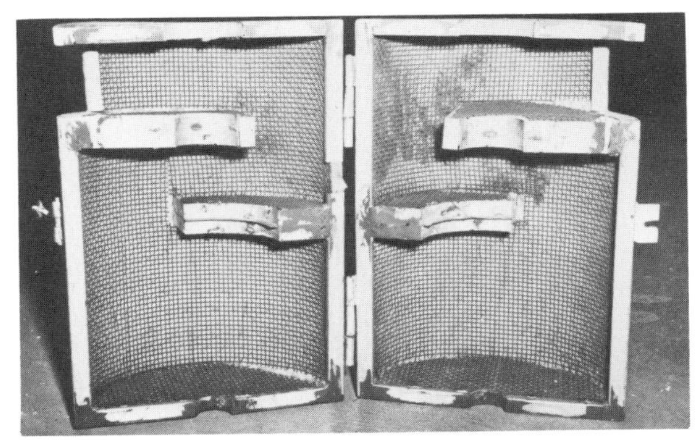

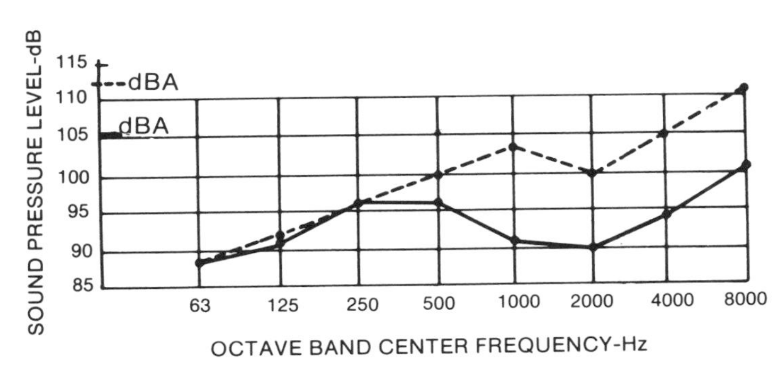

Figure 12.72 — Noise levels with and without mufflers on the primary air intakes of 54 furnace oil burners.

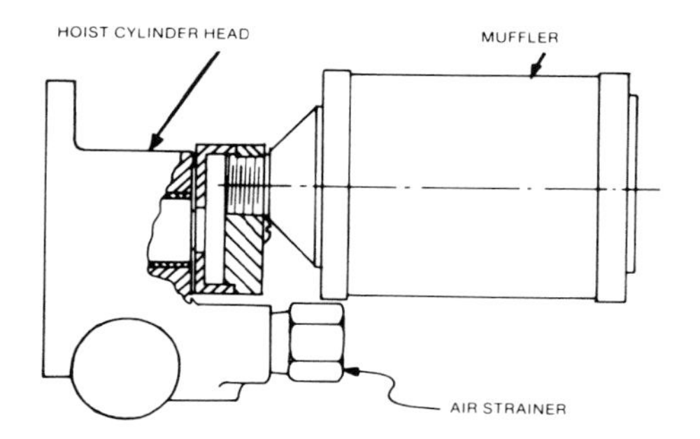

Figure 12.73 — Muffler for air hoist exhaust.

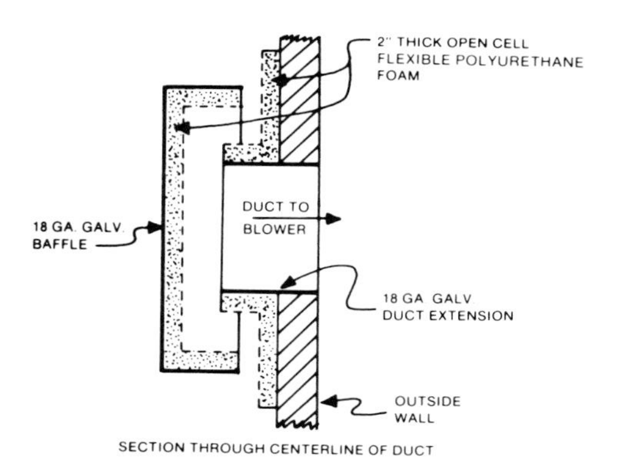

Figure 12.74 — Room air vent muffler.

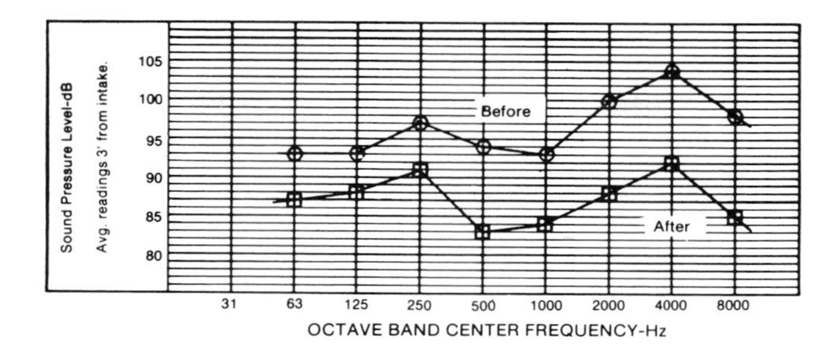

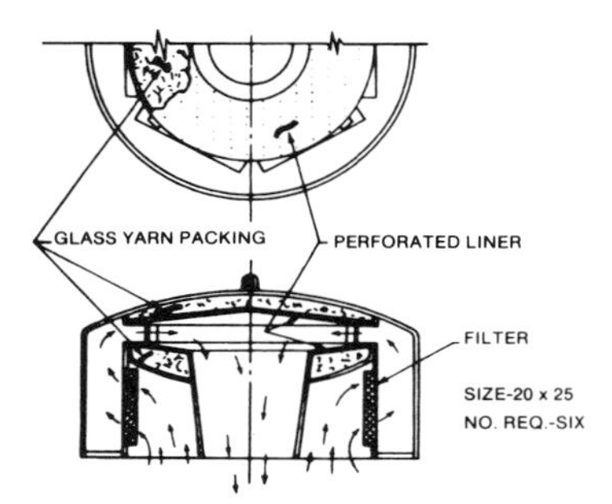

Figure 12.75 — Noise levels with and without compressor intake muffler.

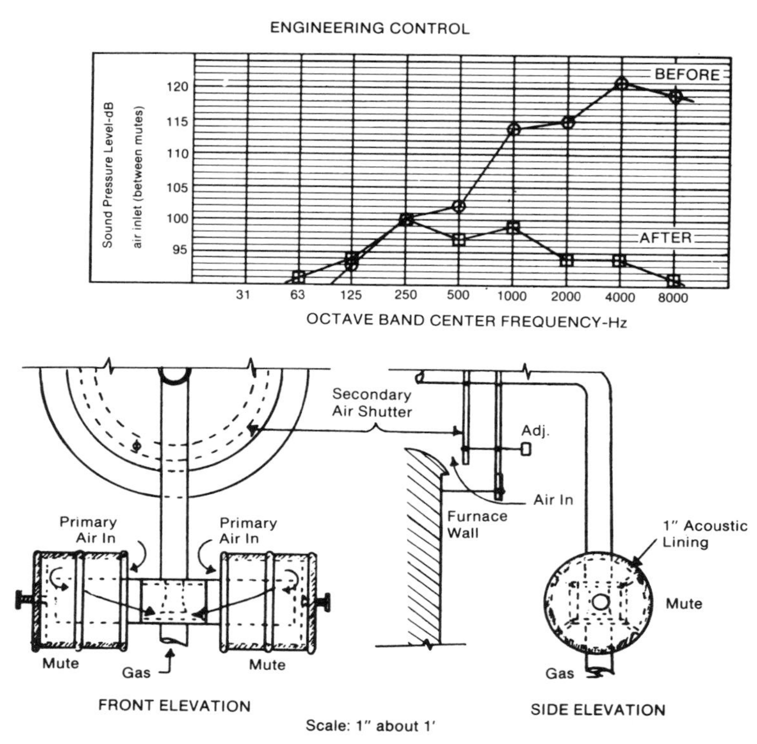

Figure 12.76 — Noise levels before and after aspirating air mutes were installed on a four-burner down-convection end-fired heater.

The design of resonant mufflers is very difficult and beyond the scope of this manual. Since they are usually effective over only a relatively narrow frequency range, it is safer to use an absorptive muffler having good performance over a broader frequency range. However, there are occasions where the noise is narrow-band and not likely to change in frequency. Such mufflers should be purchased on a performance guarantee to assure satisfactory performance.

A pneumatic conveying system handling synthetic fiber fluff discharged into a filter-bag separator as shown by the sketch in Figure 12.78. Blower noise traveled with the air and the product and was objectionable where it discharged in the filter area. An absorptive muffler was undesirable because of the possibility of snagging and plugging. A resonant muffler provided the following values of noise reduction:

OB Ctr. Freq. (Hz)	63	125	250	500	1000
Noise Reduction (dB)	12	23	13	11	10

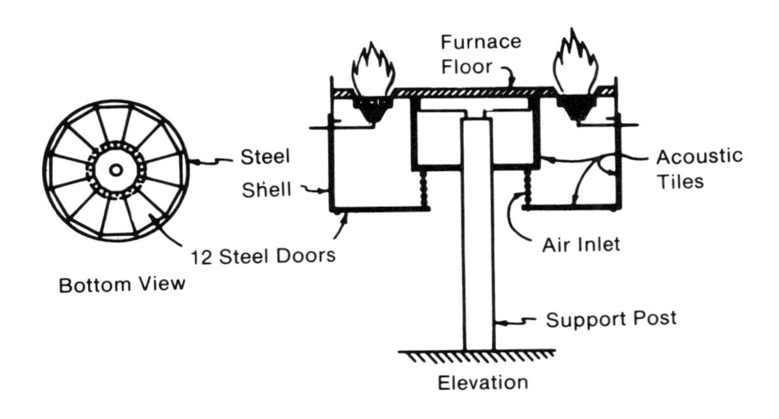

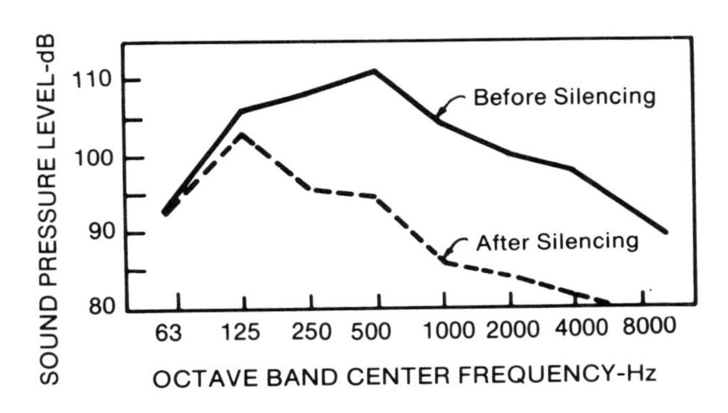

Figure 12.77 — Noise levels before and after installation of a refinery heater burner muffler.

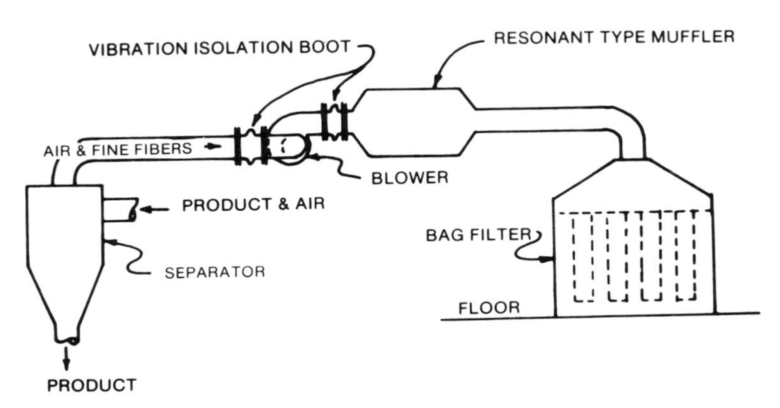

Figure 12.78 — Resonant type muffler.

Summary

In this chapter, engineering controls have been discussed. Controlling noise by purchasing quieter equipment is the best solution to a noise problem. Replacement of existing equipment, processes or materials with quieter equipment or processes is a good alternative to other engineering controls. Noise control at the source by modifying the sound source through the use of damping treatments and vibration isolation as well as reducing fluid flow and turbulence are also effective solutions in many situations. Finally, the use of mufflers, enclosures, directivity and acoustical absorption can be applied to a noise problem in selected situations.

The best solution to any noise problem is the one that is actually installed and maintained.

Acknowledgments

The authors are indebted to the many individuals who have contributed to the material in this chapter since it was first introduced in 1958. Unfortunately, AIHA's records do not provide sufficient information that detailed acknowledgements can be made. However, we wish to recognize the efforts of the following individuals:

Thomas B. Bonney
James H. Botsford
Lewis S. Goodfriend
Vaughn H. Hill
William C. Janes
Stanley H. Judd
Stanley E. Phil

References

AIMA (1974). *Bulletin of the Acoustical and Insulating Materials Association,* Park Ridge, IL.

ANSI S1.11 (1966) (R1971). "American National Standard Specification for Octave, Half-Octave, and Third-Octave Band Filter Sets," American National Standards Institute, New York, NY.

ANSI S1.4 (1983). "American National Standard Specification for Sound Level Meters," American National Standards Institute, New York, NY.

ASA STDS INDEX 3 (1985). "Index to Noise Standards, third edition," Acoust. Soc. of Am., New York, NY.

ASHRAE (1981). *Handbook, Fundamentals Volume,* Ch 7, ASHRAE, New York, NY.

Beranek, L.L. (1971). (Ed.), *Noise & Vibration Control,* McGraw-Hill, New York, NY

Beranek, L.L., Blazier, W.E., and Figwer, J.J. (1971). "Preferred Noise Criterion (PNC) Curves and Their Application to Rooms," *J. Acoust. Soc. Am.* 50(5), 1223-1228.

Bishop, D.E. (1957). "Use of Partial Enclosures to Reduce Noise in Factories," *Noise Contr.* 3,65.

Crocker, M. J. and Price, A.J. (1975). *Noise and Noise Control Volume I,* CRC Press, Inc., Cleveland, OH.

Fehr, R. D. (1951). "The Reduction of Industrial Machine Noise," *Proceedings, Second National Noise Abatement Symposium* 2,93.

Harris, C.M. (1979). (Ed.), *Handbook of Noise Control, 2nd ed.,* McGraw-Hill, New York, NY.

Harris, C.M. and Crede, C.E. (1961). (Eds.), *Shock and Vibration Handbook, Vol. II,* McGraw-Hill, New York, NY.

Haung, B. and Rivin, E.I. (1985). "Noise and Air Consumption of Blow-Off Nozzles," *Sound and Vibration* 19(7), 26-33.

Ingard, U. (1973). Unpublished data reported by Vaughn H. Hill in "Control of Noise Exposure," Chapter 37 of NIOSH (1973).

Irwin, J.D. and Graf, E.R. (1979). *Industrial Noise and Vibration Control,* Prentice-Hall, Englewood Cliffs, NJ.

Kuntz, H.L. (1985). Personal communication.

Lilley, D.T. (1985). Personal communication.

Lord, H.W., Gatty, W.S., and Evensen, H.A. (1980). *Noise Control for Engineers,* McGraw-Hill, New York, NY.

Mariner, T. and Park, A.D. (1956). "Sound Absorbing Screens," *Noise Contr.* 2,22.

NIOSH (1973). *The Industrial Environment - Its Evaluation and Control,* U.S. Dept. HEW, PHS, CDC, NIOSH.

NIOSH (1975). *Compendium of Materials for Noise Control,* U.S. Dept. HEW, Publication No. 75-165.

NIOSH (1978). *Industrial Noise Control Manual, Rev. Ed.,* DHEW (NIOSH) Publication No. 79-117.

OSHA (1980). "Noise Control - A Guide For Workers and Employers," OSHA 3048.

Peterson, A.P.G. (1980). *Handbook of Noise Measurement, 9th ed.,* General Radio Co., Concord, MA.

Rettinger, M. (1957). "Noise Level Reductions of Barriers," *Noise Contr.* 3,50.

Thomson, W.T. (1981). *Theory of Vibration with Applications,* Prentice-Hall, Englewood Cliffs, NJ.

13 Workers' Compensation

Allen L. Cudworth

Contents

Page

Introduction 523
Workers' Compensation History 524
Hearing Loss — Workers' Compensation 525
Impairment *vs.* Disability 526
Handicap 526
Impairment Calculation 526
Correction for Age 529
Summary of Impairment Considerations 530
Waiting Period 530
Date of Injury 531
Apportionment 532
Hearing Aids 532
Tinnitus 532
Duration and Level of Exposure 532
Regulation 533
Important Considerations in Handling Claims 533
The Validation of a Claim 533
Testimony and Records 534
Handling Claims 535
Summary 536
References 536

Introduction

As discussed elsewhere in this manual, there are many reasons for seeking to control noise exposures. One that is sometimes underestimated is the cost of noise-induced hearing impairment (HI), an occupational disease now scheduled or otherwise covered in the Workers' Compensation Law (WCL) of most states. The compensation is provided via Workers'

Compensation (WC) coverage, an insurance program required of employers in all states. It is imperative that everyone having responsibilities in hearing conservation programs become acquainted with the legal aspects of the industrial noise problem.

Noise exposures may result in liability under various codes. Loss of property values or injury resulting from negligent control of noise may result in legal liability to the noise producer or the source manufacturer. Laws and regulations provide added stimulus for engineering control and they sometimes specify mandatory design criteria. The current trends indicate that there are compelling legal and economic reasons, in addition to the obvious humanitarian ones, to control industrial noise.

Workers' Compensation History

At one time, occupational injury and disease were accepted as risks of employment. It was at the turn of the century that a concept emerged whereby it became the employer's obligation to compensate his employees for any injury incurred during, and resulting from, the course of their employment. In this country, the first workers' compensation law was enacted in 1908 (FECA 1908). Now compensation acts are universal.

Originally, compensation was to be paid only when a job-related injury resulted in income loss to the employee; however, in recent years, the trend is toward recognizing loss of function as a basis for an award. That is, if an employee suffers some loss of bodily function, regardless of whether it resulted in any loss of wages, the worker may be entitled to WC.

Although the loss of function concept appeared earlier, it was not until 1948 that occupational hearing loss became a significant factor in WC. In that year, a precedent was set in New York State in the decision of the case of Slawinski v. J.H. Williams & Co. (Slawinski v. J.H. Williams, 1948). The Supreme Court of New York decided that Slawinski was entitled to an award for partial loss of hearing even though he had not been disabled, *i.e.,* prevented from earning full wages at his regular employment.

Similarly, in 1953, the Wisconsin Supreme Court (Green Bay Drop Forge v. Industrial Commission, 1953) upheld a Wisconsin Industrial Commission ruling that an employee was entitled to compensation for partial hearing loss even though the Wisconsin law, as did the New York law, required disability before the HI could be compensated. Further confirmation of this concept was given in decisions reached under the United States Longshoremen's and Harbor Workers' Compensation Act in 1955 (Longshoremen's and Harbor Workers' Compensation Act, 1927) and in Missouri in 1959.

In 1959 and 1961, the Maryland courts discounted the above precedents and ruled that no compensation would be paid in the absence of disability and wage loss. Ultimately on April 14, 1967, the Maryland legislature made occupational hearing loss compensable without consideration of disability or wage loss.

Noise-induced hearing loss is usually interpreted as an occupational disease as contrasted to traumatic injury. Hearing impairment from trauma is usually associated with explosions, head injuries, *etc.* and has always been covered by WC. The "gradual injury" concept was applied to a hearing loss case in Georgia in 1962 (Shipman v. Employers Mutual Liability Insurance, 1962). The Georgia court ruled that an employee's noise-induced hearing loss resulted from the cumulative effect of a succession of injuries caused by each daily noise insult on the ear. By this reasoning, it held that the loss was compensable under the accidental injury provisions of the Georgia law.

In certain jurisdictions, where there are no specific provisions for hearing loss, awards are sometimes made on the basis of liberal interpretations of the general WCL.

Finally, damages may also be awarded in a common-law action. In such cases, the employee alleges negligence on the part of the employer or the insurance carrier and seeks a remedy outside the WC provisions. This illustrates yet another possible means by which an employee may be compensated for occupational hearing loss.

The latest survey of state WC acts (Table 13.1) shows that only a few jurisdictions do not specifically cover loss of hearing resulting from noise exposure. What is not readily apparent from Table 13.1 is that the administration of the acts varies widely, so that the difficulty in processing a case may have a profound effect on the amount of compensation actually paid to injured workers.

In summary, awards for occupational hearing loss may be made under WC: as an occupational disease covered by specific or general provisions, as a "gradual injury" covered under specific accidental injury provisions, or as either an injury or an occupational disease covered by general provisions of a workers' compensation law. In addition, recovery may be had in a common law action.

Hearing Loss - Workers' Compensation

An important provision of WCL is the definition of disability, because it is the degree of disability that determines the amount of compensation that will be paid. Disability is generally based on the impairment of the ability to understand speech. As noted below, there is a distinction between "disability" and "impairment," terms that *should not* be used interchangeably.

IMPAIRMENT *VS.* DISABILITY

"Impairment" denotes a medical condition that affects one's personal abilities as compared to non-impaired or normal characteristics. "Disability" is related to actual or presumed reduction in ability to remain employed at full wages. Permanent impairment is, therefore, a contributing factor to, but not necessarily an indication of, the extent of a patient's permanent disability within the meaning of the WCL. It can be seen then that, although they are related, the terms "impairment" and "disability" are not synonymous.

HANDICAP

In addition to the terms "impairment" and "disability," one will find reference to hearing "handicap." This term is used to describe a person's ability to understand speech under everyday circumstances. Because impaired hearing in one ear is less handicapping than binaural impairment, the term handicap normally implies consideration of the hearing loss in both ears.

IMPAIRMENT CALCULATION

Many methods have been used over the years to estimate hearing impairment. None were found to be completely satisfactory for a variety of reasons. Realizing this deficiency, the Subcommittee on Noise of the American Academy of Ophthalmology and Otolaryngology undertook the task of developing a reasonable formula. The formula was published in 1959(AAOO, 1959) and in 1961 was approved by the American Medical Association. The formula has received wide acceptance and is still included in some WCLs. In 1969 a revised audiometric threshold zero level was adopted and this led to the 1971 AMA-Guides to the Evaluation of Permanent Impairment (AMA, 1971).

Briefly, the AMA-AAOO method describes impairment as an inability to understand everyday speech under everyday conditions. On this basis, the AAOO formula considered only those frequencies which were believed to be important to the hearing of speech. Specifically, hearing threshold levels were determined by pure tone air conduction audiometry at frequencies of 500, 1000 and 2000 Hz. If the arithmetic average of the hearing levels at these three frequencies was 25 dB or less, no impairment was said to

TABLE 13.1 (cont.) - Hearing Loss S

JURISDICTION	Is occupational hearing loss compensable?	Is minimum noise exposure required for filing?	Schedule in weeks for one ear.	Schedule in weeks for both ears.	Maximum compensation (one ear).	Maximum compensation (both ears).	Hearing impairment formula.	Waiting period	Is deduction made for presbycusis?	Is award made for tinnitus?	Provision for hearing aid.
Montana	Yes	No	40	200	$ 5,720	$ 28,600	AAOO-59	6 mo.	Yes	No	Yes
Nebraska	Yes	No	50	100	$ 10,000		ME	No	No		
Nevada	Yes	No					ME	No	Yes	No	Yes
New Hampshire	Yes	No	52	214	$ 13,320	$ 54,612	ME	No	No	Yes	Yes
New Jersey	Yes	Yes	prop.	200	$ 4,320	$ 25,000	See Comments	4 wks	No	Yes	Yes
New Mexico	No	No	40	160	$ 11,945	$ 44,795	ME	No	No	Yes	
New York	Yes	Yes	60	150	$ 8,100	$ 20,250	AAO-79	3 mo.	No	No	No
North Carolina	Yes	No	prop.	150	$ 19,600	$ 42,000	AAOO-59	6 mo.	Yes	No	No
North Dakota	Yes	Yes	50	200	$ 3,000	$ 12,000	AAO-79	No	Yes	No	Yes
Ohio	Yes	No	25	125	$ 4,425	$ 22,125	ME	No	No	No	Yes
Oklahoma	Yes	No	100	300	$ 16,300	$ 48,900	ME	No	No	Yes	Yes
Oregon	Yes	Yes	60	192	$ 6,000	$ 19,200	See Comments	No	Yes	P	Yes
Pennsylvania	Yes	No	60	260	$ 20,820	$ 90,220	ME	10 wks	No	No	No
Rhode Island	Yes	No	17	100	$ 5,400	$ 18,000	AAOO-59	6 mo.	Yes	No	No
South Carolina	Yes	No	80	165	$ 22,962	$ 47,358	See Comments	No	No	Yes	Yes
South Dakota	Yes	Yes	prop.	150		$ 38,100	AAO-79	6 mo.	Yes	No	Yes
Tennessee	Yes	No	75	150	$ 14,175	$ 28,350	ME	No	No	P	Yes
Texas	Yes	No		150		$ 30,450	AAO-79	No	No	Yes	Yes
Utah	Yes	Yes	prop.	100		$ 21,500	AAO-79	6 mo.	Yes	No	Yes
Vermont	Yes	No	52	215	$ 14,456	$ 59,770	ME	No	No	No	P
Virginia	No										
Washington	Yes	No			$ 4,800	$ 28,800	AAO-79	No	No	No	Yes
West Virginia	Yes	No	100	260	$ 21,420	$ 55,692	AAO-79	No	No	P	Yes
Wisconsin	Yes	Yes	36	216	$ 5,940	$ 35,640	AAO-79	14 days	No	Yes	P
Wyoming	Yes	No	40	80	$ 9,260	$ 18,520	ME	No	No	No	Yes
U.S. Department of Labor	Yes	Yes	52	200			AAO-79	No	No	No	Yes
Longshoremen	Yes	No	52	200	$ 30,952	$119,048	AAO-79	No	No	P	Yes

*Data compiled by E.H. Berger and Loretah D. Rowland from appendix by M.S. Fox in AAO-HNS (1982), Page (1985), U.S. Chamb should only be used as a guide. Consult local WC administrators for confirmation of current statutes.

Abbreviations used in this table are: **AAOO-59**: avg. of 500, 1000, 2000 Hz >25dB; **AAO-79**: avg. of 500, 1000, 2000, 3000 Hz >25 dB; physician; **prop.**: proportionate to compensation for 100% bilateral hearing loss; **P**: possible, depending upon the medical eviden

...tutes in the United States*

Credit for improvement with hearing aid.	COMMENTS
No	Deduction for presbycusis, ½ dB for each year over 40.
	Permanent total loss of hearing compensated as permanent total disability.
No	Degree of disability determined using AAOO-59 as applied to AMA guidelines for whole man impairment.
No	
No	High level of claims activity. Uses 1000, 2000, and 3000 Hz average with a 30 dB low fence. Penalty for willful failure to wear HPDs.
	Compensation only for injuries resulting from trauma.
No	High level of claims activity. Compensation for tinnitus if accompanied by hearing loss.
	Penalty for willful failure to wear HPDs.
Yes	
	Must be permanent total hearing loss in one or both ears.
No	
Yes	High level of claims activity. Formula uses average of all frequencies 500-6000 Hz.
	Award only for total permanent hearing impairment.
No	Uses 500, 1000, and 3000 Hz average with a 25 dB low fence.
No	Statute revised 1986. Penalty for willfull failure to wear HPDs.
No	Medical opinion based on current AMA guidelines.
No	
No	
No	
	Re 1985 Virginia Supreme Court decision work-related hearing loss is not compensable since it is "an ordinary disease of life." As of 1986 new legislation is pending.
No	High level of claims activity.
No	State's commission anticipates 1500-2000 new claims with an average award of $17,000 in 1985.
No	High level of claims activity.
No	
No	
No	

...r of Commerce (1985), WC regulations (when available), and phone calls to WC offices around the country. This table

...**IOSH**: avg. 1000, 2000, 3000 >25 dB; **ME**: medical evidence, hearing loss formula up to the discretion of the consulting ...e.

exist. This value of beginning impairment is referred to as the "low fence." The concept of a low fence was developed to allow for the finding that the ability to understand everyday speech was not appreciably affected in persons having hearing thresholds above zero but less than the low fence value. Every decibel that the average hearing level exceeded 25 dB (calculated for each ear separately) constituted a 1½ percent impairment for that ear. A 92-dB average hearing level represented total impairment. Binaural impairment or "total man" impairment was determined by a weighted average in which the percentage of impairment in the better ear was multiplied by a factor of five. Although there appears to be no scientific basis for a binaural impairment calculation using a 5 to 1 weighting of the better ear, it is clear that equal weighting is not appropriate. Total loss of hearing in one ear would not cause a 50% impairment in the ability to understand speech under any conditions. All threshold values were to be stated in terms of ANSI-1969 reference threshold zero values. *Appropriate corrections must be made* for audiometric levels if the hearing thresholds for early audiograms were obtained on older audiometers calibrated according to the ASA-1951 standard.

In 1979 the AAO further revised its guide for the evaluation of hearing impairment so that more importance is attached to hearing threshold levels at high frequencies. The new definition of impairment is based on the average loss over four frequencies, 500, 1000, 2000, and 3000 Hz while the level of 25 dB was retained for the point of beginning impairment. Because noise exposure usually causes more change at 3000 Hz than at the lower frequencies, the effect of including high-frequency levels in the calculation of impairment is to increase the estimated degree of impairment for a given amount of hearing loss.

CORRECTION FOR AGE

It is generally recognized that activities of living, even without occupational noise exposure, result in some deterioration in the ability to hear. This age effect, often termed presbycusis (but which also includes sociocusis and nosoacusis - cf. Chapter 5) is characterized by a gradual but significant loss of hearing at all frequencies, with greater losses above 1000 Hz. It is not clear how age effect and noise-induced hearing loss interact, and no correction process is described in the 1979 Guide for the Evaluation of Hearing Handicap (AAO 1979). However, in the calculation of impairment, the establishment of a deduction in the form of a low fence has the effect of offsetting the expected age effect. For most WCLs, any impairment is compensable as long as an occupational factor in any way increases the impairment caused by the exposure.

The compensation provisions in some jurisdictions, however, do take into account an age factor. At least one jurisdiction deducts ½ dB from the person's average hearing level for every year of the person's age beyond 40.

SUMMARY OF IMPAIRMENT CONSIDERATIONS

Although there are many factors that affect the awards under WC, the basic consideration in determining the amount of compensation associated with hearing loss is the impairment formula. The application of such formulae varies from state to state. In about half of the states the choice of formula is left up to the examining physician who may tend to rely on the latest AMA guidelines. Other states specify an impairment formula in their WCL and those may remain unchanged for many years.

Although there are surveys (see Table 13.1) that attempt to identify the salient features of each state's position on hearing loss claims, the reader is cautioned to make certain that any such listing is up-to-date for the particular area involved.

Given the hearing loss inventory of a particular group, it is possible to calculate the potential claims value should a specified percentage of the exposed population seek compensation (Berger, 1985). Using risk tables given in the AAO guide (AAO, 1982) it is possible to estimate WC liability from noise exposure information.

As an example of the application of the impairment formulae, Table 13.2 compares the percentage impairment for a typical audiogram showing noise induced hearing loss, calculated using four different formulae. It can be seen that the newer addition of 3000 Hz in the AAO (1979) equation increases the impairment by a significant amount. Thus for a given population of noise-exposed persons, the application of this equation will result in a larger percentage with impaired hearing as well as larger amounts of impairment.

Waiting Period

This important provision in WC requires that impairment be both permanent and occupationally related. In the process associated with sensory damage from noise exposure, there is a temporary threshold shift (TTS) effect that disappears upon termination of the exposure. Although there is considerable variation in recovery time depending on the level and duration of the noise insult, most of the temporary effect is gone within 48 hours in a quiet environment. For economic reasons, longer recovery periods have been adopted in some states and an employee may not be able to pursue a claim until 6 months of nonexposure have passed. Such provisions usually do not apply to traumatically-induced hearing loss.

TABLE 13.2
Hearing Impairment Formulas and Example Calculations

Formula	Audiometric Frequency Used	Method of Calculation	Low Fence	High Fence	% Per Decibel Loss	Better Ear Correction
AAOO-1971	500, 1000, 2000	Average	25 dB	92 dB	1.5	5/1
AAO-1979	500, 1000, 2000 3000	Average	25 dB	92 dB	1.5	5/1
New Jersey	1000, 2000, 3000	Average	30 dB	97 dB	1.5	5/1
Illinois	1000, 2000, 3000	Average	30 dB	85 dB	1.82	1/1

Example Calculation From Audiogram Below

Freq. Hz	500	1000	2000	3000	4000	8000
Right Ear	15	25	45	55	65	45
Left Ear	30	45	60	85	90	70

For AAO-1979 Formula

Ear Average Threshold Level $= \frac{500,\ 1000,\ 2000,\ 3000}{4}$

Right Ear $\frac{15 + 25 + 45 + 55}{4} = 35\text{dB}$

Left Ear $\frac{30 + 45 + 60 + 85}{4} = 55\text{dB}$

To obtain % impairment, subtract 25 dB and multiply by 1.5
Right Ear = 35 − 25 = 10 × 1.5 = 15%
Left Ear = 55 − 25 = 30 × 1.5 = 45%

Binaural Impairment (BI) $= \frac{5 \times \text{Better Ear} + 1 \times \text{Poorer Ear}}{6}$

$\text{BI} = \frac{5 \times 15 + 45}{6} = \frac{75 + 45}{6} = 20\%$

Comparison of BI for Example Audiogram

Formula	AAOO-1971	AAO-1979	New Jersey	Illinois
BI	8.7%	20%	22.8%	40.8%

Although the issue has been raised, most hearing loss provisions under WCL do not address the question of personal hearing protection and the waiting period. It is not generally clear whether the wearing of adequately fitted hearing protection constitutes removal from exposure, although in at least one state, the law specifically adopts such a provision.

Date of Injury

Most WC acts were established to compensate employees for traumatic injury, and contain a provision encouraging timely filing of claims. The provision, called a statute of limitations, also applies to HL claims and requires that a date of injury be established. As with other considerations in WC law there is considerable variation between states. In some states the definition of date of injury is "the last day exposed," while in others it is the date the employee became aware of HI in a personal sense.

The statute of limitations also varies from state to state, and can be as short as 30 days to as long as 5 years.

Apportionment

The WC system was established on the principle that the employer at the time of the employee's injury should pay for the loss of earnings of the employee. However, in a case of HI where several employers might be involved in the HL leading to impairment, a new approach was necessary. This led to the inclusion of an apportionment clause in the acts of many states. For example, if the last employer at the date of injury can establish that the employee began employment with HI then prior employers or a special fund may be tapped for any HI present at the time of employment.

Apportionment does have the effect of encouraging preplacement audiometry and can be a significant loss factor in the purchase of manufacturing operations. The last survey of WCL by the U.S. Chamber of Commerce in 1981 indicates that apportionment in the form of a deduction for a preexisting loss is in effect in most of the states.

Hearing Aids

In most states, hearing impairment is evaluated without cognizance of the effect a hearing aid or other prosthesis might have on the claimant's ability to understand speech. Some states may include a prosthesis as part of the compensation, but in general, the award for HI is based on a pure tone audiogram in the absence of such a device.

Tinnitus

Although tinnitus (ringing or head noises) often accompanies noise-induced hearing loss, it may occur in the absence of noise exposure. As shown in Table 13.1, tinnitus is recognized as part of the HI consideration in a number of states and the award for compensation may be modified by its presence. There are no objective measurements for tinnitus and the award is based on the subject's symptoms.

Duration and Level of Exposure

Last but certainly not least is the factor of noise exposure in hearing loss compensation. As in the case of the other factors above, there is considerable variation in how this factor is treated. Many states include a provision that excludes a noise-induced HI claim where the occupational noise exposure is below a specified level. Other states include a provision that the noise exposure duration must exceed a minimum number of days in order to file a valid claim. Both provisions may be included in the WCL or accompanying operating procedures. The frequency spectrum characteristics as determined by A- or C-scale weighting may also be addressed in the WC act.

In a small number of states, exposure that results because of the employee's willful failure to use hearing protection is excluded as the basis for a noise-induced HI claim.

Regulation

There are both state and federal regulations that set allowable noise exposure levels for many employees. A relatively small number of states have specific codes or regulations setting allowable noise exposure limits, but all employers engaged in interstate commerce should take note of the federal noise regulation promulgated under the Occupational Safety and Health Act.

The specific actions required under the federal act will not be described in this chapter. The reader is referred to the current version of the Regulation (CFR1910.95) and the Hearing Conservation Amendment effective March 8, 1983 (Appendix I). A concise portrayal of the amendment was recently presented by Thunder and Lankford (1985).

Although some states have adopted the OSHA limits in their WC acts, there is no legal relationship between OSHA standards and WC actions. A valid HI claim may exist in the absence of an OSHA citation, and conversely, an OSHA citation does not validate a claim for HI.

Important Considerations In Handling Claims

In the broadest sense, a valid HI claim exists when it can be shown that there is permanent impairment and that it arose out of and in the course of employment. A number of considerations go into validating claims when they arise in industry. In questioning the validity of a claim, testimony and records become extremely important.

THE VALIDATION OF A CLAIM

The following is a check list of considerations to be examined.

1. Has a hearing loss been established? In the absence of a pre-employment audiogram, it is assumed that normal hearing was present prior to employment. The most recent audiometric data should be complete and professionally obtained by a clinical audiologist certified by the American Speech-Language-Hearing Association (ASHA) or licensed by the state. All audiometric records should be compared to determine consistency and detect possible fluctuating hearing levels; such records may be invalid.

2. Is the type of hearing loss consistent with excessive exposure to noise? Monaural and conductive losses do not usually result from noise exposure.
3. Are adequate noise monitoring records available to demonstrate that excessive noise exposure existed in sufficient degree and duration to result in hearing loss — *i.e.,* consistent exposure to noise levels of 85 dBA or greater during the work history?
4. Could the hearing loss be attributed to nonoccupational causes? Noise exposures resulting from a variety of hobbies or spare time activities may result in a portion of the hearing loss. A detailed personal case history must be obtained by the audiologist or physician.
5. If exposed, was the employee provided with properly fitted protection and instructed in its use? In some states the willful misuse of protection invalidates a claim for HI.

The answers to these questions should allow a determination of whether or not the hearing impairment is occupationally related and resulted from or was aggravated by conditions at work.

TESTIMONY AND RECORDS

Although most HI claims will be handled without controversy, either party has the right to a hearing to establish the validity and amount of the award. In such a hearing, the records that establish hearing loss and noise exposure become evidence that will be weighed.

The hearing is usually an adversary process, with the normal rules of evidence submission prevailing. The weight of the evidence is strongly affected by how it is obtained, its completeness, adherence to accepted standards for accuracy, and professional backup. The importance of completeness and accuracy cannot be overemphasized (Gasaway, 1985). A valid audiogram should include test results at all frequencies used in the evaluation of impairment plus those recommended for hearing conservation purposes. The audiogram should represent the permanent hearing threshold measured under acceptable background noise conditions. A certified audiometric technician under the supervision of a trained professional should obtain the audiometric data. Test results should be reviewed by a supervising professional and validated by the professional. Records of audiometer calibration and background noise conditions should also be complete and of sufficient frequency to prove that accurate testing was done.

The establishment of noise exposure or lack thereof requires similar record keeping and professional supervision. In addition to valid records of noise exposure and hearing status, a good hearing conservation program includes records of medical history and employee activities. These records should also be complete and professionally supervised.

From a legal sense, the validity of records is established through testimony.

HANDLING CLAIMS

Although the purpose of this chapter is to provide guidance in the legal aspects of hearing loss claims, the actual process of filing a claim or preparing a defense should be referred to persons skilled in such activities. Often such matters are carried out by attorneys or the representatives of insurance carriers. There are, however, steps that management can take that will eliminate some of the uncertainties and expedite the handling of valid claims.

1. In establishing an audiometric program, plan for the disposition of those persons showing abnormal hearing levels:
    a. In preplacement programs, a decision should be made as to when an employee is referred for a diagnostic work-up and how the results will be acted upon. If any HI exists at the time of preplacement testing, a diagnostic audiological test will more accurately define the level of impairment. Future audiograms will be used to determine if any further HI has occurred.

       A baseline audiogram by itself does not establish occupational hearing loss but may establish that a hearing impairment or medical condition exists that should be investigated through referral.
    b. In follow-up testing, a decision should be made as to the follow-up procedures for employees showing hearing loss. Will they be referred to their personal physicians or provided access to a qualified consultant? The insurance carrier and professional supervisor should be consulted in this decision.

In general, the employer and employee are best served when well-trained medical and audiological support is provided by the employer. Non-occupational hearing loss problems are identified and proper treatment is more conveniently accomplished through in-house medical support. There may also be greater potential for cost containment through in-house support. Having trained professional support should also provide for the anticipated requirements for referral required by OSHA.

Summary

Hearing impairment resulting from occupational exposure is covered under each state's WCL but is defined and accommodated in a variety of ways. Each state's jurisdiction is independent in its interpretation and handling of HI claims, but certain general rules apply. It is important that hearing conservationists be aware of these rules so that any hearing impairment found in exposed employees will be properly documented and appropriate steps taken. This chapter describes the features of HI and suggests the conditions that should be addressed in establishing the validity of a claim. HI is quite prevalent in working populations, even in those with no known noise exposure, and with liberalization of WC benefits it becomes increasingly important to identify the true source of hearing damage so that appropriate steps may be taken to reduce exposures and limit compensation to valid claims.

References

American Academy of Ophthalmology and Otolaryngology (1959). "Committee on Conservation of Hearing: Guide for the Evaluation of Hearing Impairment," *Trans. Am. Acad. Opthalmol. Otolaryngol. 63,* 236-238.

American Academy of Otolaryngology (1979). "Committee on Hearing and Equilibrium and the American Council of Otolaryngology Committee on the Medical Aspects of Noise: Guide for the Evaluation of Hearing Handicap," *J. Am. Med. Assoc. 241(19),* 2055-2059.

American Academy of Otolaryngology - Head and Neck Surgery (1982). "Guide for Conservation in Hearing in Noise," Rochester, MN.

American Medical Association (1971). "Guides to the Evaluation of Permanent Impairment." Committee on Rating Mental and Physical Impairment, Chicago IL.

Berger, E.H. (1985). "EARLog #15 - Workers' Compensation for Occupational Hearing Loss," *Sound and Vibration 19(2),* 16-18.

Federal Employees Compensation Act, 1908.

Gasaway, D.C. (1985). "Documentation: The Weak Link in Audiometric Monitoring Programs," *Occup. Health Saf. 54(1),* 28-33.

Green Bay Drop Forge v. Industrial Commission (1953). 265 W.S. 38, 60 N. W. 2d 409.

Longshoremen's and Harbor Workers' Compensation Act (1927). 33 U.S.C. 1901 *et seq.*

Page, J. (1985). "Workers' Compensation Update," *ASHA 27(10),* 66.

Shipman v. Employers Mutual Liability Ins. and Lockheed Corp. (1962). 125 SE(2d) 72.

Slawinski, v. J.H. Williams (1948). 208 N.Y. 546, 81 N.E. 2d 93 Aff g 273 Appellate Division, S825, 75, N.Y.S. 2d 888.

Thunder, T.D. and Lankford, J.E. (1985). "An Easy-to-Use Diagram of OSHA's Noise Exposure Regulation," *Sound and Vibration 19(2),* 24-28.

U.S. Chamber of Commerce (1985). *Analysis of Workers' Compensation Laws,* Washington, D.C.

Noise and Hearing Conservation Manual, edited by
E.H. Berger, W.D. Ward, J.C. Morrill and L.H. Royster

Appendix I Department of Labor Occupational Noise Exposure Standard

[Code of Federal Regulations, Title 29, Chapter XVII, Part 1910, Subpart G, 36 FR 10466, May 29, 1971; Amended 48 FR 9776-9785, March 8, 1983]

Occupational Noise Exposure

† 1910.95

(a) Protection against the effects of noise exposure shall be provided when the sound levels exceed those shown in Table G-16 when measured on the A scale of a standard sound level meter at slow response. When noise levels are determined by octave band analysis, the equivalent A-weighted sound level may be determined as follows:

Figure G-9

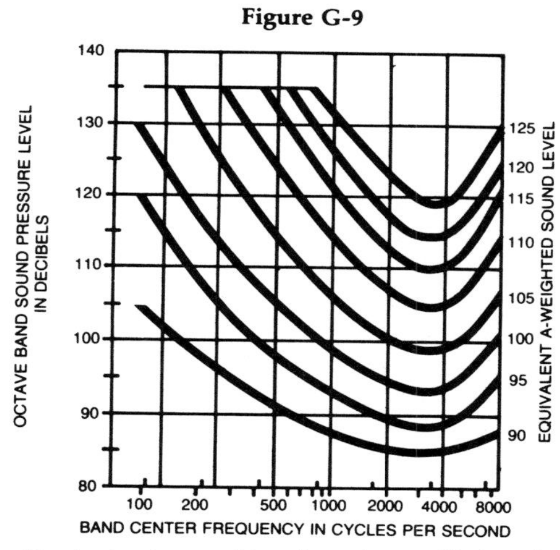

Figure G-9 — Equivalent sound level contours. Octave band sound pressure levels may be converted to the equivalent A-weighted sound level by plotting them on this graph and noting the A-weighted sound level corresponding to the point of highest penetration into the sound level contours. This equivalent A-weighted sound level, which may differ from the actual A-weighted sound level of the noise, is used to determine exposure limits from Table 1.G-16.

(b) (1) When employees are subjected to sound exceeding those listed in Table G-16, feasible administrative or engineering controls shall be utilized. If such controls fail to reduce sound levels within the levels of Table G-16, personal protective equipment shall be provided and used to reduce sound levels within the levels of the table.

(2) If the variations in noise level involve maxima at intervals of 1 second or less, it is to be considered continuous.

TABLE G-16—PERMISSIBLE NOISE EXPOSURES

Duration per day, hours	*Sound level dBA slow response*
8	90
6	92
4	95
3	97
2	100
1½	102
1	105
½	110
¼ or less	115

Table G-16 — When the daily noise exposure is composed of two or more periods of noise exposure of different levels, their combined effect should be considered, rather than the individual effect of each. If the sum of the following fractions: $C_1/T_1 + C_2/T_2 + \ldots + C_n/T_n$ exceeds unity, then, the mixed exposure should be considered to exceed the limit value. C_n indicates the total time of exposure at a specified noise level, and T_n indicates the total time of exposure permitted at that level.

Exposure to impulsive or impact noise should not exceed 140 dB peak sound pressure level.

Hearing Conservation Amendment*

(c) Hearing conservation program. (1) The employer shall administer a continuing, effective hearing conservation program, as described in paragraphs (c) through (o) of this section, whenever employee noise exposures equal or exceed an 8-hour time-weighted average sound level (TWA) of 85 decibels measured on the A scale (slow response) or, equivalently, a dose of fifty percent. For purposes of the hearing conservation program, employee noise exposures shall be computed in accordance with Appendix A and Table G-16a, and without regard to any attenuation provided by the use of personal protective equipment.

(2) For purposes of paragraphs (c) through (n) of this section, an 8-hour time-weighted average of 85 decibels or a dose of fifty percent shall also be referred to as the action level.

(d) Monitoring. (1) When information indicates that any employee's exposure may equal or exceed an 8-hour time-weighted average of 85 decibels, the employer shall develop and implement a monitoring program.

(i) The sampling strategy shall be designed to identify employees for inclusion in the hearing conservation program and to enable the proper selection of hearing protectors.

(ii) Where circumstances such as high worker mobility, significant variations in sound level, or a significant component of impulse noise make area monitoring generally inappropriate, the employer shall use representative personal sampling to comply with the monitoring requirements of this paragraph unless the employer can show that area sampling produces equivalent results.

(2) (i) All continuous, intermittent and impulsive sound levels from 80 decibels to 130 decibels shall be integrated into the noise measurements.

(ii) Instruments used to measure employee noise exposure shall be calibrated to ensure measurement accuracy.

(3) Monitoring shall be repeated whenever a change in production process, equipment or controls increases noise exposures to the extent that:

(i) Additional employees may be exposed at or above the action level; or

---

*In a unanimous eight-judge decision in *Forging Industry Association vs. Secretary of Labor (No. 83-1420),* the U.S. Court of Appeals for the Fourth Circuit Court, upheld the Secretary of Labor's promulgation of the hearing conservation amendment to the occupational noise exposure standard, and reversed the November 7, 1984, ruling by a three-judge panel which had found the amendment to be invalid. The appeal was argued June 3, 1985, and decided September 23 of the same year.

(ii) The attenuation provided by hearing protectors being used by employees may be rendered inadequate to meet the requirements of paragraph (j) of this section.

(e) Employee notification. The employer shall notify each employee exposed at or above an 8-hour time-weighted average of 85 decibels of the results of the monitoring.

(f) Observation of monitoring. The employer shall provide affected employees or their representatives with an opportunity to observe any noise measurements conducted pursuant to this section.

(g) Audiometric testing program. (1) The employer shall establish and maintain an audiometric testing program as provided in this paragraph by making audiometric testing available to all employees whose exposures equal or exceed an 8-hour time-weighted average of 85 decibels.

(2) The program shall be provided at no cost to employees.

(3) Audiometric tests shall be performed by a licensed or certified audiologist, otolaryngologist, or other physician, or by a technician who is certified by the Council of Accreditation in Occupational Hearing Conservation, or who has satisfactorily demonstrated competence in administering audiometric examinations, obtaining valid audiograms, and properly using, maintaining and checking calibration and proper functioning of the audiometers being used. A technician who operates microprocessor audiometers does not need to be certified. A technician who performs audiometric tests must be responsible to an audiologist, otolaryngologist or physician.

(4) All audiograms obtained pursuant to this section shall meet the requirements of Appendix C: *Audiometric Measuring Instruments.*

(5) *Baseline audiogram.* (i) Within 6 months of an employee's first exposure at or above the action level, the employer shall establish a valid baseline audiogram against which subsequent audiograms can be compared.

(ii) *Mobile test van exception.* Where mobile test vans are used to meet the audiometric testing obligation, the employer shall obtain a valid baseline audiogram within 1 year of an employee's first exposure at or above the action level. Where baseline audiograms are obtained more than 6 months after the employee's first exposure at or above action level, employees shall wearing [*sic*] hearing protectors for any period exceeding six months after first exposure until the baseline audiogram is obtained.

(iii) Testing to establish a baseline audiogram shall be preceded by at least 14 hours without exposure to workplace noise. Hearing protectors may be used as a substitute for the requirement that baseline audiograms be preceded by 14 hours without exposure to workplace noise.

(iv) The employer shall notify employees of the need to avoid high levels of non-occupational noise exposure during the 14-hour period immediately preceding the audiometric examination.

(6) *Annual audiogram.* At least annually after obtaining the baseline audiogram, the employer shall obtain a new audiogram for each employee exposed at or above an 8-hour time-weighted average of 85 decibels.

(7) *Evaluation of audiogram.* (i) Each employee's annual audiogram shall be compared to that employee's baseline audiogram to determine if the audiogram is valid and if a standard threshold shift as defined in paragraph (g) (10) of this section has occurred. This comparison may be done by a technician.

(ii) If the annual audiogram shows that an employee has suffered a standard threshold shift, the employer may obtain a retest within 30 days and consider the results of the retest as the annual audiogram.

(iii) The audiologist, otolaryngologist, or physician shall review problem audiograms and shall determine whether there is a need for further evaluation. The employer shall provide to the person performing this evaluation the following information:

(A) A copy of the requirements for hearing conservation as set forth in paragraphs (c) through (n) of this section;

(B) The baseline audiogram and most recent audiogram of the employee to be evaluated;

(C) Measurements of background sound pressure levels in the audiometric test room as required in Appendix D: *Audiometric Test Rooms.*

(D) Records of audiometer calibrations required by paragraph (h) (5) of this section.

(8) *Follow-up procedures.* (i) If a comparison of the annual audiogram to the baseline audiogram indicates a standard threshold shift as defined in paragraph (g) (10) of this section has occurred, the employee shall be informed of this fact in writing, within 21 days of the determination.

(ii) Unless a physician determines that the standard threshold shift is not work related or aggravated by occupational noise exposure, the employer shall ensure that the following steps are taken when a standard threshold shift occurs:

(A) Employees not using hearing protectors shall be fitted with hearing protectors, trained in their use and care, and required to use them.

(B) Employees already using hearing protectors shall be refitted and retrained in the use of hearing protectors and provided with hearing protectors offering greater attenuation if necessary.

(C) The employee shall be referred for a clinical audiological evaluation or an otological examination, as appropriate, if additional testing is necessary or if the employer suspects that a medical pathology of the ear is caused or aggravated by the wearing of hearing protectors.

(D) The employee is informed of the need for an otological examination if a medical pathology of the ear that is unrelated to the use of hearing protectors is suspected.

(iii) If subsequent audiometric testing of an employee whose exposure to noise is less than an 8-hour TWA of 90 decibels indicates that a standard threshold shift is not persistent, the employer:

(A) Shall inform the employee of the new audiometric interpretation; and

(B) May discontinue the required use of hearing protectors for that employee.

(9) *Revised baseline.* An annual audiogram may be substituted for the baseline audiogram when, in the judgment of the audiologist, otolaryngologist or physician who is evaluating the audiogram:

(i) The standard threshold shift revealed by the audiogram is persistent; or

(ii) The hearing threshold shown in the annual audiogram indicates significant improvement over the baseline audiogram.

(10) *Standard threshold shift.* (1) As used in this section, a standard threshold shift is a change in hearing threshold relative to the baseline audiogram of an average of 10 dB or more at 2000, 3000, and 4000 Hz in either ear.

(ii) In determining whether a standard threshold shift has occurred, allowance may be made for the contribution of aging (presbycusis) to the change in hearing level by correcting the annual audiogram according to the procedure described in Appendix F: *Calculation and Application of Age Correction to Audiograms.*

(h) Audiometric test requirements. (1) Audiometric tests shall be pure tone, air conduction, hearing threshold examinations, with test frequencies including as a minimum 500, 1000, 2000, 3000, 4000, and 6000 Hz. Tests at each frequency shall be taken separately for each ear.

(2) Audiometric tests shall be conducted with audiometers (including microprocessor audiometers) that meet the specifications of, and are maintained and used in accordance with, American National Standard Specification for Audiometers, S3.6 - 1969.

(3) Pulsed-tone and self-recording audiometers, if used, shall meet the requirements specified in Appendix C: *Audiometric Measuring Instruments.*

(4) Audiometric examinations shall be administered in a room meeting the requirements listed in Appendix D: *Audiometric Test Rooms.*

(5) *Audiometer calibration.* (i) The functional operation of the audiometer shall be checked before each day's use by testing a person with known, stable hearing thresholds, and by listening to the audiometer's output to make sure that the output is free from distorted or unwanted sounds. Deviations of 10 decibels or greater require an acoustic calibration.

(ii) Audiometer calibration shall be checked acoustically at least annually in accordance with Appendix E: *Acoustic Calibration of Audiometers.* Test frequencies below 500 Hz and above 6000 Hz may be omitted from this check. Deviations of 15 decibels or greater require an exhaustive calibration.

(iii) An exhaustive calibration shall be performed at least every two years in accordance with sections 4.1.2; 4.1.3; 4.1.4.3; 4.2; 4.4.1; 4.4.2; 4.4.3; and 4.5 of the American National Standard Specification for Audiometers, S3.6-1969. Test frequencies below 500 Hz and above 6000 Hz may be omitted from this calibration.

(i) Hearing protectors. (1) Employers shall make hearing protectors available to all employees exposed to an 8-hour time-weighted average of 85 decibels or greater at no cost to the employees. Hearing protectors shall be replaced as necessary.

(2) Employers shall ensure that hearing protectors are worn:

(i) By an employee who is required by paragraph (b)(1) of this section to wear personal protective equipment; and

(ii) By any employee who is exposed to an 8-hour time-weighted average of 85 decibels or greater, and who:

(A) Has not yet had a baseline audiogram established pursuant to paragraph (g) (5) (ii); or

(B) Has experienced a standard threshold shift.

(3) Employees shall be given the opportunity to select their hearing protectors from a variety of suitable hearing protectors provided by the employer.

(4) The employer shall provide training in the use and care of all hearing protectors provided to employees.

(5) The employer shall ensure proper initial fitting and supervise the correct use of all hearing protectors.

(j) Hearing protector attenuation. (1) The employer shall evaluate hearing protector attenuation for the specific noise environments in which the protector will be used. The employer shall use one of the evaluation methods described in Appendix B: *Methods for Estimating the Adequacy of Hearing Protection Attenuation.*

(2)Hearing protectors must attenuate employee exposure at least to an 8-hour time-weighted average of 90 decibels as required by paragraph (b) of this section.

(3) For employees who have experienced a standard threshold shift, hearing protectors must attenuate employee exposure to an 8-hour time-weighted average of 85 decibels or below.

(4) The adequacy of hearing protector attenuation shall be reevaluated whenever employee noise exposures increase to the extent that the hearing protectors provided may no longer provide adequate attenuation. The employee [*sic*] shall provide more effective hearing protectors where necessary.

(k) Training program. (1) The employer shall institute a training program for all employees who are exposed to noise at or above an 8-hour time-weighted average of 85 decibels, and shall ensure employee participation in such a program.

(2) The training program shall be repeated annually for each employee included in the hearing conservation program. Information provided in the training program shall be updated to be consistent with changes in protective equipment and work processes.

(3) The employer shall ensure that each employee is informed of the following:

(i) The effects of noise on hearing;

(ii) The purpose of hearing protectors, the advantages,.disadvantages, and attenuation of various types, and instructions on selection, fitting, use, and care; and

(iii) The purpose of audiometric testing, and an explanation of the test procedures.

(l) Access to information and training materials. (1) The employer shall make available to affected employees or their representatives copies of this standard and shall also post a copy in the workplace.

(2) The employer shall provide to affected employees any informational materials pertaining to the standard that are supplied to the employer by the Assistant Secretary.

(3) The employer shall provide, upon request, all materials related to the employer's training and education program pertaining to this standard to the Assistant Secretary and the Director.

(m) Recordkeeping. - (1)*Exposure measurements.* The employer shall maintain an accurate record of all employee exposure measurements required by paragraph (d) of this section.

(2) *Audiometric tests.* (i) The employer shall retain all employee audiometric test records obtained pursuant to paragraph (g) of this section.

(ii) This record shall include:

(A) Name and job classification of the employee;

(B) Date of the audiogram;

(C) The examiner's name;

(D) Date of the last acoustic or exhaustive calibration of the audiometer; and

(E) Employee's most recent noise exposure assessment.

(F) The employer shall maintain accurate records of the measurements of the background sound pressure levels in audiometric test rooms.

(3) *Record retention.* The employer shall retain records required in this paragraph (m) for at least the following periods.

(i) Noise exposure measurement records shall be retained for two years.

(ii) Audiometric test records shall be retained for the duration of the affected employee's employment.

(4) *Access to records.* All records required by this section shall be provided upon request to employees, former employees, representatives designated by the individual employee, and the Assistant Secretary. The provisions of 29 CFR 1910.20 (a) - (e) and (g) - (i) apply to access to records under this section.

(5) *Transfer of records.* If the employer ceases to do business, the employer shall transfer to the successor employer all records required to be maintained by this section, and the successor employer shall retain them for the remainder of the period prescribed in paragraph (m) (3) of this section.

(n) Appendices. (1) Appendices A, B, C, D and E to this section are incorporated as part of this section and the contents of these Appendices are mandatory.

(2) Appendices F and G to this section are informational and are not intended to create any additional obligations not otherwise imposed or to detract from any existing obligations.

(o) Exemptions. Paragraphs (c) through (n) of this section shall not apply to employers engaged in oil and gas well drilling and servicing operations.

(p) Startup date. Baseline audiograms required by paragraph (g) of this section shall be completed by March 1, 1984.

APPENDIX A: NOISE EXPOSURE COMPUTATION

This Appendix is Mandatory

I. Computation of Employee Noise Exposure

(1) Noise dose is computed using Table G-16a as follows:

(i) When the sound level, L, is constant over the entire work shift, the noise dose, D, in percent, is given by: $D = 100\ C/T$ where C is the total length of the work day, in hours, and T is the reference duration corresponding to the measured sound level, L, as given in Table G-16a or by the formula shown as a footnote to that table.

(ii) When the workshift noise exposure is composed of two or more periods of noise at different levels, the total noise dose over the work day is given by:

$$D = 100\ (C_1/T_1 + C_2/T_2 + \ldots + C_n/T_n),$$

where C_n indicates the total time of exposure at a specific noise level, and T_n indicates the reference duration for that level as given by Table G-16a.

(2) The eight-hour time-weighted average sound level (TWA), in decibels, may be computed from the dose, in percent, by means of the formula: $TWA = 16.61 \log_{10}(D/100) + 90$. For an eight-hour workshift with the noise level constant over the entire shift, the TWA is equal to the measured sound level.

(3) A table relating dose and TWA is given in Section II.

TABLE G-16a

A-weighted sound level, L (decibel)	Reference duration, T (hour)
80	32
81	27.9
82	24.3
83	21.1
84	18.4
85	16
86	13.9
87	12.1
88	10.6
89	9.2
90	8
91	7.0
92	6.1
93	5.3
94	4.6
95	4
96	3.5
97	3.0
98	2.6
99	2.3
100	2
101	1.7
102	1.5
103	1.3
104	1.1
105	1
106	0.87
107	0.76

TABLE G-16a — Continued

A-weighted sound level, L (decibel)	Reference duration, T (hour)
108	0.66
109	0.57
110	0.5
111	0.44
112	0.38
113	0.33
114	0.29
115	0.25
116	*0.22*
117	*0.19*
118	*0.16*
119	*0.14*

A-weighted sound level, L (decibel)	Reference duration, T (hour)
120	*0.125*
121	*0.11*
122	*0.095*
123	*0.082*
124	*0.072*
125	*0.063*
126	*0.054*
127	*0.047*
128	*0.041*
129	*0.036*
130	*0.031*

In the above table the reference duration, T, is computed by

$$T = \frac{8}{2^{(L-90)/5}}$$

where L is the measured A-weighted sound level.

II. Conversion Between "Dose" and "8-Hour Time-Weighted Average" Sound Level

Compliance with paragraphs (c)-(r) of this regulation is determined by the amount of exposure to noise in the workplace. The amount of such exposure is usually measured with an audiodosimeter which gives a readout in terms of "dose." In order to better understand the requirements of the amendment, dosimeter readings can be converted to an "8-hour time-weighted average sound level." (TWA).

In order to convert the reading of a dosimeter into TWA, see Table A-1, below. This table applies to dosimeters that are set by the manufacturer to calculate dose or percent exposure according to the relationships in Table G-16a. So, for example, a dose of 91 percent over an eight hour day results in a TWA of 89.3 dB, and, a dose of 50 percent corresponds to a TWA of 85 dB.

If the dose as read on the dosimeter is less than or greater than the values found in Table A-1, the TWA may be calculated by using the formula:

$$TWA = 16.61 \log_{10}(D/100) + 90$$ where

TWA = 8-hour time-weighted average sound level and

D = accumulated dose in percent exposure.

TABLE A-1
Conversion from "Percent Noise Exposure" or "Dose" to "8-Hour Time-Weighted Average Sound Level" (TWA)

Dose or percent noise exposure	TWA	Dose or percent noise exposure	TWA
10	73.4	109	90.6
15	76.3	110	90.7
20	78.4	111	90.8
25	80.0	112	90.8
30	81.3	113	90.9
35	82.4	114	90.9
40	83.4	115	91.1
45	84.2	116	91.1
50	85.0	117	91.1
55	85.7	118	91.2
60	86.3	119	91.3
65	86.9	120	91.3
70	87.4	125	91.6
75	87.9	130	91.9
80	88.4	135	92.2
81	88.5	140	92.4
82	88.6	145	92.7
83	88.7	150	92.9
84	88.7	155	93.2
85	88.8	160	93.4
86	88.9	165	93.6
87	89.0	170	93.8
88	89.1	175	94.0
89	89.2	180	94.2
90	89.2	185	94.4
91	89.3	190	94.6
92	89.4	195	94.8
93	89.5	200	95.0
94	89.6	210	95.4
95	89.6	220	95.7
96	89.7	230	96.0
97	89.8	240	96.3
98	89.9	250	96.6
99	89.9	260	96.9
100	90.0	270	97.2
101	90.1	280	97.4
102	90.1	290	97.7
103	90.2	300	97.9
104	90.3	310	98.2
105	90.4	320	98.4
106	90.4	330	98.6
107	90.5	340	98.8
108	90.6	350	99.0

TABLE A-1 — Continued

Dose or percent noise exposure	TWA	Dose or percent noise exposure	TWA
360	99.2	690	103.9
370	99.4	700	104.0
380	99.6	710	104.1
390	99.8	720	104.2
400	100.0	730	104.3
410	100.2	740	104.4
420	100.4	750	104.5
430	100.5	760	104.6
440	100.7	770	104.7
450	100.8	780	104.8
460	101.0	790	104.9
470	101.2	800	105.0
480	101.3	810	105.1
490	101.5	820	105.2
500	101.6	830	105.3
510	101.8	840	105.4
520	101.9	850	105.4
530	102.0	860	105.5
540	102.2	870	105.6
550	102.3	880	105.7
560	102.4	890	105.8
570	102.6	900	105.8
580	102.7	910	105.9
590	102.8	920	106.0
600	102.9	930	106.1
610	103.0	940	106.2
620	103.2	950	106.2
630	103.3	960	106.3
640	103.4	970	106.4
650	103.5	980	106.5
660	103.6	990	106.5
670	103.7	999	106.6
680	103.8		

APPENDIX B: METHODS FOR ESTIMATING THE ADEQUACY OF HEARING PROTECTOR ATTENUATION

This Appendix is Mandatory

For employees who have experienced a significant threshold shift [*sic*], hearing protector attenuation must be sufficient to reduce employee exposure to a TWA of 85 dB. Employers must select one of the following methods by which to estimate the adequacy of hearing protector attenuation.

The most convenient method is the Noise Reduction Rating (NRR) developed by the Environmental Protection Agency (EPA). According to EPA regulation, the NRR must be shown on the hearing protector package. The NRR is then related to an individual worker's noise environment in order to assess the adequacy of the attenuation of a given hearing protector. This Appendix describes four methods of using the NRR to determine whether a particular hearing protector provides adequate protection within a given exposure environment. Selection among the four procedures is dependent upon the employer's noise measuring instruments.

Instead of using the NRR, employers may evaluate the adequacy of hearing protector attenuation by using one of the three methods developed by the National Institute for Occupational Safety and Health (NIOSH), which are described in the "List of Personal Hearing Protectors and Attenuation Data," HEW Publication No. 76-120, 1975, pages 21-37. These methods are known as NIOSH methods #1, #2 and #3. The NRR described below is a simplification of NIOSH method #2. The most complex method is NIOSH method #1, which is probably the most accurate method since it uses the largest amount of spectral information from the individual employee's noise environment. As in the case of the NRR method described below, if one of the NIOSH methods is used, the selected method must be applied to an individual's noise environment to assess the adequacy of the attenuation. Employers should be careful to take a sufficient number of measurements in order to achieve a representative sample for each time segment.

Note.—*The employer must remember that calculated attenuation values reflect realistic values only to the extent that the protectors are properly fitted and worn.*

When using the NRR to assess hearing protector adequacy, one of the following methods must be used:

(i) When using a dosimeter that is capable of C-weighted measurements:

(A) Obtain the employee's C-weighted dose for the entire workshift, and convert to TWA (see Appendix A, II).

(B) Subtract the NRR from the C-weighted TWA to obtain the estimated A-weighted TWA under the ear protector.

(ii) When using a dosimeter that is not capable of C-weighted measurements, the following method may be used:

(A) Convert the A-weighted dose to TWA (see Appendix A).

(B) Subtract 7 dB from the NRR.

(C) Subtract the remainder from the A-weighted TWA to obtain the estimated A-weighted TWA under the ear protector.

(iii) When using a sound level meter set to the A-weighting network:

(A) Obtain the employee's A-weighted TWA.

(B) Subtract 7 dB from the NRR, and subtract the remainder from the A-weighted TWA to obtain the estimated A-weighted TWA under the ear protector.

(iv) When using a sound level meter set on the C-weighting network:

(A) Obtain a representative sample of the C-weighted sound levels in the employee's environment.

(B) Subtract the NRR from the C-weighted average sound level to obtain the estimated A-weighted TWA under the ear protector.

(v) When using area monitoring procedures and a sound level meter set to the A-weighing [*sic*] network:

(A) Obtain a representative sound level for the area in question.

(B) Subtract 7 dB from the NRR and subtract the remainder from the A-weighted sound level for that area.

(vi) When using area monitoring procedures and a sound level meter set to the C-weighting network:

(A) Obtain a representative sound level for the area in question.

(B) Subtract the NRR from the C-weighted sound level for that area.

APPENDIX C: AUDIOMETRIC MEASURING INSTRUMENTS

This Appendix is Mandatory

1. In the event that pulsed-tone audiometers are used, they shall have a tone on-time of at least 200 milliseconds.

2. Self-recording audiometers shall comply with the following requirements:

(A) The chart upon which the audiogram is traced shall have lines at positions corresponding to all multiples of 10 dB hearing level within the intensity range spanned by the audiometer. The lines shall be equally spaced and shall be separated by at least ¼ inch. Additional increments are optional. The audiogram pen tracings shall not exceed 2 dB in width.

(B) It shall be possible to set the stylus manually at the 10-dB increment lines for calibration purposes.

(C) The slewing rate for the audiometer attenuator shall not be more than 6 dB/sec except that an initial slewing rate greater than 6 dB/sec is permitted at the beginning of each new test frequency, but only until the second subject response.

(D) The audiometer shall remain at each required test frequency for 30 seconds (± 3 seconds). The audiogram shall be clearly marked at each change of frequency and the actual frequency change of the audiometer shall not deviate from the frequency boundaries marked on the audiogram by more than ± 3 seconds.

(E) It must be possible at each test frequency to place a horizontal line segment parallel to the time axis on the audiogram, such that the audiometric tracing crosses the line segment at least six times at that test frequency. At each test frequency the threshold shall be the average of the midpoints of the tracing excursions.

APPENDIX D: AUDIOMETRIC TEST ROOMS

This Appendix is Mandatory

Rooms used for audiometric testing shall not have background sound pressure levels exceeding those in Table D-1 when measured by equipment conforming at least to the Type 2 requirements of American National Standard Specification for Sound Level Meters, S1.4-1971 (R1976), and to the Class II requirements of American National Standard Specification for Octave, Half-Octave, and Third-Octave Band Filter Sets, S1.11-1971 (R1976).

TABLE D-1
Maximum Allowable Octave-Band Sound Pressure Levels for Audiometric Test Rooms

Octave-band center frequency (Hz)	500	1000	2000	4000	8000
Sound pressure level (dB)	40	40	47	57	62

APPENDIX E: ACOUSTIC CALIBRATION OF AUDIOMETERS

This Appendix is Mandatory

Audiometer calibration shall be checked acoustically, at least annually, according to the procedures described in this Appendix. The equipment necessary to perform these measurements is a sound level meter, octave-band filter set, and a National Bureau of Standards 9A coupler. In making these measurements, the accuracy of the calibration equipment shall be sufficient to determine that the audiometer is within the tolerances permitted by American Standard Specification for Audiometers, S3.6-1969.

(1) Sound Pressure Output Check

A. Place the earphone coupler over the microphone of the sound level meter and place the earphone on the coupler.

B. Set the audiometer's hearing threshold level (HTL) dial to 70 dB.

C. Measure the sound pressure level of the tones that [sic] each test frequency from 500 Hz through 6000 Hz for each earphone.

D. At each frequency the readout on the sound level meter should correspond to the levels in Table E-1 or Table E-2, as appropriate, for the type of earphone, in the column entitled "sound level meter reading."

(2) Linearity Check

A. With the earphone in place, set the frequency to 1000 Hz and the HTL dial on the audiometer to 70 dB.

B. Measure the sound levels in the coupler at each 10-dB decrement from 70 dB to 10 dB, noting the sound level meter reading at each setting.

C. For each 10-dB decrement on the audiometer the sound level meter should indicate a corresponding 10 dB decrease.

D. This measurement may be made electrically with a voltmeter connected to the earphone terminals.

(3) Tolerances

When any of the measured sound levels deviate from the levels in Table E-1 or Table E-2 by ± 3 dB at any test frequency between 500 and 3000 Hz, 4 dB at 4000 Hz, or 5 dB at 6000 Hz, an exhaustive calibration is advised. An exhaustive calibration is required if the deviations are 15 dB or greater at any test frequency.[1]

TABLE E-1
Reference Threshold Levels for Telephonics—TDH-39 Earphones

Frequency, Hz	Reference threshold level for TDH-39 earphones, dB	Sound level meter reading, dB
500	11.5	81.5
1000	7	77
2000	9	79
3000	10	80
4000	9.5	79.5
6000	15.5	85.5

[1]As amended June 28, 1983 in CFR 48(125), p. 29687.

TABLE E-2
Reference Threshold Levels for Telephonics—TDH-49 Earphones

Frequency, Hz	Reference threshold level for TDH-49 earphones, dB	Sound level meter reading, dB
500	13.5	83.5
1000	7.5	77.5
2000	11	81.0
3000	9.5	79.5
4000	10.5	80.5
6000	13.5	83.5

APPENDIX F: CALCULATIONS AND APPLICATION OF AGE CORRECTIONS TO AUDIOGRAMS

This Appendix is Non-Mandatory

In determining whether a standard threshold shift has occurred, allowance may be made for the contribution of aging to the change in hearing level by adjusting the most recent audiogram. If the employer chooses to adjust the audiogram, the employer shall follow the procedure described below. This procedure and the age correction tables were developed by the National Institute for Occupational Safety and Health in the criteria document entitled "Criteria for a Recommended Standard . . . Occupational Exposure to Noise," [(HSM)-11001].

For each audiometric test frequency:

(i) Determine from Tables F-1 or F-2 the age correction values for the employee by:

(A) Finding the age at which the most recent audiogram was taken and recording the corresponding values of age corrections at 1000 Hz through 6000 Hz;

(B) Finding the age at which the baseline audiogram was taken and recording the corresponding values of age corrections at 1000 Hz through 6000 Hz.

(ii) Subtract the values found in step (i)(B) from the value found in step (i)(A).[1]

(iii) The differences calculated in step (ii) represented that portion of the change in hearing that may be due to aging.

---

[1]See footnote on previous page.

Example: Employee is a 32-year-old male. The audiometric history for this right ear is shown in decibels below.

Employee's age	Audiometric test frequency (Hz)				
	1000	2000	3000	4000	6000
26	10	5	5	10	5
*27	0	0	0	5	5
28	0	0	0	10	5
29	5	0	5	15	5
30	0	5	10	20	10
31	5	10	20	15	15
*32	5	10	10	25	20

The audiogram at age 27 is considered the baseline since it shows the best hearing threshold levels. Asterisks have been used to identify the baseline and most recent audiogram. A threshold shift of 20 dB exists at 4000 Hz between the audiograms taken at ages 27 and 32.

(The threshold shift is computed by subtracting the hearing threshold at age 27, which was 5, from the hearing threshold at age 32, which is 25). A retest audiogram has confirmed this shift. The contribution of aging to this change in hearing may be estimated in the following manner:

Go to Table F-1 and find the age correction values (in dB) for 4000 Hz at age 27 and age 32.

	Frequency (Hz)				
	1000	2000	3000	4000	6000
Age 32	6	5	7	10	14
Age 27	5	4	6	7	11
Difference ...	1	1	1	3	3

The difference represents the amount of hearing loss that may be attributed to aging in the time period between the baseline audiogram and the most recent audiogram. In this example, the difference at 4000 Hz is 3 dB. This value is subtracted from the hearing level at 4000 Hz, which in the most recent audiogram is 25, yielding 22 after adjustment. Then the hearing threshold in the baseline audiogram at 4000 Hz (5) is subtracted from the adjusted annual audiogram hearing threshold at 4000 Hz (22). Thus the age-corrected threshold shift would be 17 dB (as opposed to a threshold shift of 20 dB without age correction).

TABLE F-1
Age Correction Values in Decibels for Males

Years	Audiometric Test Frequencies (Hz)				
	1000	2000	3000	4000	6000
20 or younger	5	3	4	5	8
21	5	3	4	5	8
22	5	3	4	5	8
23	5	3	4	6	9
24	5	3	5	6	9
25	5	3	5	7	10
26	5	4	5	7	10
27	5	4	6	7	11
28	6	4	6	8	11
29	6	4	6	8	12
30	6	4	6	9	12
31	6	4	7	9	13
32	6	5	7	10	14
33	6	5	7	10	14
34	6	5	8	11	15
35	7	5	8	11	15
36	7	5	9	12	16
37	7	6	9	12	17
38	7	6	9	13	17
39	7	6	10	14	18
40	7	6	10	14	19
41	7	6	10	14	20
42	8	7	11	16	20
43	8	7	12	16	21
44	8	7	12	17	22
45	8	7	13	18	23
46	8	8	13	19	24
47	8	8	14	19	24
48	9	8	14	20	25
49	9	9	15	21	26
50	9	9	16	22	27
51	9	9	16	23	28
52	9	10	17	24	29
53	9	10	18	25	30
54	10	10	18	26	31
55	10	11	19	27	32
56	10	11	20	28	34
57	10	11	21	29	35
58	10	12	22	31	36
59	11	12	22	32	37
60 or older	11	13	23	33	38

TABLE F-2
Age Correction Values in Decibels for Females

Years	Audiometric Test Frequencies (Hz)				
	1000	2000	3000	4000	6000
20 or younger ...	7	4	3	3	6
21	7	4	4	3	6
22	7	4	4	4	6
23	7	5	4	4	7
24	7	5	4	4	7
25	8	5	4	4	7
26	8	5	5	4	8
27	8	5	5	5	8
28	8	5	5	5	8
29	8	5	5	5	9
30	8	6	5	5	9
31	8	6	6	5	9
32	9	6	6	6	10
33	9	6	6	6	10
34	9	6	6	6	10
35	9	6	7	7	11
36	9	7	7	7	11
37	9	7	7	7	12
38	10	7	7	7	12
39	10	7	8	8	12
40	10	7	8	8	13
41	10	8	8	8	13
42	10	8	9	9	13
43	11	8	9	9	14
44	11	8	9	9	14
45	11	8	10	10	15
46	11	9	10	10	15
47	11	9	10	11	16
48	12	9	11	11	16
49	12	9	11	11	16
50	12	10	11	12	17
51	12	10	12	12	17
52	12	10	12	13	18
53	13	10	13	13	18
54	13	11	13	14	19
55	13	11	14	14	19
56	13	11	14	15	20
57	13	11	15	15	20
58	14	12	15	16	21
59	14	12	16	16	21
60 or older	14	12	16	17	22

APPENDIX G: MONITORING NOISE LEVELS NON-MANDATORY INFORMATIONAL APPENDIX

This appendix provides information to help employers comply with the noise monitoring obligations that are part of the hearing conservation amendment.

What is the purpose of noise monitoring?

This revised amendment requires that employees be placed in a hearing conservation program if they are exposed to average noise levels of 85 dB or greater during an 8 hour workday. In order to determine if exposures are at or above this level, it may be necessary to measure or monitor the actual noise levels in the workplace and to estimate the noise exposure or "dose" received by employees during the workday.

When is it necessary to implement a noise monitoring program?

It is not necessary for every employer to measure workplace noise. Noise monitoring or measuring must be conducted only when exposures are at or above 85 dB. Factors which suggest that noise exposures in the workplace may be at this level include employee complaints about the loudness of noise, indications that employees are losing their hearing, or noisy conditions which make normal conversation difficult. The employer should also consider any information available regarding noise emitted from specific machines. In addition, actual workplace noise measurements can suggest whether or not a monitoring program should be initiated.

How is noise measured?

Basically, there are two different instruments to measure noise exposures: the sound level meter and the dosimeter. A sound level meter is a device that measures the intensity of sound at a given moment. Since sound level meters provide a measure of sound intensity at only one point in time, it is generally necessary to take a number of measurements at different times during the day to estimate noise exposure over a workday. If noise levels fluctuate, the amount of time noise remains at each of the various measured levels must be determined.

To estimate employee noise exposures with a sound level meter it is also generally necessary to take several measurements at different locations within the workplace. After appropriate sound level meter readings are obtained, people sometimes draw "maps" of the sound levels within different areas of the workplace. By using a sound level "map" and information on employee locations throughout the day, estimates of individual exposure levels can be developed. This measurement method is generally referred to as *area* noise monitoring.

A dosimeter is like a sound level meter except that it stores sound level measurements and integrates these measurements over time, providing an average noise exposure reading for a given period of time, such as an 8-hour workday. With a dosimeter, a microphone is attached to the employee's clothing and the exposure measurement is simply read at the end of the desired time period. A reader may be used to read-out the dosimeter's measurements. Since the dosimeter is worn by the employee, it measures noise levels in those locations in which the employee travels. A sound level meter can also be positioned within the immediate vicinity of the exposed worker to obtain an individual exposure estimate. Such procedures are generally referred to as *personal* noise monitoring.

Area monitoring can be used to estimate noise exposure when the noise levels are relatively constant and employees are not mobile. In workplaces where employees move about in different areas or where the noise intensity tends to fluctuate over time, noise exposure is generally more accurately estimated by the personal monitoring approach.

In situations where personal monitoring is appropriate, proper positioning of the microphone is necessary to obtain accurate measurements. With a dosimeter, the microphone is generally located on the shoulder and remains in that position for the entire workday. With a sound level meter, the microphone is stationed near the employee's head, and the instrument is usually held by an individual who follows the employee as he or she moves about.

Manufacturer's instructions, contained in dosimeter and sound level meter operating manuals, should be followed for calibration and maintenance. To ensure accurate results, it is considered good professional practice to calibrate instruments before and after each use.

How often is it necessary to monitor noise levels?

The amendment requires that when there are significant changes in machinery or production processes that may result in increased noise levels, remonitoring must be conducted to determine whether additional employees need to be included in the hearing conservation program. Many companies choose to remonitor periodically (once every year or two) to ensure that all exposed employees are included in their hearing conservation programs.

Where can equipment and technical advice be obtained?

Noise monitoring equipment may be either purchased or rented. Sound level meters cost about $500 to $1,000, while dosimeters range in price from about $750 to $1,500. Smaller companies may find it more economical to rent equipment rather than to purchase it. Names of equipment suppliers may be found in the telephone book (Yellow Pages) under headings such as: "Safety Equipment," "Industrial Hygiene," or "Engineers-Acoustical." In addition to providing information on obtaining noise monitoring equipment, many companies and individuals included under such listings can provide professional advice on how to conduct a valid noise monitoring program. Some audiological testing firms and industrial hygiene firms also provide noise monitoring services. Universities with audiology, industrial hygiene, or acoustical engineering departments may also provide information or may be able to help employers meet their obligations under this amendment.

Free, on-site assistance may be obtained from OSHA-supported state and private consultation organizations. These safety and health consultative entities generally give priority to the needs of small businesses. Write OSHA for a listing of organizations to contact for aid.

APPENDIX H: AVAILABILITY OF REFERENCED DOCUMENTS

Paragraphs (c) through (o) of 29 CFR 1910.95 and the accompanying appendices contain provisions which incorporate publications by reference. Generally, the publications provide criteria for instruments to be used in monitoring and audiometric testing. These criteria are intended to be mandatory when so indicated in the applicable paragraphs of Section 1910.95 and appendices.

It should be noted that OSHA does not require that employers purchase a copy of the referenced publications. Employers, however, may desire to obtain a copy of the referenced publications for their own information.

The designation of the paragraph of the standard in which the referenced publications appear, the titles of the publications, and the availability of the publications are as follows:

Paragraph designation	Referenced publication	Available from—
Appendix B	"List of Personal Hearing Protectors and Attenuation Data," HEW Pub. No. 76-120, 1975. NTIS-PB267461.	National Technical Information Service, Port Royal Road, Springfield, VA 22161.
Appendix D	"Specification for Sound Level Meters," S1.4-1971 (R1976).	American National Standards Institute, Inc., 1430 Broadway, New York, NY 10018.
§ 1910.95(k)(2), appendix E	"Specifications for Audiometers," S3.6-1969.	American National Standards Institute, Inc., 1430 Broadway, New York, NY 10018.
Appendix D	"Specification for Octave, Half-Octave and Third-Octave Band Filter Sets," S1.11-1971 (R1976).	Back Numbers Department, Dept. STD, American Institute of Physics, 333 E. 45th St., New York, NY 10017; American National Standards Institute, Inc., 1430 Broadway, New York, NY 10018.

The referenced publications (or a microfiche of the publications) are available for review at many universities and public libraries throughout the country. These publications may also be examined at the OSHA Technical Data Center, Room N2439, United States Department of Labor, 200 Constitution Avenue, NW., Washington, D.C. 20210, (202) 523-9700 or at any OSHA Regional Office (see telephone directories under United States Government—Labor Department).

APPENDIX I: DEFINITIONS

These definitions apply to the following terms as used in paragraphs (c) through (n) of 29 CFR 1910.95.

Action Level—An 8-hour time-weighted average of 85 decibels measured on the A-scale, slow response, or equivalently, a dose of fifty percent.

Audiogram—A chart, graph, or table resulting from an audiometric test showing an individual's hearing threshold levels as a function of frequency.

Audiologist—A professional, specializing in the study and rehabilitation of hearing, who is certified by the American Speech-Language-Hearing Association or licensed by a state board of examiners.

Baseline audiogram—The audiogram against which future audiograms are compared.

Criterion sound level—A sound level of 90 decibels.

Decibel (dB)—Unit of measurement of sound level.

Hertz (Hz)—Unit of measurement of frequency, numerically equal to cycles per second.

Medical pathology—A disorder or disease. For purposes of this regulation, a condition or disease affecting the ear, which should be treated by a physician specialist.

Noise dose—The ratio, expressed as a percentage, of (1) the time integral, over a stated time or event, of the 0.6 power of the measured SLOW exponential time-averaged, squared A-weighted sound pressure and (2) the product of the criterion duration (8 hours) and the 0.6 power of the squared sound pressure corresponding to the criterion sound level (90 dB).

Noise dosimeter—An instrument that integrates a function of sound pressure over a period of time in such a manner that it directly indicates a noise dose.

Otolaryngologist—A physician specializing in diagnosis and treatment of disorders of the ear, nose and throat.

Representative exposure—Measurements of an employee's noise dose or 8-hour time-weighted average sound level that the employers deem to be representative of the exposures of other employees in the workplace.

Sound level—Ten times the common logarithm of the ratio of the square of the measured A-weighted sound pressure to the square of the standard reference pressure of 20 micropascals. Unit: decibels (dB). For use with this regulation, SLOW time response, in accordance with ANSI S1.4-1971 (R1976), is required.

Sound level meter—An instrument for the measurement of sound level.

Time-weighted average sound level—That sound level, which if constant over an 8-hour exposure, would result in the same noise dose as is measured.

Noise and Hearing Conservation Manual, edited by
E.H. Berger, W.D. Ward, J.C. Morrill and L.H. Royster

Appendix II Annotated Listing of Noise and Hearing Conservation Films and Videotapes

Elliott H. Berger

This is a list of movies and videocassettes on hearing and noise. All pricing and associated information are current as of January 1985. Please verify before ordering.

The films have been informally rated using the following scheme:

F - Fair
G - Good
VG - Very Good
E - Excellent
NA - Not available for review
* - Not rated due to association of the author with the producer

Contents
(alphabetical by film title)

1. (NA) A Brief Colloquy on Acoustic Diffraction (12 min.)
Pennsylvania State University
2. (G) Can You Hear Me? (14 min.)
BNA Communications, Inc.
3. (G) Death Be Not Loud (26 min.)
CRM - McGraw-Hill
4. (NA) Demonstrations in Acoustics (240 min.)
University of Maryland
5. (F) Ear Protection and Noise (12 min.)
International Medifilms
6. (VG) The Ears and Hearing (22 min.)
Encyclopaedia Britannica
7. (G) For Good Sound Reasons (15 min.)
Willson Products Division
8. (G) Hear: It Takes Two (20 min.)
International Medifilms

9.	(G)	Hearing Conservation Creative Media Development, Inc.	(19 min.)
10.	(G)	Hearing Conservation International Film Bureau	(22 min.)
11.	(G)	Hearing: The Forgotten Sense International Medifilms	(18 min.)
12.	(F)	How to Beat the Racket Tappi Press	(20 min.)
13.	(*)	How to Use Expandable Foam Earplugs E-A-R Division, Cabot Corporation	(6 min.)
14.	(G)	Industrial Noise OSHA Office of Information	(10 min.)
15.	(*)	It's Up to You E-A-R Division, Cabot Corporation	(12 min.)
16.	(*)	Less Than a Minute E-A-R Division, Cabot Corporation	(6 min.)
17.	(E)	Let's Hear It University of Toronto	(28 min.)
18.	(*)	Listen Up with Norm Crosby E-A-R Division, Cabot Corporation	(17 min.)
19.	(G)	Listen While You Can International Film Bureau	(21 min.)
20.	(*)	The National Hearing Quiz E-A-R Division, Cabot Corporation	(28 min.)
21.	(VG)	Nice to Hear Bilsom International Inc.	(10 min.)
22.	(VG)	Noise Film Communicators	(9 min.)
23.	(VG)	Noise International Film Bureau	(22 min.)
24.	(F)	Noise Destroys Industrial Training Systems Corp.	(12 min.)
25.	(G)	Noise and Its Effects on Health Film Fair Communications	(20 min.)
26.	(G)	Noise Polluting the Environment Encyclopaedia Britannica	(16 min.)
27.	(VG)	The Noise was Deafening International Film Bureau	(21 min.)
28.	(VG)	Noise? You're in Control! Educational Resources Foundation	(14 min.)
29.	(NA)	The Process of Holography Pennsylvania State University	(10 min.)

30.	(F)	Protect Your Hearing David Clark Company	(16 min.)
31.	(G)	Quiet Please University of Hartford	(20 min.)
32.	(VG)	Simple Harmonic Motion Pennsylvania State Universtiy	(17 min.)
33.	(E)	SOS Bilsom International, Inc.	(14 min.)
34.	(G)	Sound Advice Industrial Training Systems Corp.	(17 min.)
35.	(F)	Sound Field in Rectangular Enclosures Pennsylvania State University	(14 min.)
36.	(G)	Sound in the Seventies General Motors	(22 min.)
37.	(VG)	Sound of Sound American Optical	(17 min.)
38.	(F)	Sound Thinking TWA	(18 min.)
39.	(E)	Stick It in Your Ear Colorado Hearing and Speech Center	(15 min.)

Summaries and Order Information (alphabetical by producer)

AMERICAN OPTICAL **(617) 765-9711**
14 MECHANIC STREET **EXT. 2949**
SOUTHBRIDGE, MA 01550
ATTN: JEAN HILL

Sound of Sound (17 min.) $382.50

A comprehensive review of the effects of noise and its hazards, interspersed with interviews of employees who work in noisy environments, many of whom have lost their hearing because they didn't wear hearing protection. The film creates some interesting metaphors such as "taking the color out of vision" is like "taking the sound out of sound." The workers who are interviewed speak candidly of their occupational deafness — their loneliness and frustration, and how hearing loss negatively affects their enjoyment of life. They strongly support the film's basic message: "Keep the hearing you've got. Wear the proper hearing protection."

This film has won the Highest Honors of the National Committee on Films for Safety, The Film Festival Award for Medicine and Health (24th Nat. Conf. Pub. Relations Society of America), and the Gold Medal of the International Film and TV Festival of New York. It may be rented for $15.00 or purchased for $382.50 plus UPS shipping and tax. Rental cost applies toward purchase.

BILSOM INTERNATIONAL, INC. **(703) 620-3950**
11800 SUNRISE VALLEY DRIVE
SUITE 820
RESTON, VA 22091
ATTN: DEBBIE WELLS

Nice to Hear (10 min.) $200.00

This film includes some very good nature shots, has good ear animation, includes an extensive motivational segment and describes noise and hearing and hearing protection devices.

SOS (14 min.) $250.00

This film begins with the standard discussion of the effect of noise on hearing by following a typical worker's social interactions on the job and off the job. The last half of the film is an excellent comical animation sequence of the hair cells of the cochlea. These singing hair cells discuss how the inner ear works, how it gets hurt, and why the standard excuses for not wearing HPDs are inappropriate and inaccurate.

Slide cassettes (consisting of 45 slides) are available of each film for $80.00. Bilsom also offers a distributor's discount purchase price for each of the above films: Nice to Hear - $150.00 and SOS - $175.00 Distributor's cost for the slide cassettes is $60.00. The films produced by Bilsom may be previewed by payment of $30.00 in advance for each film. Rental cost is credited if film is purchased.

BNA COMMUNICATIONS, INC. **(301) 948-0540**
9439 KEY WEST AVENUE
ROCKVILLE, MD 20850
ATTN: CUSTOMER RELATIONS

Can You Hear Me? (14 min.) $355.00

Motivational segment on ear adapted to nature, not noise. Mediocre explanation of hearing, noise, and noise measurements. Short segments showing E-A-R Plugs in use.

Film may be rented for 3 days for $120 or one week (5 working days) for $160 plus shipping and tax. Film may be purchased for $355 plus shipping and tax. Up to one week's rental can be applied to purchase price if ordered within 30 days after renting. Film is also available for preview for $40.00 which can be applied to purchase price as well.

COLORADO HEARING AND SPEECH CENTER (303) 322-1871
INDUSTRIAL DIVISION
4280 HALE PARKWAY
DENVER, CO 80220
ATTN: BARBARA CAIN

Stick It In Your Ear (15 min.) $200.00

This film is a standout! It is unique in the way in which it approaches the subject matter of educating and motivating employees about noise and hearing conservation. It consists of nine short motivational skits, cartoons, and informational tidbits that are often quite entertaining and thus capable of capturing and holding a viewer's attention. This film is about as "snooze-proof" as a safety film can be.

The topics include the importance of good hearing both on and off the job; not letting your ears "grow old" before their time; you don't get used to noise, you "get deaf"; how hearing protectors help you hear better in noisy environments; how a hearing loss can be a social liability; and how hearing protection may be the only solution for a noise problem. The film would be better off without discussion of this last topic, since the way in which it is dealt with, and the implication that is made that noise controls are likely to lead to job elimination, is inappropriate and unnecessary to the theme of the film.

Another minor problem with the film is the explanation of why HPDs allow one to hear better in noise. The film indicates that this is because HPDs block the noise and let the speech through, which in fact is seldom the case. Rather, HPDs reduce the overall level of both noise and speech present at the ear so that the speech can be better discriminated and understood.

Film may be rented for one week for $50.00 plus postage, which can be applied to the purchase price of $200.00.

CREATIVE MEDIA DEVELOPMENT, INC. (503) 223-6794
710 S.W. NINTH AVENUE
PORTLAND, OR 97205

Hearing Conservation (19 min.) $260.00

This videotaped slide presentation (90 frames) with no animation or action footage is designed especially for "line-level" employees. The goal of the program is to impart an understanding of hearing loss cause and prevention with the intent of integrating that understanding into the overall hearing conservation program. The film approaches this task in a simplified, well organized, unimaginative fashion.

The film reviews how we hear, how hearing is damaged by noise and other agents, the effects of deafness, how we measure sound (with examples), audiometry, basic engineering controls, the advantages and disadvantages of types of hearing protectors and how to fit and use earmuffs and earplugs. In general the information is clear and accurate, with the most obvious error being the poor audio examples that are used to illustrate the range of sound levels that are discussed. Additionally, there is too much stress placed on cleanliness with regard to the fitting and use of hearing protectors.

This videotape can be customized (prior to purchase) by having Creative Media replace certain of the frames with slides of the plant in question. In spite of this the action films that are available are generally more exciting and professional looking than a videotaped set of slides. The program is accompanied by a short manual, most of which is the script for the film.

The purchase price is $260.00 for ¾″, ½″ VHS, ½″ Beta videotape or as a slide-tape program. Preview copies are available for a fee or $25.00 for a period of fourteen days.

CRM — MCGRAW-HILL **(619) 453-5000**
P.O. BOX 641
DEL MAR, CA 92014

Death Be Not Loud (26 min.) $435.00

This film concentrates on environmental noise. It begins with anecdotal interviews with New York City residents, then focuses on noise in suburbia, especially airport noise. It includes interviews with Dr. Rosen re: noise and blood pressure and animal studies. Also interviewed are Drs. Lebo and Oliphant re: music and PTS. The film illustrates how to monitor noise through examples such as a Connecticut highway noise monitoring project. A visit is made to a sleep shop where a variety of methods to aid sleep are demonstrated, including personal hearing protection. Also included is an example of successful noise reduction on Boeing jet engines.

The film presents a very biased view of extra-auditory effects of noise, implying that these effects are confirmed and serious. Furthermore, the producers suggest that noise control methods are simple (which they are not) and that there is no excuse for a noisy society.

This film may be purchased for $435.00 or rented for a 3 day period at a cost of $33.00. If purchased within 90 days, the rental cost may be applied toward the purchase price.

DAVID CLARK COMPANY **(617) 756-6216**
360 FRANKLIN STREET
WORCESTER, MA 01604

Protect Your Hearing (16 min.) $150.00

This narrative-type presentation employs action footage, animation, and cartoons to discuss the hazards of noise exposure, provide an adequate explanation of pitch and sound level, illustrate the anatomy and functioning of the ear, briefly discuss sound surveys and audiometry, and review HPDs, albeit with a decided bias against earplugs.

The hearing protection fitting section of the film only demonstrates how to use premolded earplugs (V-51R and SMR) and the David Clark line of earmuffs. It inaccurately states that earplugs should be "fitted by a doctor or other medically trained person," a statement apparently intended to steer the user into selecting the "easy-to-wear" earmuffs.

Additional inaccuracies in the film:

1) Presents a version of the "bone-conduction myth," the commonly repeated fallacy concerning the importance of the mastoid areas in determining bone-conduction sensitivity in an acoustic field. This erroneous line of reasoning is intended to suggest that an earmuff, which covers and therefore "protects" the mastoid from sound, can provide better attenuation. This is untrue.

2) Suggests that HPDs allow one to hear better in noise since "distracting background noise is filtered out." The fact that HPDs do allow listeners with normal acuity to hear more easily in noise has nothing to do with selective filtering, but rather with reducing the overall sound levels present at the cochlea.

3) Labels the frequencies below 20 Hz as "subsonic," a term usually reserved for velocity of motion of objects, as in air speed. The correct term is "infrasonic."

4) Recommends the Stop Gap for reducing the loss in attenuation due to eyeglasses. Laboratory tests indicate that the reverse is true, *i.e.* although the Stop Gap may enhance earmuff comfort when wearing glasses, it tends to reduce attenuation even further than the glasses alone.

E-A-R DIVISION, CABOT CORPORATION **(317) 872-1111**
7911 ZIONSVILLE ROAD
INDIANAPOLIS, IN 46268
ATTN: V. L. MINKNER

It's Up To You (12 min.)

$120.00 film
$ 25.00 ½″ VHS
$ 28.00 ¾″ U-matic

Take a trip through a 20-foot ear. Motivational/informational film describing noise and its hazards, and hearing protectors and how to use them.

Less Than A Minute (6 min.)
$ 80.00 film
$ 25.00 ½″ VHS
$ 28.00 ¾″ U-matic

Provides a good demonstration of how to properly use E-A-R Plugs. Includes some comments on noise, the value of hearing protection, and a glimpse of E-A-R Division, Cabot Corporation headquarters and laboratory.

How to Use Expandable Foam Earplugs (6 min.)
$ 50.00 film
$ 25.00 ½″ VHS
$ 28.00 ¾″ u-matic

E-A-R Division's Manager of Acoustical Engineering demonstrates how to correctly use E-A-R Plugs. Close-ups and animation are used to illustrate the process. Suggestions on evaluating fit and avoiding common problems are also discussed. This is a training film for E-A-R Plug users and especially those wishing to train users.

The National Hearing Quiz (28 min.)
$ 35.00 ½″ VHS
$ 35.00 ¾″ U-matic

Don Wescott, a noted narrator of educational documentaries, is the host for this question and answer film, primarily intended for public television and general education. Basic acoustics, hearing, hearing damage and hearing protection are illustrated through the use of interesting analogies and visual images.

Listen Up With Norm Crosby (17 min.)

A cast of actors and celebrities (Ed Asner, Bill Murray, Larry Brown, Charlene Tilton, Hank Aaron and others) discuss noise, its pervasiveness in society, its dangers, and how to protect ourselves from it. Contains some fine scenes with dramatic impact that illustrate the stupidity of not wearing hearing protection when exposed to harmful high level noise.

All of the above films are available in 16 mm or videotape format for a two week loan from E-A-R Division, at no charge, or for purchase at the prices shown. When prices and formats are not listed the item is unavailable for purchase from E-A-R. However, the following films may be purchased elsewhere.

The National Hearing Quiz is available for purchase in 16 mm film from Walter J. Klein Company , Ltd., Box 2087, Charlotte, NC 28211.

Listen Up With Norm Crosby is available for purchase from Better Hearing Institute, 1530 K Street, NW, suite 700, Washington, DC 20005 (202-638-7577). Purchase price is $285.00 for 16 mm film and $125.00 for videotape. This film may be ordered with a purchase order.

EDUCATIONAL RESOURCES FOUNDATION **(800) 845-8822**
P.O. DRAWER L
COLUMBIA, SC 29250

Noise? You're in Control (14 min.) $340.00

This is a motivational film to get employees to use hearing protection. It was produced to assist companies in complying with the training requirements of the OSHA Noise Standard. The film is not a complete training program on its own, since it lacks details on the attenuation and fitting of HPDs and a complete discussion of audiometric testing. These deficiencies are pointed out in the very good Leader's Implementation Guide that is supplied with the Film. The guide discusses how to use the film both for new hires and for experienced employees. It also contains an outline and script of the film.

The movie was filmed in numerous industrial environments representing a variety of potential noise problems.

OUTLINE OF FILM CONTENT

Intro: Hearing protection is important.
Why? Because noise is powerful!
Noise attacks hearing.
Protection can be achieved with hearing protection devices.
What is hearing loss?
Basics of measuring sound.
How do you know if there is a noise problem?
What will the company do about the problem?
What types of hearing loss can noise cause?
How much of a handicap is a noise induced hearing loss?
Can someone with normal hearing imagine how it feels?
Summary: How do we control hearing loss?
Conclusion.

This film is available in 16mm film, ¾″ U-matic videocassette, ½″ VHS videocassette, and ½″ Betamax videocassette for a purchase price of $340.00 or a 5-day rental of $75.00. If purchased within 30 days after renting, the rental fee can be applied to the purchase price. This film may be ordered with a purchase order number.

ENCYCLOPAEDIA BRITANNICA EDUCATION CORPORATION **(312) 321-6695**
425 N. MICHIGAN AVENUE
CHICAGO, IL 60611
ATTN: CUSTOMER SERVICE

The Ears and Hearing (22 min.) $380.00

This film describes the structure and function of the human ear and demonstrates the process by which sound waves from the environment are converted into electrochemical energy and perceived as sound. Two causes of deafness are reviewed and ways in which these malfunctions can be corrected are outlined. Excellent photography of a functioning middle ear and of middle ear surgery are included.

Noise Polluting the Environment (16 min.) $285.00

Another noise and its effects film. There are others on the list that are more informative and more entertaining.

These films can be previewed free only with the intent of purchasing. If you intend to buy them, the previewed film has to be returned. You can buy the films using a purchase order number. Use the film's product number, (The Ears and Hearing) 2834, (Noise Polluting the Environment) 3045, when ordering. These films may be rented for 10% of the list price.

FILM COMMUNICATORS **(800) 423-2400**
11136 WEDDINGTON STREET
NORTH HOLLYWOOD, CA 91601

Noise (9 min.) $295.00 film
$265.00 videotape

A unique and refreshingly different approach to safety films. This short workplace "drama" produced by Film Australia for the (Australian) Dept. of Science and Technology features Bryan Brown (star of "The Thorn Birds" and "A Town Like Alice") as a new employee in a small metal fabrication shop. A hearing conservation program has just been initiated and has been unsuccessful in motivating the employees, and in overcoming the peer influence that deems it unmanly to protect one's hearing. Bryan and another new employee refrain from using hearing protectors, in spite of their own personal concerns, due to the peer pressure.

The drama unfolds as Bryan observes, along with the audience, the kind of life he could expect to live unless he protects his hearing. He learns of the significant social handicap that hearing impairment can cause. The "happy" ending is that Bryan and the other employees in the shop begin to wear hearing protection.

This film is directed towards the motivational aspect of hearing conservation. It does not explain decibels, nor hair cells, nor the myriad of details that are normally incorporated into the educational component of the program. This information can be presented later after we have the attention and interest of the employee. The sound track effectively complements the film and heightens its impact, without the use of dialog, thus making it suitable regardless of the primary language of the audience.

Film/videotape is available for a 2-day executive preview for $30 (applicable to purchase) or a 1-week rental is available for $80.00.

FILM FAIR COMMUNICATIONS **(213) 877-3191**
10900 VENTURA BLVD.
P.O. BOX 1728
STUDIO CITY, CA 91604

Noise and Its Effects on Health (20 min.) $325.00

This is a documentary type film discussing noise and its effects on health. It is presented in a dry, technical vein, interspersed with man-on-the-street interviews with some zany characters. It explains decibels and sound, and discusses noise and its relation to stress and mental health. Considerable time is spent on aircraft/airport noise. Scientists and legislators are also interviewed (Karl Kryter, Sen. Tunney, etc.). The film makes a strong statement for socio-environmental noise control.

This film may be rented for 3 days for $25.00 plus shipping charges. The purchase price is $325.00. You can submit a purchase order for either the rental or purchase price, and the rental fee can be applied to the purchase price. The film may also be ordered by calling collect.

GENERAL MOTORS TECHNICAL CENTER **(313) 575-1626**
INSTRUCTIONAL REGULATIONS
ENVIRONMENTAL ACTIVITIES STAFF
WARREN, MI 48090
ATTN: PAUL PATAKY

Sound in the Seventies (22 min.) FREE LOAN

This movie was prepared by the Environmental Activities Staff of General Motors to illustrate significant factors in the contribution of motor vehicles to environmental noise. The movie also puts into perspective the task of noise control in our society.

The first segment of the film describes in general terms the measurement of noise in decibels or dBA. It explains how the sound level varies with the distance from the source to the listener, and it illustrates the addition of sound energies from multiple sources. This segment also illustrates annoyance which has no absolute unit of measurement.

The film describes some typical facilities and tests used to isolate and reduce noise of various vehicle components and it identifies the components which are the major contributors to the total motor vehicle noise.

Truck noise is explored using the Society of Automotive Engineers (SAE) truck test procedure. Examples of trucks experimentally modified with various noise reduction treatments are demonstrated.

A sequence of the movie portrays the phenomena associated with tire noise. It is shown that at certain vehicle speeds, the noise emitted by tires becomes predominant over the noise from other contributing components. Some tire noises are demonstrated and discussed, including the GM specification tire.

A segment of the movie describes passenger vehicle noise and the SAE test procedure.

Finally, various aspects of community noise reduction are discussed. It is shown that noise can be reduced in the initial planning of roadways, by landscaping, by construction of road berms, and by proper maintenance and operation of our motor vehicles.

The film may be requested from above address. Please specify magnetic or optical soundtrack.

INDUSTRIAL TRAINING SYSTEMS CORP. (609) 234-2600
823 EAST GATE DRIVE
MOUNT LAUREL, NJ 08054

Noise Destroys (12 min.) $324.00

This videotaped slide presentation (61 frames) with no animation or action footage is designed as an employee and supervisory training and motivational program on hearing conservation. An outline and leader's guide are provided. The program examines the nature of sound, explains the effect of over-exposure to loud noise upon hearing ability, discusses the consequences of a permanent loss of hearing, and outlines the elements of a hearing conservation program.

The most unfortunate piece of information that is included in the film is the concept that regular 8-hr., 90-dBA noise exposures are safe for the "average" worker. In fact, such exposure limits are based on socio-economic factors rather than on simply protecting hearing. Employees should be made aware that levels of 85, and even as low as 80 dBA, can be hazardous to some people.

The outline of the film follows:

Introduction—examples of hearing handicap, hearing loss is preventable

The Nature of Sound—sound vs. noise, common misconceptions (you get used to noise)
Effect of Over-Exposure to Loud Noise—temporary hearing loss, permanent loss and its consequences
Elements of a Hearing Conservation Program—protective equipment, noise measurement, engineering controls, administrative controls, audiometric testing
Conclusion—wear hearing protection

Sound Advice (17 min.) $324.00

This videotaped slide presentation (105 frames) with no animation or action footage is designed as an employee training program on hearing conservation. An outline and leader's guide are provided. The program examines the cause of hearing loss, explains how the ear functions, shows how noise affects the ear, and outlines the ways in which a hearing conservation program can prevent hearing loss caused by excessive noise.

The material is generally clear and accurate, with adequate if not exciting visuals. The most notable deficiency is in the area of demonstrating correct use of hearing protectors. Although the discussion on earmuffs is quite adequate, the presentation on inserting non-expandable (*i.e.* premolded) and expandable (*i.e.* foam) plugs is unclear, and in fact provides an incorrect demonstration of the use of E-A-R Plugs (in spite of the fact that the accompanying leader's guide provides correct instructions).

The outline of the film follows:

Introduction—hearing, prevalence of loss, its prevention
Ear Functions—outer, middle, and inner
Causes of Hearing Loss—wax, foreign objects, diseases, medications, severe blows, aging, loud noise
Noise-Induced Hearing Loss—how sound is measured, effects of noise
Hearing Conservation—measurement, engineering controls, administrative controls, protectors, audiometry
Conclusion—wear hearing protectors

Both titles are available in videotape (½″ or ¾″) or slide format for a cost of $324.00. 1-week rental available for $100.00. 5-day preview available for $20.00, cost can be applied to rental or purchase price.

INTERNATIONAL FILM BUREAU, INC. **(312) 427-4545**
332 S. MICHIGAN AVENUE
CHICAGO, IL 60604

Listen While You Can (21 min.) $425.00

This film is directed at the layman in order to provide information and motivation to protect hearing. A simulation of noise-induced hearing loss demonstrates that it involves more than reduced sensitivity. To help understand this, the film employs animation to define sound and to illustrate the construction of the ear. The animation of a functioning ear is presented very well. This leads to the types of ear damage (some of which do not heal) and the kinds of noise conditions that are dangerous. Tests for measuring hearing ability and means of protecting the ear from damage are shown. The main point of the film is that noise damage is a sure route to incurable deafness, but that such deafness is unnecessary because there are simple precautions for avoiding it.

This film was produced by Stewart Hardy Films for Ministry of Defense (Navy Dept.), United Kingdom. The American version was prepared by International Film Bureau, Inc.

Hearing Conservation (22 min.) $425.00

A typical dangerous noise environment is used to study the problems of noise-induced hearing loss, noise reduction, and hearing conservation. First, noise must be measured and evaluated; the film defines noise, explains the factors determining its hazard, and indicates damage risk criteria. The two types of noise reduction are then considered and compared: reduction at source and direct protection of ears. Finally, the film illustrates the discovery and measurement of noise-induced hearing loss. The emphasis in this film is on preventing hearing loss, and if it has already occurred, on discovering it before it affects normal hearing of speech. This film was produced by Stewart Hardy Films for Ministry of Defense (Navy Dept.), United Kingdom.

NOTE: Both of these films incorrectly state that "muffs offer better protection than plugs."

Noise (22 min.) $425.00

This film is a comprehensive well-executed primer on noise and hearing. It contains very good descriptions (with animation) of sound, frequency, level, and how the ear works. Understandable explanations of "3 dB trading," A-weighting, dosimetry and audiometry follow, with comments regarding typical presbycusic vs. noise induced audiograms.

The last segment of the film reviews source/path/receiver concepts with some poorly thought out examples that, in general, are not of the caliber of the preceding portion of the film. Produced by the National Coal Board, England.

The above 3 films can each be rented for a 3 day period for $45.00 or $60.00 for a week. The 3 day rental fee of $45.00 can be applied towards the purchase price.

The Noise Was Deafening (21 min.) $525.00

This film on hearing loss and noise reduction emphasizes the irreversible effects of excessive noise on hearing loss, shows how noise surveys can be conducted, and points out some of the methods for noise reduction. Using animation, the physiology of the ear is explained. Frequency, pitch, dB, and A-weighting are described and illustrated with animation. A noise survey is conducted and the measurements are recorded on a chart of the work areas. Recommendations are made for reducing noise levels including mechanical reduction, job rotation, and hearing protection. Many real world problems are brought up and discussed. Suggestions are made for implementing all phases of a hearing conservation program except for audiometric testing. This film will be of interest to both management and employees as they become more aware of the implications of hearing loss and the means for reducing the noise which can cause it. Produced by Millbank Films Limited.

The above film can be rented for a 3 day period for $55.00 or $75.00 for a 5 day period.

INTERNATIONAL MEDIFILMS **(213) 851-4555**
3393 BANHAM BLVD.
LOS ANGELES, CA 90068
ATTN: MS. CHRIS KEVIN

Hearing, the Forgotten Sense (18 min.) $350.00

The purpose of this film is to orient management, supervisory personnel, and employees toward noise situations and to provide incentive for all levels of employment to prevent personal hearing loss. It applies not only to work situations, but to home activities as well.

In this film, the audience participates in a non-technical hearing acuity test. This test shows how hearing levels vary and the need for hearing conservation. This film contends that hearing is literally the "forgotten sense." It instructs the audience that hearing loss is painless and shows no visible signs of damage. In its attack on carelessness and complacency, at work and home, this film becomes an ideal reinforcement for "Hear: It Takes Two."

Hear: It Takes Two (20 min.) $400.00

The purpose of this film is to motivate employees to use hearing protection provided by management. It makes each employee aware of their individual responsibility in this area and shows how hearing loss affects the worker, his family, and his work. It verifies that hearing loss goes undetected until damage is permanent.

This film consists primarily of interviews with many different noise exposed persons who contend that ear protection is an acceptable and practical method of preventing industrial hearing loss. Actual workers tell of their hearing problems and how these problems occurred. They discuss actual problems in private and work life occasioned by hearing loss. At the end of the film, a short sequence discusses the danger signs of excessive noise exposure and how to protect against it.

Ear Protection and Noise (12 min.) $280.00

The purpose of this film is to illustrate the danger of prolonged exposure to high level noise. It makes the employee aware of the cumulative effect of industrial noise on the ear mechanism and instructs on the care and use of ear protectors. Aram Glorig was the consultant for this film.

This film is dated; however, it does do a good job of pointing out indications of hearing damage and how to fit protectors. It stresses five points about hearing protection: 1 - snug fit, 2 - aid in perception, 3 - won't cause infection, 4 - plugs need reseating, and 5 - must be worn. It should be noted that, based on more recent studies on the use of hearing protection devices, some of the points made by this film must be corrected. As an example, if an employee exhibits significant hearing loss, the wearing of hearing protection may actually reduce the ability to hear warning signals, etc.

The above films can be previewed with a rental charge of:

$50.00 for "It Takes Two"
$50.00 for "Ear Protection and Noise"
$50.00 for "Hearing, the Forgotten Sense"

The rental charge can be deducted from the price if you purchase the film. The film can be ordered with a purchase order. A $5.00 handling charge (in addition to shipping charges) will be added to all purchases and rentals.

OSHA OFFICE OF INFORMATION **(202) 523-8615**
U.S. DEPARTMENT OF LABOR
ROOM N-3637
WASHINGTON, DC 20210
ATTN: MR. DANIEL HAM

Industrial Noise (OSHA Spectrum Special, 10 min.) FREE LOAN

The principal intent of this film is to demonstrate the feasibility and value of engineering noise controls in an industrial environment. The example that OSHA uses is 13 wire drawing machines at the Ray Magnet Co. in Virginia. The noise levels are 95 dBA.

The film begins with employee interviews to demonstrate the noxious and hazardous effects of daily occupational exposure to high noise levels. The film reviews CFR 1910.95 paragraphs a) and b) which mandate engineering noise controls when the TWA is greater than 90 dBA. In this regard the film is not current since no mention is made of the Hearing Conservation Amendment or the acceptability of hearing protection devices as a control measure.

The film describes how the machines were redesigned by replacement of the gear drive mechanisms with belt drives and addition of new bearing lubrication systems. The cost was $10,000/machine for a measured noise reduction of 10 dBA. The employees and management are then further interviewed to substantiate the success of the controls in terms of a better work environment and more efficient machinery operation.

This videotape is available on a free-loan basis from OSHA in both ½" VHS and ¾" U-matic formats.

THE PENNSYLVANIA STATE UNIVERSITY (814) 865-6364
APPLIED RESEARCH LABORATORY
P.O. BOX 30
STATE COLLEGE, PA 16801
ATTN: BARBARA CROCKEN

Simple Harmonic Motion (17 min.) $200.00

High-school/college level tutorial which explains the concepts of phase, amplitude and frequency and demonstrates simple harmonic motion using real-world objects.

Sound Field in Rectangular Enclosures (14 min.) $180.00

Very technical, moderately informative computer generated film. The title tells it all. Color is used to represent pressure in a 3-dimensional simulation of normal modes, where the modes are depicted separately and in various combinations.

A Brief Colloquy on Acoustic Diffraction (12 min.) $160.00

Presents theoretical diffraction of sound by a barrier with various degrees of absorption, and identifies the shadow boundary and other concepts.

The Process of Holography (10 min.) $150.00

College level tutorial explaining how plane and spherical waves combine to form a wavefront, and the cause of the 3 components in the reconstruction.

These four computer-generated films can be previewed for a fee of $15.00. Since only a limited number of preview copies are available, it is advisable to make preview requests for particular dates as far in advance as possible.

When ordering please specify: film; optical or magnetic sound; if preview, dates needed; if purchase, number of copies. Payment must be made by check in advance to The Graduate Program in Acoustics, Penn State University.

These films have been reviewed in greater detail by Conrad Hemond [J. Acoust. Soc. Am. 65(5), 1352].

TAPPI PRESS **(404) 446-1400**
ONE DUNWOODY PARK
ATLANTA, GA 30338

How to Beat the Racket (20 min.) $559.95

Primarily a motivation film that compares not wearing HPDs to taking a stupid chance. Contains a mediocre explanation of the hearing mechanism and explanation of NIPTS. No animation. Presents the "bone conduction myth." Presents short interviews illustrating problems associated with hearing loss.

This film is by the Pulp and Paper Industry Technical Association. It can be obtained by a check sent to the above company, using stock No. 01-03-S002. Members of Tappi Press may purchase the film for $375.17; non-members may purchase the film for $559.95. This film is not available for preview or rental.

TWA **(816) 891-4966**
GROUND SAFETY DEPT., 1-135 MCI
P.O. Box 20126
KANSAS CITY, MO 64195
ATTN: SKULI GUDMUNDSSON

Sound Thinking (18 min.) $425.00 Videotape

The subject of this film revolves around an employee who during the course of his annual audiogram discusses how he came to start religiously wearing his hearing protectors when exposed to potentially harmful industrial noise. Until recently he did not, and he partially attributes that to his upbringing — his dad was a pressman who never wore hearing protection. In fact, as a result of that his father is significantly hearing-impaired today. Incidents resulting from that hearing loss are what motivated his son to start protecting his hearing.

Through flashbacks we learn how hearing loss has affected his father in day-to-day and social situations, especially how his grandson almost drowned as a result of his inability to hear the child's cries for help. The happy ending is that dad now wears a hearing aid and the son now wears his earmuffs.

The film then goes on to discuss three types of hearing protectors - earplugs, earmuffs, and semi-aural devices and illustrates how to fit and use them. It concludes with a couple of mediocre audio demonstrations of hearing loss and the limitations of hearing aids.

The film proceeds at a fairly slow pace, has only a moderately interesting story line and audio-visual demonstrations, and the quality of the video filming tends toward the amateurish.

The videotape may be purchased in ¾" U-Matic, ½" VHS or BETA III formats for $425.00. Preview copies are available at no charge for a period of five days. For an additional fee this videotape can be tailored to meet specific hearing protection needs (*i.e.*, different hearing protection devices or adding company logo to the film). Fee would be determined by changes in videotape.

UNIVERSITY OF HARTFORD **(203) 243-4786**
COLLEGE OF ENGINEERING
200 BLOOMFIELD AVENUE
WEST HARTFORD, CONN 06117
ATTN: DR. CONRAD J. HEMOND, JR.

Quiet Please (20 min.) FREE LOAN

A basic acoustics film that begins with what is sound and how we hear. Describes source-path-receiver and decibels. Comments upon and illustrates industrial noise, noise vs. health, community noise, jet noise, truck noise and general noise control procedures.

Film available on a Free Loan basis from the above address.

UNIVERSITY OF MARYLAND **(301) 454-3520**
LECTURE-DEMONSTRATION FACILITY
DEPARTMENT OF PHYSICS AND ASTRONOMY
COLLEGE PARK, MD 20742
ATTN: DR. RICHARD E. BERG

Demonstrations in Acoustics (240 min.) $60.00

This set of four 60-minute, color videocassettes was designed for use in an elementary course on acoustics for non-science majors. The demonstrations were designed to illustrate acoustical phenomena at the introductory, non-mathematical level. However, because of the wide range of topics treated and the uniqueness of some of the demonstrations, the tapes can be used with success in upper level science and engineering classes as well. Due to the striking nature of some of the demonstrations, many of the tapes are excellent for "open house" situations where non-scientists and laypeople can watch the tapes at their leisure.

The set consists of 29 titled segments, each containing one or more demonstrations, some with explanations, tables and applications. The shortest segment is approximately three minutes and the longest is about twelve minutes. Each segment contains some discussion or description of the equipment and principles illustrated. Over sixty individual demonstrations are presented.

The tapes will be made available in any format: 60-minute ¾" U-matic cassettes and 60-minute Beta or VHS cassettes are most common, and may be ordered directly. Other formats may be available, but check before ordering. All tapes will be reproduced at a cost of $15 per hour or $60 for the four-hour set. Please advise to whom the invoice should be sent; you will be billed by the Department of Physics and Astronomy. Send blank tapes in a re-usable shipping case. UPS is the best method of shipment, insured for $200.00. Ship the package to the above address.

UNIVERSITY OF TORONTO **(416) 978-6302**
INSTRUCTIONAL MEDIA SERVICES
FACULTY OF MEDICINE
TORONTO, ONTARIO, CANADA M5S 1A1

"Let's Hear It" with Dr. Peter Alberti (28 min.) $420.00

Dr. Alberti is Professor of Otolaryngology at Mount Sinai Hospital, University of Toronto.

The film begins by discussing basics of sound and how we hear. It shows some excellent scanning electron microscope photos of hair cells in a normal and in a noise damaged ear. It points out that hearing protectors *do not* perform as well in industrial environments as manufacturer's data would indicate.

The film follows the hearing conservation program of one industry. It examines the various phases of the program such as identification, training and fitting. It demonstrates the use of foam and premolded plugs and earmuffs.

The purchase price is $420.00 for 16 mm film and $378.00 for ¾" video cassette. Payment may be made in the form of a check or a purchase order. Film may be rented from City Films, 542 Gordon Baker Road, Willowdale, Ontario M2H 3B4 (416) 499-1400 for a 3-day period for $50.00 plus shipping and U.S. Customs charges of $25.00 (if shipped to the United States).

WILLSON SAFETY PRODUCTS **(215) 376-6161**
P.O. BOX 622
READING, PA 19603
ATTN: DONNA MAURER

For Good Sound Reasons (15 min.) $150.00

The film begins by Pop introducing Son to a new job at the plant. He uses some of the old false expressions often heard, such as "your ears will toughen up," or "wearing earmuffs is for sissies" and presents the incorrect "bone conduction myth." The film discusses noise, hearing, and hearing protection.

The film can be ordered with a purchase order number for $150.00 or it may be borrowed with a deposit of $150.00 which is refunded when the film is returned.

Subject Index

absorption coefficient (see sound absorption coefficient)
acoustic reflex 179
acoustic trauma 182
action level 539
addition of decibels (see decibels, addition of)
administrative control of noise exposure (see noise control)
age correction 190, 192ff, 284, 309, 529, 554ff
age, effect on hearing (also see presbyacusis) 192, 296
age-corrected hearing thresholds 191ff, 309
age-related threshold level (ARTL) 296ff
ambient noise (also see background noise) 235ff, 244
analyzers, frequency 58ff, 91
annoyance 8, 49, 237, 276, 419
apportionment (see workers' compensation)
area noise survey (see noise surveys, area)
atresia 220
attenuation 27
audiogram 12, 234ff
  age correction 190, 192ff, 284, 309, 529, 554ff
  analysis 280, 541
  annual 273, 541
  baseline 272, 540
  baseline revision 542
  collapsing ear canal 220, 222, 225, 277
  criteria for interpretation 281
  earphone placement 276
  evaluation 256, 541, 554
  fluctuations (see audiogram, variability)
  learning effect 299, 301ff
  referral 280ff
  review 283, 284, 294
  significant threshold shift 307ff
  standard threshold shift (STS) 13, 15, 281, 283, 307ff, 541, 542
  use in employee education/motivation 247, 389, 400, 402
  validity 253, 261ff, 283
  variability 249, 261ff, 304ff
audiologist 271, 283, 540
audiometer 246ff
  Bekesy (see self-recording)
  calibration of 264ff, 315, 543, 552ff
    acoustic 266
    coupler 242, 266
    exhaustive 268
    functional (biological) 265
    OSHA regulations 268, 540ff
    records 268
    shifts in 300
  computer-controlled 257ff
  manual 246ff
  microprocessor (see computer-controlled)
  self-recording 250ff, 551ff
    scoring 256, 552
  test environment for 235ff, 541, 552
audiometric data base analysis (ADBA) 293ff
  benefits 294, 295, 314
  education/motivation 295
  evaluation criteria 300
  individual audiogram review 294
  HCP effectiveness 294, 309, 314
  learning effect 299, 301, 310
  population comparisons 294, 296, 300, 309
  purpose 295
  STS criteria 307
  test-retest statistics 304, 310
  variability-related statistics 312
audiometric technician 141, 269ff, 395, 540
audiometric test booths 237ff, 541, 552
audiometric zero 188ff, 553ff

Note: abbreviated terms are indexed under their complete designation. See list of abbreviations on pp. xiii-xiv.

audiometry (also see audiogram; audiometer) 12
computer-controlled 257ff
instructions for 251, 254, 275
manual 246ff
problems in 277
self-recording 250ff
auditory fatigue (see temporary threshold shift)
aural history (see employee, history)
averaging, exponential 43, 66
average level (also see sound, average levels) 63, 80
A-weighted sound level (dBA; see weighting filters)

background noise (also see ambient noise) 76, 144ff, 543, 552
barriers 497, 500
baseline audiogram 272, 540
basilar membrane 179
biological check of audiometer (see audiometer, calibration)
boilermakers' ear 2
bone conduction (BC) 246, 323ff
B-weighting (see weighting filters)

calibration
accuracy 117
audiometer 264ff, 543, 552ff
dosimeter 81, 117
records of 268
sound level meter 56, 76, 117
calibrators 56ff, 114, 267
canal caps (see hearing protection devices, semi-aural)
center frequency 21, 59
cerumen 219, 221, 358
cholesteatoma 222
cochlea 179
coincidence frequency 459
collapsing ear canal 220, 222, 225, 277
communication interference (also see hearing protection devices, speech discrimination) 7, 281
community noise analyzer 45
compensation (see workers' compensation)
contributing noise sources (see noise, contributing source)
correction for background noise 76
Council for Accreditation in Occupational Hearing Conservation (CAOHC) 271, 389, 540
crest factor 53
criterion sound level (see dose, criterion sound level)
critical intensity 213
cutoff sound level, cutoff level (see instrumentation, dosimeter, threshold)
C-weighting (see weighting filters)

damage risk criteria 2, 210ff
damping 443, 451ff, 458ff
coincidence frequency 459
constrained layer 458, 463
critical damping constant 451
free layer 458
loss factor 451
materials 459
ratio 451ff
data recording forms
audiometric 245, 249, 270, 279
sound levels 118, 119, 121-124, 136, 151, 426
day-night average level (L_{dn}) 64
decibel 22
addition of 28, 76, 141
subtraction of 144
definitions, OSHA standard 561
degrees of freedom 163
diffuse sound field 31
diplacusis 199
direct sound field 31
directivity 423, 467
factor 25
index (DI) 26
dizziness 288
dominant noise source (see noise, dominant source)
dose 67ff, 546
measurement (also see instrumentation, dosimeter; monitoring) 67, 102ff, 126, 138, 546ff
area 86, 126ff, 558
criterion sound duration 67
criterion sound level 67
exchange rate 65ff, 153, 209ff
partial 67, 126
personal 86, 559
trading ratio 65, 66
dosimeter (see instrumentation)

double hearing protection 352
ducts, lined 508ff
dynamic balance 445
dynamic characteristics, meter (see sound level meter, meter dynamics)

ear 177ff
  inner 179
  middle 178
  outer (see external ear)
ear anatomy (also see external ear) 177ff, 218ff
  basilar membrane 179
  cochlea 179
  eustachian tube 179
  hair cells 180ff, 183
  mastoid 219
  middle ear muscles 179
  organ of Corti 180
  ossicles 179
  pinna 178
  Reissner's membrane 179
  tectorial membrane 180
ear canal (see external ear)
ear disease or anomaly 220ff
  atresia 220
  cholesteatoma 222
  foreign body 221
eardrum 177, 219ff
earmuffs (also see hearing protection) 333ff
  cushions 336, 338
  effect of eyeglasses on 334, 569
  headbands 333ff
earphones 75, 89, 241ff
  circumaural 241
  noise-attenuating 241, 277
  recreational 102, 356
  supra-aural 242
earplugs (also see hearing protection) 338ff
  custom molded 346ff
  fiberglass 345
  foam 346
  formable 343ff
  insertion of 339
  premolded 340ff
  sizes 342
  V-51R 340
  wax 345
ear wax 219, 221, 358

education 14, 385ff
  use of visual aids in (see visual aids)
educator, selection of 394ff
effects of noise (see noise, effects of)
electromagnetic interference 90
employee
  counseling 280ff, 285
  history 226, 279
  notification of 540, 544
  training and education (see education; see motivation)
enclosure (see noise control)
engineering control 11, 418, 420ff
equal-energy theory (see noise-induced hearing loss)
equal-loudness contours 47ff, 186
equal-TTS theory (see noise-induced hearing loss)
equivalent sound level (L_{eq}) 63
eustachian tube 179
evaluation of HCP (see hearing conservation program, evaluation)
evaluation of external ear 217ff
  benefits 227
exchange rate 65ff, 153, 209
exhaust noise 429, 437ff
exponential averaging 42ff
exposure (see noise exposure)
exposure classification scheme (see noise exposure)
external ear 177, 217ff
  abnormalities 220ff
  auricle 218
  canal 177, 218
  cerumen 219, 221, 358
  concha 219, 342
  dimensions 219
  eardrum (see tympanic membrane)
  helix 219
  infection (see otitis)
  lobule 219
  occluding wax (see cerumen)
  otitis (also see ear disease) 220
  pinna 177, 218
  sebaceous glands 219
  tragus 219
  tympanic membrane 177, 219ff
external otitis (see external ear, otitis)

FAST response 42ff, 79
fatigue, auditory 6, 13, 272

films, educational 394, 563ff
filters 21
  weighting (see weighting filters)
Fletcher-Munson curves 47ff, 186
fluctuating noise, effects of 208ff
forms, data recording (see data recording forms)
free field 25, 30
frequency
  band-edge (see frequency, cutoff)
  center 21, 59
  coincidence 459
  cutoff 21
  forcing 450
  natural 447ff
  resonance 448, 453
  surge 454
frequency analyzer 58ff, 91
functional check 265

government regulations 3ff, 67, 100ff, 211ff, 537ff
  hearing conservation amendment 4, 539ff
  hearing protectors 370, 543, 549
graphic level recorder (GLR) 44

handicap (see workers' compensation)
hearing aid 405, 532
hearing conservation amendment (see government regulations)
hearing conservation program (HCP) 3, 9ff, 385ff, 539ff
  audiometric testing 12, 234
  benefits 9, 227, 235, 286, 290, 295, 397, 400, 409
  effectiveness 15, 309ff
  evaluation of 15ff, 293ff, 383
    criteria 300
  key individual 385, 395
hearing damage 181, 200, 204
hearing disability (see workers' compensation)
hearing impairment (HI, see workers' compensation)
hearing level (HL) 188
hearing loss (also see noise-induced hearing loss) 6, 181
  conductive 182, 287
  neural 182
  rate of change of 299ff
  sensory 182, 286
  susceptibility to (see susceptibility)
hearing loss claims (see workers' compensation)
hearing measurement (see audiometer)
hearing protection devices 13, 320ff, 543ff
  abuse and modification 335, 341, 345, 350, 352, 407
  active 355
  air/acoustical leak 321, 334ff
  amplitude sensitive 355
  attenuation 13, 14, 325ff
    interlaboratory variability in 331
    OSHA procedure 364
  bone conduction 323, 352, 569
  calculating protection 326ff, 543, 549ff
  circumaural (see earmuffs)
  communication (also see hearing protection, speech discrimination) 368
  communication headsets 356
  derating attenuation 328, 331, 362ff
  dual (muffs and plugs) 352
  ear infection (also see external ear) 220ff, 358
  earmuffs (see earmuffs)
  earplugs (see earplugs)
  education and training 14, 391, 399, 541
  evaluation of effectiveness (also see NRR) 359, 365
  fitting 13, 325, 332ff, 342, 391, 541
  fourteen-hour quiet 273, 540
  hygiene 358
  labeling 371
  laboratory test methods 325
  localization with 369
  maximum sound attenuation 323, 352, 569
  misuse 407
  noise reduction rating (NRR) 328ff, 550
  nonlinear 355
  occlusion effect 324, 342, 346, 351
  real-world attenuation 13, 103, 314, 328, 359ff
  recreational earphones 102, 356
  selection 13, 332, 364, 372
  semi-aural 350ff
  seven-dB correction 330, 364
  speech discrimination 355, 365ff, 569

standard deviation 326, 361
standards 370
ultrasound attenuation 326
use in baseline audiometry 272
hearing threshold level (HTL) 184, 188
hygiene 358

impact noise (see noise, impact)
impairment, hearing (see workers' compensation)
impulse noise (see noise, impulse)
industrial-noise-induced permanent threshold shift (INIPTS) 204ff
from non-steady noise 208ff
from pure tones 213
from steady noise 204ff
from very high levels 213
industrial-noise-exposed population (INEP) 204, 298
inner ear 179
insertion loss (IL) 26
instructions, test 251, 254, 275
instrumentation 41ff, 54ff
accuracy 46, 50, 87ff, 114, 117
bandwidth 59
community noise analyzer 45
crest factor 53
dosimeter 44, 81ff, 112, 133ff
accuracy 114
artifact 84
limitations 87, 112
microphone placement 83, 133
protection 85
threshold (cutoff) level 66, 83
electromagnetic interference 90
exponential averaging (SLOW and FAST) 42ff
frequency analyzer 58ff, 91
graphic level recorder (GLR) 44
integrating/averaging SLM 45, 61
noise floor 73
octave-band analyzer 59, 147
otoscope 223
precision SLM (see sound level meter)
real time analyzer 59
selection 70ff, 112
sound level meter (see sound level meter)
spectrum analyzer 58ff
tape recorder 61
integrating/averaging SLM 45, 61
intensity, sound 22

key individual 385, 395

learning effect 299, 301ff
lobule 219
localization 289, 369
loss factor (see damping)
loudness 186
loudness level 48, 186

machines
enclosures 477ff, 486ff
engineering modification (see noise control)
mountings 432, 446ff, 453, 480
malingering 256, 258
masking 7, 235
mastoid 219
medical referrals 280, 283, 286ff
meter dynamics 42ff, 79, 120
microphones 54ff, 70ff, 74ff
directivity 72
dummy 91
orientation 70, 83, 133, 135
placement 70, 83, 88, 113
protection 72, 85
random incidence 72
types 55, 72
windscreen 75, 84
minimum audible field (MAF) 184
minimum audible pressure (MAP) 184, 189
monitoring 86, 540, 558
area vs. personal 86, 88, 125ff
motivation 395ff, 402ff
ADBA 295, 315, 393, 406, 409
employee input 407
individual audiograms 247, 389, 402
management 397, 410ff
peers 406
permanent and temporary hearing loss 391, 404ff
rewards 408
movies 394, 563ff
muffler 429, 433, 475, 508ff

natural frequency 447ff
noise 20
ambient 72, 76, 235, 244
community 64
continuous 538
contributing source 144

control (see noise control)
dominant source 141ff
effects of
on annoyance 8, 62
on cardiovascular system 9
on communication 7
on irritability 9
on non-auditory functions 9
on performance 8
on safety 8
on sleep 9
on stress reaction 9
impact/impulse 54, 85, 88, 214, 435, 441, 538
noise control 418ff
administrative 11ff, 418, 538
barriers 497, 500
engineering 11, 418, 420ff
damping (see damping)
ducts, lined 508
enclosures 477ff, 486ff
fans 430
fluid flow 468
hearing protection devices (see hearing protection devices)
lagging 475, 494
leaks 490
muffler 429, 433, 475, 508ff
multiple wall enclosure 497
nozzles 469
operator isolation 429
quieter materials 443
path modification 477ff
plant design 422ff
room absorption 502
source modification 11, 444ff
source placement 428ff
source replacement 430ff
vibration isolation 432, 446ff, 453, 480
noise dose (see dose)
noise dosimeters (see instrumentation)
noise exposure 10, 102, 201
classification scheme 102
computation 546ff
history of 280
variability 150
noise floor (also see ambient noise) 73
noise-induced hearing loss (also see industrial-noise-induced PTS) 1, 6, 201, 204ff
equal-energy theory 65, 153, 209
equal-TTS theory 210
permanent threshold shift (NIPTS) 200ff
susceptibility 201ff
noise measurement 10, 38ff, 70ff
noise radiation (see radiation, sound)
noise reduction (NR) 26, 478, 486, 505
noise reduction rating (NRR, see hearing protection devices)
noise regulations (see government regulations)
noise sampling procedures 133ff, 152
noise silencer 434
noise survey 10, 99ff
area 10, 86, 88, 125
basic 107, 115ff
detailed 132ff
engineering 141ff
forms (see data recording forms)
general (see basic)
political considerations in 99
preparation for 108ff
report preparation 169
sampling 133ff, 152
special 146
updating 172
noise survey instrumentation
choice of 70ff, 112
nonauditory effects of noise (see noise, effects of)
non-industrial-noise-exposed population (NINEP, also see audiometric data base analysis) 191ff, 296ff
normal hearing 189
nosoacusis 6, 190

occluding wax 219, 221, 358
occlusion effect 324, 342, 346, 351
occupational hearing conservationist 269
Occupational Safety and Health Administration (OSHA) 537ff
hearing conservation amendment (see government regulations)
octave band (OB) 20, 59, 60
one-third octave band 21, 59, 60
otoscope 223
otoscopic examination 217ff, 280
in audiometry 280
in fitting HPDs 217, 220ff
outer ear (see external ear)

paracusis 199
peak sound level 47, 53
permanent hearing loss (see noise-induced hearing loss)
permanent threshold shift (PTS, also see industrial-noise-induced PTS, noise-induced hearing loss) 6, 200, 207ff
permissible exposure level (PEL) 3, 68, 537ff
phons 48, 186
pink noise 328
pinna 177, 218
pitch 19, 199
population comparisons 296
precision sound level meter (see sound level meter)
preferred noise criteria (PNC) 486
presbyacusis, presbycusis 6, 190, 284, 296
pressure equalization (PE) tubes 221
protectors (see hearing protection devices)
pure tone 20

race, effect on hearing 202, 297ff
radiation, sound (also see directivity) 25, 30, 420
radiation efficiency 444, 467
real-ear attenuation at threshold (REAT) 325
recording forms, data (see data recording forms)
records 15, 534, 545
  calibration 268
  noise level 172, 545
recreational headsets 102, 356
recruitment 258
reference populations 191ff, 296ff
referral 226 280ff
regulations (see government regulations; see workers' compensation)
resonance (also see frequency)
  ear canal 178
  isolator 447ff
reverberant sound field 31, 468
room absorption 31ff, 423, 502
room constant 31, 423, 480, 485
root mean square value (RMS) 23, 50

sabins 484, 505
safety 7, 282, 288, 392, 411, 420
sampling (see statistical techniques)
sebaceous glands 219
sensitivity, auditory (see threshold, auditory)
sex, effect on hearing 202, 297ff
significant threshold shift 307ff
silencer, noise 434
single number rating 328ff, 550
SLOW response (see sound level meter, meter dynamics)
sociacusis, sociocusis 6, 190
sone 188
sound (also see noise) 19, 39
  amplitude 20
  average levels 63ff
    L_{dn}: day-night average level 64
    L_{DOD}: Department of Defense equivalent sound level 65
    L_{eq}: equivalent continuous sound level 63ff
    L_{OSHA}: Occupational Safety and Health Administration equivalent sound level 65
    TWA: time-weighted average 66, 102ff, 126, 547
  frequency 19
  intensity 22
  measurement 38
  pitch 19, 199
  power, power level 22, 30, 423
  pressure, pressure level 19, 23, 30, 423
    reference 23
  pure tone 20
  speed 20
  wavelength 20
sound absorption coefficient (also see sabins) 27, 32, 479ff, 502ff
sound exposure level (SEL) 68
sound level contours 140
sound level meter (SLM) 40ff, 112, 135
  accuracy 114
  averaging 45
  basic 40ff
  calibration 56, 76ff, 114
  crest factor 53ff, 85
  impulse dynamics 46
  integrating 45
  measurement tolerance 50
  meter dynamics 42ff, 79, 120
  orientation (see microphone orientation)
  peak sound level 47, 53

precision 46
pulse range 53
slow response 42ff, 79, 120, 539
tolerances 50
true peak 47
types 46
sound power level 22, 30, 423
sound pressure level (SPL) 19, 23, 30, 423
sound survey (see noise survey)
sound transmission loss (STL) 27
source modification (see noise control)
specification, equipment noise 424
spectrum analyzers 58ff
speech discrimination (SD, also see hearing protection devices, speech discrimination) 7, 281
speech interference level (SIL) 7
speech perception 200
speed of sound 20
standard threshold shift (STS, also see audiogram) 283ff
standing waves 71
statistical techniques 150ff
chi-square 159
confidence intervals 155ff, 163ff
degrees of freedom (df) 163
mean 155, 158, 326
normal distribution 159
sampling 137, 152
standard deviation 158, 326
tolerance limits 157, 166
subtraction of decibels 144
susceptibility (also see race, see sex) 201ff
effect of age on 203
effect of gender on 202
effect of pigmentation on 202

tape recorder 61
temporary hearing loss (see temporary threshold shift)
temporary threshold shift (TTS) 6, 13, 210ff, 272
auditory fatigue 6
in audiometric data base analysis 304, 314
in baseline audiograms 272ff
in employee motivation 405
in evaluation of HPD effectiveness 314
test booths, audiometric 237ff, 541, 552
test-retest comparisons (see audiometric data base analysis)
testimony 534
threshold, auditory 184, 188
minimum audible field (MAF) 184
minimum audible pressure (MAP) 184, 189
threshold shift (see audiogram, see noise-induced hearing loss, see temporary threshold shift)
time constants 43, 46
time-weighted average (TWA) 66, 102ff, 126, 547
tinnitus 199, 256, 258, 277, 287, 532
trading ratio 65, 66
tragus 219
training
of audiometric test personnel 269ff
of ear examiners 223
programs 269, 397ff, 544
transmissibility 450ff
transmission loss (TL) 478ff
turbulence 75, 84, 468

vibration isolation 432, 446ff, 453, 480
visual aids in education 393ff, 563ff
movies 394, 563ff
slides 393
variety 394
videotapes 394, 563ff

Walkman (see recreational headsets)
Walsh-Healy Act 3, 211
wavelength 20
weighting filters 47ff
A-weighting (dBA) 24, 42, 48, 204
B-weighting 48
C-weighting (dBC) 48, 50
wind screen 75, 84
workers' compensation (WC) 2, 101, 523ff
apportionment 532
disability 525
handicap 198, 526
hearing loss claims 269, 285, 391, 410, 533
impairment 526
impairment claims 533ff
handling 535
testimony and records 534
validation 533
waiting period 530
impairment formulae 526
state regulations 527, 533